Laboratory DNA Science

OF RELATED INTEREST
From the Benjamin/Cummings Series in the Life Sciences

General Biology

N. A. Campbell
Biology, Fourth Edition (1996)

N. A. Campbell, L. G. Mitchell, and J. B. Reece
Biology: Concepts and Connections (1994)

J. Dickey
Laboratory Investigations for Biology (1995)

J. G. Morgan and M. E. B. Carter
Investigating Biology: A Laboratory Manual for Biology, Second Edition (1996)

Plant Biology

M. G. Barbour, J. H. Burk, and W. D. Pitts
Terrestrial Plant Ecology, Second Edition (1987)

J. D. Mauseth
Plant Anatomy (1988)

L. Taiz and E. Zeiger
Plant Physiology (1991)

Biochemistry and Cell Biology

W. M. Becker, J. B. Reece, and M. F. Poenie
The World of the Cell, Third edition (1996)

R. F. Boyer
Modern Experimental Biochemistry, Second Edition (1993)

C. K. Mathews and K. E. van Holde
Biochemistry, Second Edition (1996)

G. L. Sackheim
Chemistry for Biology Students, Fifth Edition (1995)

Molecular Biology and Genetics

L. E. Hood, I. L. Weissman, W. B. Wood, and J. H. Wilson
Immunology, Second Edition (1984)

J. D. Watson, N. H. Hopkins, J. W. Roberts, J. A. Steitz, and A. M. Weiner
Molecular Biology of the Gene, Fourth Edition (1987)

Microbiology

I. E. Alcamo
Fundamentals of Microbiology, Fourth Edition (1994)

R. M. Atlas and R. Bartha
Microbial Ecology: Fundamentals and Applications, Third Edition (1992)

J. G. Cappuccino and N. Sherman
Microbiology: A Laboratory Manual, Fourth Edition (1996)

T. R. Johnson and C. L. Case
Laboratory Experiments in Microbiology, Brief Version, Fourth Edition (1995)

G. J. Tortora, B. R. Funke, and C. L. Case
Microbiology, An Introduction, Fifth Edition (1995)

Anatomy and Physiology

E. N. Marieb
Essentials of Human Anatomy and Physiology, Fourth Edition (1994)

E. N. Marieb
Human Anatomy and Physiology, Third Edition (1995)

E. N. Marieb and J. Mallatt
Human Anatomy (1992)

A. P. Spence
Basic Human Anatomy, Third Edition (1991)

Benjamin/Cummings offers a series of multimedia products designed for use in introductory anatomy and physiology. Highlights include:

A.D.A.M. Standard (1994)

A.D.A.M.-Benjamin/Cummings Interactive Physiology: Cardiovascular Module (1995)

A.D.A.M.-Benjamin/Cummings Interactive Physiology: Muscular Module (1995)

A.D.A.M.-Benjamin/Cummings Interactive Physiology: Respiratory Module (1996)

Laboratory DNA Science

An Introduction to Recombinant DNA Techniques and Methods of Genome Analysis

Mark V. Bloom
DNA Learning Center, Cold Spring Harbor Laboratory

Greg A. Freyer
College of Physicians and Surgeons, Columbia University

David A. Micklos
DNA Learning Center, Cold Spring Harbor Laboratory

Illustrations by Susan Zehl Lauter

The Benjamin/Cummings Publishing Company, Inc.
Menlo Park, California • Reading, Massachusetts
New York • Don Mills, Ontario • Wokingham, U.K. • Amsterdam
Bonn • Paris • Milan • Madrid • Sydney • Singapore • Tokyo
Seoul • Taipei • Mexico City • San Juan, Puerto Rico

Sponsoring Editor: Catherine Pusateri
Associate Editor: Kimberly Viano
Editorial Assistant: Lisa Woo-Bloxberg
Production Editor: Judith Hibbard
Visual Communications Manager: Don Kesner
Composition and Film Manager: Lillian Hom
Permissions Editor: Marty Granahan
Senior Manufacturing Coordinator: Merry Free Osborn
Marketing Manager: Larry Swanson
Copy Editor: Jan deProsse
Text Designer: Mark Ong
Illustrators: Val Felts and Susan Zehl Lauter
Compositor: Side by Side Studios
Film House: Banta Digital Services
Printer: Banta, Menasha
Cover Designer: Yvo Riezebos
Cover Illustration: Richard Wagner, UCSF Graphics

Library of Congress Cataloging-in-Publication Data

Bloom, Mark V.

Laboratory DNA Science : an introduction to recombinant DNA techniques and methods of genome analysis / Mark V. Bloom, Greg A. Freyer, David A. Micklos.

p. cm.

Includes bibliographical references and index.

ISBN 0-8053-3040-2

1. Recombinant DNA—Laboratory manuals. I. Freyer, Greg A. II. Micklos, David A. III. Title

QH442.B59 1995

574.87'3282—dc20 95-35419

CIP

1 2 3 4 5 6 7 8 9 10—BAM— 99 98 97 96 95

The cover was printed using soy-based inks.

The Benjamin/Cummings Publishing Company, Inc.
2727 Sand Hill Road
Menlo Park, CA 94025

Foreword

It has been almost ten years since I became aware of the clandestine relationship between Greg Freyer, a postdoctoral fellow in my laboratory, and Dave Micklos, the then newly appointed fund-raiser for Cold Spring Harbor Laboratory. I would occasionally find Greg working in the cold room in the evening on projects that were definitely unrelated to his main postdoctoral research. It transpired that he and Dave had embarked on an ambitious project to enlighten the country's biology teachers about the wonders of DNA. They wanted to spread the fun far beyond its normal province of academic research institutions. Their audacious plan was to devise a modern biology curriculum for teachers, who could then expose their students to the mysteries of DNA. Thus began the imaginative journey to the current volume, *Laboratory DNA Science*.

My own involvement in this project has been minimal, merely providing a small amount of bench space and sharing some of our reagents in the early days. Greg's exceptional technical skills led him to develop special vectors designed for teaching and challenging experiments that might actually work (in the hands of novice students). Dave was the first guinea pig; he had little previous laboratory experience but possessed great drive and an infectious enthusiasm to see the project succeed. Dave joined Cold Spring Harbor Laboratory in 1982 as Development Director, ostensibly to raise money for the institution. Jim Watson was therefore quite surprised when he returned from a sabbatical in 1984 to discover that Dave had gone off to raise money for his own project, now known as the DNA Learning Center.

Today, we stand in the midst of a whirlwind of technological change. Just as computers are changing the way we communicate, biotechnology is offering bright prospects for improving our health care and food supply. People must learn about DNA and its potential if they are to draw their own conclusions about the benefits of biotechnology. Yet, many of our major public policy decisions are being made by individuals with little or no knowledge of science. The traditional biology curriculum is only slowly incorporating the new ideas of molecular biology that provide the foundation for the biotechnology industry. Sadly, most of our adult population, who receive their higher education only through the medium of television, know little about DNA and its wonderful properties. It is time for change.

Some of our leading scientists are calling for their colleagues to help educate society, but most are just talking. However, in 1990, Greg and Dave did much more than talk when they produced their first book, *DNA Science*. This was the forerunner of the present volume. With this new book, Mark Bloom, Greg Freyer, and Dave Micklos have expanded their horizons. They have produced a first-rate textbook that includes experiments to fascinate, challenge, and explain much of the molecular biology that underpins biotechnology. It is aimed at an audience of college students who are anxious to leave behind the traditional teachings of biology and are curious about the new age. I am pleased to have played even a small role at the inception of this project, and I wish the readers well in all their future experiments, whether they are taken from this book or devised independently.

Richard J. Roberts
New England Biolabs

Preface

Time flies. It hardly seems possible that ten years have passed since we began developing *Laboratory DNA Science*. These 23 laboratories had their origin in an unlikely collaboration at Cold Spring Harbor Laboratory between Dave Micklos, then Development Director, and Greg Freyer, then a postdoctoral fellow in Richard Roberts' lab group. It all began with Greg preparing gels of *Lambda* restriction digests for Dave to show to students who came for tours of the Laboratory. Students were fascinated with those simple "DNA fingerprints" of the *Lambda* virus, and in 1984 we began to develop a lab curriculum that could be used to introduce students to molecular genetics. Greg's original plan to make and analyze a recombinant plasmid was contained in a slim "manuscript" consisting of 10–12 pages of lab notes that were comprehensible only to a bench scientist. Happily, in early 1985, Rich Roberts gave Dave one half-counter of bench space where the labs could be fleshed out before being tested with teaching faculty from local Long Island schools.

The working title of the first ten labs was *Recombinant DNA for Beginners*, which was taken from Graham Nash's record album, *Songs for Beginners*. This album was composed of simple songs that nearly any beginning guitar player or vocalist could easily tag along with—and in some sense, feel like a pop musician. Our course was likewise intended to be a collection of relatively simple experiments that could make any student feel like a molecular biologist.

When the first labs and a mini-text were formally published in 1990, we opted for a simpler title, *DNA Science*. This aphorism for molecular genetics—looking at biology through the lens of DNA—was coined in casual conversation by Cold Spring Harbor Laboratory President James D. Watson. Mark Bloom joined the author group to lead the development of *Laboratory DNA Science,* which focuses entirely on practical work and extends the original lab sequence with 13 additional laboratories that cover gene library construction, hybridization, and polymerase chain reaction. We believe these to be the most thoroughly tested laboratories available today for teaching molecular genetics. Each experiment incorporates insights from our own instruction of 2,000 teaching faculty at training workshops across the country and from the 30,000 students in our teaching laboratory at the DNA Learning Center.

Albert Einstein's comment that "God is subtle but not malicious" applies equally well to molecular genetics. On the face of it, this field must seem almost maliciously abstract to even bright and motivated students. Doing experiments is the probably the only way to reduce this abstraction and to give students a sense of the subtle beauty of molecular mechanics. For better or worse, bands on gels, colonies on plates, and dots on filter paper are the major methods of inference in molecular genetics—the ways of knowing molecules. Only when these methods are used to obtain predictable results will students begin to believe that experiments can, indeed offer a window on the unseen molecular world.

Teachers sometimes try to put the best spin on a failed experiment by saying that students can learn as much from failure as from success. In fact, in students' eyes, a failed experiment can mean a failure of that mode of inference. A failed experiment throws students back into the abstractive quandary in which they began. We believe that experiments should work for students. Therefore, we have struggled to adapt and refine these laboratories to increase the chances for success.

Nothing has been left to chance. Greg even engineered the teaching plasmids pAMP, pKAN, and pBLU specifically for the course. Restriction digests of these plasmids yield fragments of markedly different sizes, thus making gel interpretation straightforward. Being derivatives of the pUC19 expression vector, these plasmids transform well, are highly amplified in *E. coli*, and give consistent yields in plasmid preparations. pBLU was developed to simplify colorimetric screening of *E. coli* transformed with recombinant plasmids. This plasmid includes the entire *lac*Z gene but excludes *lacI* repressor activity; thus, color development requires neither a complementing cell line nor IPTG in the growth medium.

Our extensive contacts within the biological research community keep us abreast of emerging techniques. While Doug Hanahan was at Cold Spring Harbor Laboratory, he introduced us to the colony transformation method, which obviates the need for liquid culturing of mid-log cells in simple experiments. Prescott Deininger, of the LSU Medical Center, and Mark Batzer, of the Lawrence Livermore National Laboratory, introduced us to the *Alu* insertion polymorphisms, which proved to be perfect for educational demonstrations of human DNA typing. Collaborations with scientists at Cetus, Perkin-Elmer, and Roche Molecular Systems were key to the development of the PCR experiments; scientists at Boehringer Mannheim aided with colorimetric detection used in hybridization experiments. The entire lab sequence is supported by quality-assured reagents and kits available from the Carolina Biological Supply Company. (See Appendix 1 for ordering information.)

The development of *DNA Science* is one of the first examples of bench molecular biologists making a substantive commitment to help bring biology education into the gene age. Although this educational role is now accepted and legitimized at high levels of the biological research establishment, this was not so in 1984—especially at a "pure science" place like Cold Spring Harbor Laboratory. At that time, research biologists were generally thought to have only one responsibility—to the bench. Rich Roberts' assistance was almost without precedent: He freely allowed Greg to participate in the project, donated enzymes and lab supplies, and provided lab space for Dave.

Of course, none of this would have been possible at all without the strong support of Jim Watson, Dean of DNA, who has maintained Cold Spring Harbor as a place of "great dreams and rigorous thinking." We also owe a special debt of gratitude to our guardian angels, Mary Jean and Henry Harris, whose generosity and friendship have sustained us for many years.

So here it is. Our ten-year effort to make a "foolproof" laboratory course in molecular genetics. One word of caution: These are simple laboratories but not trivial ones. Complex biochemistry is going on in the background; the failure of any one component can bring the whole system down. When your labs are working, don't change a thing. You'd be surprised to discover that even a small substitution can confound results—sometimes in very subtle ways. *E. coli* may seem as predictable as a light switch, but please believe us when we say that this lowly creature can fool you from time to time. Stay on your toes and have fun.

Mark Bloom, Greg Freyer, and Dave Micklos

Acknowledgments

We would like to thank our editors, Cathy Pusateri and Kim Viano, and the others at Benjamin/Cummings who worked so hard to keep their promise to publish *Laboratory DNA Science* in only ten months. Our thanks also to the collaborators who provided advice, technical support, equipment, and reagents:

Mark Batzer, Lawrence Livermore National Laboratory

Donald Comb, New England Biolabs, Inc.

Prescott Deininger, Louisiana State University Medical Center

Bob Diller and Dieter Schluter, Brinkmann Instruments, Inc.

Ray Gladden, Carolina Biological Supply Company

Lou Hosta, Parke Flick, Jack Chase, and Tom Mann, U.S. Biochemicals, Inc.

Rick Martin, Jim Pease, Dennert Ware, and Brian Holaway, Boehringer Mannheim Corporation

Cindy Rapheal, Valerie Erdman, Larry Haff, and Fenton Williams, The Perkin-Elmer Corporation

Spencer Teplin, Cold Spring Harbor Laboratory

Tom White, Stan Rose, Ellen Daniell, and Mike Zoccoli, Roche Molecular Systems, Inc.

Finally, special thanks to the professors who helped us to strengthen the scientific accuracy and teaching effectiveness of the protocols presented in *Laboratory DNA Science:*

Riaz Ahmad, University of Central Oklahoma

Jane E. Aloi, Saddleback College

G. Douglas Crandall, Emmanuel College

Don DeRosa, Boston University School of Medicine

Robert Dorrance, Herkimer County Community College

Steve Goodwin, University of Massachusetts at Amherst

Alice Grier Lee, Washington and Jefferson College
Tracie M. Jenkins, Mercer University
Michael P. Kolotila, Northern Essex Community College
Craig W. Maki, United States Air Force Academy
Innocent N. Mbawuike, Baylor University, College of Medicine
John W. Moon, Jr., Harding University Main Campus
Elain C. Rubenstein, Skidmore College
Shirley M. Russo, Quisigamond Community College
Nelson Samuel, California Baptist College
Mark F. Sanders, University of California - Davis
Daniel C. Scheirer, Northeastern University
Eric Stavney, Foothill College
Wendy L. Whyte, BioData Incorporated
H. Patrick Wooley, East Central College

For my parents, Leonard and Hope.
M. B.

For my wife, Joadie, and my three sons, Kurt, Eric, and Alec.
G. F.

For Dana Kuhn and Alec Spencer, my links to past and future.
D. M.

Contents

Laboratory DNA Science

Laboratory Safety

Individuals should use this manual only in accordance with prudent laboratory safety precautions and under the supervision of a person familiar with such precautions. Use of this manual by unsupervised or improperly supervised individuals could result in serious injury. Instructors and students are urged to pay particular attention to specific cautions placed throughout this manual and to carefully follow the instructions contained in these caution boxes.

Instructors should be familiar with and follow all national, state, local, or institutional regulations or practices pertaining to the use and disposal of materials used in this manual. Alternative procedures for the handling and disposal of laboratory waste should supersede those suggested in this manual when local requirements, conditions, or practices so dictate.

DNA RECOMBINATION

The Recombinant DNA Advisory Committee (RAC) of the National Institutes of Health (NIH) is the formal arbiter of research involving DNA recombination and transfer. In the wake of the debate over the potential dangers presented by recombinant DNA in the late 1970s, the RAC published guidelines to govern research and to rate the risks of various types of experiments. In 1982, the RAC exempted from any type of regulation experiments using plasmids and DNA from prokaryotic systems within prokaryotic hosts [1]. This acknowledged that recombinant-DNA experiments using DNA from closely related bacterial strains essentially entail no danger or ethical dilemma. In recent years, regulations have been further relaxed to exempt certain types of experiments that recombine DNA from different species.

The recombinant-DNA experiments in this course use only DNA sequences from *E. coli* and its indigenous plasmids and viruses (*Lambda*); the plasmids pAMP, pKAN, and pBLU were constructed using only prokaryotic sequences. Thus, the laboratories in this book— including those suggested for further research—were specifically designed to fall within the least controversial category of recombinant-DNA experiments. Instructors who would like to expand their teaching to experiments involving interspecific mixing of DNA should familiarize themselves with current NIH guidelines.

E. COLI AND RESISTANCE TRANSFER

E. coli is part of the normal bacterial flora of the large intestine, and it is rarely associated with severe illness in healthy individuals. Nonetheless, one might be concerned about the consequences of accidentally ingesting *E. coli* bacteria, especially those that have been transformed with plasmids. Research has shown that the *E. coli* strains MM294 and JM101 used in this laboratory sequence are unlikely to grow inside human intestines and are incapable of transferring plasmid DNA to *E. coli* living there.

MM294, JM101, and other *E. coli* strains commonly used in research today are derivatives of wild-type strain K-12, which was originally isolated in 1922 from the feces of a diphtheria patient at Stanford University. The work of Edward Tatum in the 1940s popularized the use of K-12 in biochemical and genetic studies. Research in the 1970s showed that strain K-12 cannot effectively colonize the human gut, probably because of genetic changes accumulated during decades of in vitro culture [2]. Furthermore, K-12 strains have lost the O antigen domain of the lipopolysaccharide that composes the outer membrane. The O antigen is thought to be necessary for infection in mammals [3].

Plasmid DNA can be transferred from one bacterial cell to another during conjugation. Transport requires a mobility protein (encoded by the plasmid-borne *mob* gene) and a specific site (*nic*) on the plasmid. The *mob* protein nicks the plasmid at *nic* and then attaches to the nicked strand to conduct the plasmid through a mating channel into a recipient cell. Most newer plasmids—including pAMP, pKAN, and pBLU—lack both the *mob* gene and the *nic* site and thus cannot be mobilized for transport [4,5].

The prelab notes in Laboratory 2 contain detailed instructions for the responsible handling and disposal of *E. coli,* which should be followed throughout this laboratory course.

ETHIDIUM BROMIDE STAINING

Ethidium bromide is the most rapid and sensitive means for straining DNA; however, it is a mutagen by the Ames microsome assay and a suspected carcinogen. Methylene blue is presumably a safer stain and can be used in simple demonstrations of electrophoresis; however, its lack of sensitivity makes it inappropriate for most experiments in this book.

With responsible handling, as we discussed in the prelab notes in Laboratory 3, a dilute staining solution of ethidium bromide poses little risk to experimenters or to the environment. The greatest risk is of inhaling ethidium bromide powder when mixing a 5 mg/ml stock solution; therefore, we strongly recommend purchasing a premixed 5 mg/ml solution from a supplier. The stock solution should be handled according to the instructions contained in the Material Safety Data Sheet provided by the supplier. The 5 mg/ml stock solution is diluted to make a staining solution with a final concentration of 1 μgm/ml, according to the procedure described in Appendix 2.

REFERENCES

1. National Institutes of Health. 1982. Guidelines for research involving recombinant DNA molecules. Federal Register 47/167: 38051.
2. Bachman, Barbara. 1987. Derivations and genotypes of some mutant derivatives of Escherichia coli K-12. In Escherichia coli *and* Salmonella typhimurium: *Cellular and molecular biology* (ed. F. C. Neidhardt), vol. 2, p. 1190. American Society for Microbiology, Washington, D.C.
3. Raetz, Christian R. H. 1987. Structure and biosynthesis of Lipid A in *Escherichia coli*. In Escherichia coli *and* Salmonella typhimurium: *Cellular and molecular biology* (ed. F. C. Neidhardt), vol. 1, p. 498. American Society for Microbiology, Washington, D.C.
4. Luria, Salvador E., and Joan L. Suit. 1987. Colicins and *Col* plasmids. In Escherichia coli *and* Salmonella typhimurium: *Cellular and molecular biology* (ed. F. C. Neidhardt), vol. 2, pp. 1620-1621. American Society for Microbiology, Washington, D.C.
5. Sambrook, Joseph, Edward Fritsch, and Thomas Maniatis. 1989. Molecular cloning: A laboratory manual, vol. I, p. 1.5. Cold Spring Harbor Laboratory Press, Cold Spring Harbor, New York.

Basic Techniques

LABORATORY 1

Measurements, Micropipetting, and Sterile Techniques

This laboratory introduces the micropipetting and sterile pipetting techniques used throughout the course. Mastery of these techniques is important for good results in all of the experiments that follow. Most of the laboratories are based on *microchemical* protocols that use very small volumes of DNA and reagents. These require the use of an adjustable micropipettor (or microcapillary pipet) that measures as little as one microliter (μl), one millionth of a liter.

Many experiments require growing *Escherichia coli* in a culture medium that also provides an ideal environment for other microorganisms. Therefore, it is important to maintain sterile conditions, to minimize the chances of contaminating an experiment with foreign bacteria or fungi. *Sterile conditions* must be maintained whenever living bacterial cells are to be reused in further cultures. Use sterilized materials for everything that comes in contact with a bacterial culture: nutrient media, solutions, pipets, micropipet tips, inoculating loops and cell spreaders, flasks, culture tubes, and plates.

Remember this rule of thumb: Use sterile technique if live bacteria are needed at the end of a manipulation (general culturing and transformations). Sterile technique is not necessary if the bacteria are destroyed by the manipulations in the experiment or when solutions for DNA analysis (plasmid isolation, DNA restriction, and DNA ligation) are being used.

II. Small-Volume Micropipettor

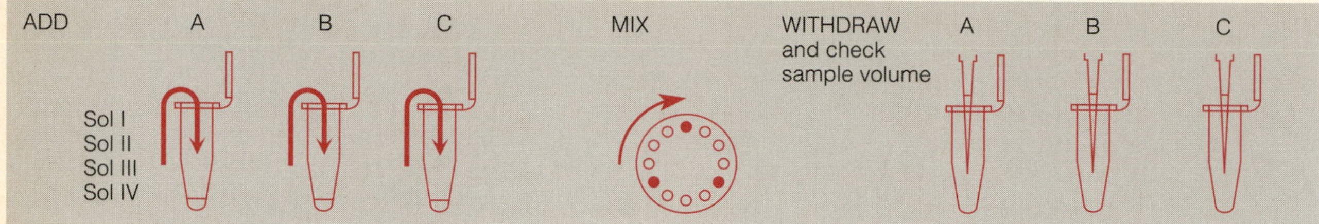

III. Large-Volume Micropipettor

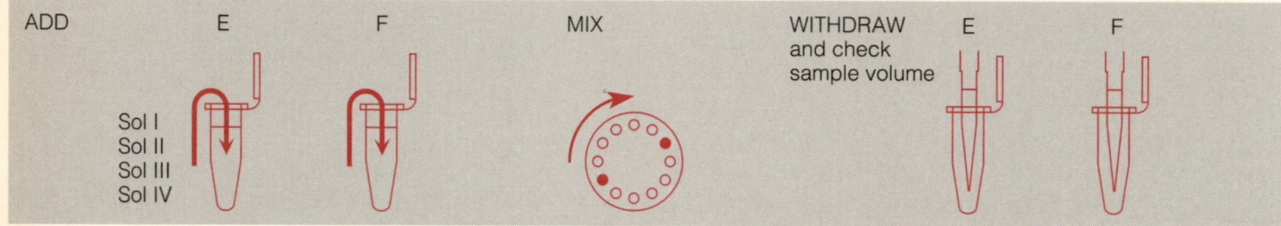

IV. Sterile Use of 10-ml Standard Pipet

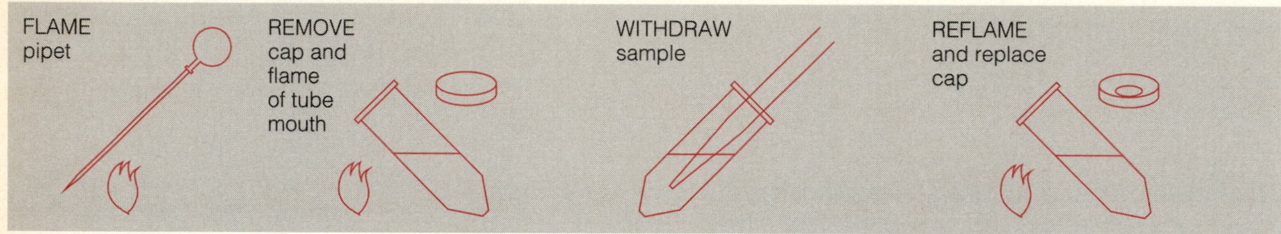

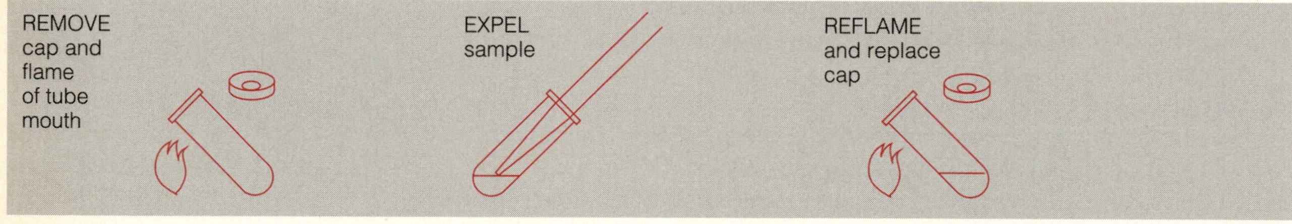

PRELAB NOTES

Metric Conversions

Become familiar with metric units of measurement and their conversions. We will concentrate on liquid measurements based on the liter, but the same prefixes (milli- and micro-) also apply to dry measurements based on the gram. The two most useful units of liquid measurement in molecular biology are the *milli*liter (ml) and *micro*liter (μl).

1 ml = 0.001 liter	1,000 ml = 1 liter
1 μl = 0.000001 liter	1,000,000 μl = 1 liter

Digital Micropipettors

A digital micropipettor is essentially a precision pump fitted with a disposable tip. The volume of air space in the barrel is adjusted by screwing the plunger farther in or out of the piston, and the volume is displayed on a digital readout. Depressing the plunger displaces the specified volume of air from the piston; releasing the plunger creates a vacuum, which draws an equal volume of fluid into the tip. The withdrawn fluid is then expelled by depressing the plunger again.

The volume range of digital micropipettors varies from one manufacturer to another. A small-volume micropipettor (with a range of 0.5–10 μl or 1–20 μl) and a large-volume micropipettor (100–1,000 μl) are used most frequently. A mid-range micropipettor (10–100 μl or 20–100 μl) is used less frequently, but is especially useful for prelab preparation.

Take the following precautions when using a digital micropipettor:

- Never rotate the volume adjustor beyond the upper or lower range of the pipet, as stated by the manufacturer.
- Never use the micropipettor without the tip in place; this could ruin the piston.
- Never invert or lay the micropipettor down with a filled tip; fluid could run back into the piston.
- Never let the plunger snap back after withdrawing or expelling fluid; this could damage the piston.
- Never immerse the barrel of the micropipettor in fluid.
- Never flame the tip of the micropipettor.
- Never reuse a tip that has been used to measure a different reagent.

Microcapillary Pipets

Microcapillary pipets are an inexpensive alternative to adjustable micropipettors. These disposable glass capillary tubes come in various sizes that include the range of volumes used in this course. Several types of inexpensive micropipet aids are available. A thumbscrew micrometer or wire plunger is much easier to control than a rubber bulb. An easily controllable pipet bulb can be made at no extra cost by simply tying a knot in the length of latex tubing that is usually provided with the capillary pipets.

Under conditions of high static electricity, capillary pipetting can be very difficult or even impossible. A reagent droplet can have a greater static attraction to the glass capillary than to the polypropylene test tube to which it must be transferred. Even under ideal circumstances, microcapillary pipets are more difficult to master than adjustable micropipettors. Allow yourself time to become competent with them before attempting any experiments.

Transfer Pipets

Small polypropylene transfer pipets are handy because they have an integrated bulb. The smallest size, which holds a *total* volume of approximately 1 ml, has a thin tip that can be used to measure amounts in microliters. Before using the transfer pipet, calibrate it with a digital micropipet or microcapillary pipet. Pressing on the pipet barrel, rather than the bulb, creates less air displacement and makes measuring small volumes easier.

Sterile 10-ml Pipets

Presterilized, disposable 10-ml plastic pipets are a convenient option for bacterial work and are supplied either in bulk pack or individually wrapped. Bulk-packed pipets should be opened immediately before use. To dispense, cut one corner of the plastic wrapper at the end opposite the pipet tips; avoid touching and contaminating the opening of the wrapper. Tap the bag to push the pipet end through the cut opening. Reclose the wrapper with tape to keep it sterile for future use. To use individually wrapped pipets properly, peel back only enough of the wrapper to expose the wide end of the pipet and affix the end into the pipet aid or bulb. Immediately before use, completely peel back the wrapper.

To Flame or Not to Flame

Scientists disagree about whether it is necessary to flame pipets and the mouths of tubes as part of sterile technique. Flaming kills some of the microbes that tend to accumulate on tube rims. It also warms the air at the mouth of the container, creating an outward convection current that prevents microorganisms from falling into the container. Even so, the effect of flaming may be primarily psychological, when fresh sterile supplies are used and manipulations are done quickly. Especially when individually wrapped supplies are used, flaming can be omitted without fear of compromise to sterility. Flaming is not recommended for presterilized plasticware, because it melts easily.

MEASUREMENTS, MICROPIPETTING, AND STERILE TECHNIQUES

Reagents	Supplies and Equipment
1 ml of Solution I, colored	100–1000-µl micropipettor + tips
1 ml of Solution II, colored	0.5–10-µl micropipettor + tips
1 ml of Solution III, colored	10-ml pipet
1 ml of Solution IV, colored	pipet aid or bulb
25 ml of Solution V, colored	50-ml conical tube
	15-ml culture tube
	1.5-ml tubes
	beaker for waste/used tips
	burner flame (optional)
	microfuge (optional)
	permanent marker
	test tube rack

I. General Use of Digital Micropipettors

(10 minutes)

1. Rotate the volume adjustor to the desired setting. Note the change in plunger length as the volume changes. Be sure to properly locate the decimal point when reading the volume setting.
2. *Firmly* seat a proper-sized tip on the end of the micropipettor.
3. When withdrawing or expelling fluid, always hold the tube firmly between your thumb and forefinger. Hold the tube nearly at eye level to observe the change in the fluid level in the pipet tip. Do not pipet with the tube in the test tube rack. Do not have another person hold the tube while you are pipetting.
4. Each tube must be held in the hand during each manipulation. Open the top of the tube by flipping up the tab with your thumb. During manipulations, grasp the tube body (rather than the lid), to provide greater control and to avoid contamination of the mouth of the tube.
5. For best control, grasp the micropipettor in your palm and wrap your fingers around the barrel; work the plunger (piston) with the thumb. Hold the micropipettor almost vertical when filling it.
6. Most digital micropipettors have a two-position plunger with friction "stops." Depressing to the first stop measures the desired volume. Depressing to the second stop introduces an additional volume of air to blow out any solution remaining in the tip. Notice these friction stops; they can be felt with the thumb.
7. To withdraw the sample from a reagent tube:
   a. Depress the plunger to *first stop* and hold it in this position. Dip the tip into the solution to be pipetted, and draw fluid into the tip by *gradually* releasing the plunger. Be sure that the tip remains in the solution while you are releasing the plunger.

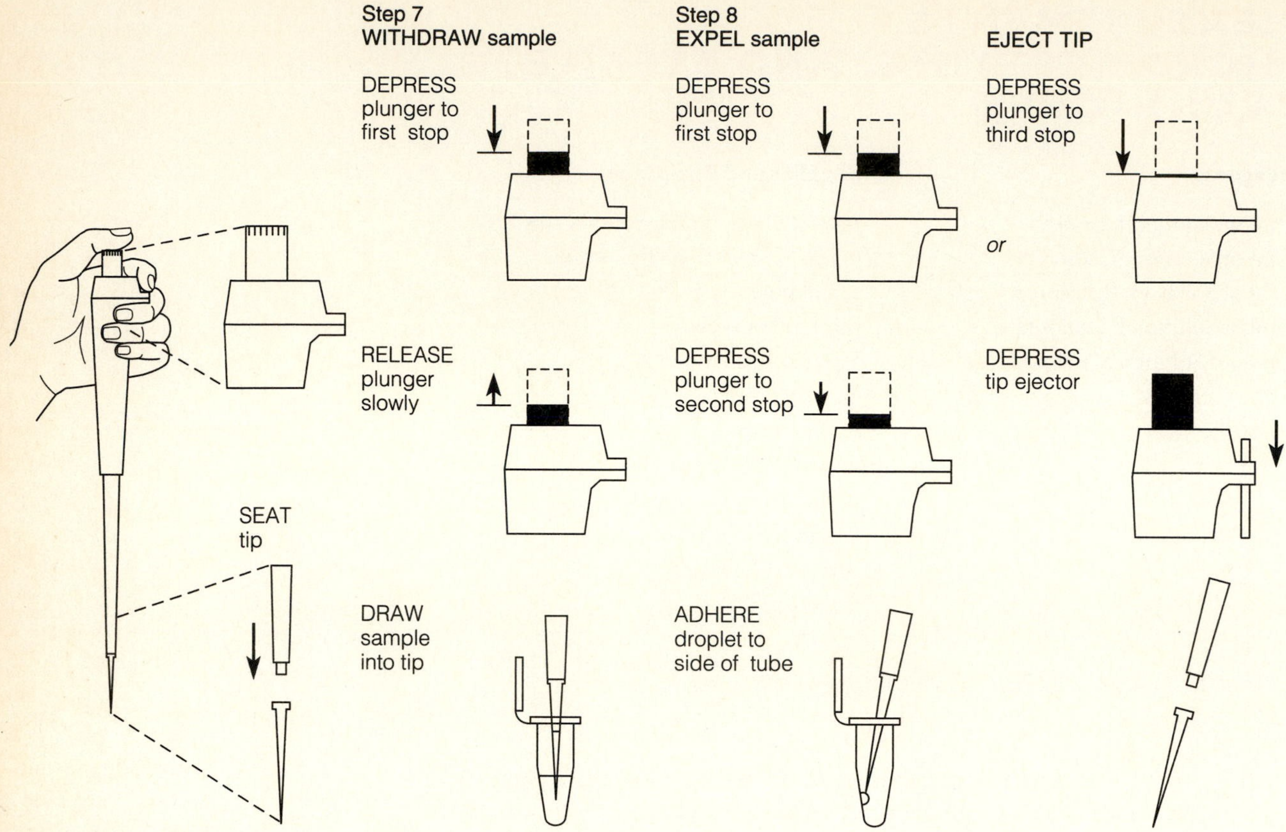

Use of Digital Micropipet (steps 7 and 8)

b. Slide the pipet tip out along the inside wall of the reagent tube to dislodge any excess droplets adhering to the outside of the tip.

c. Check that there is no air space at the very end of the tip. To avoid future pipetting errors, learn to recognize the approximate levels to which particular volumes fill the pipet tip.

d. If you notice air space at the end of the tip or air bubbles within the sample in the tip, carefully expel the sample back into its supply tube. Coalesce the sample by sharply tapping the tube on the bench top or pulsing it in a microfuge.

8. To expel the sample into a reaction tube:

a. Touch the tip of the pipet to the inside wall of the reaction tube into which the sample will be emptied. This creates a capillary effect that helps draw fluid out of the tip.

b. *Slowly* depress the plunger to the first stop to expel the sample. Depress to second stop to blow out the last bit of fluid. Hold the plunger in the depressed position.

c. Slide the pipet out of the reagent tube with the measurement plunger depressed, to avoid sucking any liquid back into the tip.

d. Manually remove or eject the tip into a beaker kept on the lab bench for this purpose. The tip is ejected either by depressing the plunger beyond the second stop or by depressing a separate tip-ejection button.

9. To prevent cross-contamination of reagents:
   a. Always add appropriate amounts of a single reagent sequentially to all reaction tubes.
   b. Release each reagent drop onto a new location on the inside wall, near the bottom of the reaction tube. In this way, the same tip can be used to pipet the reagent into each reaction tube.
   c. Use a *fresh tip* for each new reagent to be pipetted.
   d. If the tip becomes contaminated, switch to a new one.

II. Practice with Small-Volume Micropipettor

(15 minutes)

This exercise simulates setting up a reaction, using a micropipettor with a range of 0.5–10 µl or 1–20 µl.

1. Use a permanent marker to label three 1.5-ml tubes A, B, and C.
2. Use the matrix below as a checklist while adding solutions to each reaction tube.

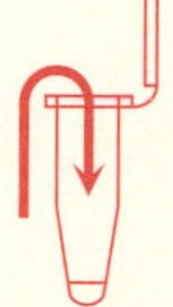

Tube	Sol. I	Sol. II	Sol. III	Sol. IV
A	4 µl	5 µl	1 µl	—
B	4 µl	5 µl	—	1 µl
C	4 µl	4 µl	1 µl	1 µl

3. Set the micropipettor to 4 µl, and add Solution I to each reaction tube.
4. Use a *fresh tip* to add the appropriate volume of Solution II at a clean spot on reaction tubes A, B, and C.
5. Use a *fresh tip* to add 1 µl of Solution III to tubes A and C.
6. Use a *fresh tip* to add 1 µl of Solution IV to tubes B and C.
7. Close the tops. Pool and mix reagents, using one of the following methods:
   a. Sharply tap the tube bottom on the bench top. Be certain that all of the drops have pooled into one drop at the bottom of the tube.

   or

   b. Place in the microfuge and apply a short pulse of several seconds. Make sure that reaction tubes are placed in a *balanced* configuration in the microfuge rotor. Spinning tubes in an unbalanced position will damage the microfuge motor.

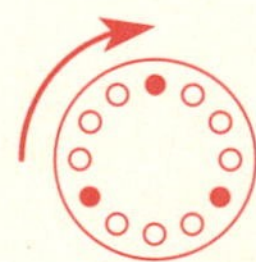

An empty 1.5-ml tube can be used to balance a sample with a volume of 20 µl or less.

8. A total of 10 µl of reagents was added to each reaction tube. To check the accuracy of your measurements, set the pipet to 10 µl and very carefully withdraw the solution from each tube.

   a. Is the tip barely filled?

   or

   b. Does a small volume of fluid remain in the tube? This indicates an overmeasurement.

   or

c. After withdrawing all fluid, is an air space left in the end of the tip? This indicates an undermeasurement. (The actual volume of fluid can be determined by simply rotating the volume adjustment to expel air and push fluid to the very end of the tip. Then, read the volume directly.)

9. If several measurements were inaccurate, repeat the exercise to obtain nearly perfect results.

III. Practice with Large-Volume Micropipettor

(10 minutes)

This exercise simulates a bacterial transformation or plasmid preparation, for which a 100–1000-µl micropipettor is used. It is far easier to mismeasure when using a large-volume micropipettor. If the plunger is not released slowly, an air bubble may form or solution may be drawn into the piston.

1. Use a permanent marker to label two 1.5-ml reaction tubes E and F.
2. Use the matrix below as a checklist while adding solutions to each reaction tube.

Tube	Sol. I	Sol. II	Sol. III	Sol. IV
E	100 µl	200 µl	150 µl	550 µl
F	150 µl	250 µl	350 µl	250 µl

3. Set the micropipettor to add appropriate volumes of solutions I–IV to reaction tubes E and F. Follow the same procedure as for a small-volume pipettor.
4. A total of 1,000 µl of reactants was added to each reaction tube. To check the accuracy of your measurements, set the micropipettor to 1,000 µl and carefully withdraw the solution from each tube.

a. Is the tip barely filled?

or

b. Does a small volume of fluid remain in the tube?

or

c. After withdrawing all fluid, is an air space left in the end of the tip?

5. If your measurements were inaccurate, repeat the exercise to obtain nearly perfect results.

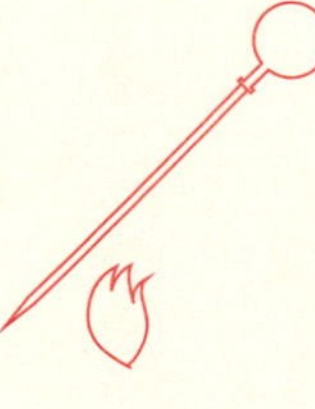

IV. Practice Sterile Use of 10-ml Standard Pipet

(10 minutes)

The following directions include flaming the pipet and mouth of the tube. It is probably best to first learn to flame and then omit flaming when safety or the situation dictates. These directions also assume one-person pipetting, which is rather difficult. The process is much easier when working as a team: One person handles the pipet while the other removes and replaces caps on tubes.

The key to successful sterile technique is to work quickly and efficiently. Before beginning, clear off the lab bench and arrange tubes, pipets, and culture medium within easy reach. Locate the burner in a central position on the lab bench to avoid reaching over a flame.

Loosen caps of tubes so they are ready for easy removal. Remember that the longer a top is off a tube, the greater the chance of microbe contamination. Do not place a sterile cap on a nonsterile lab bench.

CAUTION

Always use a pipet aid or bulb to draw solutions up the pipet. Never pipet solutions using mouth suction: This method is not sterile and can be dangerous.

Nonsterile pipets can be used for this practice exercise.

1. Light burner flame.
2. Set pipet aid to 5 ml or depress pipet bulb.
3. Select a sterile 10-ml pipet, and insert the end into a pipet aid or bulb. Remember to handle only the large end of the pipet; avoid touching the lower two-thirds.
4. Remove the remaining wrapper, and quickly pass the lower two-thirds of the pipet cylinder several times through a flame. Be sure to flame any portion of the pipet that will enter the sterile container. The pipet should become warm but not hot enough to melt a plastic pipet or to cause a glass pipet to crack when immersed in the solution to be pipetted.
5. Hold a 50-ml conical tube containing Solution V in your free hand, and remove the tube cap, using the little finger of the hand holding the pipet aid or bulb. Do not place the cap on the lab bench.
6. Quickly pass the mouth of the conical tube through the burner flame several times; be careful not to melt the plastic pipet.
7. Withdraw 5 ml of Solution V from the conical tube.
8. Reflame the mouth of the tube, and replace the top.
9. Remove the top of the sterile 15-ml culture tube with the little finger of the hand holding the pipet. Quickly flame the mouth of the tube.
10. Expel fluid into the culture tube. Reflame the mouth of the tube and replace the top.

This expels contaminated air and prepares a vacuum to withdraw fluid.

When using an individually wrapped pipet, be careful to open the end of the wrapper opposite the pipet tip. Unwrap only enough of the pipet to attach the end into a pipet aid or bulb.

RESULTS AND DISCUSSION

Inaccurate pipeting and improper sterile techniques are chief contributors to poor laboratory results. If you are still uncomfortable with micropipettors or sterile techniques, take the time now for additional practice. These techniques will soon become second nature to you.

1. Why must tubes be balanced in a microfuge rotor?
2. Use the rotor diagrams below to show how you would balance 1–12 tubes (for the 12-place rotor) and 1–16 tubes (for the 16-place rotor). When balancing an odd number of tubes, first try to make a balanced triangle of three tubes, and then add balanced pairs. Which number of tubes cannot be balanced in the 12- or 16-place rotors?

12-Place Rotor

16-Place Rotor

3. What common error in handling a micropipettor can account for pipetting too much reagent into a tube? What errors account for underpipetting?
4. When is it necessary to use sterile technique?
5. What does flaming accomplish?

FOR FURTHER RESEARCH

Devise a method to determine the percentage of error in micropipetting. Then, determine the percentage of error when you purposely make pipetting mistakes using small- and large-volume micropipettors.

LABORATORY 2

Bacterial Culture Techniques

This laboratory contains most of the culture techniques that are used throughout the course. In Part A, *Escherichia coli* cells are streaked onto LB agar plates so that single cells are isolated from one another. Each cell then reproduces to form a visible colony composed of genetically identical clones. Streaking cells to obtain individual colonies is usually the first step in genetic manipulations of microorganisms. Using cells derived from a single colony minimizes the chance of using a cell mass that has been contaminated with a foreign microorganism. To demonstrate antibiotic resistance, the growth of wild-type *E. coli* and of an *E. coli* containing an ampicillin-resistance gene are compared, using LB medium containing ampicillin. The resistant strain contains the plasmid pAMP, which produces an enzyme that destroys the ampicillin in the medium, thus allowing these cells to grow.

Part B provides a protocol for growing small-scale suspension cultures of *E. coli* that reach stationary phase with overnight incubation. Overnight cultures are used for purification of plasmid DNA and for inoculating mid-log cultures. When growing *E. coli* strains that contain a plasmid, it is best to maintain selection for antibiotic resistance by using LB broth containing an appropriate antibiotic. Strains containing an ampicillin-resistance gene such as pAMP should be grown in LB broth with ampicillin.

Part C provides a protocol for culturing *E. coli* to mid-log phase, the growth stage in which cells can be rendered most competent to uptake plasmid DNA. The protocol begins with an overnight suspension culture of *E. coli,* which contains a high proportion of healthy cells capable of further reproduction. The object is to subculture (reculture) a small volume of the overnight culture in a larger volume of fresh nutrient broth. This "resets" the culture to zero growth where, after a short lag phase, the cells enter the log-growth phase. As a general rule, one volume of overnight culture (the *inoculum*) is added to 100 volumes of fresh LB broth in an Erlenmeyer flask. To provide good aeration for bacterial growth, the flask volume should be at least four times the total culture volume.

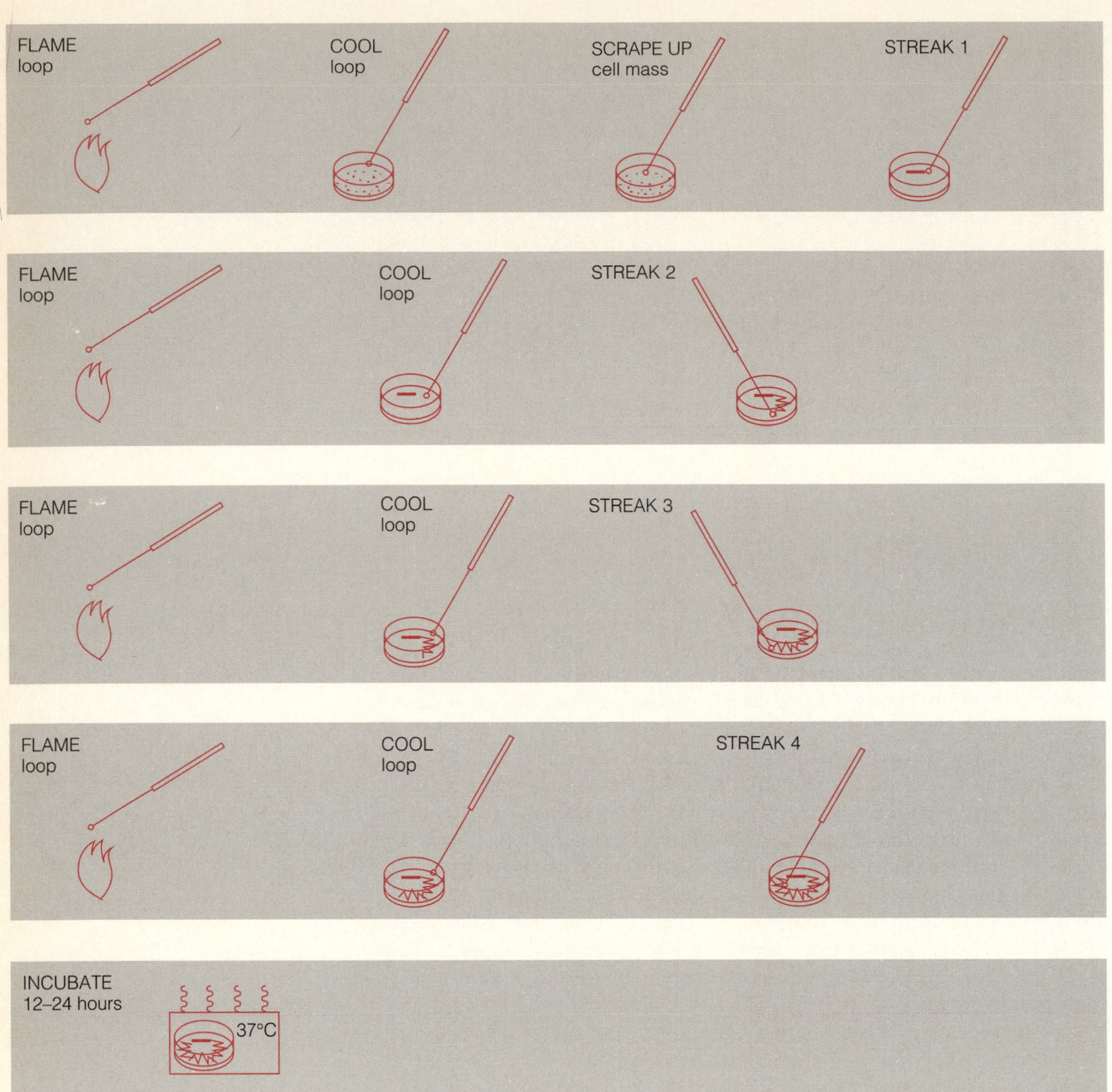

Laboratory 2/Part A
Isolation of Individual Colonies

PRELAB NOTES

E. coli Strains

All protocols involving bacterial growth, transformation, and plasmid isolation have been tested and optimized with two *E. coli* strains: MM294, developed in the laboratory of Matthew Messelson at Harvard University, and JM101, developed in the laboratory of Joachim Messing at the Max Planck Institute in Munich and later at the University of California, Davis. Laboratories 5–10 make use of MM294 and plasmids pAMP and pKAN (and their derivatives). Laboratories 14–21 make use of JM101 and the plasmid pBLU and its derivatives. Other strains commonly used for molecular biological studies should give comparable results. However, growth properties of other *E. coli* strains in suspension culture may differ significantly. For example, the time needed to reach mid-log phase and the cell number represented by specific optical densities differ from one strain to another.

Nutrient Agar

Almost any rich nutrient agar can be used for plating cells, although we prefer LB (Luria-Bertani) agar. Presterilized, ready-to-pour agar is a great convenience. It only needs to be melted in a microwave oven or boiling water bath, cooled to approximately 60°C, and poured onto sterile culture plates.

CAUTION

To prevent boiling over, the agar container should be no more than half full. Loosen the cap to prevent bottle from exploding.

Ampicillin

Plasmids having ampicillin resistance are most commonly used for cloning DNA sequences in *E. coli.* Ampicillin is very stable in agar plates, thresholds for selection are relatively broad, and contaminants are infrequent. Despite its stability, ampicillin, like most antibiotics, is inactivated by prolonged heating. Therefore, it is important to allow the agar solution to cool until the container can be held comfortably in the hand (about 60°C) before adding the antibiotic. Use sodium salt, which is very soluble in water, instead of free acid form, which is difficult to dissolve.

Prewarming the Incubator

Inexpensive classroom incubators are notorious for fluctuating temperatures and miscalibrated thermostats. Therefore, always prewarm the incubator. Make temperature adjustments several days before use. It is important to make these adjustments after the incubator has been warmed for at least 12 hours, *with the door closed:* Heat will build up during the 12–24 hour incubation. When in doubt, err on the cool side. *E. coli* will grow very well (though more slowly) several degrees below 37°C but can be killed by temperatures several degrees above 37°C.

Responsible Handling and Disposal of *E. coli*

E. coli is a commensal organism of *Homo sapiens,* and a normal part of the bacterial flora of the human gut. It is not considered pathogenic and is rarely associated with illness in healthy individuals. Furthermore, K-12-derived *E. coli* strains—including MM294, JM101, and the most commonly used laboratory strains—are ineffective in colonizing the human gut and lack the cell-surface "O antigen" responsible for virulence. Adherence to simple guidelines for handling and disposal makes working with *E. coli* a nonthreatening experience.

1. To avoid contamination, always reflame the inoculating loop or cell spreader one final time before placing it on the lab bench.
2. Keep your nose and mouth away from the tip end, when pipetting suspension culture, to avoid inhaling any aerosol that might be created.
3. Do not overincubate plates. Because a large number of cells are to be inoculated, *E. coli* is generally the only organism that will appear on plates incubated for 12–24 hours. However, with a longer incubation, contaminating bacteria and slower-growing fungi can arise. If the plates cannot be observed following the initial incubation, refrigerate them to retard the growth of contaminants.
4. Collect for treatment bacterial cultures *and* tubes, pipets, and micropipettor tips that have come into contact with the cultures. Disinfect these materials as soon as possible after use. Contaminants, often odorous and sometimes potentially pathogenic, can be readily cultured over a period of several days at room temperature. Disinfect bacteria-contaminated materials in one of two ways:
   a. Autoclave the materials at 121°C for 15 minutes. Tape three or four culture plates together and loosen the tube caps before autoclaving. Collect contaminated materials in a "bio bag" or heavy-gauge trash bag; seal bag before autoclaving. Dispose of autoclaved materials in the regular garbage.

   or

   b. Treat the materials with a solution containing 5,000 parts per million (ppm) available chlorine (10% bleach solution). Immerse contaminated pipets, tips, and tubes (open) directly into a sink or tub containing the bleach solution. Plates should be placed in a sink or tub and flooded with bleach solution. Allow the materials to stand in the bleach solution for 15 minutes or longer. Then, drain any excess bleach solution, seal the materials in a plastic bag, and dispose of them in the regular garbage.
5. Wipe down the lab bench with soapy water, 10% bleach solution, or a disinfectant (such as Lysol) at the end of the lab session.
6. Wash your hands before leaving the laboratory.

ISOLATION OF INDIVIDUAL COLONIES

Cultures and Plates	**Supplies and Equipment**
MM294 culture	inoculating loop
MM294/pAMP culture	"bio bag" or heavy-duty trash bag
2 LB agar plates	10% bleach or disinfectant
2 LB + ampicillin (LB/amp) plates	burner flame
	37°C incubator
	permanent marker

Streak and Incubate Plates

(15 minutes)

Plan out your manipulations before beginning to streak the plates. Organize the lab bench to allow yourself plenty of room, and work quickly. If you are working from a stab or slant culture, loosen the cap before starting.

Make it a habit to mark the plate bottoms; these always stay attached to the agar!

1. Use a permanent marker to label the *bottom* of each agar plate with your name and the date. Each plate will have been previously marked to indicate whether it is plain LB agar (LB) or LB agar + ampicillin (LB/amp).
2. Select the two LB plates. Mark one plate –pAMP, for cells without plasmid, and the other +pAMP, for cells with plasmid.
3. Select the two LB/amp plates. Mark one –pAMP, for cells without plasmid, and the other +pAMP, for cells with plasmid.
4. Hold the inoculating loop like a pencil, and sterilize the loop in the burner flame until it glows red hot. Then pass the entire wire section (up to the handle) through the flame.

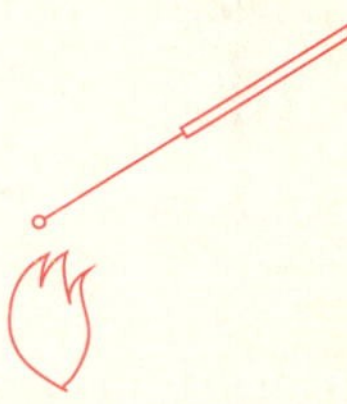

5. Cool the inoculating loop for 5 seconds. *To avoid contamination, do not set the loop on the lab bench.*
6. If you are working from a culture plate:
   a. Remove the lid from the *E. coli* culture plate with your free hand. *Do not place the lid on the lab bench.* Hold the lid face down just above the culture plate, to help prevent contaminants from falling onto the plate or lid.
   b. Stab the inoculating loop several times into a clean area of the agar to cool the loop.

   c. Use the loop tip to scrape up a *visible* cell mass from one or several colonies. Be careful not to gouge the agar. Replace the culture plate lid, and proceed to step 7.

   or

   If you are working from a stab culture:

   a. Grasp the bottom of the vial containing the *E. coli* culture between the thumb and two fingers of your free hand. Remove the vial cap, using the little finger of the same hand that holds the inoculating loop. *Avoid touching the rim of the cap.*
   b. Quickly pass the mouth of the vial several times through the burner flame.
   c. Stab the inoculating loop into the side of the clean agar several times to cool it.

d. Stab the loop several times into the area of the culture where bacterial growth is apparent. Remove the loop, flame the mouth of the vial, and replace the cap. Proceed to step 7.

7. Select the LB –pAMP plate, and lift the lid only enough to perform streaking, as shown below. *Do not set the lid on the lab bench.*

   a. *Streak 1:* Glide the tip of the inoculating loop back and forth across the agar surface to make a streak across the top of the plate. Avoid gouging the agar. Replace the lid between streaks.

   b. *Streak 2:* Reflame the inoculating loop, and then cool it by stabbing it into a clean area of the agar. Draw the loop tip once through the first streak, and without lifting the loop, make a tight zigzag streak across one-fourth of the agar surface. Replace the plate lid.

   c. *Streak 3:* Reflame the inoculating loop, and cool it in the agar, as before. Draw the loop tip once through the last lines of the second streak, and make another zigzag streak in an adjacent quarter of the plate—*without retouching the previous streak.*

   d. *Streak 4:* Reflame the loop and cool it. Draw the tip once through the third streak, and make a final zigzag streak in the remaining quarter of the plate.

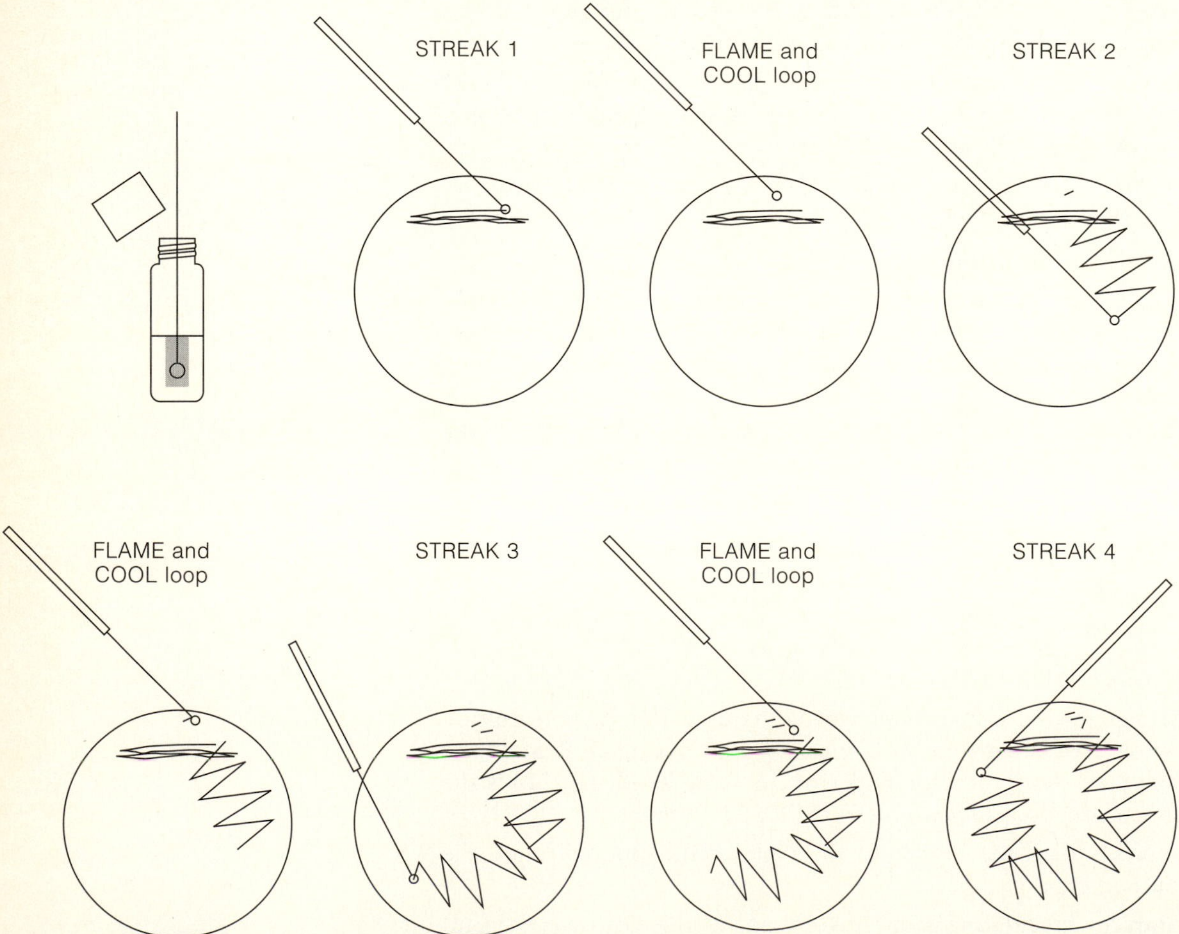

Streaking to Isolate Individual Colonies (step 7)

8. Repeat steps 4–7 to streak *E. coli* onto the LB/amp –pAMP plate.
9. Repeat steps 4–7 to streak *E. coli* pAMP onto the LB +pAMP plate.
10. Repeat steps 4–7 to streak *E. coli* pAMP onto the LB/amp +pAMP plate.
11. Reflame the loop, and allow it to cool before placing it on the lab bench. Make it a habit to always flame the loop one last time before incubation.
12. Place plates upside down in a 37°C incubator, and incubate them for 12–24 hours.
13. Take the time for proper cleanup.
    a. Segregate bacterial cultures for proper disposal.
    b. Wipe down the lab bench with soapy water, a 10% bleach solution, or disinfectant (such as Lysol) at the end of the lab session.
    c. Wash your hands before leaving the laboratory.

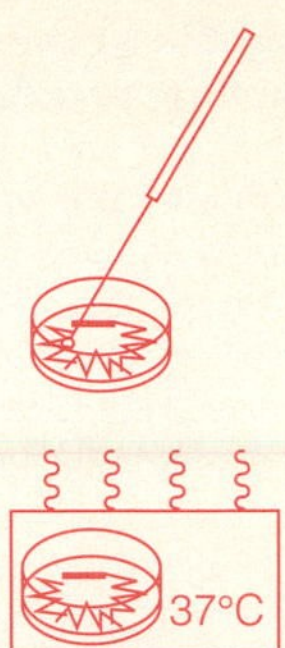

During incubation, water vapor may condense on the upper surface of the plate, form droplets, and fall onto the lower surface. The plates are inverted to prevent water droplets from falling onto the agar and causing *E. coli* colonies to run together. Condensation will be most noticeable in plates poured while the medium is still quite hot.

RESULTS AND DISCUSSION

This laboratory demonstrates a method for streaking bacteria to single colonies. It also introduces antibiotics and plasmid-borne resistance to antibiotics, topics that will be important in several laboratories that follow.

There are two classes of antibiotics: *bacteriostats,* which prevent cell growth, and *bacteriocides*, which kill cells outright. The antibiotics used in this course, ampicillin and kanamycin, are both bacteriocides, although their modes of action are quite different. Ampicillin, a derivative of penicillin, blocks synthesis of the peptidoglycan layer that lies between *E. coli's* inner and outer cell membranes. Thus, ampicillin does not affect existing cells with intact cell envelopes but kills dividing cells as they synthesize new peptidoglycan. Kanamycin (which will be introduced in later laboratories) is a member of the aminoglycoside family of antibiotics, which block protein synthesis by covalently modifying the bacterial ribosome. Thus, kanamycin kills both dividing and quiescent cells.

The ampicillin-resistance gene carried by the plasmid pAMP produces a protein, β-lactamase, that disables the ampicillin molecule. β-Lactamase cleaves a specific bond in the β-lactam ring, a four-membered ring in the ampicillin molecule that is essential to its antibiotic action. β-lactamase not only disables ampicillin within the bacterial cell, but also disables it in the surrounding medium when it leaks through the cell envelope. The kanamycin-resistance gene of the plasmid pKAN produces the enzyme kanamycin phosphotransferase, which prevents kanamycin from interacting with the ribosome.

Antibiotic-Resistant Growth

After 12–24 hours of incubation, colonies should range in diameter from 0.5 to 3 millimeters (mm). If you are unable to observe the plates on the day after streaking, store them at 4°C to arrest *E. coli* growth and to slow the growth of any contaminating microbes. Wrap them in Parafilm or plastic wrap to retard drying.

Peptidoglycan Biosynthesis, Ampicillin Action, and Ampicillin Resistance
(1) A transpeptidase removes an alanine residue (A) from a pentapeptide (PEP^5). The resulting tetrapeptide (PEP^4) is joined to the peptide bridge (B) to cross-link two adjacent polysaccharide chains (P). (2) The β–lactam ring of ampicillin structurally mimics the peptide bridge and irreversibly binds the transpeptidase, making it unavailable for peptidoglycan synthesis. (3) The pAMP resistance protein, β-lactamase, cleaves the β-lactam ring of ampicillin, making it unable to bind the transpeptidase.

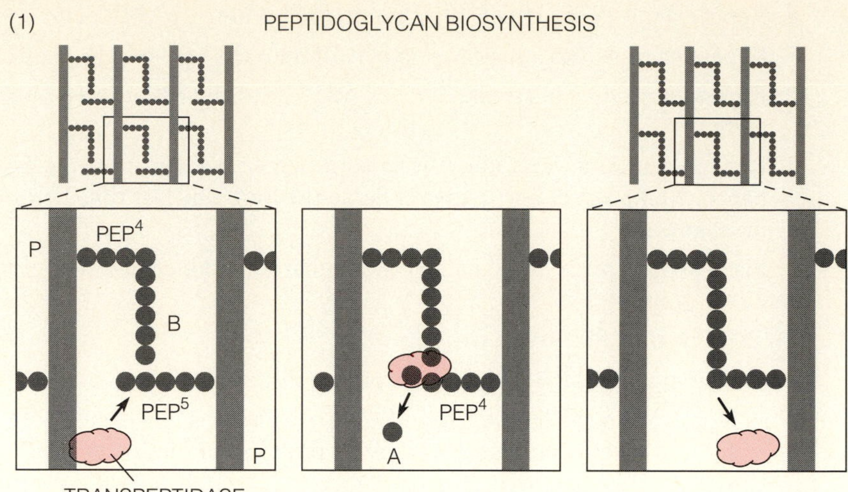

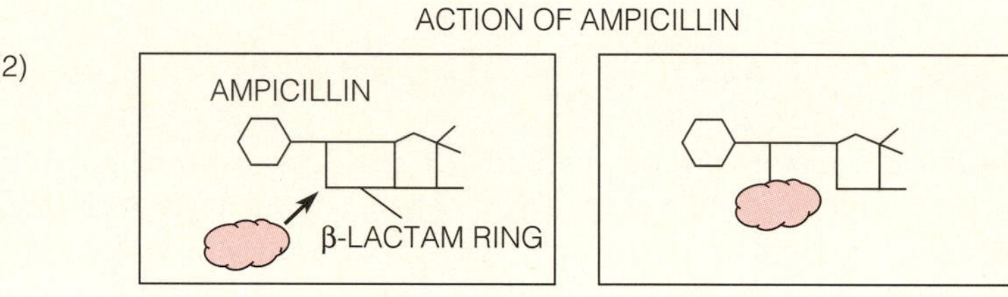

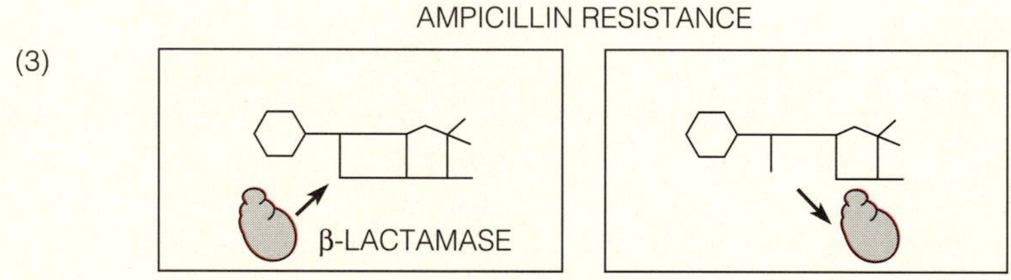

Observe the plates, and use the matrix below to record which plates have bacterial *growth* and which have *no growth*. On plates with growth, distinct, individual colonies should be observed within one of the streaks.

On the LB/amp plate, growth must be observed in the secondary streak to count as antibiotic-resistant growth. In a heavy inoculum, nonresistant cells in the primary streak can be isolated from the antibiotic on a bed of other nonresistant cells. Although these cells may be growing on top of one another, *they will not have expanded* into the medium outside of the width of the original streak.

	Transformed Cells +pAMP	**Wild-Type Cells −pAMP**
LB/amp	experiment	negative control
LB	positive control	positive control

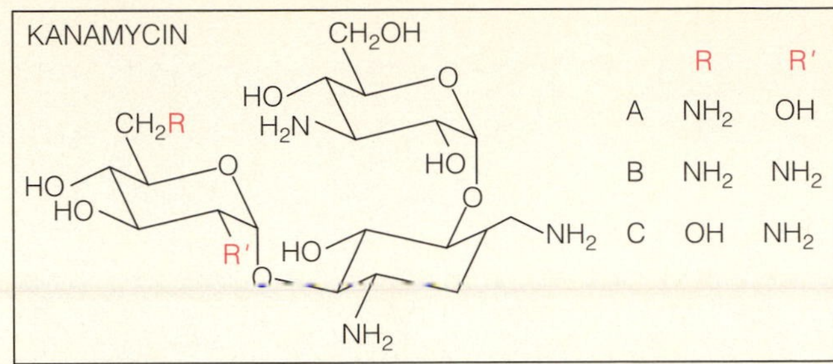

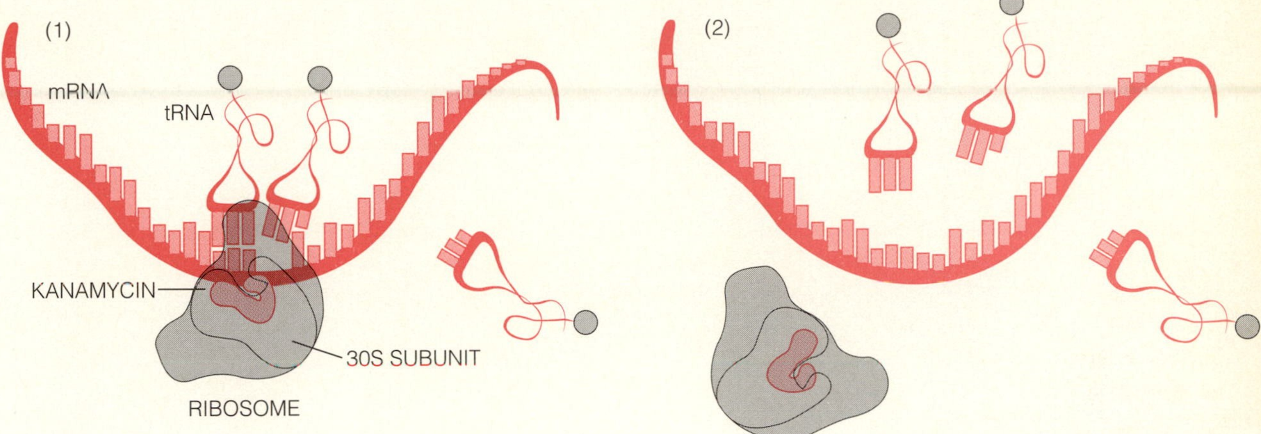

Kanamycin Action
Kanamycin poisons protein synthesis by irreversibly binding the 30S subunit of the ribosome. (1) The kanamycin/ribosome complex initiates protein synthesis by binding mRNA and the first tRNA. (2) However, the second tRNA is not bound, and the ribosome-mRNA complex dissociates. The pKAN resistance protein, from the aminoglycoside family, is a phosphotransferase that adds phosphate groups to the kanamycin molecule, thus blocking its ability to bind the ribosome. Other resistance proteins from this family can be acetyltransferases, which add acetyl groups to the antibiotic. Three forms of the phosphotransferase (A, B, or C) vary by the group present at R and R′, as indicated in the figure.

On the LB/amp +pAMP plate, tiny "satellite" colonies may be observed radiating from the edges of large, well-established colonies. These satellite colonies are ampicillin-sensitive cells that begin to grow in an "antibiotic shadow," where the antibiotic in the media has been broken down by the large resistant colony. Satellite colonies are generally a sign that the antibiotic has been weakened either by being added to medium that has not been cooled enough, by long-term storage of more than 30 days, or by overincubation.

1. Were the results as you expected? Explain the possible causes for variations from the expected results.
2. In step 7:
   a. What is the reason for the zigzag streaking pattern?
   b. Why is the inoculating loop resterilized between each new streak?
   c. Ideally, why should a new streak intersect the previous one at only a single point?
3. Describe the appearance of a single *E. coli* colony. Why can it be considered genetically homogeneous?
4. Upcoming laboratories use cultures of *E. coli* cells derived from one or several discrete parental colonies isolated by streaking. Why is it important to use this type of culture in genetic experiments?
5. *E. coli* strains containing the plasmid pAMP are resistant to ampicillin. Describe how this plasmid functions to bring about resistance.

Inoculate and Incubate Overnight Culture

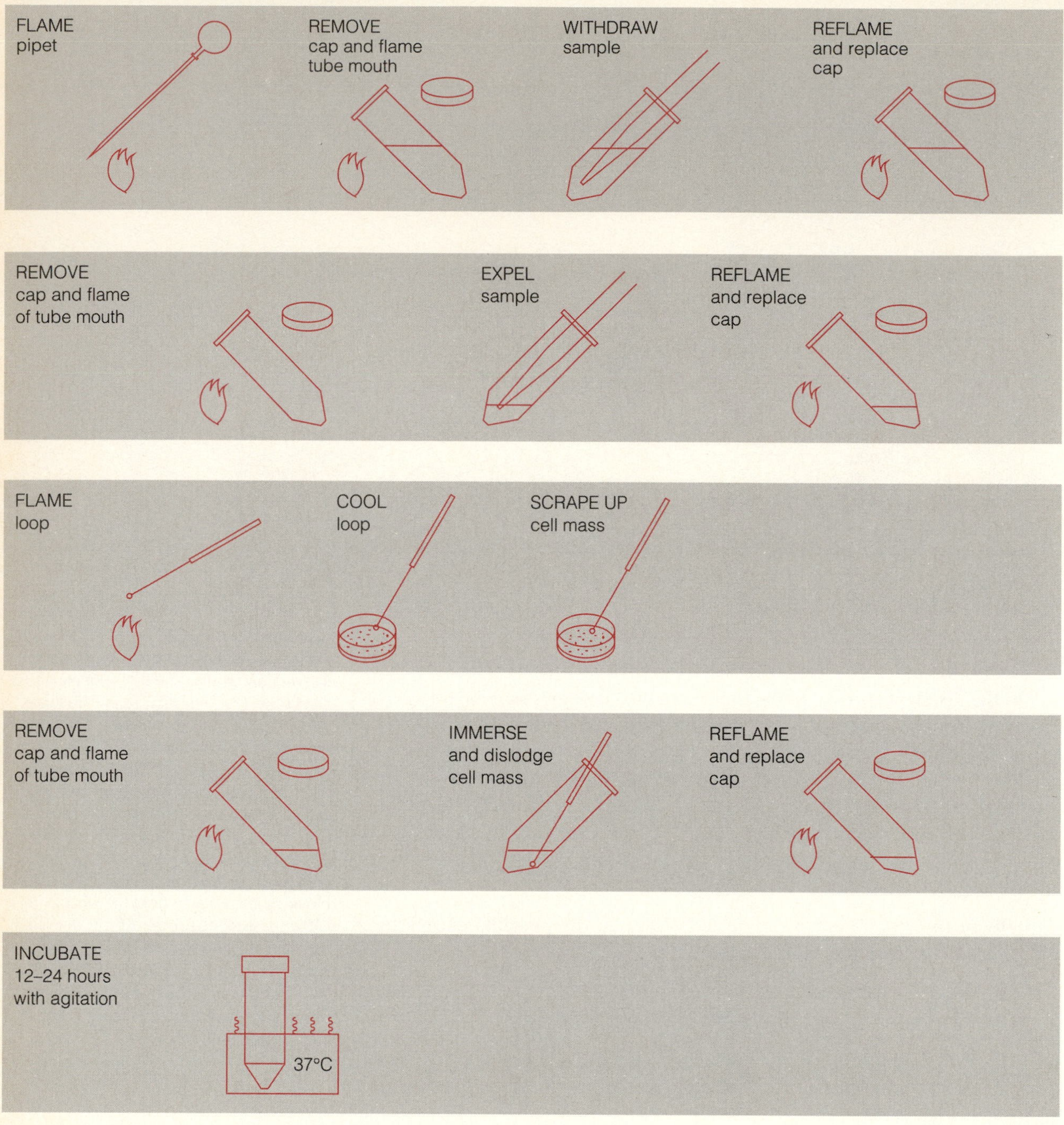

Laboratory 2/Part B
Overnight Suspension Culture

PRELAB NOTES

Review the Prelab Notes in Laboratory 2A, "Isolation of Individual Colonies."

Although it is best to grow an overnight culture with shaking (aeration), it is not essential when growing cells for purifying plasmid or for inoculating a larger culture. For these purposes, suspensions can be incubated, without shaking, in a rack within a 37°C incubator. Since cultures grow more slowly without aeration, incubate the suspension for 2–3 days to obtain an adequate number of cells.

Ironically, a 15-ml culture tube is not the best choice for *E. coli* growth. A 50-ml conical tube is preferable for growing overnight cultures, because it provides greater surface area for aeration. When culturing transformed strains for plasmid preparations, it is best (but not essential) to use antibiotic media to maintain selection. It is also prudent to inoculate a backup overnight culture, just in case the first one was not properly inoculated.

Overnight Suspension Culture

Culture and Media	Supplies and Equipment
E. coli plate	inoculating loop
LB broth (w/appropriate antibiotic)	10-ml standard pipet
	50-ml conical tube, sterile
	pipet aid or bulb
	10% bleach or disinfectant
	burner flame
	permanent marker
	37°C shaking water bath *or* 37°C dry shaker (*or* dry shaker in 37°C incubator)

Inoculate and Incubate Overnight Culture

(10 minutes)

Review "Practice Sterile Use of 10-ml Standard Pipet," Section IV in Laboratory 1, "Measurements, Micropipetting, and Sterile Techniques." Think sterile! A pipet should be considered contaminated whenever the tip or lower shaft comes into contact with anything in the environment—lab bench, hands, or clothing. When contamination is suspected, discard the pipet, and start again with a fresh one. Plan out the steps to be performed, organize your lab bench, and work quickly.

1. Label a sterile 50-ml tube with your name and the date.
2. Use a 10-ml pipet to sterilely transfer 5 ml of LB broth into the tube.
   a. Attach pipet aid or bulb to the pipet. Briefly flame the pipet cylinder.
   b. Remove the cap of the LB bottle, using the little finger of the hand holding the pipet bulb. Flame the mouth of the LB bottle.
   c. Withdraw 5 ml of LB broth. Reflame the mouth of the bottle, and replace the cap.

If working as a team, one partner can handle the pipet, and the other handles the tubes and caps.

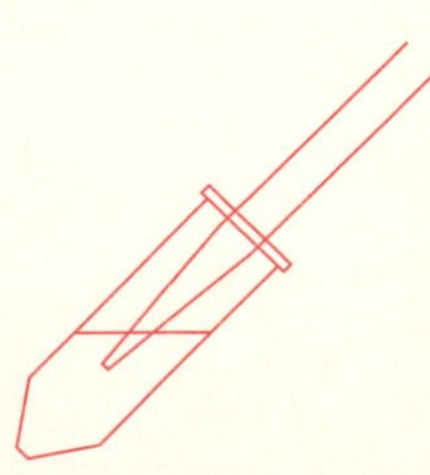

Pipet flaming can be eliminated if individually wrapped pipets are used.

d. Remove the cap of the sterile 50-ml culture tube. Briefly flame the mouth of the culture tube.

e. Expel the sample into the tube. Briefly reflame the mouth of the tube, and replace the cap.

3. Locate a well-defined colony 1–4 mm in diameter on a freshly streaked plate.
4. Sterilize the inoculating loop in the burner flame until it glows red hot. Then pass the entire wire section (up to the handle) through the flame.

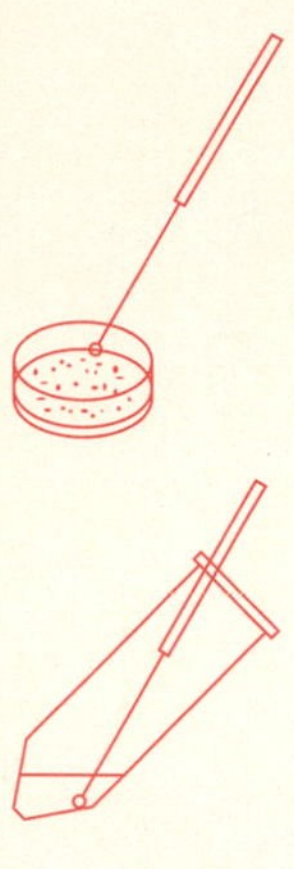

5. Cool the tip of the loop by stabbing it several times into a clean area of the agar.
6. Use the loop to scrape up a visible cell mass from one selected colony.
7. Sterilely transfer the colony into the culture tube:
   a. Remove the cap of the culture tube, using the little finger of the hand holding the loop.
   b. Briefly flame the mouth of the culture tube.
   c. Immerse the tip of the loop in the broth, and agitate it to dislodge the cell mass.
   d. Briefly reflame the mouth of the culture tube, and replace the cap.
8. Reflame the loop before setting it on the lab bench.
9. Loosely replace the cap of the tube to allow air to flow into the culture. Affix a loop of tape over the cap to prevent it from becoming dislodged during shaking.

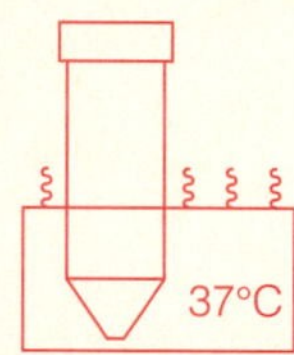

Cultures for plasmid preparations can be incubated at 37°C, without shaking, for 2 days or longer. Following initial incubation, the culture can be stored at room temperature for several days, until you are ready to use it.

10. Incubate the culture for 12–24 hours at 37°C, preferably with continuous agitation.
11. Take the time for proper cleanup.
    a. Segregate for proper disposal bacterial cultures *and* tubes and pipets that have come into contact with cultures.
    b. Wipe down the lab bench with soapy water, a 10% bleach solution, or disinfectant (such as Lysol).
    c. Wash your hands before leaving the laboratory.

RESULTS AND DISCUSSION

E. coli has simple nutritional requirements and grows slowly on a minimal medium containing (1) an energy source such as glucose, (2) salts such as NaCl and $MgCl_2$, (3) the vitamin biotin, and (4) the nucleoside thymidine. *E. coli* synthesizes all necessary vitamins and amino acids from these precursors. It grows rapidly in a complete medium, such as LB, in which yeast extract and hydrolyzed milk protein (casein) provide a ready supply of vitamins and amino acids.

A liquid bacterial culture goes through a series of growth phases. For approximately 30 minutes following inoculation, there is a *lag phase* during which there is no cell growth. The bacteria begin dividing rapidly during *log phase*, when the number of cells doubles every 20–25 minutes. As nutrients in the media are depleted, cells stop dividing and enter the *stationary phase*, with a concentration of approximately 10^9 cells/ml. During the *death phase*, waste products accumulate and the cells begin to die.

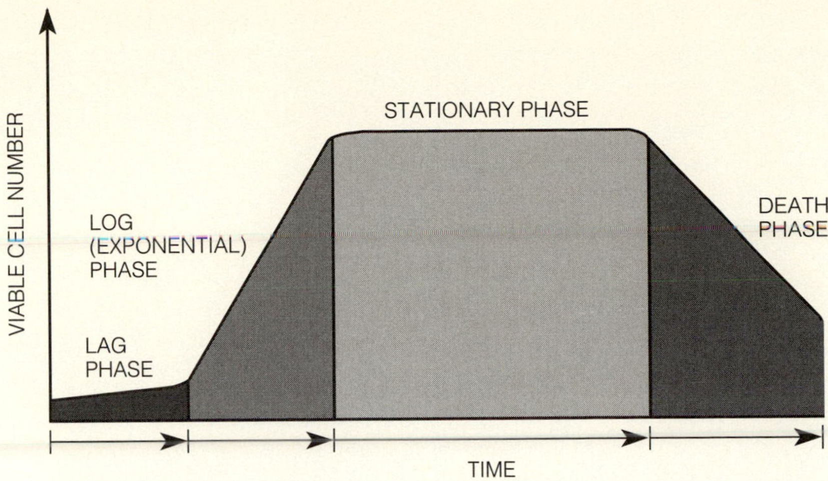

***E. Coli* Growth Curve**

Optimum growth in liquid culture is achieved with continuous agitation, which aerates the cells, facilitates the exchange of nutrients, and flushes away waste products of metabolism. It can safely be assumed that a culture in complete medium has reached stationary phase, following overnight incubation with continuous shaking.

A culture in the stationary phase will look very cloudy and turbid. Discard any overnight culture where vigorous growth is not evident. Expect less growth in cultures incubated for several days *without continuous shaking*. To gauge growth, shake the tube to suspend the cells that have settled at the bottom of the tube.

1. Why is 37°C the optimum temperature for *E. coli* growth?
2. Why is it ideal to provide continuous shaking for a suspension culture? Give two reasons.
3. What growth phase is reached by a suspension of *E. coli* following overnight shaking at 37°C?
4. Approximately how many *E. coli* cells are in a 5-ml suspension culture at stationary phase?
5. If a spectrophotometer can read accurately within the range of 0.1–1.0 optical density units, then how can you obtain the absorbance of a stationary culture (which may have an A_{550} of 8.0)?

Inoculate and Incubate Culture

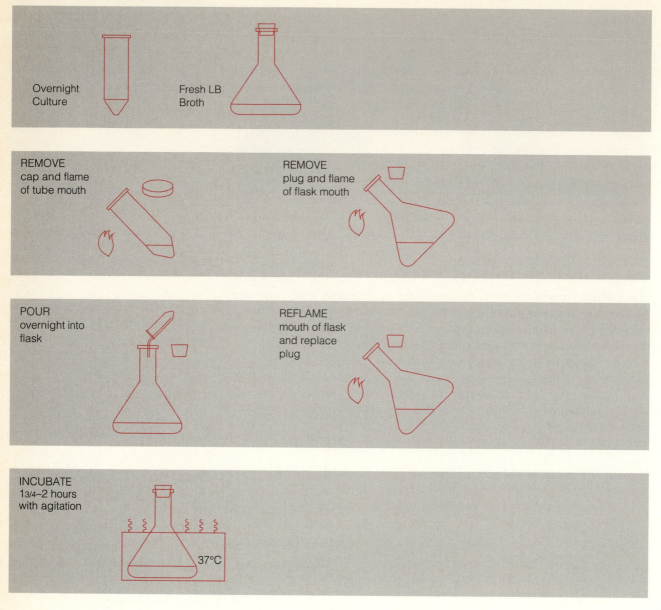

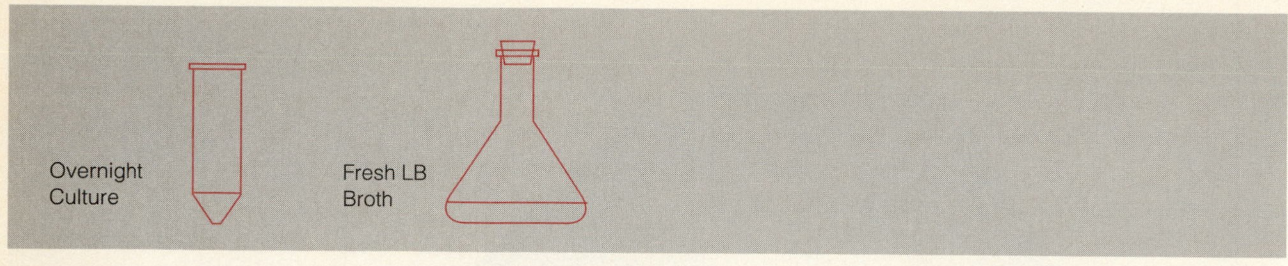

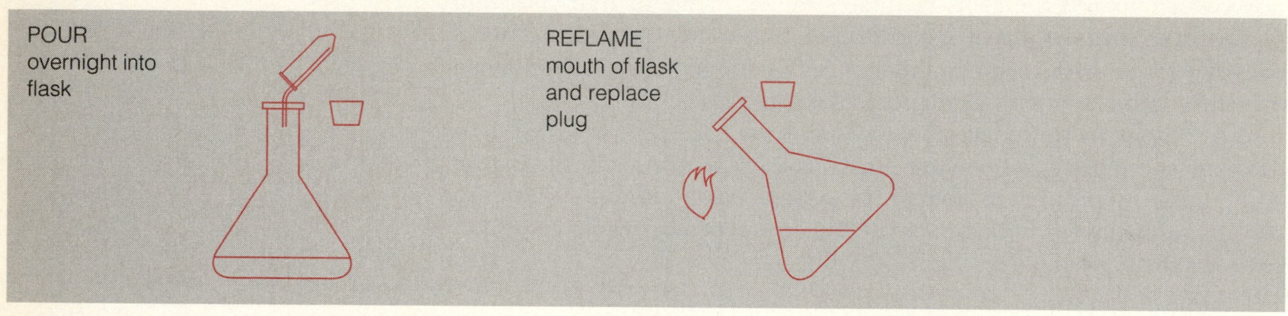

Laboratory 2/Part C
Mid-Log Suspension Culture

PRELAB NOTES

Competent Cell Yield

In Laboratories 8, 15, and 20, "the Classic Procedure for Preparing Competent Cells," each experiment requires 10 ml of mid-log cells, which yields 1 ml of competent cells. If competent cells are being prepared in large quantity for group use, remember that the ratio of mid-log cells to competent cells is 10:1. A 100-ml mid-log culture will yield 10 ml of competent cells, sufficient for fifty 200-μl transformations. If you are preparing cells for freezer storage, aliquot fifty 200-μl samples in sterile tubes—be sure to include 15% sterile glycerol.

Sterile Technique

Scrupulous sterile technique must be used when preparing overnight and mid-log cultures. No antibiotic is used, so any contaminant will multiply as cells are repeatedly manipulated or stored for future use.

Culture Aeration

Proper temperature, aeration, and nutrient exchange—as provided by a shaking incubator—are essential to achieve vigorous, predictable cell growth. A shaking water bath or temperature-controlled dry shaker is most commonly used, but one economical alternative is to place a small platform shaker inside a 37°C incubator. To provide a large surface-to-volume ratio for aeration, culture cells in an Erlenmeyer flask with a volume of LB broth that is no more than one-fourth the total flask volume.

Timing of Culture

Inoculate a mid-log culture 2–4 hours before you plan to perform the laboratory. Using the protocol below, *E. coli* strains MM294 and JM101 will reach mid-log phase after 90–110 minutes of incubation. Cells can either be used immediately or stored on ice for as long as 2 hours before beginning the laboratory.

The amount of time a culture takes to reach mid-log phase is likely to be affected by any change in protocol. For example, a culture inoculated with an overnight culture that was grown without shaking will take longer to reach mid-log phase than one inoculated with shaken cells. A culture begun by inoculating into LB broth prewarmed to 37°C will reach mid-log phase more quickly than one begun at room temperature. Different strains of *E. coli* display different growth properties. Strains MM294 and JM101 can also exhibit different growth properties in a nutrient broth other than LB broth.

Mid-Log Suspension Culture

Cultures and Media

MM294 overnight culture
LB broth, sterile

Supplies and Equipment

500-ml Erlenmeyer flask, sterile
10-ml pipets, sterile (optional)
10% bleach or disinfectant
burner flame
37°C shaking water bath (or 37°C dry shaker *or* dry shaker + 37°C incubator)
spectrophotometer (optional)

Inoculate and Incubate Culture

(2 hours, including incubation)

The time estimate in step 4 is based on inoculation of LB at room temperature. Cells will reach the mid-log phase more quickly if the overnight culture is inoculated into LB that is prewarmed to 37°C.

1. Sterilely transfer 2.5 ml of overnight culture into 250 ml of LB broth at *room temperature.*
2. If you are using a 1-ml overnight culture:
   a. Remove the cap from the overnight culture tube, and flame the mouth. *Do not place the cap down on the lab bench.*
   b. Remove the plug from the flask, and flame the mouth.
   c. Pour the entire overnight culture into the flask. Reflame the mouth of the flask, and replace the plug. Swirl the culture to mix it.

   or

   If you are transferring only a portion of a larger overnight culture:
   a. Flame the pipet cylinder.
   b. Remove the cap from the overnight culture tube, and flame the mouth of the tube.
   c. Withdraw 1 ml of overnight suspension. Reflame the mouth of the overnight culture tube, and replace the cap.
   d. Remove the plug from the flask, and flame the mouth.
   e. Expel the overnight sample into the flask. Reflame the mouth of the flask, and replace the plug. Swirl the culture to mix it.

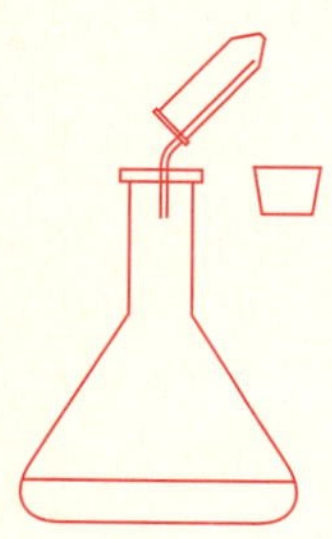

3. Incubate at 37°C with continuous shaking.
4. *If a spectrophotometer is available:* Approximately 1 hour after inoculating, sterilely withdraw a sample of culture, and measure the absorbance (optical density at 550 mm). Repeat this procedure at approximately 20-minute intervals until the culture reaches an OD_{550} of 0.3–0.4 (90–110 minutes for MM294 or JM101).

   or

   If a spectrophotometer is not available: You can safely assume that an MM294 or JM101 culture has reached OD_{550} 0.3–0.4 after 1.75 hours of incubation with continuous shaking (*based on inoculation of the LB broth at room temperature*).

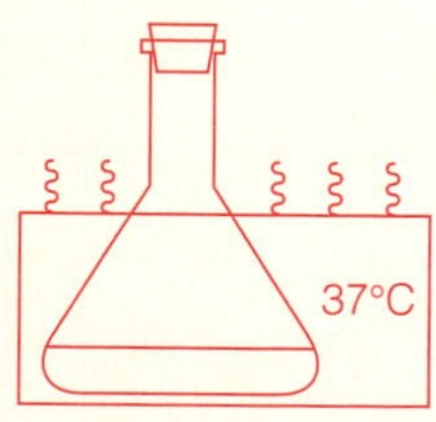

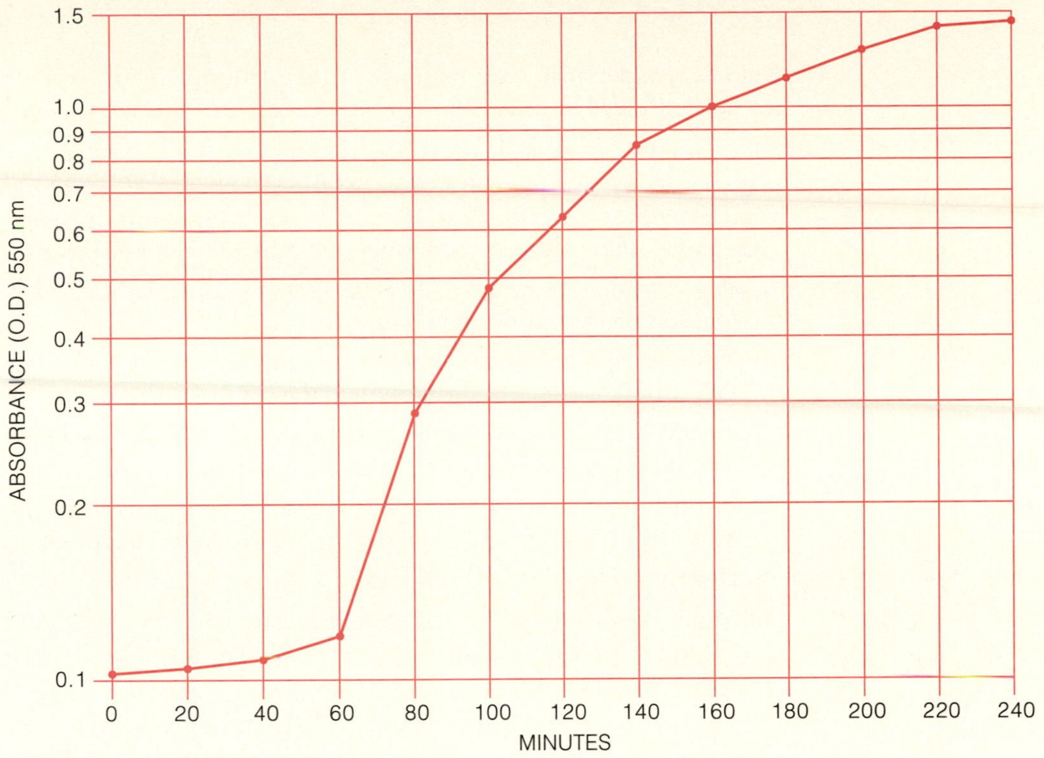

Growth Curve for *E. coli* Strain MM294

5. Store the culture flask or 10-ml aliquots of cells on ice as long as 2 hours before beginning calcium chloride treatment. This arrests cell growth in the mid-log phase.
6. Take the time for proper cleanup.
   a. Segregate for proper disposal bacterial cultures *and* tubes, pipets, especially those that have come into contact with the cultures.
   b. Disinfect the overnight culture, tubes, and pipets with 10% bleach or disinfectant (such as Lysol).
   c. Wipe down the lab bench with soapy water, a 10% bleach solution, or disinfectant.
   d. Wash your hands before leaving the laboratory.

RESULTS AND DISCUSSION

1. What variables influence the length of time needed for an *E. coli* culture to reach mid-log phase?
2. What are the disadvantages of beginning a mid-log culture from a colony scraped off of a plate, as opposed to an inoculum of an overnight culture?

FOR FURTHER RESEARCH

The following experiments can be started in the morning by one experimenter and continued by others throughout the day, until late afternoon.

1. Plot an OD_{550} growth curve for a particular strain of *E. coli.* Inoculate a 250-ml mid-log culture, as described above. Determine the optical density of samples sterilely withdrawn at 20-minute intervals, beginning at time zero and continuing for as many hours as possible.
   a. Plot a graph of time (x axis) *versus* OD^{550} (y axis). Label the lag, log, stationary, and death phases.
   b. What is the slope of the curve at a point that corresponds to an OD^{550} of 0.3? Describe the growth of the culture at this point.
2. Correlate OD^{550} to viable cell number for a particular strain of *E. coli.* Inoculate a 250-ml mid-log culture, as described above.
   a. At time zero and at 20-minute intervals for 4 hours, sterilely withdraw 1-ml aliquots, and store them on ice to arrest cell growth.
   b. Determine the OD^{550} of each aliquot.
   c. Make a 10^2 dilution of each aliquot by sterilely mixing 10 µl of aliquot with 990 µl of fresh LB broth. Then, make three serial dilutions, as follows:

      10^4 dilution = 10 µl of 10^2 culture + 990 µl of LB.

      10^5 dilution = 100 µl of 10^4 culture + 900 µl of LB.

      10^6 dilution = 100 µl of 10^5 culture + 900 µl of LB.
   d. Spread 100 µl of each dilution onto an LB agar plate, for a total of three plates for each time point (aliquot). *Label each plate bottom with the time point and dilution.* Invert the plates, and incubate for 12–24 hours at 37°C.
   e. For each time point, select a dilution plate that has 30–300 colonies. Multiply the number of colonies by the dilution factor to give cell number/ml in the original aliquot.
   f. Plot a graph of time (x axis) *versus* OD^{550} (y axis). Label the growth phases. Then add time (x axis) *versus* cell number/ml to the same graph. What are the differences between the two graphs?
   g. Plot a graph of cell number/ml (x axis) *versus* OD^{550} (y axis). An OD^{550} 0.3 corresponds to what concentration of cells? What is the cell concentration at each of the following points:

      lag phase

      first third of log phase (early log)

      second third of log phase (mid-log)

      final third of log phase (late log)

      stationary phase
   h. How can you account for a discrepancy between cell number/ml measured by plating serial dilutions *versus* counting by microscopy?

DNA Restriction and Electrophoresis

This laboratory introduces the genotypic analysis of DNA using restriction enzymes and gel electrophoresis. Three samples of purified DNA from bacteriophage λ (48,502 base pairs in length) are incubated at 37°C, each with one of three restriction endonucleases: *Eco*RI, *Bam*HI, or *Hin*dIII. Each enzyme has five or more restriction sites on the λ chromosome and therefore produces six or more restriction fragments of varying lengths. A fourth sample, the negative control, is incubated without an endonuclease and remains intact.

The digested DNA samples are then loaded into wells of an 0.8% agarose gel. An electrical field applied across the gel causes the DNA fragments to move from their origin (the sample well) through the gel matrix toward the positive electrode. The gel matrix acts as a sieve through which smaller DNA molecules migrate faster than larger ones; restriction fragments of different sizes separate into distinct bands during electrophoresis. The characteristic pattern of the bands produced by each restriction enzyme is made visible by staining the bands with a dye that binds to the DNA molecule.

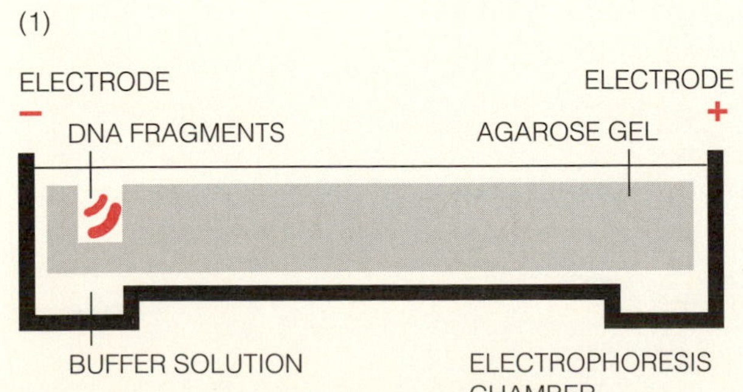

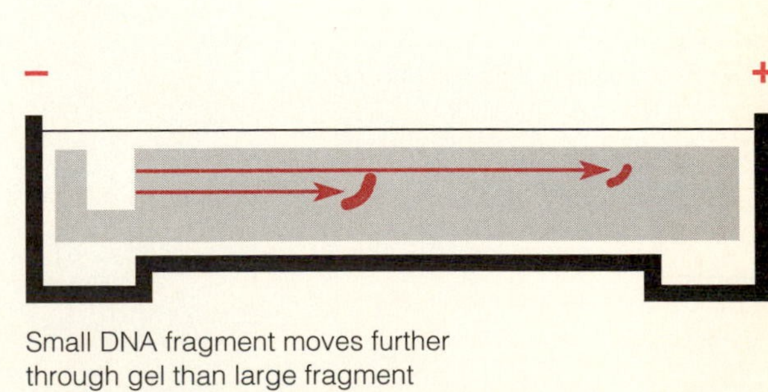

Small DNA fragment moves further through gel than large fragment

Agarose Gel Electrophoresis of DNA Fragments
(Art concept developed by Lisa Shoemaker.)

I. Set Up Restriction Digests

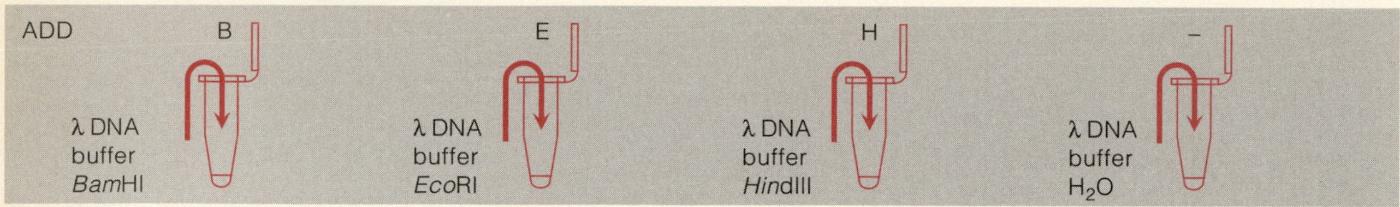

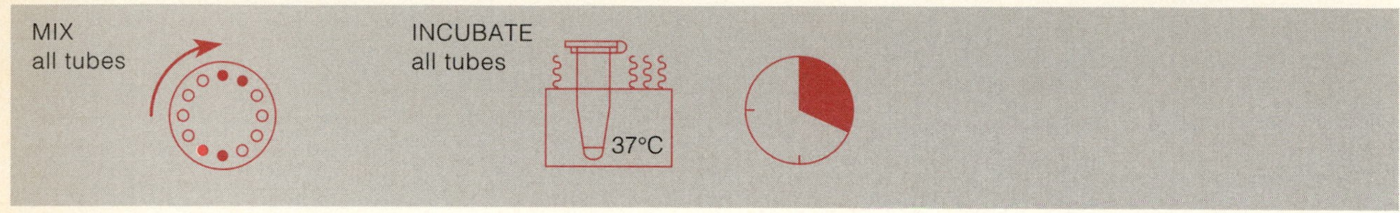

II. Cast 0.8% Agarose Gel

III. Load Gel and Electrophorese

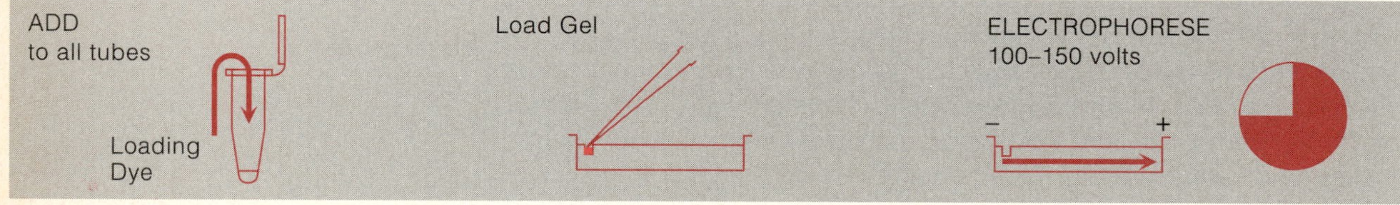

IV. Stain Gel and View

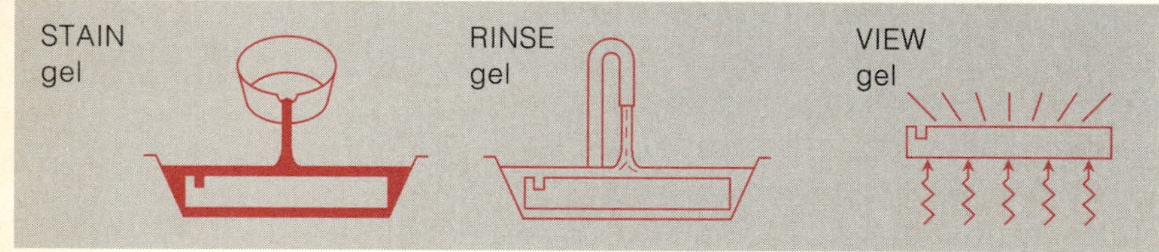

V. Photograph Gel

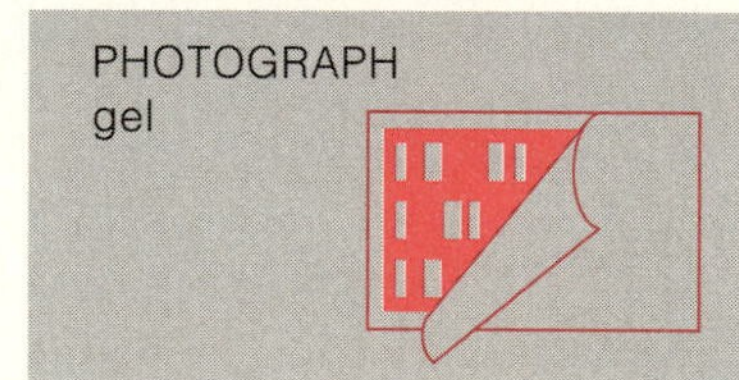

Laboratory 3
DNA Restriction and Electrophoresis

PRELAB NOTES

Storing and Handling Restriction Enzymes

Restriction enzymes, like many enzymes, are most stable at cold temperatures, and lose activity if they are warmed for any length of time. Because maintaining these enzymes in good condition is critical to the success of the experiments in this course, follow the guidelines below for handling them.

1. Always store enzymes in a *non-frost-free* freezer that maintains a constant temperature of –10° to –20°C. *Non-frost-free* freezers typically develop a layer of frost around the chamber, which acts as an efficient insulator and helps maintain a constant temperature. Frost-free freezers, on the other hand, go through freeze-thaw cycles that would subject enzymes to repeated warming and subsequent loss of enzymatic activity. If a frost-free freezer must be used, store the enzymes in their Styrofoam shipping container within the freezer. The container will help to maintain a constant temperature during the thaw cycle.
2. Remove restriction enzymes from the freezer directly onto crushed or cracked ice in an insulated ice bucket or cooler. Make certain that the tubes are pushed down into the ice and not just sitting on top of it. Keep the enzymes on ice at all times during handling, and return them to the freezer immediately after use.
3. When a large shipment of an enzyme is received, split it into several smaller aliquots of 50–100 μl in 1.5-ml tubes. Use a permanent marker on tape to clearly identify aliquots by enzyme type, concentration in units/μl, and date received. Use up one aliquot before starting another.
4. Keep aliquots of enzymes, buffer, and DNA in a cooler filled with ice before lab and also for dispensing; thus, unused aliquots will remain fresh.
5. Although it is good technique to set up restriction digests on ice, it is much simpler to set up reactions in a test tube rack at room temperature. Little loss of enzyme activity will occur during the brief time needed to set up the reaction.

Storing DNA and Restriction Buffer

Purified DNA is generally stored in the refrigerator (at approximately 4°C). DNA can be kept in the freezer (at approximately –20°C) for long-term storage of several months or longer. However, it is not advisable to store DNA in the freezer during times of active use: Ice crystals formed during repeated freezing will nick and shear DNA over time. Freeze damage is especially relevant to plasmid DNA used for transformations; nicked or linearized plasmid does not transform as well as the supercoiled form. Restriction buffer is best stored frozen and is not affected by freeze-thaw cycles.

Buffers

Many types of buffers are used in this course: restriction buffer, ligation buffer, PCR buffer, electrophoresis buffer, hybridization buffer, and various wash buffers. Each has a different chemical composition and use. Always double-check to ensure that you are using the proper buffer.

Tris-Borate-EDTA (TBE) electrophoresis buffer can be reused several times. Collect the used buffer, and store it in a large carboy. If different gels will be run over a period of several days, store the buffer in an electrophoresis chamber with the cover in place to retard evaporation. Before reusing buffer that has been stored in an electrophoresis chamber, rock the chamber back and forth to mix the buffer at either end. This reequilibrates ions that accumulate at either end during electrophoresis.

Groups of restriction enzymes operate under various salt and pH conditions. For optimal activity, several different buffers would be needed for the enzymes used in this course. Wherever possible, we use a "compromise" restriction buffer—a universal buffer that is a compromise between the conditions preferred by various enzymes.

All buffers are used at a final working concentration of 1×. Rely on the standard $C_1V_1 = C_2V_2$ formula to determine how much buffer to add to obtain a 1× solution:

(vol. buffer)	(conc. of buffer) = (total vol. of reaction)	(1X buffer)
(1 µl)	(10× buffer) = 10 µl	(1×)
(5 µl)	(2× buffer) = 10 µl	(1×)

For convenience, we use 2× restriction buffer whenever possible in 10 µl reactions: It saves a pipetting step to add water to bring the reaction up to 10 µl total volume. It is also easier and more accurate to pipet 5 µl than to pipet 1 µl. Compare a typical restriction reaction using 2× versus 10× restriction buffer:

	2X Buffer	10X Buffer
DNA	4 µl	4 µl
enzyme	1 µl	1 µl
buffer	5 µl	1 µl
water	—	4 µl
total solution	**10 µl**	**10 µl**

Diluting DNA

DNA for near-term use can be diluted with distilled or deionized water. However, it should be diluted with Tris-EDTA (TE) buffer for long-term storage. EDTA in the buffer binds divalent cations, such as Mg^{++}, that are necessary cofactors for DNA-degrading nucleases. Always dilute DNA to the concentration specified by the protocol.

1. Determine the total volume of DNA required by multiplying the number of experiments times the total volume of DNA per experiment, including overage.

   (10 experiments) (20 µl DNA) = 200 µl DNA

2. Plug this number into the $C_1V_1 = C_2V_2$ formula, along with the desired final DNA concentration and the concentration of the stock DNA. Solve for V_1, the volume of stock DNA needed in the dilution.

   (C_1 stock DNA) (V_1) = (C_2 final DNA) (V_2 total volume)

   (0.5 µg/µl) (V_1) = (0.1 µg/µl) (200 µl)

   $$(V_1) = \frac{(0.1\ \mu g/\mu l)\ (200\ \mu l)}{(0.5\ \mu g/\mu l)} = 40\ \mu l \text{ stock DNA}$$

3. Add water or TE to make the total volume of final solution.

   40 µl stock DNA + 160 µl H_20 or TE = 200 µl final solution

Pooling Reagents

Reagent aliquots often become spread in a film around the sides or caps of 1.5-ml tubes, during aliquoting and moving to and from the freezer or refrigerator and ice bucket. Use one of the following methods to pool reagent droplets to make them easier to find in the tube.

1. Spin the tubes briefly in a microfuge.
2. Spin the tubes briefly in a preparatory centrifuge, using adapter collars for 1.5-ml tubes. Alternatively, spin the tubes within a 15-ml tube, and remove them carefully.
3. Tap the tubes sharply on the bench top.

Restriction Enzyme Activity

The *unit* is the standard measure of restriction enzyme activity; it is usually defined as the amount of an enzyme needed to digest to completion 1 µg of λ DNA in a 50-µl reaction in 1 hour at 37°C. The unit strength of various restriction enzymes varies from one batch to another and also from one manufacturer to another. Typical batches of commercially available enzymes have activities in the range of 10–20 units/µl.

We suggest using endonucleases at 10 units/µl—a working concentration of approximately 1 unit per microliter of reaction mix. Although this is technically far more enzyme than is required, such overkill assures complete digestion by compensating for the following experimental conditions:

1. To save time, reaction times for restriction digests have often been shortened to 20–30 minutes. Complete digestion of the DNA would not occur during an abbreviated incubation, if the restriction enzyme was used at the standard condition of 1 unit/µg of DNA.
2. Many enzymes do not exhibit 100% activity in a compromise buffer.
3. Enzymes lose activity over time, due to imperfect handling.
4. It is easy to underpipet when measuring 1 µl of an enzyme, especially considering that the pipettor's mechanical error is greatest at the low end of its volume range.

Incubating Restriction Reactions

A constant-temperature water bath for incubating reactions can be made by maintaining a trickle of flowing tap water into a Styrofoam box.

Monitor the temperature with a thermometer. An aquarium heater can also be used to maintain a constant temperature.

The bare minimum incubation at 37°C needed for a restriction reaction to go to completion is 20 minutes. However, reactions can be incubated for several hours or even overnight, provided that you use commercial-quality DNA and enzymes that are free of extraneous nucleases that could degrade the sample. Enzymes lose their activity after several hours, and the reaction simply stops. Stop incubation whenever it is convenient; reactions may be stored in a freezer (at –20°C) until you are ready to continue. Thaw reactions before adding loading dye.

Importance of Intact Sample Wells

Very subtle damage caused to sample wells during casting and loading is one major cause of gel defects. (See "Field Guide to Electrophoresis Effects," following the "Results and Discussion" section.) Good electrophoretic separation and straight, well-defined banding require that DNA fragments of the same size enter the gel at essentially the same time against the *vertical front face* of the sample well. Anything that deforms the front face of the well or takes it out of the perpendicular to the electrical field will alter the appearance of DNA bands. The following list explains some common sources of damage to sample wells that can cause diffuse, blurred, wavy, or nicked bands.

1. *Inserting the pipet tip directly into a well when loading DNA is almost certain to excavate the face of each well. There is also a fair chance of puncturing the bottom of the well, thus allowing DNA to escape into the buffer.* Gels should always be submerged in an electrophoresis buffer before loading. Center the pipet tip on the well and lower it only until it dips through the buffer surface; then carefully expel the DNA/loading dye solution. Sucrose or glycerol incorporated in the loading dye will increase the density of the DNA solution, causing it to sink and fill the well without damage. Buffer can also lubricate and cool the newly cast gel, thus mitigating problems 2–4.
2. Wiggling or rocking the comb while removing it deforms the wells. If necessary, one partner should hold the casting tray steady, while the other pulls the comb straight up and out of the set agarose.
3. Removing the comb before the agarose has completely set causes the molten well edge to partially collapse into the well. Don't hurry when setting gel: It is impossible to tell the difference between almost-set gel and completely set gel.
4. Burrs or "flash" on a casting comb can nick the gel as the comb is removed. Hardened agarose can present the same problem. Inspect each comb, file to remove any burrs, and wash the comb to remove hardened agarose.
5. A bent comb or loose casting tray notches will take the comb out of perpendicular. Straighten the comb or tighten the notches with masking tape.

Storing Agarose Gel

Agarose gel can be cast a day or two before its use. To prevent drying, keep it covered with TBE electrophoresis buffer in an electrophoresis chamber or zip-lock plastic bag.

Electrophoresing

Hydrogen gas bubbling off of the negative electrode and oxygen gas rising from the positive electrode (products of electrolysis of water) are the first signs that current is flowing through the electrophoresis system. Shortly after, bands of loading dye should be seen moving into the gel and migrating toward the positive pole of the apparatus. The loading dye band quickly resolves into two bands of color: The faster-moving, purplish band is bromophenol blue, and the slower-moving, aqua band is xylene cyanol. Bromophenol blue migrates through a 0.8% gel at the same rate as a DNA fragment of approximately 300 base pairs (bp). Xylene cyanol migrates at a rate approximately equivalent to 9,000 base pairs. The best separation for analysis of λ and plasmid DNA is achieved when the bromophenol blue migrates 40–70 mm from the origin.

The migration of DNA through an agarose gel depends upon voltage: The higher the voltage, the faster the rate of migration. However, higher voltages accentuate imperfections in the gel—such as differences in density and thickness from one part of the gel to another. Common effects include slanted and U-shaped bands ("smiles"). Slanted effects most typically occur at the edges of the gel, where the liquid gel adheres to the casting tray to form a miniscus, making the edges of the solidified gel thicker than the center. Resistance is decreased in the thicker edges of the outermost lanes, thus allowing DNA molecules toward the outer edge of the band to migrate faster than like-sized molecules toward the inner edge of the band. Also, heat generated at high voltages can begin to melt the gel and change its sieving properties. For these reasons, avoid using more than 125 volts in a minigel system.

Ethidium Bromide Staining and Responsible Handling

Ethidium bromide is the most rapid, sensitive, and reproducible means currently available for staining DNA. However, like many natural *and* man-made substances, ethidium bromide is a mutagen, by the Ames microsome assay, and a suspected carcinogen. The protocols in this manual limit the use of ethidium bromide to a single staining procedure that can be performed by the instructor in a controlled area. Scientists often incorporate ethidium bromide into agarose gel, thus eliminating the staining step. We caution against using this shortcut in the teaching laboratory, because it inevitably leads to contamination of electrophoresis equipment and lab surfaces, and ethidium bromide contamination is very difficult to detect or monitor.

With responsible handling, the dilute solution (1 μg/ml) used for gel staining poses minimal risk. The greatest risk is incurred by inhaling ethidium bromide powder when mixing a 5 mg/ml stock solution. Therefore, we suggest purchasing a ready-mixed stock solution from a supplier. The stock solution is then diluted to make a staining solution with a final concentration of 1 μg/ml of ethidium bromide.

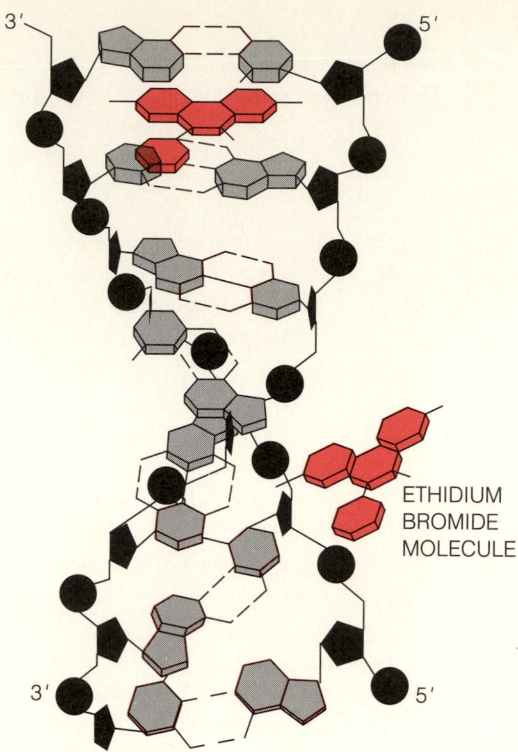

Intercalation of Ethidium Bromide into DNA Helix

CAUTION

Handling and Decontamination of Ethidium Bromide

1. Always wear disposable gloves when working with ethidium bromide solutions or stained gels.
2. Limit the use of ethidium bromide to a restricted sink area.
3. Following gel staining, use a funnel to decant as much as possible of the ethidium bromide solution into a storage container for reuse or decontamination and disposal.
4. Disable stained gels and used staining solution, according to accepted laboratory procedure. The method given below is adapted from Quillardet and Hofnung (1988).
   a. If necessary, add water to reduce the concentration of ethidium bromide to less than 0.5 mg/ml.
   b. Add 1 volume of 0.05 M $KMnO_4$, and *mix carefully*.
   c. Add 1 volume of 0.25 N HCl, and *mix carefully*.
   d Let the solution stand at room temperature for several hours.
   e. Add 1 volume of 0.25 N NaOH, and *mix carefully*.
   f. Discard the disabled solution down a sink drain. Drain disabled gels, and discard them in the regular trash.

   Caution: $KMnO_4$ is an irritant, and it is explosive. Solutions containing $KMnO_4$ should be handled in a chemical hood.
5. *One previous method of treating ethidium bromide with bleach solution was deemed inadequate and has been abandoned.*

DNA Staining with Methylene Blue

The volumes and concentrations of DNA used in all laboratories have been optimized for staining with ethidium bromide. If methylene blue staining is preferred, increase stated concentrations by three or four times for λ DNA, and increase them by two for plasmid DNA. If DNA *concentration* is increased, the volumes used in laboratories remain as stated.

Methylene blue stains have two advantages that make them attractive to educators: They are nontoxic and are visible in white light. However, these stains have the very serious disadvantages of relative insensitivity, longer staining/destaining times, and lack of reproducibility. We have found methylene blue staining extremely variable and prone to amplify errors in technique. The various proprietary stains—based upon azure components of methylene blue—offer no significant improvement.

The insensitivity of methylene blue staining makes it difficult to properly visualize λ digests, which produce fragments that differ greatly in size. (The mass of λ DNA required to visualize the lower-molecular weight bands overloads the larger bands.) The increased DNA mass needed for methylene blue staining can significantly increase the cost of experiments involving plasmid. For these reasons, methylene blue staining is not listed as an option for Laboratories 11–23.

Viewing Stained Gels

Transillumination, where light passes through the gel, allows superior viewing of gels stained with either ethidium bromide or methylene blue.

A mid-wavelength (260–360 nm) ultraviolet lamp emits light in the optimum range for illuminating ethidium-bromide-stained gels. Avoid short-wavelength lamps; the radiation they emit is very dangerous to humans and also damaging to the DNA. Long-wavelength ("black light") lamps, though safe, provide less intense illumination.

CAUTION

Ultraviolet light can damage the retina of the eye. Never look directly at an unshielded UV light source without eye protection. View only through a filter or safety glasses that absorb the harmful wavelengths.

A fluorescent light box for viewing slides and negatives provides ideal illumination for methylene-blue-stained gels. An overhead projector can also be used. Cover the surface of the light box or projector with plastic wrap to keep liquid off of the apparatus.

Photographing Gels

Photographs of DNA gels provide a permanent record of the experiment, allowing time to analyze results critically, to discover subtleties of gel interpretation, and to correct mistakes. Furthermore, time exposure can record bands that are faint or invisible to the unaided eye.

A Polaroid "gun camera," equipped with a close-up diopter lens, is used to photograph gels on either an ultraviolet or white-light transilluminator. A plastic hood extending from the front of the camera forms a minidarkroom and provides correct lens-to-subject distance. Alternatively, a close-focusing 35mm camera can be used. For UV photography, two filters are placed in front of the lens: A 23A orange is placed closest to the camera, and a 2B UV-blocking filter (clear) is placed closest to the subject. Yellow or orange filters will intensify the contrast in gels stained with methylene blue. The UV filter set described above works well and can be left in place for both ethidium-bromide and methylene-blue photography.

For UV photography, use Polaroid type 667 high-speed film (black and white). Regular Polaroid color film is less expensive for use with methylene-blue-stained gels.

Semilog Graph Paper

Restriction fragments used in the analysis range in size from about 500 to 20,000 bp. Thus, log paper with three cycles will spread the data points best.

For Further Information

The protocol presented here is based on the following published methods:

Helling, R. B., H. M. Goodman, and H. W. Boyer. 1974. Analysis of *Eco*RI fragments of DNA from lambdoid bacteriophages and other viruses by agarose-gel electrophoresis. *Journal of Virology* 14: 1235.

Sharp, P. A., B. Sugden, and J. Sambrook. 1973. Detection of two restriction endonuclease activities in *Haemophilus parainfluenzae* using analytical agarose–ethidium bromide electrophoresis. *Biochemistry* 12: 3055.

Quillardet, P. and M. Hofnung. 1988. Ethidium bromide and safety—Readers suggest alternative solutions. (Letter to editor.) *Trends in Genetics* 4: 89.

DNA RESTRICTION AND ELECTROPHORESIS

Reagents

0.1 μg/μl λ DNA
restriction enzymes:
- *Eco*RI
- *Bam*HI
- *Hin*dIII

2× restriction buffer
distilled water
loading dye
0.8% agarose
1× Tris-Borate-EDTA (TBE) buffer
1 μg/ml ethidium bromide (or 0.025% methylene blue)

for decontamination:
0.05M $KMnO_4$
0.25 N HCl
0.25 N NaOH

Supplies and Equipment

0.5–10-μl micropipettor + tips
1.5-ml tubes
aluminum foil
beaker for agarose
beaker for waste/used tips
camera and film (optional)
disposable gloves
electrophoresis box
masking tape
microfuge (optional)
Parafilm or waxed paper (optional)
permanent marker
plastic wrap (optional)
power supply
semilog graph paper
test tube rack
transilluminator (optional)
37° water bath
60°C water bath (for agarose)

I. Set Up Restriction Digests

(10 minutes; then 20+ minutes incubation)

1. Use a permanent marker to label four 1.5-ml tubes in which restriction reactions will be performed:

   B = *Bam*HI
   E = *Eco*RI
   H = *Hin*dIII
   — = no enzyme

2. Use the matrix below as a checklist while adding reagents to each reaction. Read down each column, adding the same reagent to all appropriate tubes. *Use a fresh tip for each reagent.* Refer to the detailed directions that follow.

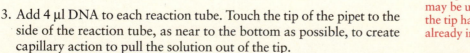

Tube	λ DNA	Buffer	*Bam*HI	*Eco*RI	*Hin*dIII	H_2O
B	4 μl	5 μl	1 μl	—	—	—
E	4 μl	5 μl	—	1 μl	—	—
H	4 μl	5 μl	—	—	1 μl	—
—	4 μl	5 μl	—	—	—	1 μl

3. Add 4 μl DNA to each reaction tube. Touch the tip of the pipet to the side of the reaction tube, as near to the bottom as possible, to create capillary action to pull the solution out of the tip.

It is unnecessary to change tips when adding the same reagent. The same tip may be used for all tubes, provided that the tip has not touched any solution already in the tubes.

4. Use a *fresh tip* to add 5 μl of restriction buffer to a clean spot on each reaction tube.

Always add buffer to the reaction tubes before adding the enzymes.

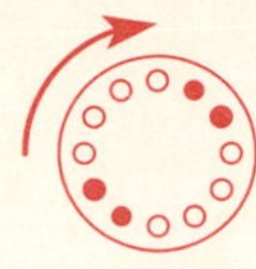

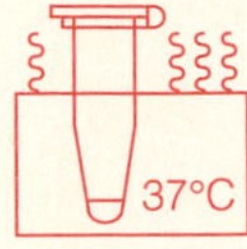

Enzymes lose activity after several hours, and the reaction stops.

5. Use *fresh tips* to add 1 µl of *Eco*RI, *Bam*HI, and *Hin*dIII to the appropriate tubes.
6. Use a *fresh tip* to add 1 µl of deionized water to the – tube.
7. Close the tube tops. Pool and mix reagents by pulsing them in a microfuge or by sharply tapping the bottom of the tube on the lab bench.
8. Place the reaction tubes in a 37°C water bath, and incubate them for 20 minutes or longer.

Stop Point

Following incubation, freeze the reactions at –20°C until you are ready to continue. Thaw the reactions before continuing to Section III, step 1.

II. Cast 0.8% Agarose Gel

(15 minutes)

Gel is cast directly in the box in some electrophoresis apparatuses.

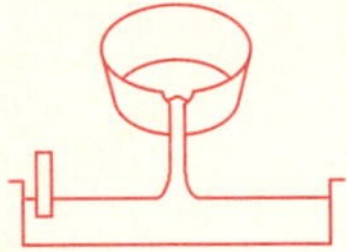

1. Seal the ends of the gel-casting tray with tape, and insert a well-forming comb. Place the gel-casting tray out of the way on the lab bench, so that the agarose to be poured in the next step can set undisturbed.
2. Carefully pour agarose solution into the casting tray to fill it to a depth of about 5 mm. The gel should cover only about ⅓ the height of the comb teeth. Use a pipet tip to move large bubbles or solid debris to the sides or end of the tray, while the gel is still liquid.
3. The gel will become cloudy as it solidifies (in about 10 minutes). *Take care not to move or jar the casting tray while the agarose is solidifying*. Touch the corner of the agarose, *away* from the comb, to test whether the gel has solidified.
4. When the agarose has set, unseal the ends of the casting tray. Place the tray on the platform of the gel box, so that the comb is at the negative (black) electrode.
5. Fill the box with TBE buffer to a level that barely covers the entire surface of the gel.
6. Gently remove the comb, pulling it straight up and out of the set agarose. Do not rock or wiggle the comb. If necessary, have your partner steady the casting tray.
7. Removal of the comb usually pulls the agarose well edges above the buffer surface; these edges appear as "dimples" around the wells. Add buffer until any dimples disappear and the buffer surface is smooth.

Too much buffer will channel the current over the top rather than through the gel, thus increasing the time required to separate DNA fragments. TBE buffer can be used several times; do not discard it. If you are using buffer remaining in an electrophoresis box from a previous experiment, rock the chamber back and forth to remix ions that have accumulated at either end. Buffer solution helps to lubricate the comb and thus prevent damage to wells.

Some gel boxes are designed so that you must remove the comb before inserting the casting tray into the box. In this case, flood the casting tray and gel surface with buffer *before* removing the comb.

Stop Point

Cover the electrophoresis box, and save the gel until you are ready to continue. The gel will remain in good condition for several days, if it has been completely submerged in buffer.

III. Load Gel and Electrophorese

(10 minutes; then 40–60 minutes electrophoresis)

1. Remove restriction digests from the 37°C water bath.
2. Add 1 µl loading dye to each reaction by either of two methods.
   a. Add 1 µl of loading dye to each reaction tube. Close the tops of the tubes, and mix the solution by tapping the bottom of the tube on the lab bench, pipetting in and out, or pulsing it in a microfuge. Make sure that the tubes are placed in a *balanced* configuration in the rotor.

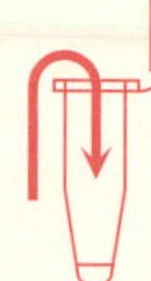

   or

   b. Place four individual droplets of loading dye (1 µl each) on a small square of Parafilm or waxed paper. Withdraw the contents from one reaction tube, and mix it with a loading dye droplet by pipetting in and out. Immediately load the dye mixture, according to step 2. Repeat this successively, with a clean tip, for each reaction.

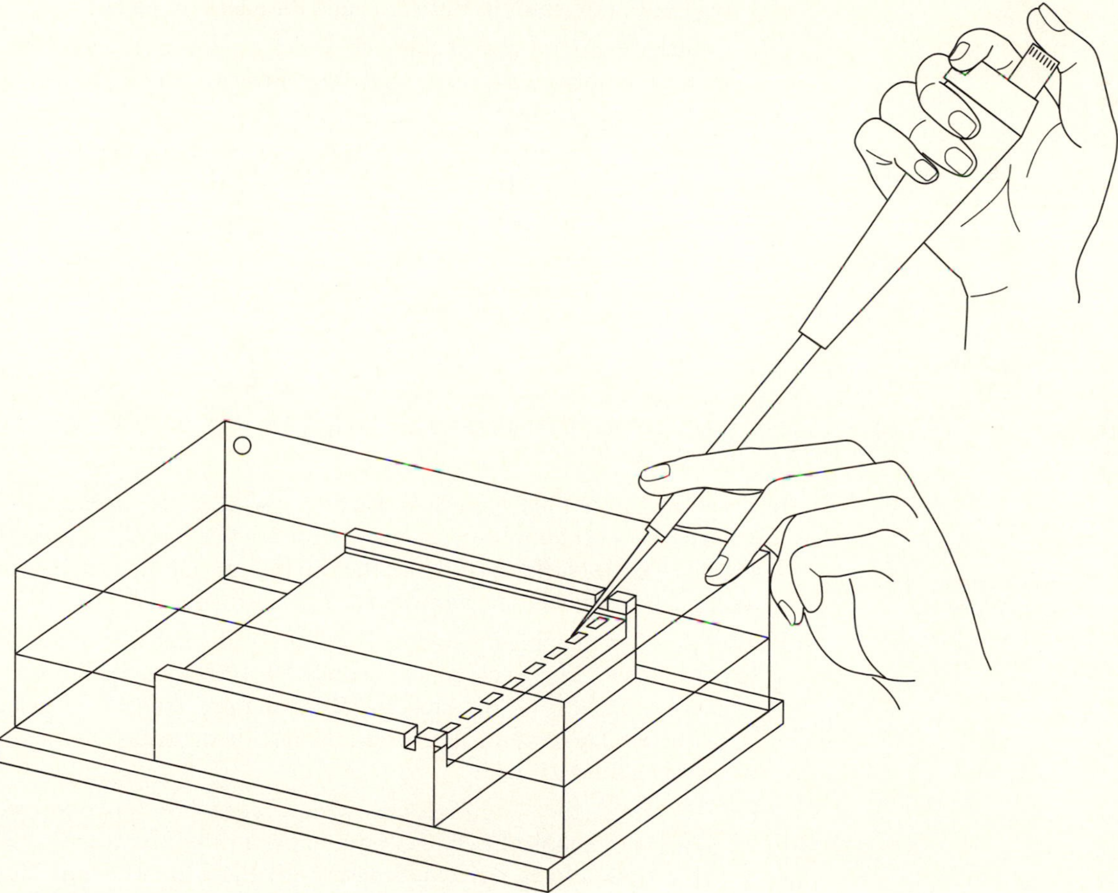

Hand Positions for Loading on Agarose Gel (step 2)

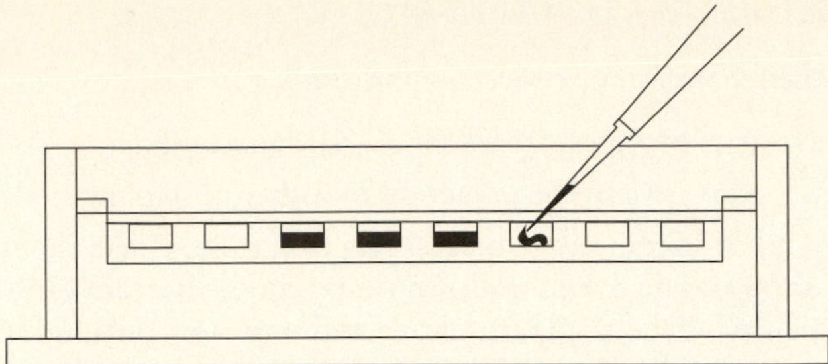

Placing a piece of dark construction paper beneath the gel box will help to make the wells more visible.

3. Use a micropipettor to load 10 µl from each reaction tube into a separate well in the gel, as shown in the diagram below. Use a *fresh tip* for each reaction.
   a. Steady the pipet over the well, using two hands.
   b. If there is air in the end of the tip, carefully depress the plunger to push the sample to the end of the tip. (An air bubble ejected into the well can form a "cap" over the well, thus causing DNA/loading dye to flow into the buffer around the edges of the well.)
   c. Center the pipet tip over the well, dip the tip in only enough to pierce the buffer surface, and gently depress the pipet plunger to slowly expel the sample.

Take care *not* to insert the pipet tip directly into the well; this will damage or puncture the well. Sucrose in the loading dye weighs down the DNA sample, causing it to sink to the bottom of the well.

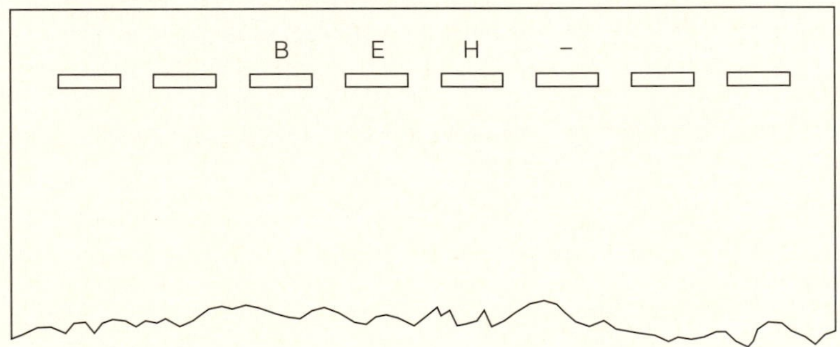

4. Close the top of the electrophoresis box, and connect the electrical leads to a power supply—anode to anode (red-red) and cathode to cathode (black-black). Make sure that both electrodes are connected to the same channel of power supply.
5. Turn the power supply on, and set it to 100–150 volts. The ammeter (if present) should register approximately 50–100 milliamperes. Current flow can also be detected by observing gas bubbles released from the electrode wires. If current cannot be detected, check the connections, and try again.
6. Electrophorese for 40–60 minutes. Good separation has occurred when bromophenol blue bands have moved 40–70 mm from the wells. If time allows, electrophorese until the bromophenol blue bands are near the end of the gel. *Stop* electrophoresis *before* the bromophenol blue band runs off the end of the gel.
7. Turn off the power supply, disconnect leads from inputs, and remove the top of the electrophoresis box.

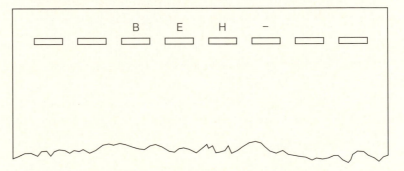

Alternatively, set the power supply on a lower voltage, and run the gel for several hours. When you are running two gels off of the same power supply, the current will be double that needed for a single gel at the same voltage.

8. Carefully remove the casting tray from the electrophoresis box, and slide the gel into a disposable weigh boat or other shallow tray. Label the staining tray with your name and the date.

Stop Point

The gel can be stored overnight, covered with a small volume of buffer in a staining tray or a zip-lock plastic bag, for viewing and photographing the next day. However, over a longer period of time, the DNA will diffuse through the gel, and the bands will become indistinct or entirely disappear.

9. Stain and view the gel, using one of the methods described in Sections IVA and IVB.

If desired, staining may be performed by an instructor in a controlled area, when students are not present.

IVA. Stain Gel with Ethidium Bromide and View

(10–15 minutes)

CAUTION

Review the section on "Ethidium Bromide Staining and Responsible Handling." Wear disposable gloves when staining, viewing, and photographing the gel and also during cleanup. Confine all staining to a restricted sink area.

1. Flood gel with an ethidium bromide solution (1 μg/ml), and allow them to stain for 5–10 minutes.
2. Following staining, use a funnel to decant as much ethidium bromide solution as possible from the staining tray back into a storage container.
3. Rinse the gel and the tray under running tap water.
4. If desired, gel can be destained in tap water or distilled water for 10 minutes or longer, to help remove background ethidium bromide.
5. View under an ultraviolet transilluminator or other UV source.

Staining time and the "background" of unbound ethidium bromide markedly increase in thicker gels.

Staining solution may be reused to stain 15 or more gels. When the staining time increases markedly, refresh the solution with 5 mg/ml stock solution according to the recipe in Appendix 1 or else disable it according to the procedure outlined on page 38.

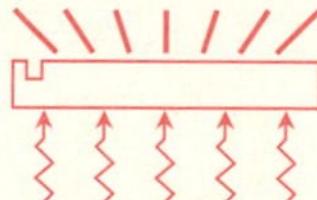

CAUTION

Ultraviolet light can damage your eyes. Never look directly at an unshielded UV light source without eye protection. View only through a filter or safety glasses that absorb the harmful wavelengths.

Band intensity and contrast increase dramatically when a gel is destained for 15–30 minutes in tap water. More simply, rinse and drain the gels, stack the staining trays, cover the top gel, and let the gels set overnight at room temperature.

6. Take the time for proper cleanup.
   a. Wipe down the camera, the transilluminator, and the staining area.
   b. Decontaminate gel and any staining solution that will not be reused.
   c. Wash your hands before leaving the laboratory.

IVB. Stain Gel with Methylene Blue and View

(30+ minutes)

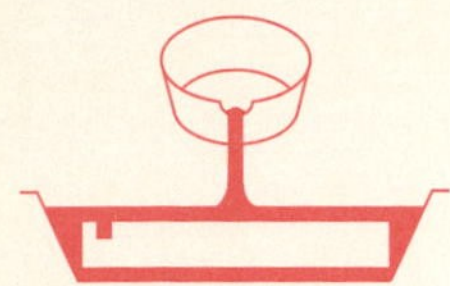

The staining solution may be reused to stain 15 or more gels.

Destaining time is decreased by agitating and rinsing the gel in warm water. For best results, continue to destain the gel overnight in a small volume of water. (Gel can destain too much if left overnight in a large volume of water.) Cover the staining tray to retard evaporation.

1. Wear disposable gloves during the staining and cleanup.
2. Flood gel with a 0.025% methylene blue solution, and allow them to stain for 20–30 minutes.
3. Following staining, use a funnel to decant as much methylene blue solution as possible from the staining tray back into a storage container.
4. Rinse the gel in running tap water. Let it soak for several minutes in several changes of fresh water. DNA bands will become increasingly distinct as the gel destains.
5. View the gel over a light box; cover the surface with plastic wrap to prevent staining.

V. Photograph Gel

(5 minutes)

Exposure times vary according to the mass of DNA in the lanes, the level of staining, the degree of background staining, the thickness of the gel, and the density of the filter. Experiment to determine the best exposure. When possible, stop the lens down (to a higher f/number) to increase the depth of the field and the sharpness of the bands.

1. For ultraviolet (UV) photography of ethidium-bromide-stained gels: Use Polaroid high-speed type 667 film (ISO 3000). Set the camera aperture to f/8 and the shutter on 1 second. Depress the shutter once or twice for a 1–2-second exposure.

   or

   For white-light photography of methylene-blue-stained gels: Use Polaroid type 667 color film, with an aperture to f/8 and a shutter speed of 1/125th of 1 second.
2. Place your left hand firmly on top of the camera to steady it. Firmly grasp the small white tab and pull it straight out from the camera. This causes a larger tab to appear.
3. Grip the large tab at its center, and in one steady motion, pull the film straight out from the camera. This starts development.
4. Allow the film to develop for the recommended time (45 seconds at room temperature). Do not disturb the print while it is developing.
5. After the full development time has elapsed, separate the print from the negative by peeling back at the end nearest the large tab.

CAUTION

Avoid spilling caustic developing jelly on your skin or clothes. If jelly does get on your skin, wash immediately with plenty of soap and water.

6. Wait to see the results from the first photo before making other exposures.

RESULTS AND DISCUSSION

Agarose gel electrophoresis, combined with ethidium bromide staining, allows for the rapid analysis of DNA fragments. Before the introduction of this method in 1973, however, analysis of DNA molecules was a laborious task. The original separation method, involving ultracentifugation of DNA in a sucrose gradient, gave only crude size approximations and required more than 24 hours to complete.

Electrophoresis using a polyacrylamide gel in a glass tube was an improvement but could only be used to separate small DNA molecules of no more than 2,000 bp. Another drawback was that the DNA had to be radioactively labeled before analysis. Following electrophoresis, the polyacrylamide gel was cut into thin slices, and the radioactivity in each slice was determined. The amount of radioactivity detected in each slice was plotted *versus* the distance migrated, producing a series of radioactive peaks representing the DNA fragments.

DNA restriction analysis is at the heart of recombinant-DNA technology and of the laboratories in this course. The ability to cut DNA predictably and precisely enables scientists to manipulate and recombine DNA molecules at will. When discrete bands of like-sized DNA fragments are seen in one lane of an agarose gel, this shows that each of the several billion λ DNA molecules present in each restriction reaction were all cut in precisely the same place.

By convention, DNA gels are "read" from left to right, with the sample wells oriented at the top. The area extending from the well in a vertical column down the gel is called a *lane*. Thus, reading down the lane identifies fragments generated by a particular restriction reaction. Scanning across lanes identifies fragments that have comigrated the same distance down the gel and are thus of like size.

1. Why is water added to the "–" tube in Section I, step 6?
2. What is the function of a compromise restriction buffer?
3. What are the two functions of loading dye?
4. How does ethidium bromide stain DNA? How does this relate to the need to minimize ethidium bromide exposure to humans?
5. Troubleshooting electrophoresis: What would occur . . .
   a. if the gel box is filled with water instead of TBE buffer?
   b. if water is used to prepare the gel instead of TBE buffer?
   c. if the electrodes are reversed?
6. Examine the photograph of your stained gel (or view it on a light box or overhead projector). Compare your gel with the ideal gel shown on page 48, and try to account for the fragments of λ DNA in each lane. Can you account for any differences in separation and band intensity between your gel and the ideal gel?
7. Troubleshooting gels: What effect would be observed in the stained bands of DNA in an agarose gel . . .
   a. if the casting tray is moved or jarred while the agarose is solidifying in Part II, step 3?
   b. if the gel is run at very high voltage?
   c. if a large air bubble or clump of debris is allowed to set in the agarose?
   d. if too much DNA is loaded in a lane?

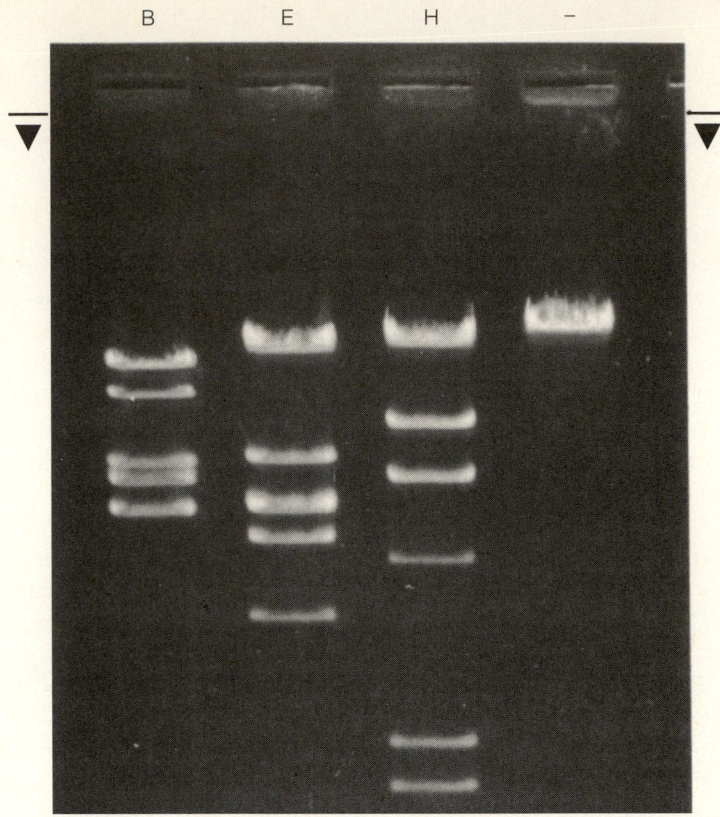

Ideal Gel

e. if there is incomplete (partial) digestion by a restriction enzyme?

8. Linear DNA fragments migrate at rates that are inversely proportional to the $\log_{10}$ of their molecular weight. (For simplicity's sake, base-pair length is substituted for molecular weight.) Use the photograph of your stained gel and a ruler to construct a graph that relates the base-pair size of each restriction fragment to the distance migrated through the gel. The "bp Map" column of the matrix on the next page gives the base pair sizes of λ DNA fragments generated by the *Hin*dIII digest (as obtained from a the restriction map on page 50):

   a. Orient your gel photo with the wells at the top and locate the H lane. Working from top to bottom, match the base-pair sizes of the *Hin*dIII fragments in the "bp Map" column of the matrix with bands that appear in the H lane. Label each band with the size of the kilobase pair (kbp). For example, 27,491 bp equals 27.5 kbp. (Alternatively, use the ideal gel shown above.)

   b. Carefully measure the distance (in mm) that each *Hin*dIII fragment migrated from the sample well. Measure from the front edge of the well to the leading edge of each band. Enter the distance migrated by each fragment into the "Distance Migrated" column of the matrix. (Alternatively, measure the distances on an overhead-projected image of methylene-blue-stained gel.)

   c. Set up semilog graph paper with the distance migrated as the x (arithmetic) axis and a log of base-pair length as the y (logarithmic) axis. Then, plot the distance migrated versus the bp length

***Hind*III**		***Eco*RI**			***Bam*HI**		
Distance Migrated	**bp Map**	**Distance Migrated**	**bp Graph**	**bp Map**	**Distance Migrated**	**bp Graph**	**bp Map**
	27,491*						
	23,130*						
	9,416						
	6,557						
	4,361						
	2,322						
	2,027						
	564†						
	125‡						

* The pair appears as a single band on the gel.
† Usually not visible in methylene-blue-stained gel.
‡ Usually runs off the end of mini-gel during electrophoresis.

for each *Hin*dIII fragment. (The first cycle of the *y* axis should begin with 100 bp and end with 1,000 bp—counting in 100-bp increments. The second cycle begins with 1,000 bp and ends with 10,000 bp. The third cycle begins with 10,000 bp, counting in 10,000-bp increments.) Next, plot the distance migrated versus the base-pair length for each *Hin*dIII fragment. Finally, connect the data points.

d. Measure the distance migrated (in mm) by each *Eco*RI and *Bam*HI fragment. Measure from the front edge of the well to the front edge of each band. Enter these distances into the appropriate "Distance Migrated" columns of the matrix.

e. Locate on the *x* axis the distance migrated by each *Eco*RI and *Bam*HI fragment. Then, use a ruler to draw a vertical line from this point to its intersection with the *Hin*dIII data line. Now, extend a horizontal line from this point to the *y* axis; this gives the bp size of the fragment. Enter the result into the appropriate "bp Graph" columns of the matrix above.

f. Use restriction maps of λ on page 50 to determine the actual bp sizes of *Eco*RI and *Bam*HI fragments, and enter these into the appropriate "bp Map" columns.

g. For which fragment sizes was your graph most accurate? For which fragment sizes was it *least* accurate? What does this tell you about the resolving ability of agarose gel electrophoresis?

9. DNA fragments of similar size will not always resolve on a gel. This is seen in lane E, where *Eco*RI fragments of 5,804 bp and 5,643 bp migrate as a single heavy band. These are referred to as a *doublet* and can be recognized because they are brighter and thicker than similarly sized singlets. What could be done to resolve the doublet fragments?

10. Determine a range of sensitivity of DNA detection by ethidium bromide (or methylene blue/azure) by comparing the *mass* of DNA in the bands of the largest and smallest detectable fragments on the gel. Use this formula to determine the mass of DNA in a given band:

$$\frac{\text{fragment bp (conc. DNA) (vol. DNA)}}{\lambda \text{ bp}}$$

For example:

$$\frac{24{,}251 \text{ bp } (0.1\ \mu\text{g}/\mu\text{l})\ (4\ \mu\text{l})}{48{,}502 \text{ bp}} = 0.2\ \mu\text{g}$$

Now, compute the mass of DNA in the largest and smallest *singlet* fragments on the gel.

11. λ DNA can exist as a circular, as well as a linear, molecule. At each end of the linear molecule is a single-stranded sequence of 12 nucleotides, called a COS site. The COS sites at each end are complementary to one another; thus, they can base pair to form a circular molecule. These complementary ends are analogous to the "sticky ends" created by some restriction enzymes. Commercially available λ DNA is likely to be a mixture of linear and circular molecules. This leads to the appearance of more bands on the gel than could be predicted from a homogeneous population of linear DNA molecules. This also causes the partial loss of other fragments. For example, the left-most *Hin*dIII site is 23,130 bp from the left end of the linear λ genome; the right-most site is 4,361 bp from the right end. The 4,361 band is faint in comparison to other bands of similar size on the gel. This indicates that a percentage of the DNA molecules are circular—combining the 4,361-bp terminal fragment with the 23,130-bp terminal fragment to produce a 27,491-bp fragment. However, the combined 27,491 bp usually runs as a doublet along with the 23,130-bp fragment from the linear molecule.
    a. Use a protractor to draw three circles about 3 inches in diameter. These represent λ DNA molecules with base-paired COS sites.
    b. Label a point at 12 o'clock on each circlet 48/0. This marks the point where the COS sites are joined.

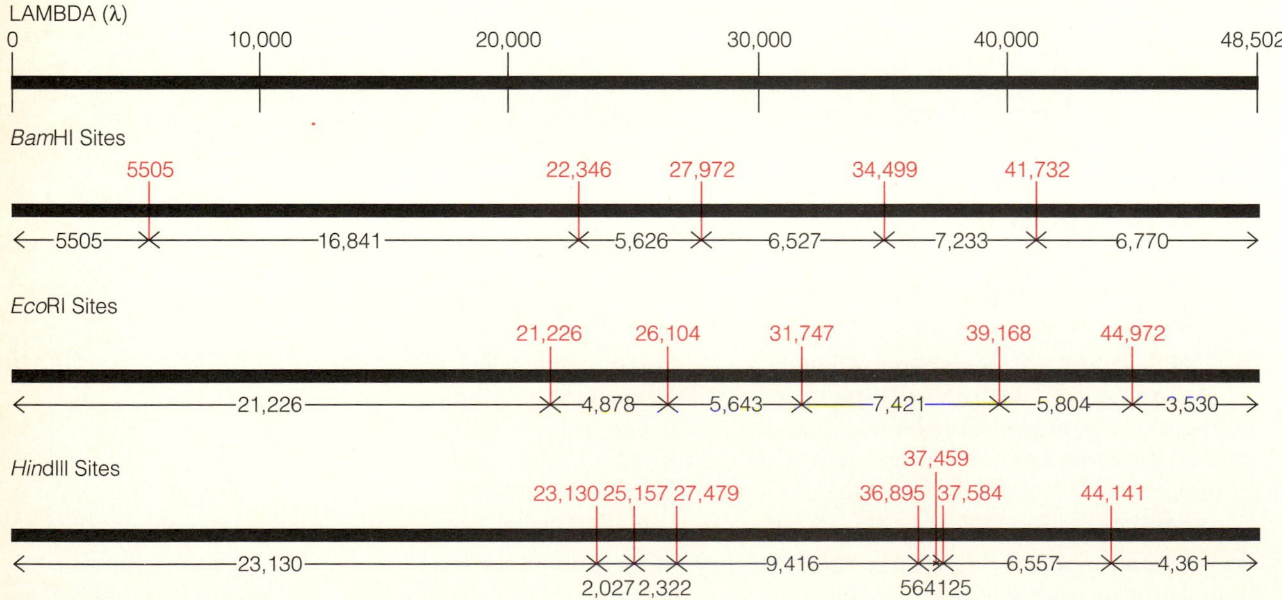

Restriction Maps of the Linear λ Genome

c. Using data from the restriction maps of the linear λ genome on page 50, make a rough map of restriction sites for *Hin*dIII on one of the circles. Note the situation described above.

d. Next make rough restriction maps of *Bam*HI and *Eco*RI sites on the two remaining circles. Note the length of each restriction fragment in bp.

e. What *Bam*HI and *Eco*RI fragments are created in the circular molecules? Can you locate each of these fragments on your gel or on the ideal gel? Why or why not?

FOR FURTHER RESEARCH

1. Preparations of λ DNA contain some circular molecules that are covalently linked at the COS sites. Other circles are only hydrogen bonded and can dissociate to form linear molecules. Heating λ DNA to 65°C for 10 minutes will linearize any noncovalent COS circlets in the preparation by breaking hydrogen bonds that hold the complementary COS sites together.

   a. Set up duplicate restriction digests of λ DNA with several enzymes. Following digestion, add loading dye to one reaction from each set, and heat at 65°C for 10 minutes, while holding the duplicates on ice. Following the heat step, quickly load all samples and electrophorese them in a 0.8% agarose gel. Relate changes in restriction patterns of heated versus unheated DNA to a restriction map of the circular λ genome, as in question 11 above.

   b. How can you estimate the approximate percentage of the linear, covalently closed circular, and hydrogen-bonded circular forms in your preparation?

2. Design and carry out a series of experiments to study the kinetics of a restriction reaction.

   a. Determine the approximate percentage of digested λ DNA at various time points (differing incubation times at 37°C).

   b. Repeat the experiments with several enzyme dilutions and several DNA dilutions.

   c. In each case, at what point does the reaction appear to be complete?

3. Test the relative stability of *Bam*HI, *Eco*RI, and *Hin*dIII at room temperature.

4. Determine the identity of an unknown restriction enzyme.

   a. Perform single digests of λ DNA with the unknown enzyme and with several known restriction enzymes. Run the restriction fragments in an agarose gel at 50 volts to produce well-spread and well-focused bands.

   b. For each fragment, plot the distance migrated *versus* base-pair size, as in question 8 above. Use a graph to determine the base-pair lengths of unknown fragments, and compare this with restriction maps of commercially available enzymes.

5. Research the steps needed to purify a restriction enzyme from *E. coli*, and characterize its recognition sequence.

Field Guide to Electrophoresis Effects

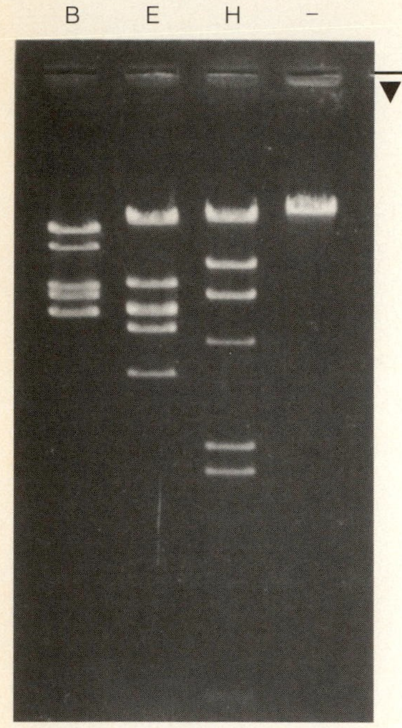

Ideal Gel

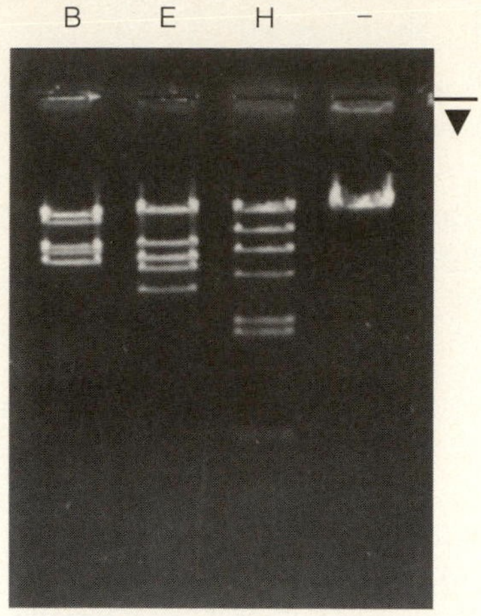

Short Run
Bands compressed; short time electrophoresing.

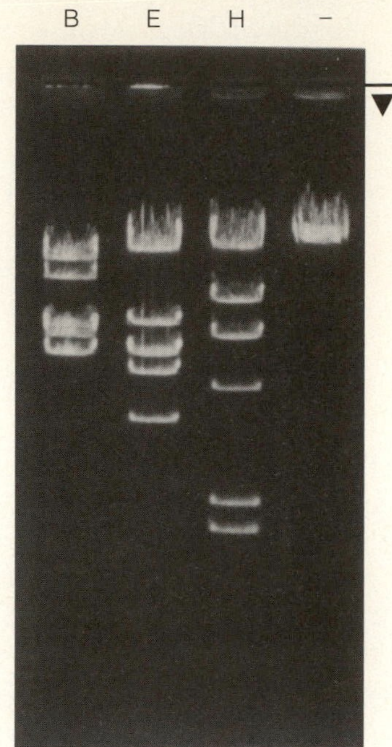

Overloaded
Bands smeared in all lanes; too much DNA in digests.

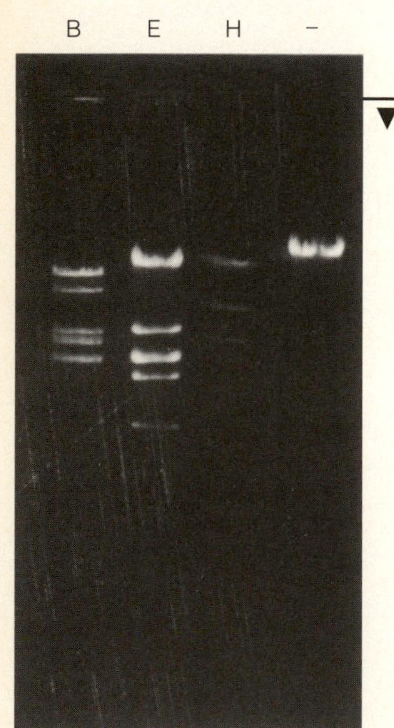

Punctured Wells
Bands faint in lanes B and H; DNA lost through hole punched in bottom of well with pipet tip.

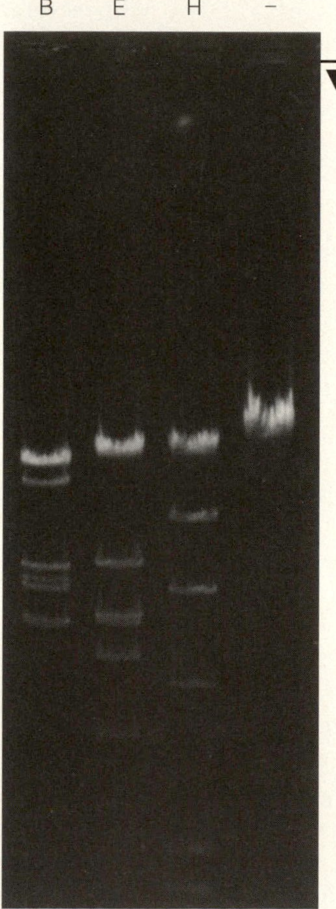

Long Run
Bands spread; long time electrophoresing.

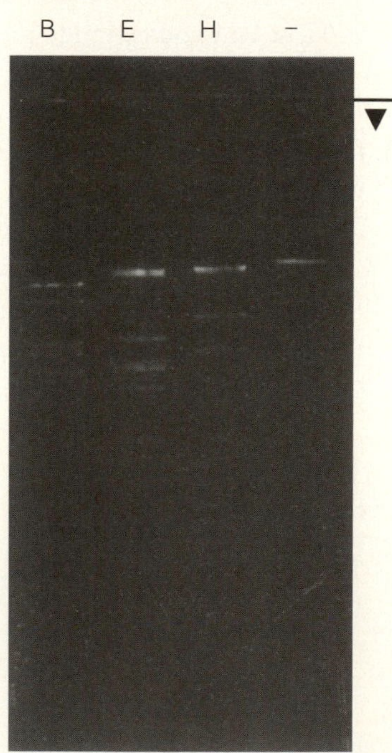

Underloaded
Bands faint in all lanes; too little DNA in digests.

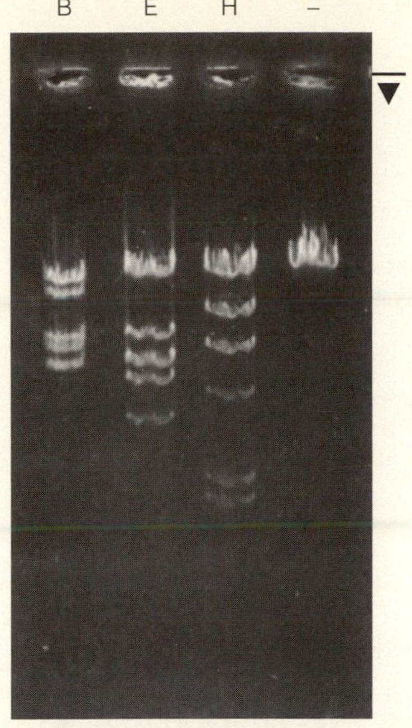

Poorly Formed Wells
Wavy bands in all lanes; comb removed before gel was completely set.

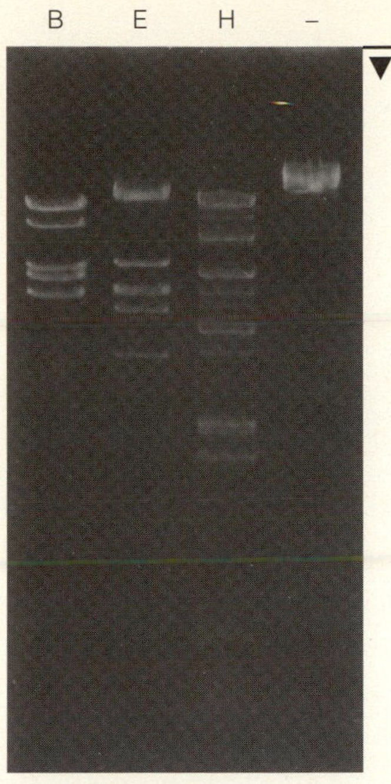

Enzymes Mixed
Extra bands in Lane H; *Bam*HI and *Hin*dIII mixed in digest.

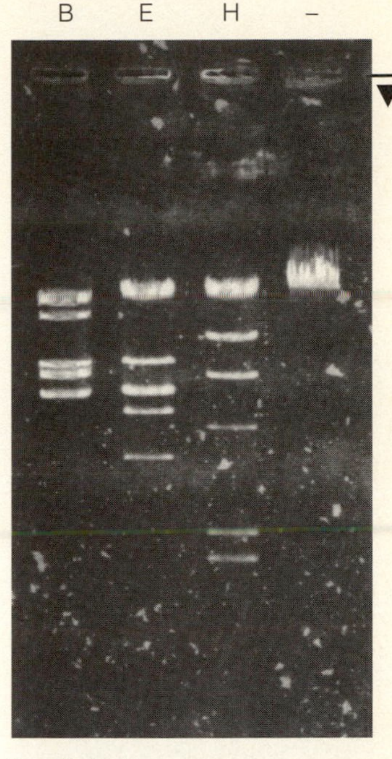

Precipitate
Precipitate in TBE buffer used to make gel.

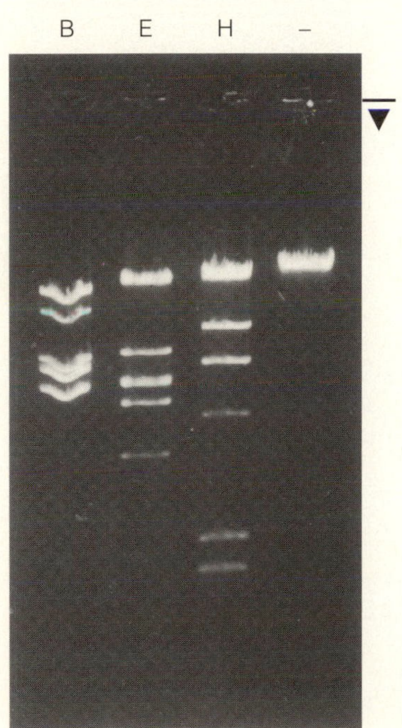

Bubble in Lane
Bump in band in lane B; bubble in lane.

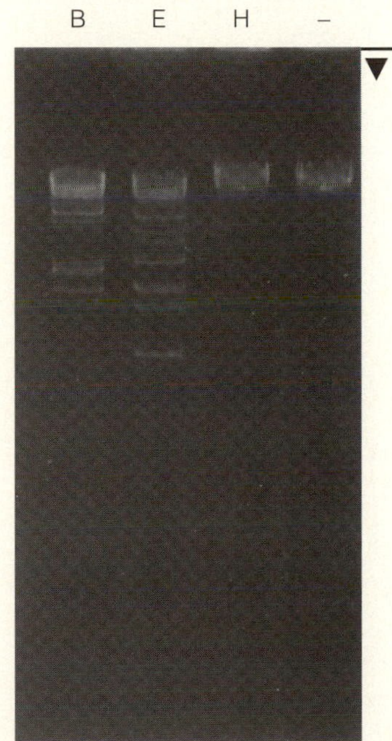

Incomplete Digest
Bands faint in lane H; very little *Hin*dIII in digest. Also, extra bands are present in lanes B and E.

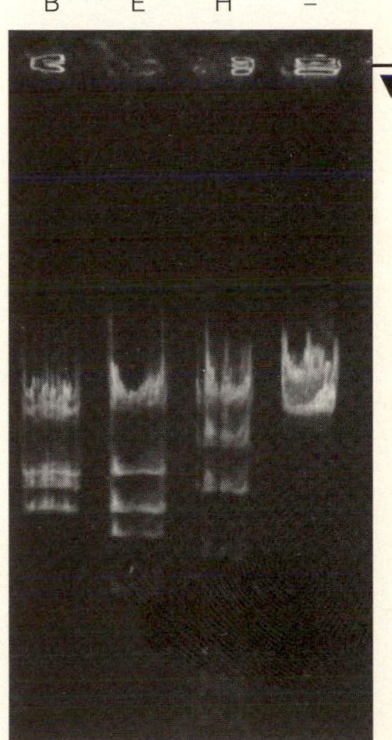

Gel Made with Water
Bands smeared in all lanes; gel made with water or wrong concentration of TBE buffer.

DNA Modification

LABORATORY 4

Effects of DNA Methylation on Restriction

In this laboratory, the *Eco*RI methylation system is used to illustrate the sequence specificity of a modifying enzyme that protects DNA from restriction enzyme digestion. M.*Eco*RI methylase adds a methyl group to the second adenine residue in the *Eco*RI recognition site, thus preventing the endonuclease from binding and cutting the DNA. *S*-adenosyl methionine (SAM), which is included in the methylation reaction, donates the methyl group that is attached to the DNA molecule by the methylase.

Two samples of λ DNA are incubated at 37°C with *Eco*RI methylase; then, one sample is incubated with *Eco*RI and the other is incubated with *Hin*dIII. These samples, along with cut and uncut controls, are electrophoresed in an agarose gel and stained. Comparison of the band patterns reveals that the methylated DNA is protected from digestion by *Eco*RI. However, methylation at the *Eco*RI site has no effect on the activity of the restriction enzyme *Hin*dIII.

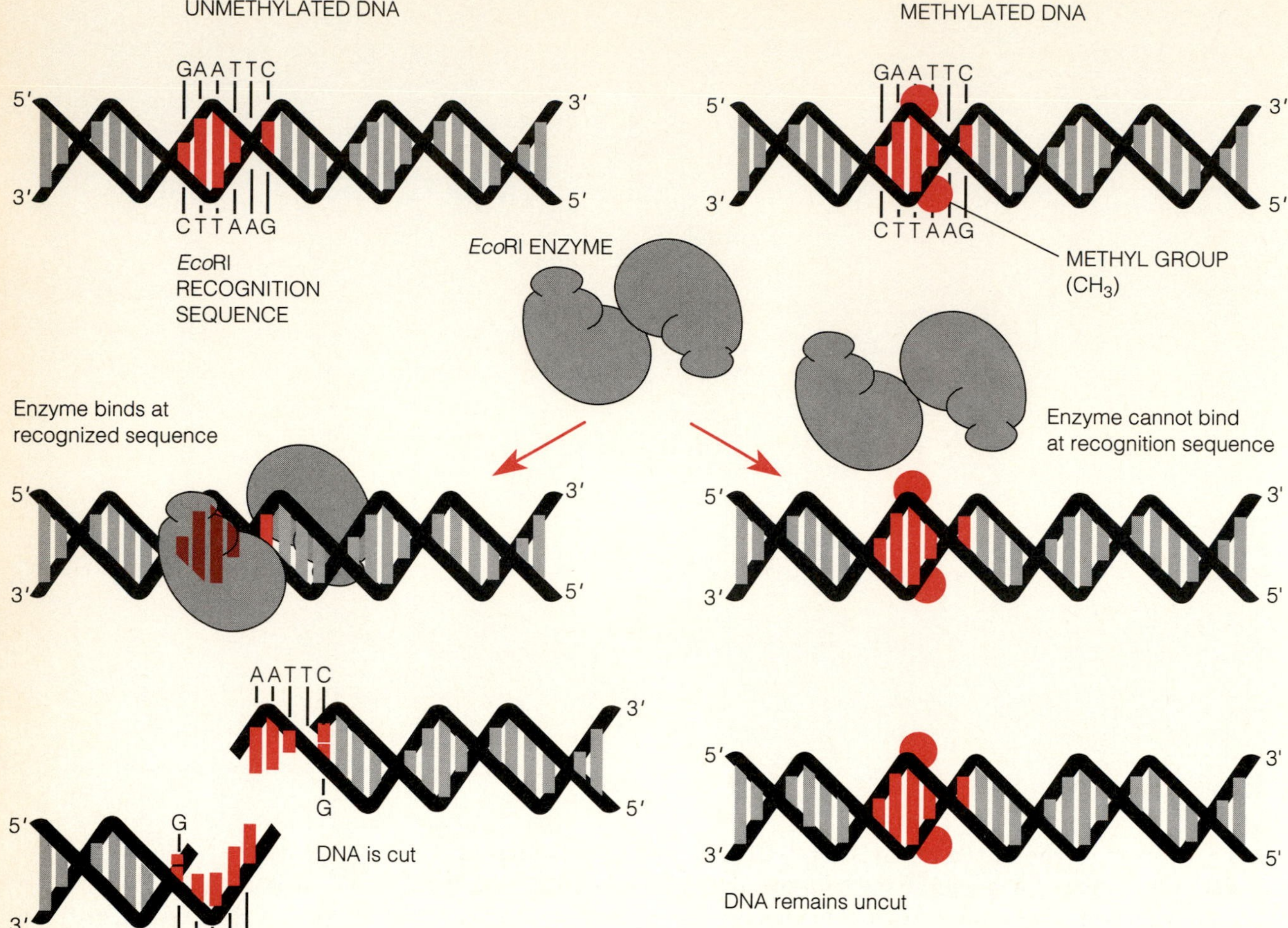

Molecular Detail of *Eco*RI Restriction-Modification
(Art concept developed by Lisa Shoemaker.)

I. Set Up Methylase Reactions

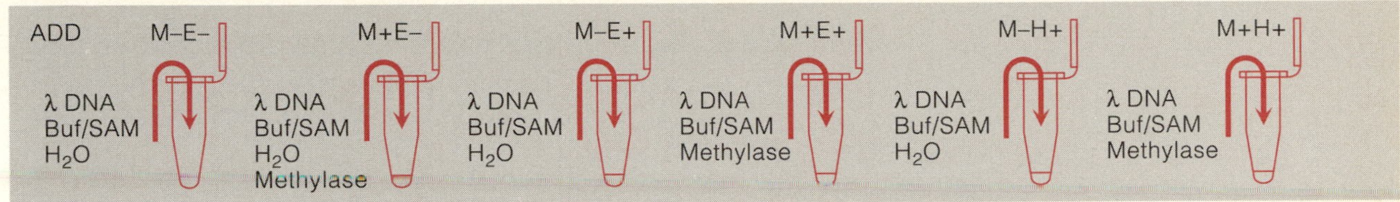

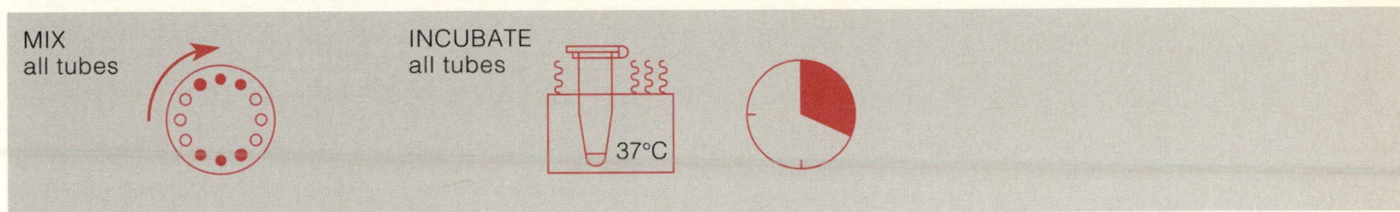

II. Cast 0.8% Agarose Gel

III. Set Up Restriction Reactions

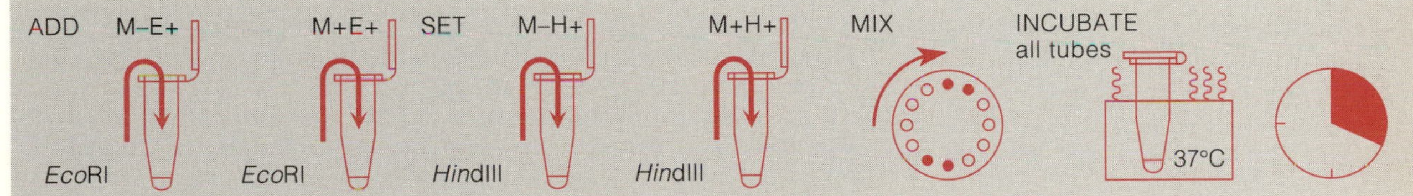

IV. Load Gel and Electrophorese

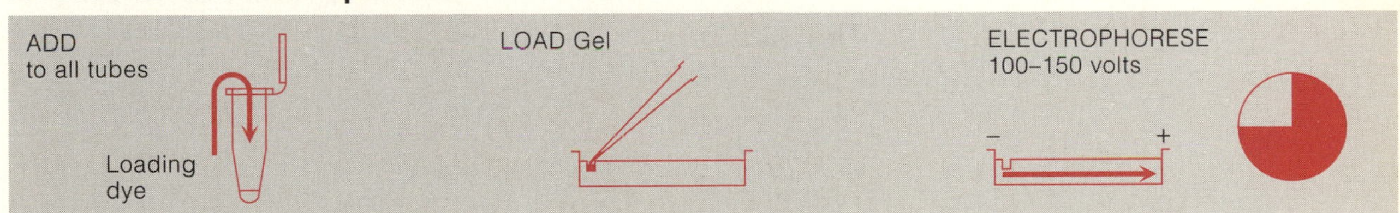

V. Stain Gel and View

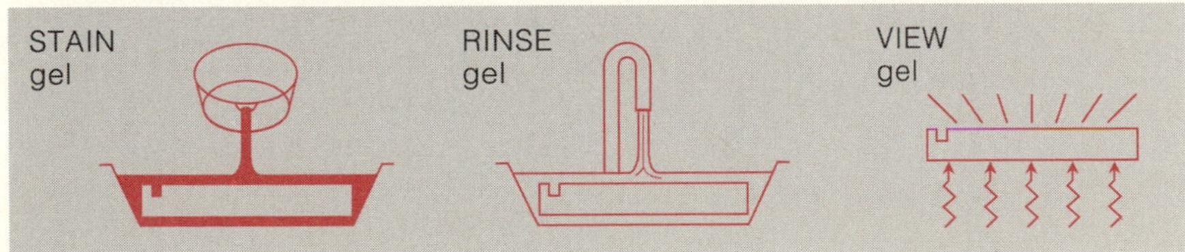

VI. Photograph Gel

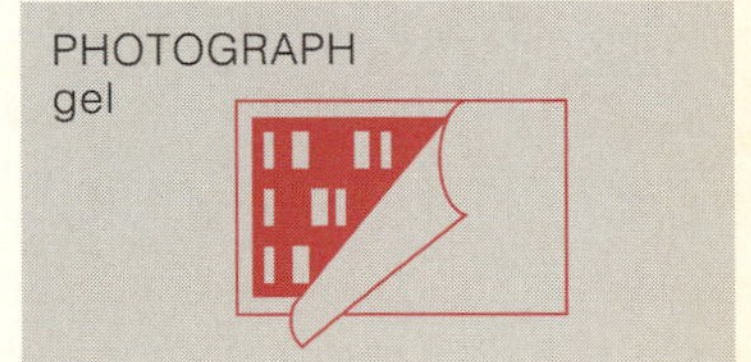

PRELAB NOTES

Review "Prelab Notes" in Laboratory 3, DNA Restriction and Electrophoresis.

S-Adenosyl Methionine

S-adenosyl methionine (SAM) is incorporated into 2× restriction buffer, so the same buffer is used for both methylation and restriction reactions. Because SAM is unstable, the buffer/SAM solution should be mixed within a day or two of the lab and discarded after use. Also, make sure to work from a fresh stock of SAM not more than several months old.

To Avoid Confusion

This laboratory has *two distinct steps* involving two similar-sounding reagents. *In the first step*, DNA is preincubated with *Eco*RI methylase. *In the second step*, the methylated DNA is incubated with *Eco*RI restriction enzyme. To avoid mishaps, do not set out the endonuclease *Eco*RI until after the methylation reactions have been set up.

For Further Information

The protocol presented here is based on the following published methods:

Greene, P. J., M. S. Poonian, A. L. Nussbaum, L. Tobias, D. E. Garfin, H. W. Boyer, and H. M. Goodman. 1975. Restriction and modification of a self-complementary octanucleotide containing the *Eco*RI substrate. *Journal of Molecular Biology* 99: 237.

Helling, R. B., H. M. Goodman, and H. W. Boyer. 1974. Analysis of *Eco*RI fragments of DNA from lambdoid bacteriophages and other viruses by agarose-gel electrophoresis. *Journal of Virology* 14: 1235.

Sharp, P. A., B. Sugden, and J. Sambrook. 1973. Detection of two restriction endonuclease activities in *Haemophilus parainfluenzae* using analytical agarose–ethidium bromide electrophoresis. *Biochemistry* 12: 3055.

EFFECTS OF DNA METHYLATION ON RESTRICTION

Reagents

0.1 µg/µl λ DNA
*Eco*RI methylase
restriction enzymes:
*Eco*RI
*Hin*dIII
2× restriction buffer/SAM
distilled water
loading dye
0.8% agarose solution
1 × Tris-Borate-EDTA (TBE) buffer
1 µg/ml ethidium bromide solution (or 0.025% methylene blue solution)

for decontamination:
0.05 M $KMnO_4$
0.25 N HCl
0.25 N NaOH

Supplies and Equipment

0.5–10-µl micropipettor + tips
1.5-ml tubes
aluminum foil
beaker for agarose
beaker for waste/used tips
camera and film (optional)
disposable gloves
electrophoresis box
masking tape
microfuge (optional)
Parafilm or waxed paper (optional)
permanent marker
plastic wrap (optional)
power supply
test tube rack
transilluminator (optional)
37°C water bath
60°C water bath (for agarose)

I. Set Up Methylase Reactions

(10 minutes; then 20+ minutes incubation)

1. Use a permanent marker to label six 1.5-ml tubes in which methylation and restriction reactions will be performed:

   M–E– = No methylase, no *Eco*RI
   M+E– = Methylase, no *Eco*RI
   M–E+ = No methylase, *Eco*RI
   M+E+ = Methylase, *Eco*RI
   M–H+ = No methylase, *Hin*dIII
   M+H+ = Methylase, *Hin*dIII

2. Use the matrix below as a checklist while adding reagents to each reaction. Read down each column, adding the same reagent to all appropriate tubes; *use a fresh pipet tip for each reagent.* Refer to the detailed instructions that follow.

Tube	λ DNA	Buffer/SAM	M.*Eco*RI methylase	H_2O
M–E–	4 µl	5 µl	—	1 µl
M+E–	4 µl	5 µl	1 µl	—
M–E+	4 µl	5 µl	—	1 µl
M+E+	4 µl	5 µl	1 µl	—
M–H+	4 µl	5 µl	—	1 µl
M+H+	4 µl	5 µl	1 µl	—

It is not necessary to change pipet tips when you are adding the same reagent. The same tip may be used for all tubes, provided that the tip has not touched any solution already in the tubes.

3. Add 4 µl of DNA to each reaction tube. Touch the tip of the pipet to the side of the reaction tube, as close to the bottom as possible, to create capillary action to pull the solution out of the tip.

Always add buffer to the reaction tubes before adding enzymes.

4. Use a *fresh tip* to add 5 µl of restriction buffer/SAM to a clean spot on each reaction tube.
5. Use a *fresh tip* to add 1 µl of *Eco*RI methylase to the appropriate tubes.

To avoid confusing methylase with the reagents in Part III, discard the methylase reagent tube after completing step 5.

6. Use a *fresh tip* to add a 1 µl of distilled water to the appropriate tubes.
7. Close the tops of the tubes. Pool and mix reagents by pulsing them in a microfuge or by sharply tapping the bottom of the tube on the lab bench.

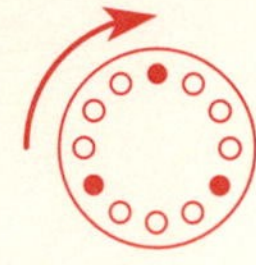

8. Place reaction tubes in a 37°C water bath, and incubate them for 20 minutes or longer.

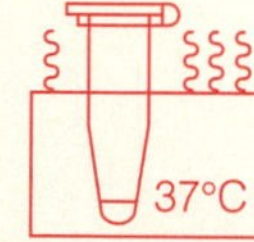

Stop Point

Following incubation, freeze the reactions at –20°C until you are ready to continue. Thaw the reactions before continuing on to Section III, step 1.

After several hours, methylase loses its activity, and the reaction stops.

II. Cast 0.8% Agarose Gel

(15 minutes)

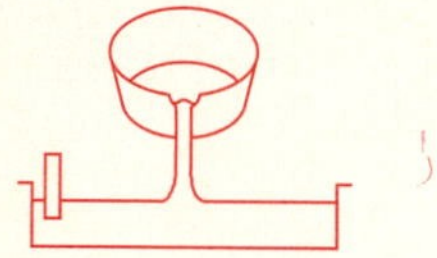

The gel is cast directly in the box in some electrophoresis apparatuses.

1. Seal the ends of a gel-casting tray with tape, and insert a well-forming comb. Place the gel-casting tray out of the way on the lab bench, so that the agarose to be poured in the next step can set undisturbed.
2. Carefully pour agarose solution into the casting tray to fill it to a depth of about 5 mm. The gel should cover only about ⅓ the height of the comb teeth. Use a pipet tip to move large bubbles or solid debris to the sides or end of the tray, while the gel is still liquid.
3. The gel will become cloudy as it solidifies (about 10 minutes). *Take care not to move or jar the casting tray while the agarose is solidifying.* Touch the corner of the agarose *away from* the comb to test whether the gel has solidified.
4. When the agarose has set, unseal the ends of the casting tray. Place the tray on the platform of the gel box, so that the comb is at the negative (black) electrode.

Too much buffer will channel the current over the top rather than through the gel, thus increasing the time required to separate the DNA fragments. TBE buffer can be reused several times; do not discard it. If you are using buffer remaining in the electrophoresis box from a previous experiment, rock the chamber back and forth to remix the ions that have accumulated at either end.

5. Fill the box with TBE buffer so that it barely covers the entire surface of the gel.

Buffer solution helps to lubricate the comb and prevent damage to the wells.

6. Gently remove the comb, pulling it straight up and out of the set agarose; do not rock or wiggle it. If necessary, have your partner steady the casting tray.

7. Removal of the comb usually pulls the agarose well edges above the buffer surface; the edges appear as "dimples" around wells. Add buffer until any dimples disappear and the buffer surface is smooth.

Some gel boxes are designed so that you must remove the comb before inserting the casting tray into the box. In this case, flood the casting tray and gel surface with buffer *before* removing the comb.

Stop Point

Cover the electrophoresis box, and save the gel until you are ready to continue. The gel will remain in good condition for several days, if it is completely submerged in buffer.

III. Set Up Restriction Reactions

(5 minutes; then 20+ minutes incubation)

1. Remove methylation reactions from the water bath.
2. Use the matrix below as a checklist while adding reagents to each reaction. Read down each column, adding the same reagent to all appropriate tubes; *use a fresh pipet tip for each reagent.* Refer to the detailed directions that follow.

Tube	EcoRI	*Hind*III
M–E–	—	—
M+E–	—	—
M–E+	1 µl	—
M+E+	1 µl	—
M–H+	—	1 µl
M+H+	—	1 µl

It is not necessary to change tips when you are adding the same reagent. The same tip may be used for all tubes, provided that the tip has not touched any solution already in the tubes.

3. Add 1 µl of *Eco*RI to the M–E+ and M+E+ tubes. Touch the tip of the pipet to the side of the reaction tube, as close to the bottom as possible, to create capillary action to pull the solution out of the tip.
4. Use a *fresh tip* to add 1 µl of *Hind*III to the M–H+ and M+H+ tubes.
5. Close the tops of the tubes. Pool and mix the reagents by pulsing them in the microfuge or by sharply tapping the bottom of the tube on the lab bench.
6. Place the reaction tubes in a 37°C water bath, and incubate the restriction reactions for 20 minutes or longer.

After several hours, enzymes lose their activity, and the reaction stops.

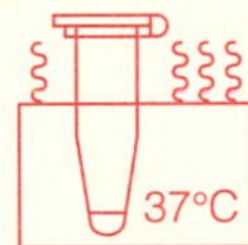

Stop Point

Following incubation, freeze the reactions at –20°C until you are ready to continue. Thaw the reactions before continuing on to Section IV, step 1.

IV. Load Gel and Electrophorese

(10 minutes; then 40–60 minutes electrophoresis)

1. Remove the restriction digests from the 37°C water bath.
2. Add 1 µl of loading dye to each reaction by one of two methods.

a. Add 1 µl of loading dye to each reaction tube. Close the tops of the tubes, and mix the solution by either tapping the bottom of the tube on the lab bench, pipetting in and out, or pulsing it in a microfuge. Make sure that the tubes are placed in a *balanced* configuration in the rotor.

or

b. Place six individual droplets of loading dye (1 µl each) on a small square of Parafilm or waxed paper. Withdraw the contents from one reaction tube, and mix it with a loading dye droplet by pipetting in and out. Immediately load the dye mixture, according to step 2. Repeat this successively, with a clean pipet tip, for each reaction.

Placing a piece of dark construction paper beneath the gel box will make the wells more visible.

3. Use a micropipettor to load 10 µl from each reaction tube into a separate well in the gel, as shown in the diagram below. Use a *fresh tip* for each reaction.

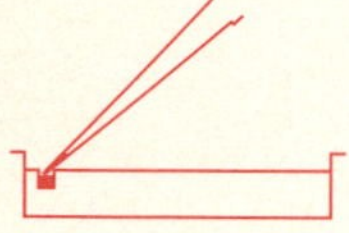

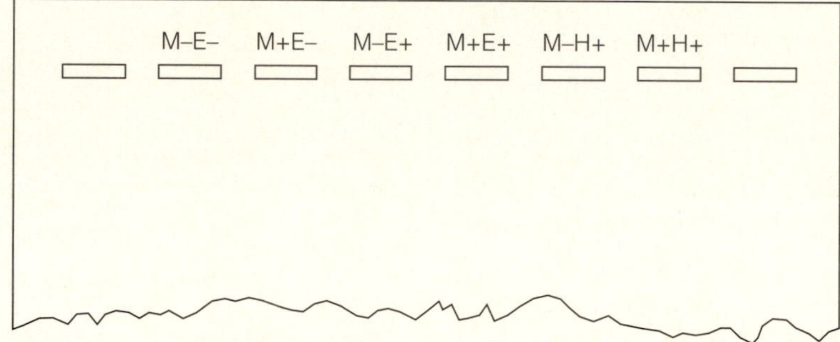

a. Steady the pipet over the well, using two hands.

b. If there is air in the end of the pipet tip, carefully depress the plunger to push the sample to the end of the tip. (An air bubble ejected into a well can form a "cap" over the well, causing DNA/loading dye solution to flow into the buffer around the edges of the well.)

c. Center the pipet tip over the well, dip the tip in only enough to pierce the buffer surface, and then gently depress the pipet plunger to slowly expel the sample.

Take care not to insert the pipet tip directly into the well; this could damage or puncture the well. Sucrose in the loading dye weighs down the DNA sample, thus causing it to sink to the bottom of the well.

4. Close the top of the electrophoresis box, and connect the electrical leads to a power supply—anode to anode (red-red) and cathode to cathode (black-black). Make sure that both electrodes are connected to the same channel of the power supply.

5. Turn the power supply on, and set it to 100–150 volts. The ammeter (if present) should register approximately 50–100 milliamperes. Current flow can also be detected by observing gas bubbles released from the electrode wires. If current is not detected, check the connections and try again.

Alternatively, set the power supply on a lower voltage, and run the gel for several hours. When you are running two gels off of the same power supply, the current is double that used for a single gel at the same voltage.

6. Electrophorese for 40–60 minutes. Good separation has occurred when the bromophenol blue bands have moved 40–70 mm from the wells. If time allows, electrophorese until the bromophenol blue bands near the end of the gel. *Stop* the electrophoresis before the bromophenol blue band runs off the end of the gel.

7. Turn off the power supply, disconnect the leads from the inputs, and remove the top of the electrophoresis box.
8. Carefully remove the casting tray from the electrophoresis box, and slide the gel into a disposable weigh boat or other shallow tray. Label the staining tray with your name and the date.

Stop Point

Gel can be stored overnight, covered with small volume of buffer in a staining tray or a zip-lock plastic bag, for viewing/photographing the next day. However, over longer periods of time, the DNA will diffuse through the gel, and the bands will either become indistinct or disappear entirely.

9. Stain and view the gel, using one of the methods described in Sections VA and VB.

Staining can be performed by an instructor in a controlled area, when students are not present.

VA. Stain Gel with Ethidium Bromide and View

(10–15 minutes)

CAUTION

Review "Ethidium Bromide Staining and Responsible Handling" in Laboratory 3. Wear disposable gloves when staining, viewing, and photographing gel and also during cleanup. Confine all staining to a restricted sink area.

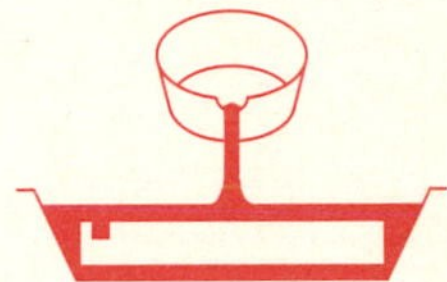

1. Flood the gel with an ethidium bromide solution (1 µg/ml), and allow them to stain for 5–10 minutes.
2. Following staining, use a funnel to decant as much ethidium bromide solution as possible from the staining tray back into a storage container.
3. Rinse the gel and the tray under running tap water.
4. If desired, the gel can be destained in tap water or distilled water for 10 minutes or longer, to help remove background ethidium bromide.
5. View the gels under an ultraviolet transilluminator or other UV source.

Staining time and the "background" of unbound ethidium bromide are markedly increased in thicker gels.

Staining solution can be reused to stain 15 gels or more. When the staining time increases markedly, spike the solution with 5 mg/ml stock solution according to the recipe in Appendix 1 or else disable the staining solution according to the procedure on page 38.

Band intensity and contrast increase dramatically if a gel is destained for 15–30 minutes in tap water. More simply, rinse and drain the gels, stack the staining trays, cover the top gel, and let it set overnight at room temperature.

CAUTION

Ultraviolet light can damage your eyes. Never look directly at an unshielded UV light source without eye protection. View only through a filter or safety glasses that absorb the harmful wavelengths.

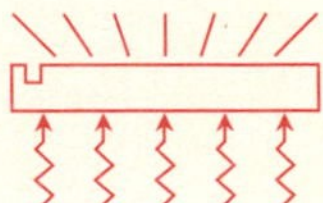

6. Take the time for proper cleanup.
   a. Wipe down the camera, transilluminator, and staining area.
   b. Decontaminate the gel and any staining solution that will not be reused.
   c. Wash your hands before leaving the laboratory.

The staining solution can be reused to stain 15 gels or more.

Destaining time is decreased by agitating and rinsing the gel in warm water, with agitation. For best results, continue to destain the gel overnight in a small volume of water. (The gel may destain too much if left overnight in a large volume of water.) Cover the staining tray to retard evaporation.

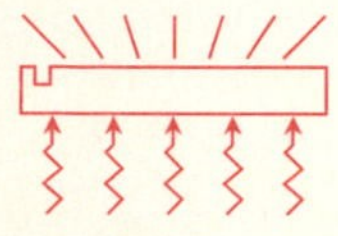

VB. Stain Gel with Methylene Blue and View

(30+ minutes)

1. Wear disposable gloves during staining and cleanup.
2. Flood the gel with 0.025% methylene blue, and allow them to stain for 20–30 minutes.
3. Following staining, use a funnel to decant as much methylene blue solution as possible from the staining tray back into a storage container.
4. Rinse the gel under running tap water. Let it soak for several minutes in several changes of fresh water. The DNA bands will become increasingly distinct as the gel destains.
5. View the gel over a light box; cover the surface with plastic wrap to prevent staining.

VI. Photograph Gel

(5 minutes)

Exposure times vary according to the mass of the DNA in the lanes, the level of staining, the degree of background staining, the thickness of the gel, and the density of the filter. Experiment to determine the best exposure. When possible, stop the lens down (to a higher f/number) to increase the depth of field and the sharpness of the bands.

1. *For ultraviolet (UV) photography of ethidium-bromide-stained gels:* Use Polaroid type 667 high-speed film (ISO 3000). Set the camera aperture to f/8 and the shutter to 1 second. Depress the shutter once or twice for a time exposure of 1–2 seconds.

   or

   For white-light photography of methylene-blue-stained gels: Use Polaroid type 667 color film, with the aperture set to f/8 and a shutter speed of 1/125th of a second.
2. Place your left hand firmly on top of the camera to steady it. Firmly grasp the small white tab and pull it straight out from camera. This will cause a larger tab to appear.
3. Grip the large tab in center, and in one steady motion, pull the film straight out from the camera. This starts development.
4. Allow the film to develop for the recommended time (45 seconds at room temperature). Do not disturb the print while it is developing.
5. After the full development time has elapsed, separate the print from the negative by peeling back at the end nearest to the large tab.

CAUTION

Avoid getting caustic developing jelly on your skin or clothing. If the jelly does get on your skin, wash immediately with plenty of soap and water.

6. Wait to see the results of your first photo before making other exposures.

RESULTS AND DISCUSSION

Each Type II restriction enzyme has a corresponding methylase that recognizes the same nucleotide sequence. However, the methylase *adds* a methyl group, rather than cutting, within the recognition sequence. This modification blocks the recognition site and prevents subsequent binding by a restriction enzyme. Within the host bacterium, methylation protects the host DNA from cleavage by its endogenous restriction enzyme. Foreign DNA that is unmethylated at the recognition site is unprotected.

The methylation reaction requires *S*-adenosyl methionine (SAM), the common methyl-donating molecule in both prokaryotes and eukaryotes. SAM is composed of the nucleoside *adenosine* and the amino acid *methionine,* as its name implies. The donation of a methyl group from the methionine portion of the molecule converts it into *S*-adenosyl homoserine.

One common occurrence in this laboratory is partial methylation, where methyl groups are added to only a fraction of the *Eco*RI sites within the λ DNA molecules. Cleavage at the unprotected sites produces a partial digest, yielding restriction fragments of varying lengths. These fragments are evidenced as additional bands in an agarose gel; the intensity of the fragments is inversely proportional to the level of DNA methylation.

1. Examine the photograph of your stained gel (or view it on a light box or overhead projector). Compare your gel to the ideal gel shown below. How can you account for differences in separation and band intensity?

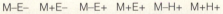

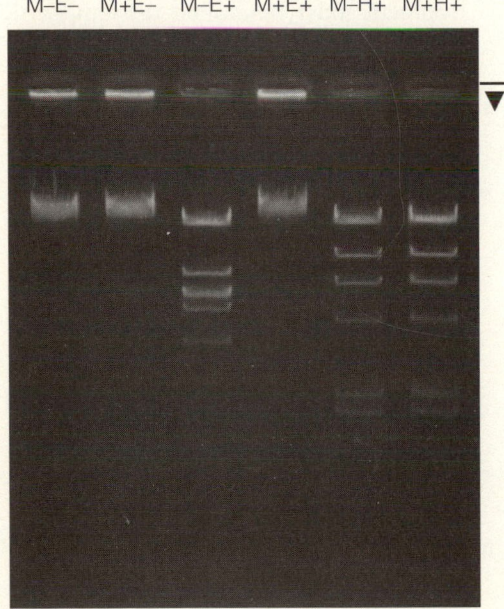

Ideal Gel

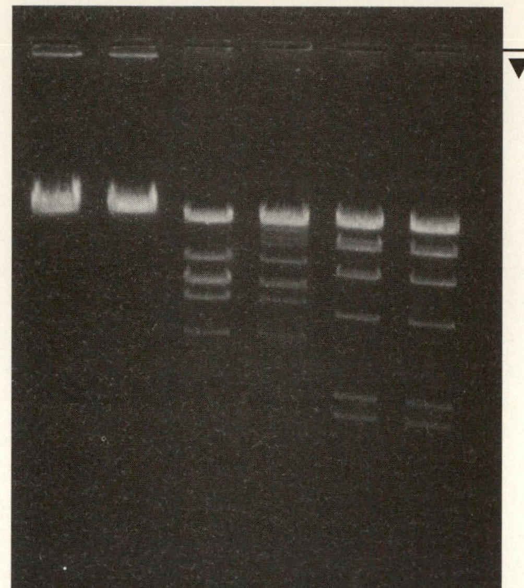

Incomplete Methylation
Faint bands in Lane M+E+ DNA partially cut by *Eco*RI.

2. What does the M+H+ control tell you about M.*Eco*RI methylation?
3. What does the M+E– control tell you about methylation?
4. What biological function do methylases perform in bacteria? What adaptive value do they have for a bacterium?
5. Experimental constraints demand that a plasmid be constructed in two digestion steps whose order cannot be reversed. In step 1, a *Bam*HI fragment is inserted into the *Bam*HI site of plasmid pAMP. In step 2, an *Eco*RI fragment must be cloned into an *Eco*RI site *within* the *Bam*HI insert. Unfortunately, the pAMP "backbone" also contains an *Eco*RI site, which is not the intended cloning site for the *Eco*RI fragment in step 2.
   a. Draw a diagram of this cloning experiment.
   b. Explain how *Eco*RI methylase could be used to solve this experimental problem.

FOR FURTHER RESEARCH

1. Design a series of experiments to study the kinetics of a methylation reaction.
   a. Determine the approximate percentages of sites that are protected at various times.
   b. Repeat the experiments with several different methylase and DNA dilutions.
   c. In each case, at what point does protection appear to be complete?

2. Design an experiment using M.*Eco*RI methylase to map the locations of *Eco*RI restriction sites in the λ genome.
3. Research the use of methylases in constructing a genomic library.
4. Research the role of DNA methylation in gene regulation in higher organisms.
5. Research the role of DNA methylation in the eukaryotic phenomenon of gene imprinting.
6. Research the role of methylation in controlling the movement of transposable elements in maize (corn).

Plasmid Transformation and Identification

LABORATORY 5

Rapid Colony Transformation of *E. coli* with Plasmid DNA

This laboratory demonstrates a rapid method of transforming *E. coli* with a foreign gene. The bacterial cells are rendered "competent" to uptake plasmid DNA containing an ampicillin-resistance gene (pAMP). Transformants are detected by their antibiotic-resistant phenotype.

Samples of *E. coli* cells are scraped off of a nutrient agar plate (LB agar) and suspended in two tubes containing a solution of calcium chloride. Plasmid pAMP is added to one cell suspension, and both tubes are incubated at 0°C for 15 minutes. Following a brief heat shock at 42°C, cooling, and addition of LB broth, samples of the cell suspensions are plated on two types of media, LB agar and LB agar plus ampicillin (LB/amp).

The plates are incubated for 12–24 hours at 37°C and then checked for bacterial growth. Only cells that have been transformed by taking up the plasmid DNA with the ampicillin-resistance gene will grow on the LB/amp plate. Subsequent division of an antibiotic-resistant cell will produce a colony of resistant clones. Thus, each colony on an ampicillin plate represents a single transformation event.

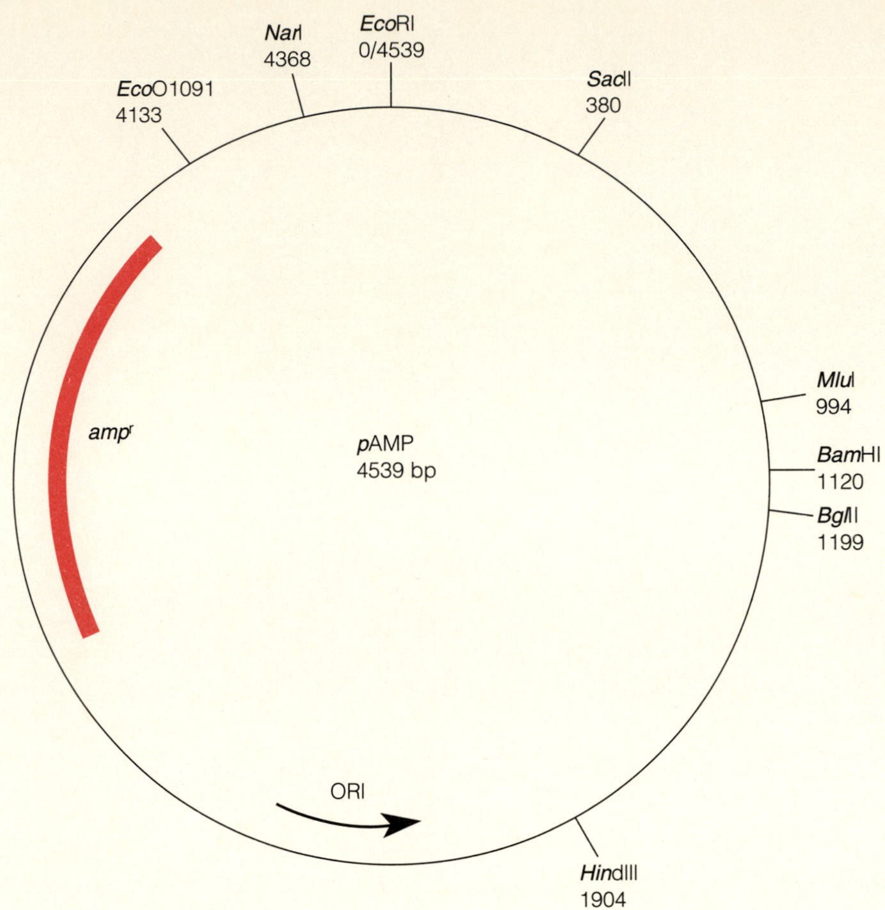
NarI
4368
EcoRI
0/4539
EcoO1091
4133
SacII
380
MluI
994
ampr
pAMP
4539 bp
BamHI
1120
BglII
1199
ORI
HindIII
1904

Transform *E. coli* with pAMP

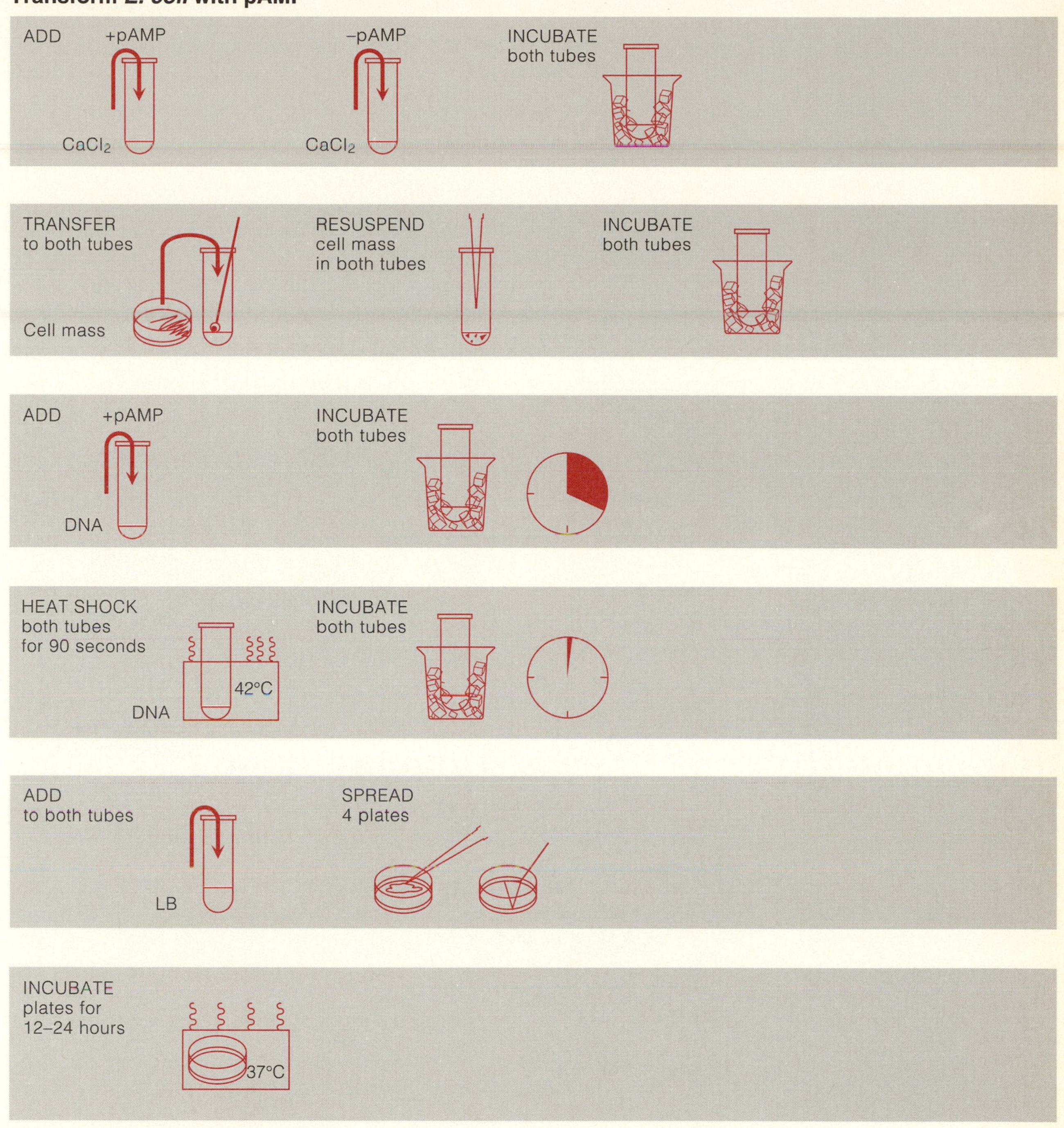

Laboratory 5
Rapid Colony Transformation of *E. coli* with Plasmid DNA

PRELAB NOTES

Review "Prelab Notes" in Laboratories 1 and 2, regarding sterile technique and *E. coli* culture.

Transformation Scheme

Most transformation protocols can be divided into these four major steps.

1. *Preincubation:* The cells are suspended in a solution of cations and incubated at 0°C. The cations are thought to complex with negatively charged phosphates of lipids in the *E. coli* cell membrane. The low temperature congeals the cell membrane, stabilizing the distribution of charged phosphates and allowing them to be more effectively shielded by the cations.

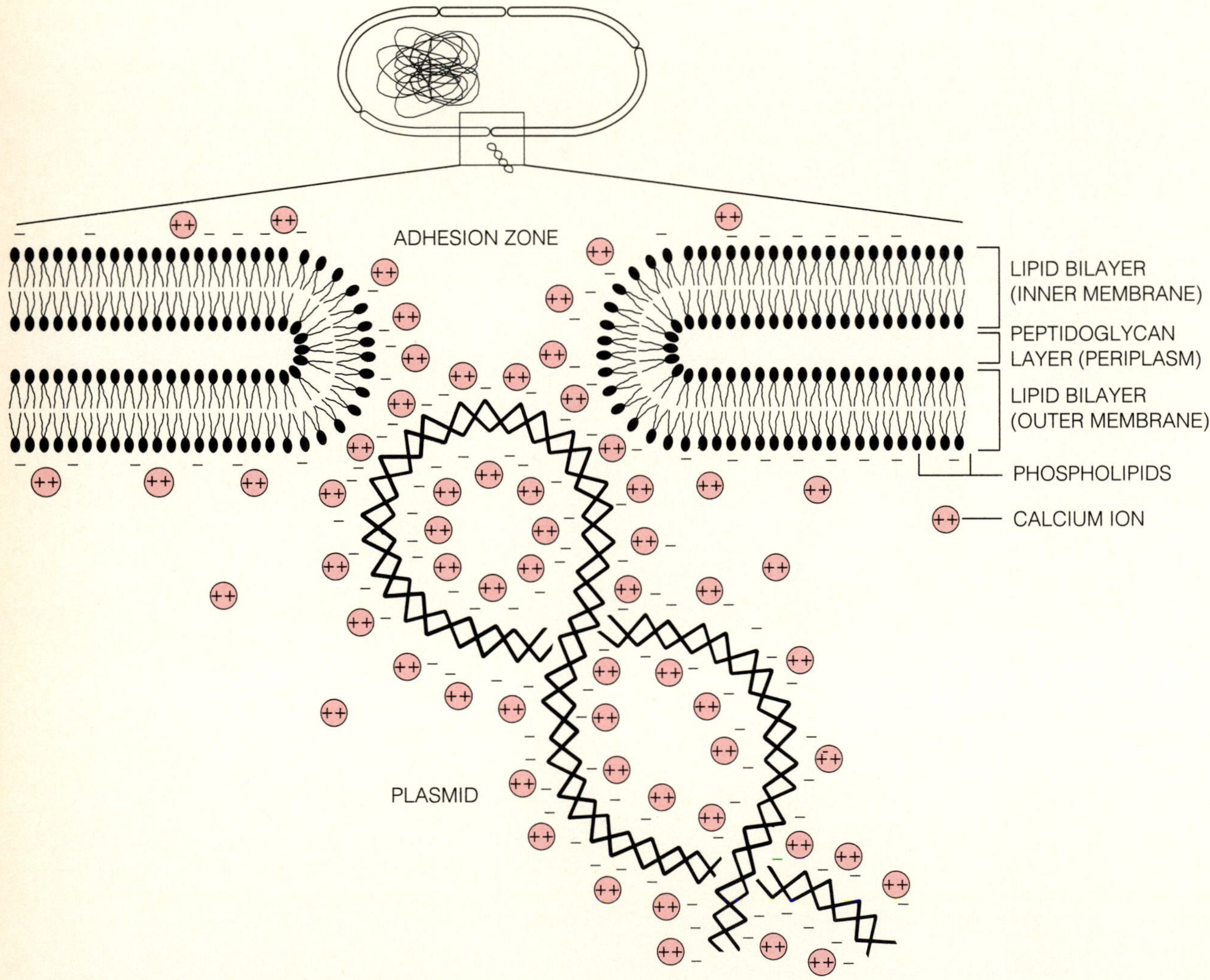

Proposed Molecular Mechanism of DNA Transformation of *E. coli*
Calcium ions (+ +) complex with negatively charged oxygens (–) to shield DNA phosphates from phospholipids at the adhesion zone.

2. *Incubation:* DNA is added, and the cell suspension is further incubated at 0°C. The cations are thought to neutralize negatively charged phosphates in the DNA and the cell membrane.
3. *Heat shock:* The cell/DNA suspension is briefly incubated at 42°C and then returned to 0°C. The rapid temperature change creates a thermal imbalance on either side of the *E. coli* membrane, which is thought to create a draft that sweeps plasmids into the cell.
4. *Recovery:* LB broth is added to the DNA/cell suspension and incubated at 37°C (ideally with shaking) before plating on selective media. Transformed cells recover from the treatment, amplify the transformed plasmid, and begin to express the antibiotic-resistance protein.

Simplifications of Colony Transformation

The classic transformation procedure used in Laboratories 8, 15, and 20 employs all of the above steps. However, only the incubation and heat shock steps are absolutely essential. To save time, the simplified colony method omits inessential preincubation and recovery steps. (If time permits, a preincubation of 5–15 minutes or a recovery of 5–30 minutes may be included.) The colony method also begins with *E. coli* colonies scraped from an agar plate, thus eliminating the need for preparing mid-log phase cells. Since liquid culturing is not used, equipment for shaking incubation and spectrophotometric analysis are not required. Therefore, this procedure entails minimal preparation time, uses less equipment, and is virtually foolproof.

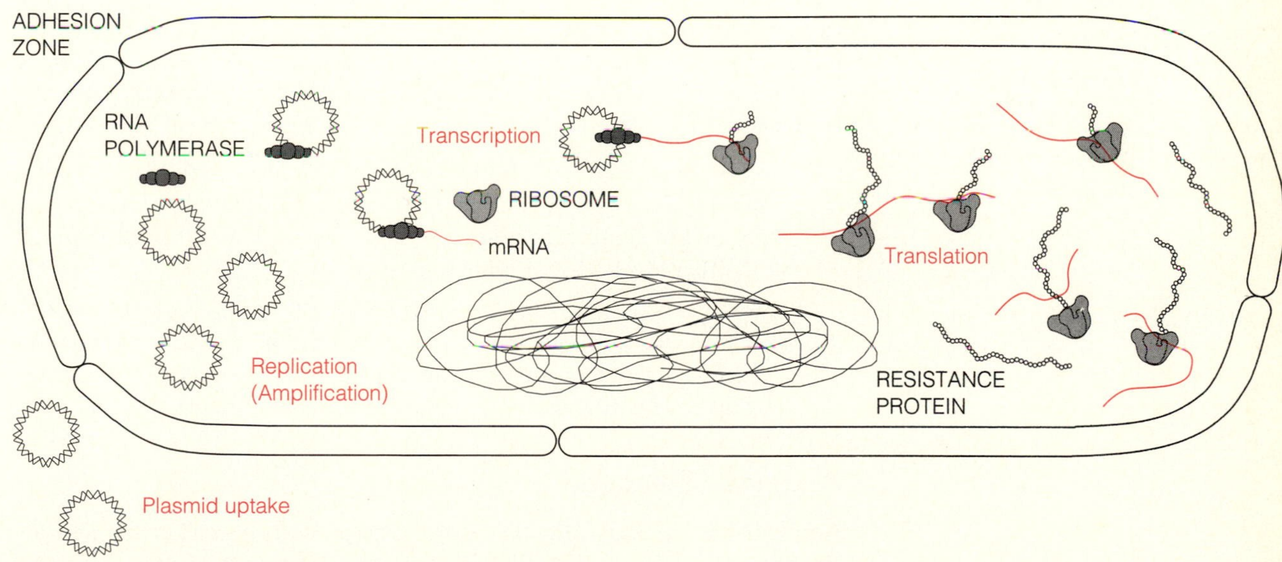

Expression of Antibiotic Resistance by Transformed *E. coli*

The Relative Inefficiency of Colony Transformation

The transformation efficiencies achieved with the colony protocol (5×10^3 to 5×10^4 colonies per microgram of plasmid) are up to 200 times less than those of the classic protocol (5×10^4 to 10^6 colonies per microgram). Colony transformation is perfectly suitable for transforming *E. coli* with purified, intact plasmid DNA. However, it will give marginal results with ligated DNA, which is composed of relaxed circular plasmid and linear plasmid DNA. These forms yield 5–100 times fewer transformants than an equivalent mass of intact, supercoiled plasmid.

Maintenance of *E. coli* Strains for Colony Transformation

The classic procedure (Laboratories 8, 15, and 20) includes several liquid culturing steps that help ensure vigorous, healthy *E. coli* cells that readily transform. However, the colony method eliminates these safeguards and uses cells cultured on solid agar. Therefore, it is important to monitor *E. coli* cells to be used in the colony method. If there is a drop in the number of transformants—from the expected 50–500 colonies per plate to essentially zero—discard your culture and obtain a fresh one.

Prolonged reculturing (passaging) of *E. coli* cells can result in a loss of competence that will make the bacterium virtually impossible to transform using the colony method. There is some evidence that loss of transforming ability in MM294 may result from exposure of the cells to temperatures below 4°C. Therefore, take care to store stab and slant cultures and streaked plates at room temperature.

Plasmid pAMP

Almost any plasmid containing a selectable antibiotic-resistance marker can be substituted for pAMP to demonstrate transformation of *E. coli* to an antibiotic-resistant phenotype. However, pAMP was constructed specifically as a teaching molecule and therefore offers advantages in other contexts:

1. pAMP is derived from a pUC expression vector that replicates to a high number of copies per cell. Therefore, yields from plasmid preparations are significantly greater than those obtained with pBR322 and other less highly amplified plasmids.
2. pAMP was designed for use with another teaching plasmid, pKAN. Each produces unique and readily recognizable restriction fragments when separated on an agarose gel. Thus, recombinant molecules formed by ligating these fragments can be easily characterized.

Antibiotic Selection

Ampicillin is the most practical antibiotic-resistance marker for demonstration purposes, especially in the rapid transformation protocol described here. Ampicillin interferes with construction of the peptidoglycan layer in the cell wall and only kills replicating cells that are assembling new cell envelopes. Nonreplicating *E. coli* with intact cell envelopes are unaffected. Thus, cells can be plated onto ampicillin-containing medium directly following heat shock, omitting the recovery step. Transformed cells can recover in the presence of ampicillin, because most express the resistance protein before replicating. Kanamycin selection, on the other hand, is less amenable to rapid transformation; a recovery step before plating is essential, because kanamycin acts quickly to kill any cells (replicating or nonreplicating) that are not actively expressing the resistance protein.

Test Tube Selection

The type of test tube used is a critical factor in achieving high-efficiency transformation and can also be important in the colony protocol. Therefore, we recommend using a presterilized 15-ml (17 mm × 100 mm) polypropylene culture tube. The critical heat shock step has been optimized for the thermal properties of a 15-ml polypropylene tube. Tubes of different material (such as polycarbonate) or a different thickness conduct heat differently. Also, the small volume of cell suspension forms a thin layer across the bottom of a 15-ml tube, thus allowing heat to be transferred quickly to all cells. A smaller tube (such as 1.5-ml) increases the depth of the cell suspension through which the heat must be conducted. Thus, *any* change in tube specifications requires recalibrating the duration of the heat shock. The "Falcon 2059" tube developed by Becton Dickinson is the standard for transformation experiments. Other quality brands are completely acceptable, but batches of tubes are occasionally contaminated with surfactants or other chemicals that can inhibit transformation. (Unfortunately, this can only be determined by experience.)

Purified Water

Extraneous salts and minerals in the transformation buffer can also affect results. Use the most highly purified water available; pharmacy-grade distilled water is recommended. You might want to obtain, from a local research center or hospital, water that has been purified through a multistage ion-exchange system, such as Milli-Q.

Eliminating Autoclaving

Calcium chloride and LB medium can be sterilized by passing each solution through a sterile filter with a pore size of 0.22 or 0.44 microns, which will exclude any contaminating microbes. Store the filtered solutions in presterilized 50-ml conical tubes. Presterilized, ready-to-pour agar need only be melted, cooled, and poured.

Presterilized Supplies

Presterilized supplies can be used to good effect in transformations; 15-ml culture tubes and individually packaged 100–1000-μl micropipet tips are handy. A 3-ml transfer pipet, marked in 250-μl gradations, can be substituted for a 100–1000-μl micropipettor with no loss of speed or accuracy. A presterilized inoculating loop, calibrated to deliver 10 μl, can be substituted for a 1–10-μl micropipettor. Eliminate flaming when you are using individually wrapped plasticware.

Technically, everything used in this experiment should be sterilized. However, it is acceptable to use clean, nonsterile plastic supplies that have been stored in a plastic bag or container. (Plasticware is manufactured by machine, so it is virtually sterile.) Aliquots of calcium chloride, LB broth, and plasmid DNA can be stored in nonsterile 1.5-ml tubes, *provided that they are used within a day or two.* Clean, nonsterile 1–10-μl micropipet tips can be used for adding DNA to cells in step 9. Antibiotic selection covers such minor lapses of sterile technique.

Starter Plates

Streaked "starter plates" of *E. coli* are the source of bacterial colonies used in this laboratory. For best results, use cells immediately following an incubation of 12–24 hours at 37°C. However, good results can also be obtained when cells are stored *at room temperature* for 1–2 additional days following the initial incubation. During this time, colonies grow to a large size and become tacky, making them easier to scrape up with the inoculating loop. A several-day-old plate can also be reincubated at 37°C for several hours or overnight before use.

For Further Information

The protocol presented here is based on the following published methods:

Cohen, S. N., A. C. Y. Chang, and L. Hsu. 1972. Nonchromosomal antibiotic resistance in bacteria: Genetic transformation of *Escherichia coli* by R-factor DNA. *Proceedings of the National Academy of Sciences* 69: 2110.

Hanahan, D. 1983. Studies on transformation of *Escherichia coli* with plasmids. *Journal of Molecular Biology* 166: 557.

Hanahan, D. Techniques for transformation of *E. coli.* 1987. In D. M. Glover (ed.), *DNA Cloning: A Practical Approach.* vol. 1. Oxford: IRL Press.

Mandel, M. and A. Higa. 1970. Calcium-dependent bacteriophage DNA infection. *Journal of Molecular Biology* 53: 159.

RAPID COLONY TRANSFORMATION OF *E. COLI* WITH PLASMID DNA

Culture, Media, and Reagents

MM294 starter culture
50 mM $CaCl_2$
0.005 µg/µl of pAMP
LB broth
2 LB plates
2 LB/amp plates

Supplies and Equipment

inoculating loop
100–1,000-µl micropipettor + tips (or 3-ml transfer pipets)
0.5–10-µl micropipettor + tips (or 10-µl calibrated loop)
2 15-ml culture tubes
beaker of crushed or cracked ice
beaker of 95% ethanol
beaker for waste/used tips
"bio-bag" or heavy-duty trash bag
10% bleach or disinfectant
burner flame
cell spreader
37°C incubator
permanent marker
test tube rack
37°C water bath (optional)
42°C water bath

Transform *E. coli* with pAMP

(45 minutes)

This entire experiment must be performed under sterile conditions. Review sterile techniques discussed in Laboratory 1, "Measurements, Micropipetting, and Sterile Techniques."

1. Use a permanent marker to label sterile 15-ml culture tubes:
   + = plus pAMP plasmid
   – = negative control (no plasmid)
2. Use a 100–1,000-µl micropipettor with a *sterile tip* (or sterile transfer pipet) to add 250 µl of $CaCl_2$ solution to each tube.
3. Place both tubes on ice.
4. Use a sterile inoculating loop to transfer several large (3-mm) colonies from the starter plate to the + tube:
   a. Sterilize the loop in the burner flame until it glows red hot. Then pass the lower half of the shaft through the flame.
   b. Stab the loop several times into the side of the agar to cool it.
   c. Gently scrape up enough colonies to form a *visible* cell mass in the eye of the inoculating loop, but be careful not to transfer any agar in the process. (Impurities in the agar can inhibit transformation.)
   d. Immerse the loop tip *directly into a CaCl2 solution* in the tube. Dislodge a cell mass by vigorously spinning the loop handle back and forth between your thumb and forefinger. Hold the tube up to the light to observe the cell mass drop off into the $CaCl_2$ solution.

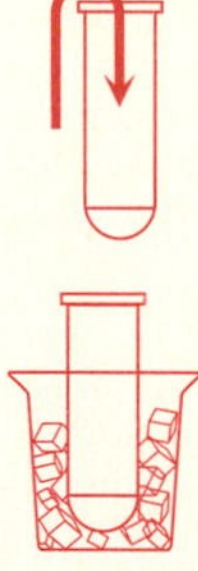

Optimally, flame the mouth of a 15-ml tube after removing and before replacing the cap.

If there are no individual colonies on the starter plate, scrape up several adjoining colonies on a streak, which will appear like beads on a string.

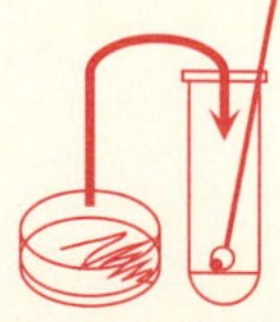

Make sure that the cell mass does not remain on the loop or on the side of the tube.

e. Reflame the loop before setting it down.

5. Immediately suspend the cells in the + tube by repeatedly pipetting in and out, using a 100–1,000-µl micropipettor with a sterile tip (or a sterile transfer pipet).

CAUTION

Keep your nose and mouth away from the end of the pipet tip when pipetting suspension culture, to avoid inhaling any aerosol that might be created.

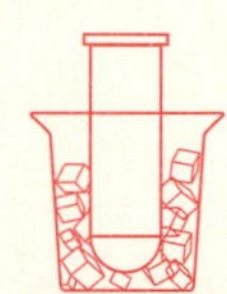

Cells become difficult to resuspend if they are allowed to clump together in the $CaCl_2$ solution. Suspending cells in the +pAMP tube first allows cells to preincubate for several minutes at 0°C while the –pAMP tube is being prepared. If time permits, both tubes can be preincubated on ice for 5–15 minutes.

a. Pipet carefully to avoid making bubbles in the suspension or splashing the suspension far up on the sides of the tube.

b. Wipe off any condensation on the tube bottom, and hold the tube up to the light to check that the suspension is homogeneous. The suspension should look slightly milky, and no visible clumps of cells should remain.

6. Return the + tube to the ice.

7. Transfer a second mass of cells to the – tube, as described in steps 4 and 5 above.

8. Return the – tube to the ice. Now both tubes should be on ice.

9. Use a 1–10-µl micropipettor (or calibrated inoculating loop) to add 10 µl of 0.005 µg/µl pAMP *directly into the cell suspension* in the + tube. Tap the tube lightly with your finger to mix the suspension. Avoid making bubbles in the suspension or splashing it up the sides of the tube.

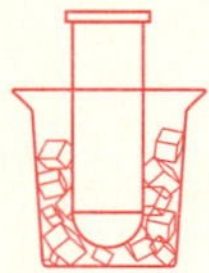

Double-check both tubes for complete suspension of cells; this is probably the most important variable for obtaining good results. If either suspension looks rather clear, add an additional cell mass and resuspend it.

10. Return the + tube to the ice. Incubate both tubes on ice for another 15 minutes.

11. While the cells are incubating on ice, use a permanent marker to label two LB plates and two LB/amp plates with your name and the date. Divide these plates into sets containing one of each plate type, and mark them as follows:

+ : Mark one LB plate and one LB/amp plate.

– : Mark one LB plate and one LB/amp plate.

Make it a habit to mark the plate bottoms, which always stay attached to the agar. To save plates, experimenters can omit *either* the +LB *or* the –LB plate.

12. Following the 15-minute incubation, heat shock the cells in both tubes. *The cells must receive a sharp and distinct shock.*
    a. Carry an ice beaker to the water bath. Remove tubes from the ice, and *immediately* immerse them in a 42°C water bath for 90 seconds.
    b. Immediately return both tubes to the ice for at least 1 more minute.

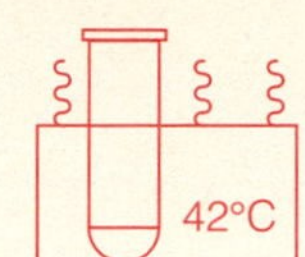

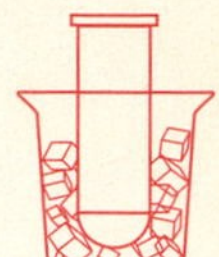

If time permits, the cells may be allowed to recover at 37°C for 5–30 minutes. Gentle shaking is also helpful.

Stop Point

An extended period on ice following heat shock will not affect the transformation. If necessary, store the + and – tubes on ice in the refrigerator (0°C) for as long as several days, until you have time to plate the cells. Do not put cell suspensions in the freezer, however.

13. Place both tubes in a test tube rack at room temperature.
14. Use a 100–1,000-µl micropipettor with a sterile tip (or sterile transfer pipet) to add 250 µl of LB broth to each tube. Tap the tubes lightly with your finger to mix the solution.

15. Use the matrix below as a checklist as + and – cells are spread on each type of plate in the following steps:

	+ Transformed Cells	– Nontransformed Cells
LB/amp	100 µl	100 µl
LB	100 µl	100 µl

Do not allow the cells to sit too long on the plates before spreading them. The object is to evenly distribute and separate the transformed cells on the plate surface so that each gives rise to a distinct colony of clones.

16. Use a 100–1,000-µl micropipettor with a sterile tip (or transfer pipet) to add 100 µl of cell suspension from the – tube onto the –LB plate and another 100 µl onto the –LB/amp plate.
17. Sterilize the cell spreader, and spread the cells over the surface of each – plate in succession.
    a. Dip the spreader into an ethanol beaker, and briefly pass it through a burner flame to ignite the alcohol. Allow the alcohol to burn off *away from the burner flame.*

Do not overheat the spreader in the burner flame; cool it before touching cell suspensions. The spreader, submerged in alcohol, is already sterile. The only purpose of the flame is to burn off the alcohol before spreading. The spreader will become too hot if it is held directly over the burner flame for several seconds; this can also kill *E. coli* cells on the plate.

CAUTION

Be extremely careful *not* to ignite the ethanol in the beaker.

    b. Lift the plate lid, like a clam shell, only enough to allow spreading.
    c. Cool the spreader by touching it to the agar surface away from the cells or to condensed water on the plate lid.
    d. Touch the spreader to the cell suspension, and gently drag it back and forth several times across the agar or membrane surface. Rotate the plate one-quarter turn, and then repeat the spreading motion.
    e. Replace the plate lid. Return the cell spreader to the ethanol *without flaming.*

Take care not to gouge the agar.

Sterile Spreading Techniques (steps 16 and 17)

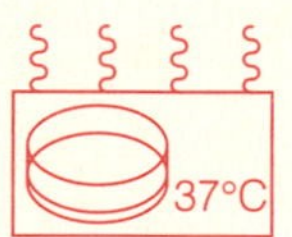

Save the +LB/amp plate as a source of a colony to begin an overnight suspension culture, if you are planning to execute Laboratory 6, "Purification and Identification of Plasmid DNA." Wrap the plate with Parafilm or plastic wrap, and refrigerate it until you are ready to use it.

18. Use a 100–1,000-µl micropipettor with a fresh, sterile tip (or a transfer pipet) to add 100 µl of cell suspension from the +AMP tube onto the +LB plate and another 100 µl onto the +LB/amp plate.
19. Repeat steps 17a–e to spread the cell suspension on the +LB and +LB/amp plates.
20. Reflame the spreader once more before setting it down.
21. Let the plates set for several minutes, to allow the cell suspensions to be absorbed.
22. Wrap the plates together with tape, place them upside down in a 37°C incubator, and incubate for 12–24 hours.
23. After the initial incubation, store the plates at 4°C to arrest *E. coli* growth and to slow the growth of any contaminating microbes.
24. Take the time for proper cleanup:
    a. Segregate, for proper disposal, culture plates and tubes, pipets, and micropipettor tips that have come into contact with *E. coli.*
    b. Disinfect overnight cell suspensions, tubes, and tips with a 10%-bleach solution or a disinfectant (such as Lysol).
    c. Wipe down the lab bench with soapy water, a 10%-bleach solution, or disinfectant.
    d. Wash your hands before leaving the laboratory.

RESULTS AND DISCUSSION

Count the number of individual colonies on the +LB/amp plate. Observe the colonies through the bottom of the culture plate, and use a permanent marker to mark each colony as it is counted. Between 50 and 500 colonies should be observed: 100 colonies equals a transformation efficiency of 10^4

colonies per microgram of plasmid DNA. (Question 3 explains how to compute transformation efficiency.)

If the plates have been overincubated or left at room temperature for several days, tiny "satellite" colonies may be observed to radiate from the edges of large, well-established colonies. Nonresistant satellite colonies grow in an "antibiotic shadow" where ampicillin has been broken down by the large resistant colony. Do not include satellite colonies in your count of transformants. A "lawn" should be observed on positive controls, where bacteria cover nearly the entire agar surface, and individual colonies cannot be discerned.

1. Record the number of colonies on each plate or, if cell growth is too dense to count individual colonies, record "lawn." Were the results as you expected? Explain possible reasons for variations from expected results.

	+ Transformed Cells	**– Nontransformed Cells**
LB/amp	experiment	negative control
LB	positive control	positive control

2. Compare and contrast the growth on each of the following pairs of plates. What does each pair of results tell you about the experiment?
   a. +LB and –LB
   b. –LB/amp and –LB
   c. +LB/amp and –LB/amp
   d. +LB/amp and +LB
3. Transformation efficiency is expressed as the number of antibiotic-resistant colonies per microgram of plasmid DNA. The object is to determine the mass of pAMP that was spread on the experimental plate and was therefore responsible for the transformants observed.
   a. Determine the *total mass (in μg) of pAMP* used in the transformation = concentration pAMP × volume pAMP used in step 9.
   b. Determine the *fraction of cell suspension spread* onto the +LB/amp plate (step 18) = volume suspension spread/*total* volume suspension (steps 2 and 14).
   c. Determine the *mass of pAMP in cell suspension spread* onto the +LB/amp plate = total mass pAMP (*a*) × fraction cell suspension spread (*b*).
   d. Express *transformation efficiency* in scientific notation as the number of colonies per μg of pAMP = number of colonies observed/mass pAMP in cell suspension spread (*c*).
4. What factors might influence transformation efficiency?
5. Your Favorite Gene (YFG) is cloned into pAMP, and *E. coli* is transformed with 0.2 μg of intact pAMP/YFG recombinant, according to the protocol above. Using the information below, calculate the number of molecules of YFG that have been cloned *in the entire culture* when the culture enters the stationary phase 200 minutes after inoculation.

a. A transformation efficiency equal to 10^6 colonies per microgram of intact pAMP is achieved.

b. pAMP has an average copy number of 100 molecules per transformed cell.

c. Following heat shock (step 12) the entire 250 µl of cell suspension is used to inoculate 25 ml of fresh LB broth. The culture is incubated, with shaking, at 37°C.

d. Transformed cells enter the log phase 60 minutes after inoculation and then begin to replicate on an average of once every 20 minutes.

6. The above transformation protocol is used with 10 µl of intact plasmid DNA at nine different concentrations. The following numbers of colonies are obtained when 100 µl of transformed cells are plated on selective medium:

0.00001 µg/µl	4 colonies
0.00005 µg/µl	12 colonies
0.0001 µg/µl	32 colonies
0.0005 µg/µl	125 colonies
0.001 µg/µl	442 colonies
0.005 µg/µl	542 colonies
0.01 µg/µl	507 colonies
0.05 µg/µl	475 colonies
0.1 µg/µl	516 colonies

a. Calculate the transformation efficiency at each concentration.

b. Plot a graph of DNA mass *versus* colonies.

c. Plot a graph of DNA mass *versus* transformation efficiency.

d. What is the relationship between the mass of DNA transformed and transformation efficiency?

e. At what point does the transformation reaction appear to be saturated?

f. What is the true transformation efficiency?

Interpretable experimental results can be obtained only when the colony transformation protocol can be repeated with reproducible results: 100–500 colonies on the +LB/amp plate.

FOR FURTHER RESEARCH

1. Design an experiment to compare the transformation efficiencies of linear *versus* circular plasmid DNAs. Keep the molecular weight constant.

2. Design a series of experiments to test the relative importance of each of the four major steps of most transformation protocols: (1) preincubation, (2) incubation, (3) heat shock, and (4) recovery. Which steps are absolutely necessary?

3. Design a series of experiments to compare the transforming effectiveness of calcium chloride *versus* salts of other monovalent (+), divalent (++), and trivalent (+++) cations—such as KCl, $MgCl_2$, and Fe_2O_3.

a. Make up 50 mM solutions of each salt.

   b. Check the pH of each solution, and buffer to approximately pH 7, when necessary.
   c. Is calcium chloride unique in its ability to facilitate transformation?
   d. Is there a consistent difference in the transforming effectiveness of monovalent *versus* divalent *versus* trivalent cations?

4. Design a series of experiments to determine saturating conditions for transformation.
   a. Transform *E. coli,* using the DNA concentrations listed in question 6 above.
   b. Plot a graph of DNA mass *versus* colonies per plate.
   c. Plot a graph of DNA mass *versus* transformation efficiency.
   d. At what mass does the reaction appear to become saturated?
   e. Repeat the experiment with concentrations clustered on either side of the presumed saturation point, to produce a fine saturation curve.

5. Repeat experiment 4 above, but instead transform with a 1:1 mixture of pAMP and pKAN at each concentration. Plate the transformants on LB/amp, LB/kan, and LB/amp+kan plates. *Be sure to include a recovery of 40–60 minutes, with shaking.*
   a. Calculate the percentage of double transformations at each mass.

$$\frac{\text{colonies LB/amp+kan plate}}{\text{colonies LB/amp plate} + \text{colonies LB/kan plate}}$$

   b. Plot a graph of DNA mass *versus* colonies per plate.
   c. Plot a graph of DNA mass *versus* the percentage of double transformations. Under saturating conditions, what percentage of bacteria are doubly transformed?

6. Plot a recovery curve for *E. coli* transformed with pKAN. Allow the cells to recover for 0–120 minutes, and plate samples at 20-minute intervals.
   a. Plot a graph of recovery time *versus* colonies per plate.
   b. At what point is antibiotic expression maximized?
   c. Can you discern a point where the cells began to replicate?

Purification and Identification of Plasmid DNA

Growth of *E. coli* on ampicillin plates demonstrates transformation to an antibiotic-resistant phenotype. Laboratory 5 showed that the observed phenotype is caused by the uptake of plasmid pAMP, a DNA molecule that is well characterized.

In experimental situations where numerous recombinant plasmids are generated by joining two or more DNA fragments, the antibiotic-resistance marker only indicates which cells have taken up a plasmid bearing the resistance gene; it does not indicate anything about the structure of the new plasmid. Therefore, it is standard procedure to isolate plasmid DNA from transformed cells and identify the molecular genotype using DNA restriction analysis. In cases where the recombinant molecules are formed by combining well-characterized fragments, restriction analysis is sufficient to confirm the structure of a hybrid plasmid. In other cases—for example, when cloning PCR (polymerase chain reaction) amplification products or products of mutagenesis reactions—it may be necessary to determine the exact nucleotide sequence of the inserted DNA fragment.

In Part A, "Plasmid Minipreparation of pAMP," a small-scale protocol is used to purify enough plasmid DNA for restriction analysis from transformed *E. coli*. Cells taken from an ampicillin-resistant colony are grown to the stationary phase in a suspension culture. The cells from 1 ml of culture are harvested and lysed, and plasmid DNA is separated from the cellular proteins, lipids, and chromosomal DNA. This procedure yields 2–5 μg of relatively crude plasmid DNA, in contrast to large-scale preparations that yield 1 mg or more of pure plasmid from a 1-liter culture.

In Part B, "Restriction Analysis of Purified pAMP," a sample of plasmid DNA isolated in Part A and a control sample of pAMP are cut with the restriction enzymes *Bam*HI and *Hin*dIII. The two samples are coelectrophoresed on an agarose gel, and the restriction fragments are stained and visualized. The purified DNA is shown to have a restriction "fingerprint" identical to that of pAMP. *Bam/Hin*d restriction fragments of the miniprep DNA comigrate with the 784- and 3,755-bp *Bam/Hin*d fragments of control pAMP. This provides genotypic proof that pAMP molecules were successively transformed into *E. coli* in Laboratory 5.

Isolate Plasmid DNA

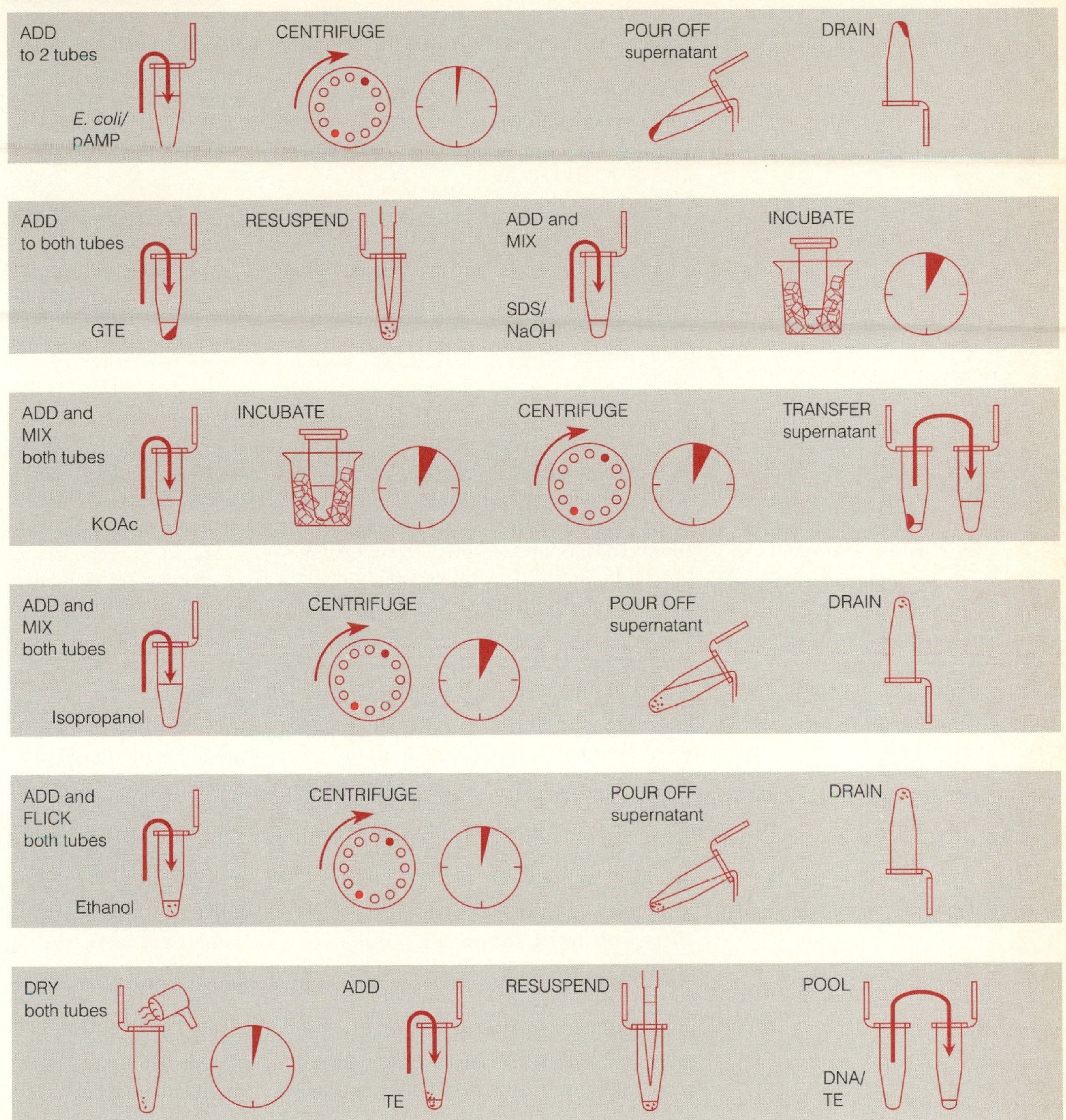

Laboratory 6/Part A
Plasmid Minipreparation of pAMP

PRELAB NOTES

Optimally, minipreps should be prepared from cells that have been recently manipulated for transformation. This completes a conceptual unit that firmly cements the genotype–phenotype relationship. Alternatively, use streaked plates of transformed *E. coli* to prepare overnight cultures.

Plasmid Selection

pAMP gives superior yields on minipreparations, compared to pBR322. pAMP is a derivative of a pUC expression vector, and it is highly amplified; more than 100 copies are present per *E. coli* cell. If you are substituting a different plasmid for miniprep purposes, select a member of the pUC family, such as pUC18 or pUC19, or another plasmid with a high copy number.

Centrifuge Requirements

A microfuge that generates at least 6,000 times the force of gravity (6000 × *g*) provides efficient and rapid purification of plasmid DNA. A number of small, inexpensive, but slower-spinning centrifuges are available for 1.5-ml tubes. Models that reach 2,000–3,000 × *g* can give reasonable plasmid minipreparations; however, the centrifuge times stated in the protocol should be doubled. A slower-spinning clinical or preparatory centrifuge cannot be substituted.

Supplies

Sterile supplies are not required for this protocol. Standard 1-ml pipets, transfer pipets, or microcapillary pipets can be used instead of micropipettors. Use good-quality, colorless 1.5-ml tubes, beginning with step 11. The walls of poor-quality tubes, especially colored ones, often contain tiny air bubbles that can be mistaken for ethanol droplets in step 19. We have observed experimenters drying DNA pellets for 15 minutes or longer, trying to rid their tubes of these phantom droplets. (Actual drying time is several minutes.)

Fine Points of Technique

Take care not to overmix reagents; excessive manipulation shears both plasmid and chromosomal DNA. The success of this protocol largely depends on maintaining chromosomal DNA in large pieces that can be differentially separated from intact plasmid DNA. Mechanical shearing reduces the yield of high-quality, supercoiled plasmid DNA and increases contamination with short-sequence chromosomal DNA. Make sure that the microfuge will be immediately available for step 13. If you are sharing a microfuge, coordinate with the other experimenters to begin steps 12 and 13 together.

Columns for Plasmid Purification

A number of types of microcolumns are available to preferentially bind plasmid DNA. In most cases, a cell lysate is run through the column, followed by one or more buffer washes. Bound plasmid is ultimately released from the column with a high-salt wash. Use of columns still requires lysis steps, typically with lysozyme or proteinase K, but saves time by eliminating alcohol precipitation. However, these columns add an expense of $0.50–1.50 per column, thus making them difficult to justify for teaching purposes.

For Further Information

The protocol presented here is based on the following published methods:

Birnboim, H. C. and J. Doly. 1979. A rapid alkaline extraction method for screening recombinant plasmid DNA. *Nucleic Acids Research* 7: 1513.

Ish-Horowicz, D. and J. F. Burke. 1981. Rapid and efficient cosmid cloning. *Nucleic Acids Research* 9: 2989.

PLASMID MINIPREPARATION OF pAMP

Culture and Reagents

E. coli/pAMP overnight culture
glucose/Tris/EDTA (GTE)
SDS/sodium hydroxide (SDS/NaOH)
potassium acetate/acetic acid (KOAc)
isopropanol
95% ethanol
Tris/EDTA (TE)

Supplies and Equipment

100–1,000-µl micropipettor + tips
0.5–10-µl micropipettor + tips
1.5-ml tubes
beaker of crushed ice
beaker for waste/used tips
10%-bleach solution or disinfectant (Lysol)
disposable gloves
hair dryer
microfuge
paper towels
permanent marker
test tube rack

Isolate Plasmid DNA

(50 minutes)

The instructions below are for making duplicate minipreps, which provide balance in the microfuge and insurance if a critical mistake is made.

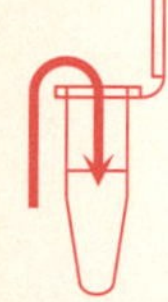
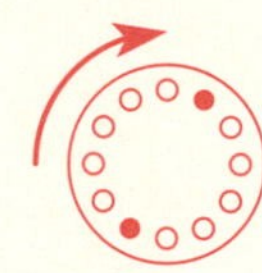

Double all centrifuge times stated in the protocol when you are using a small microfuge that generates forces of 2,000–3,000 × *g*.

1. Shake the culture tube to resuspend the *E. coli* cells.
2. Label two 1.5-ml tubes with your initials. Transfer 1,000 µl (1 ml) of *E. coli*/pAMP overnight suspension into each tube.
3. Close the caps, and place the tubes in a *balanced* configuration in the microfuge rotor. Spin for 1 minute to pellet cells.
4. Pour off the supernatant from both tubes into a waste beaker for later disinfection. *Take care not to disturb the cell pellets.* Invert the tubes, and touch their mouths to a clean paper towel, to wick off as much as possible of the remaining supernatant.

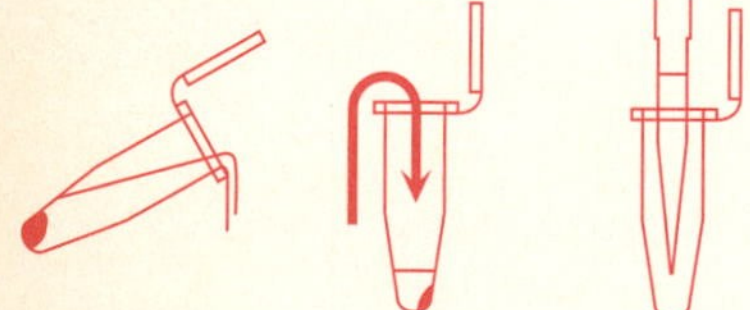

Accurate pipetting is essential to good plasmid yield. The volumes of reagents are precisely calibrated so that the sodium hydroxide added in step 6 is neutralized by the acetic acid in step 8. *Do not overmix.* Excessive agitation shears the single-stranded DNA and decreases plasmid yield.

5. Add 100 µl of *ice-cold* GTE solution to each tube. Resuspend the pellets by pipetting solution in and out several times. Hold the tubes up to the light to check that the suspension is homogeneous and that no visible clumps of cells remain.
6. Add 200 µl of room temperature SDS/NaOH solution to each tube. Close the caps, and mix the solutions by rapidly inverting the tubes about five times.
7. Stand the tubes on ice for 5 minutes. The suspension will become relatively clear.
8. Add 150 µl of *ice-cold* KOAc solution to each tube. Close the caps, and mix the solutions by rapidly inverting the tubes about five times. A white precipitate will immediately appear.
9. Stand the tubes on ice for 5 minutes.

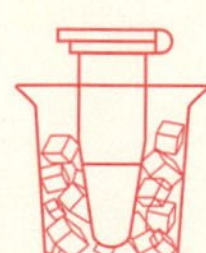

10. Place the tubes in a *balanced* configuration in the microfuge rotor, and spin them for 5 minutes to pellet the precipitate.
11. Transfer 400 µl of supernatant from each tube into two clean 1.5-ml tubes. *Avoid pipetting the precipitate.* Wipe off any precipitate clinging to the outside of the tip before expelling the supernatant. Discard old tubes containing precipitate.

Precipitate typically collects along the side of the tube, rather than forming a tight pellet.

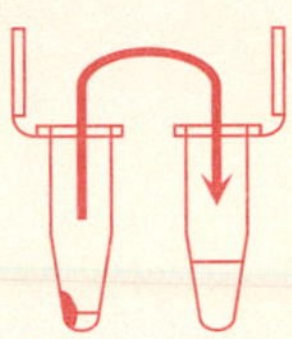

12. Add 400 µl of isopropanol to each tube of supernatant. Close the caps, and mix the solution by rapidly inverting the tubes about five times. *Stand the tubes at room temperature for only 2 minutes.*
13. Place the tubes in a *balanced* configuration in a microfuge rotor, and spin them for 5 minutes to pellet the nucleic acids. Align the tubes in the rotor so that the cap hinges point outward. The nucleic acid residue, visible or not, will collect at the bottom of the tube under the hinge, during centrifugation.

Do step 12 quickly, and make sure that the microfuge will be immediately available for step 13. The isopropanol preferentially precipitates nucleic acids rapidly; however, proteins and other cellular components remaining in the solution will also begin to precipitate in time.

14. Pour off the supernatant from both tubes. *Take care not to disturb the nucleic acid pellets.* Wick off as much as possible of the remaining alcohol on a paper towel.

The pellet will likely appear as either a tiny, teardrop-shaped smear or as small particles on the bottom of each tube. However, pellet size is not a valid predictor of plasmid yield or quality. Large pellets will be composed primarily of RNA, salt, and cellular debris carried over from the original precipitate; small pellets, or those impossible to see, often indicate a cleaner preparation.

15. Add 200 µl of 95% ethanol to each tube, and close the caps. Flick the tubes several times to wash the pellets.

Nucleic acid pellets are insoluble in ethanol and will not resuspend during washing.

Stop Point

Store the DNA in ethanol at –20°C until you are ready to continue.

16. Place the tubes in a balanced configuration in a microfuge rotor, and spin for 2–3 minutes to recollect the nucleic acid pellets.

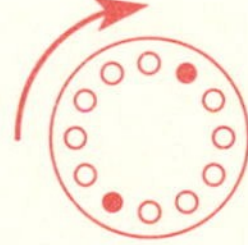

17. Pour off the supernatant from both tubes. *Take care not to disturb the nucleic acid pellets.* Wick off as much as possible of the remaining alcohol on a paper towel.
18. Dry the nucleic acid pellets, using one of following methods:
    a. Direct a stream of warm air from a hair dryer across the open ends of the tubes for about 3 minutes. *Do not blow the pellets out of the tubes.*

    or

    b. Close the caps, and pulse the tubes in the microfuge to pool the remaining ethanol. *Carefully*, use the micropipettor to draw off the ethanol. Let the pellets air dry at room temperature for 10 minutes.

Air bubbles cast in the tube wall can be mistaken for ethanol droplets.

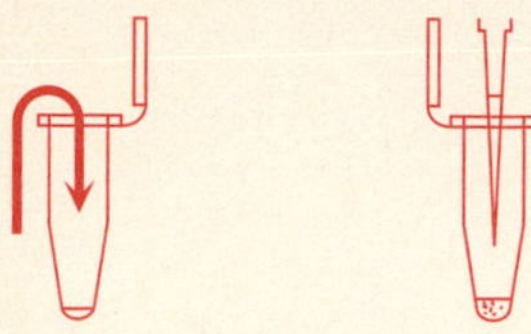

If the nucleic acid pellets are not dry, they will be difficult to resuspend in the TE buffer.

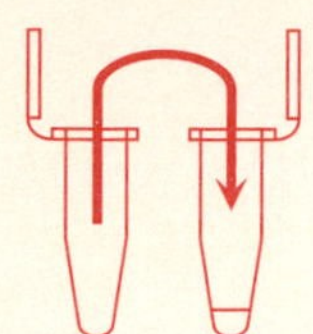

If you are using a 0.5–10-µl micropipettor, set it to 7.5 µl and pipet twice.

19. Be sure that the nucleic acid pellets are dry and that all the ethanol has evaporated before proceeding to step 20. Hold each tube up to the light and confirm that no ethanol droplets remain and that the nucleic acid pellet, if visible, appears white and flaky. Sniff the mouth of the tube; if ethanol is still evaporating, an alcohol odor will be detected.

20. Add 15 µl of TE to each tube. Resuspend the pellets by scraping them with a pipet tip and vigorously pipetting in and out. Rinse down the side of the tube several times, concentrating on the area where the pellet should have formed during centrifugation (beneath the cap hinge). Check that all DNA is dissolved and that no particles remain in the tip or on the side of the tube.
21. Pool the DNA/TE solutions into one tube.

Stop Point

Freeze the DNA/TE solution at –20°C until you are ready to continue. Thaw before using.

22. Take the time for proper cleanup.
    a. Segregate for proper disposal culture tubes, paper towels, and micropipettor tips that have come into contact with *E. coli.*
    b. Disinfect the overnight culture, pipet tips, and the supernatant from step 4 with a 10%-bleach solution or disinfectant (such as Lysol).
    c. Wipe down the lab bench with soapy water, a 10%-bleach solution, or disinfectant.
    d. Wash your hands before leaving the laboratory.

RESULTS AND DISCUSSION

The minipreparation is a simple and efficient procedure for isolating plasmid DNA. You should be familiar with the molecular and biochemical effects of each reagent used in the protocol:

- *Glucose/Tris/EDTA:* Glucose functions to maintain osmotic pressure, while the Tris buffers the cells at pH 8.0. EDTA binds divalent cations in the lipid bilayer, thus weakening the cell envelope. Following cell lysis, EDTA limits DNA degradation by binding Mg^{++} ions that are a necessary cofactor for bacterial nucleases.
- *SDS/sodium hydroxide:* This alkaline mixture lyses the bacterial cells. The detergent SDS dissolves the lipid components of the cell envelope and the cellular proteins. Sodium hydroxide denatures the chromosomal and plasmid DNA into single strands; the intact circles of plasmid DNA remain intertwined.
- *Potassium acetate/acetic acid:* The acetic acid brings the pH to neutral, allowing the DNA strands to renature. The large, disrupted chromoso-

mal strands cannot rehybridize perfectly but instead collapse into a partially hybridized tangle. At the same time, the potassium acetate precipitates the SDS from the cell suspension, along with the associated proteins and lipids. The renaturing chromosomal DNA is trapped in the SDS/lipid/protein precipitate. Only smaller plasmid DNA, fragments of chromosomal DNA, and RNA molecules escape the precipitate and remain in solution.

- *Isopropanol:* The alcohol rapidly precipitates nucleic acids but precipitates proteins slowly. Thus, a quick precipitation preferentially brings down nucleic acids.
- *Ethanol:* A wash with ethanol removes some remaining salts and SDS from the preparation.
- *Tris-EDTA:* Tris buffers the DNA solution. EDTA protects the DNA from degradation by DNases by binding the divalent cations (especially Mg^{++}) that are necessary cofactors for DNase activity.

1. Consider the three major classes of biologically important molecules: proteins, lipids, and nucleic acids. Which steps of the miniprep procedure act on proteins? on lipids? on nucleic acids?
2. Which aspect of plasmid DNA structure allows it to renature efficiently in step 8?
3. What other kinds of molecules, in addition to plasmid DNA, would you expect to be present in the final miniprep sample? How could you find out?

FOR FURTHER RESEARCH

1. Design experiments to determine which other kinds of molecules, besides plasmid DNA, may be present in the final miniprep sample.
2. Determine the approximate mass of plasmid DNA per milliliter of cells that you isolated.
   a. Set up 20-μl *Hin*dIII restriction digests of pAMP and λ DNA. Use 15 μl of your pAMP preparation and a known mass of λ DNA as a standard. The single digest of pAMP will produce a linear fragment of 4,539 bp; the λ digest will include a 4,361-bp fragment (of known mass) that can be compared to the unknown mass of pAMP.
   b. Make 1:10, 1:50, and 1:100 dilutions of the digested pAMP and λ DNA.
   c. Electrophorese equal volumes of each dilution in an agarose gel, stain them with ethidium bromide, and photograph the results.

CAUTION

Review "Ethidium Bromide Staining and Responsible Handling" in Laboratory 3. Wear disposable gloves when staining, viewing, and photographing gel and during cleanup. Confine all staining to a restricted sink area.

d. Identify a lane of the λ digest where the 4,361-bp fragment is *barely* visible; identify a lane of pAMP (4,539-bp) of equal intensity. These bands should have a nearly equivalent mass of DNA.

e. Determine the mass of λ DNA in the selected fragment, using the following formula:

$$\frac{\text{fragment bp (conc. DNA) (vol. DNA)}}{\lambda \text{ bp}}$$

f. Multiply the mass from step e by the dilution factor of the selected pAMP lane.

I. Set Up Restriction Digests

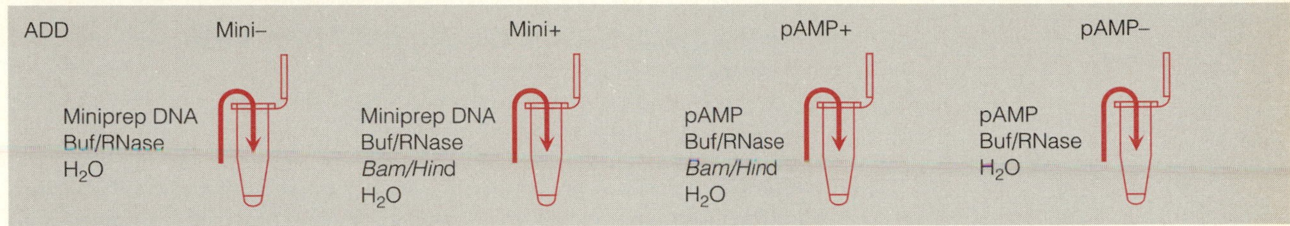

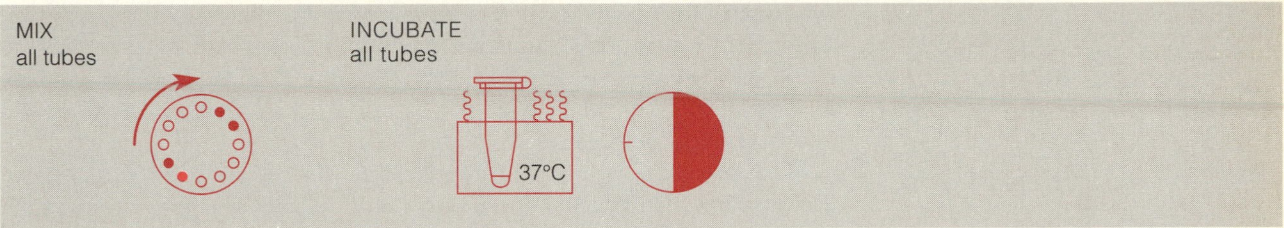

II. Cast 0.8% Agarose Gel

III. Load Gel and Electrophorese

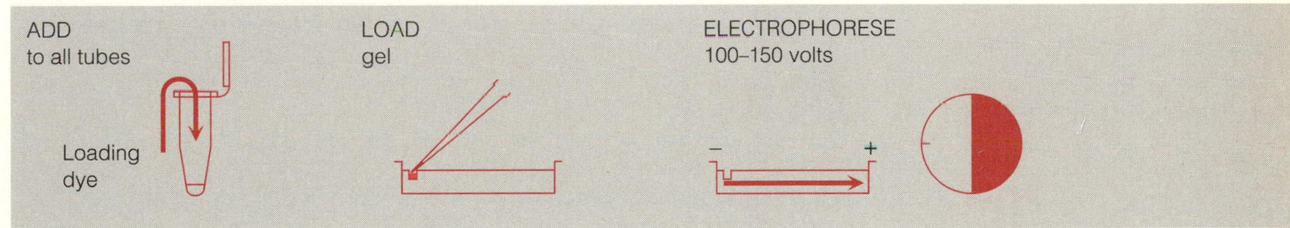

IV. Stain Gel and View

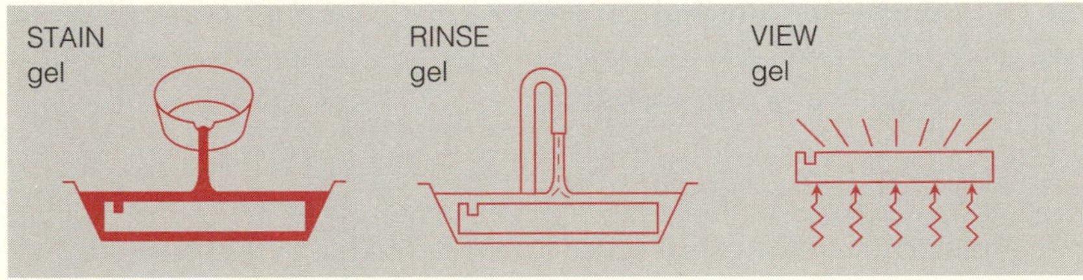

V. Photograph Gel

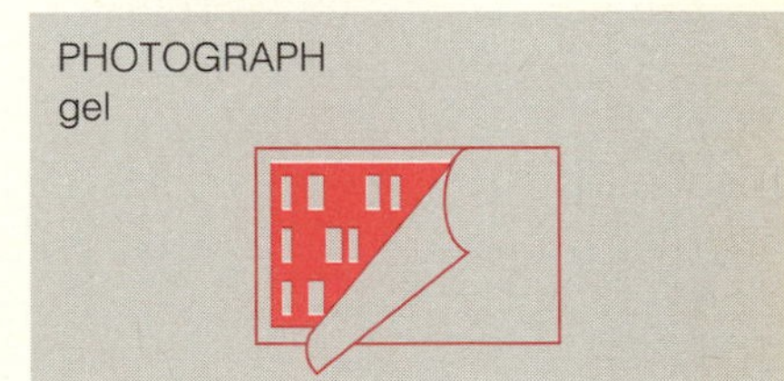

PRELAB NOTES

Review the Prelab Notes in Laboratory 3, "DNA Restriction and Electrophoresis."

Limiting DNase Activity

Miniprep DNA is impure, unlike the highly purified plasmid DNA available from commercial vendors. A significant percentage of the nucleic acid in the preparation is, in fact, RNA and fragmented chromosomal DNA. Miniprep DNA is typically contaminated with nucleases (DNases) that cleave DNA into small fragments. Residual DNases will degrade plasmid DNA if the miniprep is left for a long period of time at room temperature or even on ice. For this reason, you should store minipreps at –20°C, and thaw them just before use.

The situation is further complicated during restriction digestion. DNases and restriction endonucleases both require divalent cations, such as Mg^{++}. EDTA is incorporated in the TE buffer at a low concentration of 1 mM and chelates (binds) a large proportion of divalent cations in the DNA preparation. Although some divalent cations will likely remain free to activate DNases, a higher concentration of EDTA would also chelate Mg^{++} in the restriction buffer, thus inhibiting restriction enzyme activity. A balance is reached at an EDTA concentration that inhibits most of the contaminating DNases without significantly reducing the activity of the restriction enzymes.

Another balance must be reached. On the one hand, contaminants in the miniprep limit restriction enzyme activity: A 20-minute incubation is often insufficient for complete digestion. On the other hand, DNases are activated by Mg^{++} in the restriction buffer and will significantly degrade plasmid DNA if the restriction reaction is incubated for too long. Experience has shown that an incubation of 30–40 minutes gives optimal results.

RNase

Miniprep DNA is contaminated by large amounts of ribosomal RNA and smaller amounts of messenger RNA and transfer RNA. If this RNA is not removed from the preparation, it will obscure the DNA bands in the agarose gel. Therefore, RNase is added to the restriction digest; during incubation, the RNase digests RNA into very small fragments (of several hundred nucleotides or fewer). These RNA fragments run well ahead of the DNA fragments of interest. Use only RNase A from bovine pancreas; boil the RNase solution for 15 minutes to destroy contaminating DNases that may be present in the preparation.

For Further Information

The protocol presented here is based on the following published methods:

Helling, R. B., H. M. Goodman, and H. W. Boyer. 1974. Analysis of *Eco*RI fragments of DNA from lambdoid bacteriophages and other viruses by agarose gel electrophoresis. *Journal of Virology* 14: 235.

Sharp, P. A., B. Sugden, and J. Sambrook. 1973. Detection of two restriction endonuclease activities in *Haemophilus parainfluenzae* using analytical agarose-ethidium bromide electrophoresis. *Biochemistry* 12: 3055.

RESTRICTION ANALYSIS OF PURIFIED pAMP

Reagents

miniprep DNA/TE
0.1 μg/μl pAMP
*Bam*HI/*Hin*dIII
5× restriction buffer/RNase
distilled water
loading dye
0.8% agarose
1× Tris/Borate/EDTA (TBE) buffer
1 μg/ml ethidium bromide (or 0.025% methylene blue)

for decontamination:
0.05 M $KMnO_4$
0.25 N HCl
0.25 N NaOH

Supplies and Equipment

0.5–10-μl micropipettor + tips
1.5-ml tubes
aluminum foil
beaker for agarose
beaker for waste/used tips
camera and film (optional)
disposable gloves
electrophoresis box
masking tape
microfuge (optional)
Parafilm or waxed paper (optional)
permanent marker
plastic wrap (optional)
power supply
test tube rack
transilluminator (optional)
37°C water bath
60°C water bath (for agarose)

I. Set Up Restriction Digests

(10 minutes; then 30–40 minutes of incubation)

Refer to Laboratory 3, "DNA Restriction and Electrophoresis," for detailed instructions on setting up digests.

1. Use a permanent marker to label four 1.5-ml tubes, as shown in the matrix below. Restriction reactions will be performed in these tubes.
2. Use the matrix below as a checklist while adding reagents to each reaction. Read down each column, adding the same reagent to all appropriate tubes. *Use a fresh pipet tip for each reagent.*

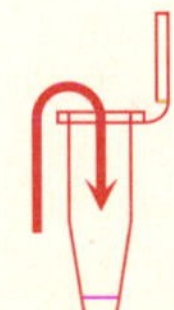

Tube	Miniprep DNA	pAMP	Buffer/ RNase	BamHI/ HindIII	H_20
Mini–	5 μl	—	2 μl	—	3 μl
Mini+	5 μl	—	2 μl	2 μl	1 μl
pAMP+	—	5 μl	2 μl	2 μl	1 μl
pAMP–	—	5 μl	2 μl	—	3 μl

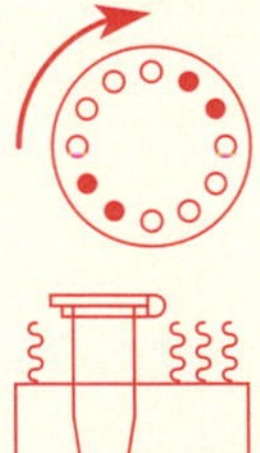

3. After adding all of the reagents, close the tops of the tubes. Pool and mix the reagents by pulsing them in a microfuge or by sharply tapping the bottom of the tube on the lab bench.
4. Place the reaction tubes in a 37°C water bath, and incubate for only 30–40 minutes.

Do not overincubate. During a longer incubation, DNases in the miniprep can degrade plasmid DNA.

Stop Point

Following incubation, freeze the reactions at –20°C until you are ready to continue. Thaw the reactions before continuing on to Section III, step 1.

II. Cast 0.8% Agarose Gel

(15 minutes)

1. Seal the ends of the gel-casting tray with tape, and insert a well-forming comb. Place the gel-casting tray out of the way on the lab bench, so that the agarose to be poured in the next step can set undisturbed.

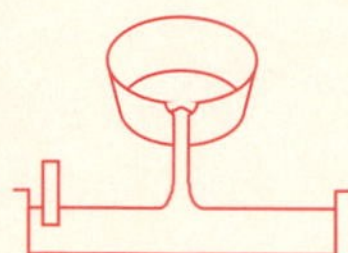

2. Carefully pour enough agarose solution into the casting tray to fill it to a depth of about 5 mm. The gel should cover only about one-third the height of the comb teeth. Use a pipet tip to move large bubbles or solid debris to the sides or end of the tray, while the gel is still liquid.
3. The gel will become cloudy as it solidifies (in about 10 minutes). *Take care not to move or jar the casting tray while the agarose is solidifying.* Touch a corner of the agarose *away* from the comb to test whether the gel has solidified.
4. When the agarose has set, unseal the ends of the casting tray. Place the tray on the platform of the electrophoresis box, so that the comb is at the negative (black) electrode.

Too much buffer will channel the current over the top rather than through the gel, thus increasing the time required to separate the DNA fragments. TBE buffer can be reused several times; do not discard it. If you are using buffer remaining in the electrophoresis box from a previous experiment, rock the chamber back and forth to remix the ions that have accumulated at either end.

5. Fill the box with TBE buffer to a level that barely covers the entire surface of the gel.
6. Gently remove the comb, pulling it straight up and out of the set agarose. Do not rock or wiggle the comb. If necessary, have your partner steady the casting tray.

Buffer solution helps to lubricate the comb and prevent damage to the wells.

7. The removal of the comb usually pulls the agarose well edges above the buffer surface; these edges appear as "dimples" around the wells. Add buffer until any dimples disappear and the buffer surface is smooth.

Stop Point

Cover the electrophoresis box, and save the gel until you are ready to continue. The gel will remain in good condition for several days if it is completely submerged in buffer.

III. Load Gel and Electrophorese

(10 minutes; then 20–40 minutes of electrophoresis)

1. Remove the restriction digests from the 37°C water bath.
2. Add 1 μl of loading dye to each reaction using one of two methods.

   a. Add 1 μl of loading dye to each reaction tube. Close the tops of the tubes, and mix the solution by either tapping the bottom of the tube on the lab bench, pipetting in and out, or pulsing in a microfuge. Make sure that the tubes are placed in a *balanced* configuration in the rotor.

   or

   b. Place four individual droplets of loading dye (1 μl each) on a small square of Parafilm or waxed paper. Withdraw the contents from one reaction tube, and mix it with a loading dye droplet by pipetting in and out. Immediately load the dye mixture, according to step 3. Repeat this successively, with a clean pipet tip, for each reaction.

3. Load 10 µl from each reaction tube into a separate well in the gel, as shown in the diagram below. Use a *fresh tip* for each reaction.
   a. Steady the pipet over the well, using two hands.
   b. If there is air in the end of the tip, carefully depress the plunger to push the sample to the end of the tip. (An air bubble ejected into the well can form a "cap" over the well, causing the DNA/loading dye solution to flow into the buffer around the edges of the well.)

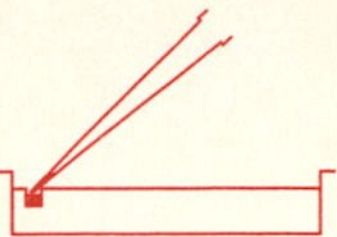

Placing a piece of dark construction paper beneath the gel box will make the wells more visible.

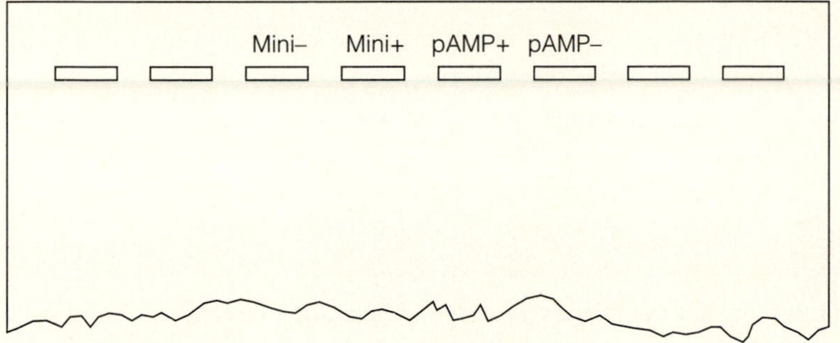

   c. Center the pipet tip over the well, dip the tip in only enough to pierce the buffer surface, and then gently depress the pipet plunger to slowly expel the sample.

Take care not to insert the pipet tip directly into the well; this could damage or puncture the well. Sucrose in the loading dye weighs down the DNA sample, causing it to sink to the bottom of the well.

4. Close the top of the electrophoresis box, connect the electrical leads to a power supply, and electrophorese at 100–150 volts for 20–40 minutes. Adequate separation has occurred when the bromophenol blue band has moved 30–40 mm from the sample wells.
5. Turn off the power supply, remove the gel-casting tray from the electrophoresis box, and transfer the gel to a disposable weigh boat (or other shallow tray) for staining.

The *Bam*HI/*Hin*dIII digest yields two bands containing small fragments of 784 bp and 3,755 bp, which are easily resolved during a short electrophoresis run. The 784-bp fragment runs behind the purplish band of bromophenol blue (equivalent to approximately 300 bp); the 3,755-bp fragment runs in front of the aqua band of xylene cyanol (equivalent to approximately 9,000 bp).

Stop Point

The gel can be stored overnight, covered with a small volume of buffer in the staining tray or in a zip-lock plastic bag, for viewing or photographing the next day. However, over a longer period of time, the DNA will diffuse through the gel, and the bands will either become indistinct or entirely disappear.

6. Stain and view the gel, using one of the methods described in Sections IVA and IVB.

Staining can be performed by an instructor in a controlled area, when students are not present.

IVA. Stain the Gel with Ethidium Bromide and View

(15 minutes)

CAUTION

Review "Ethidium Bromide Staining and Responsible Handling" in Laboratory 3. Wear disposable gloves when staining, viewing, and photographing gel and during cleanup. Confine all staining to a restricted sink area.

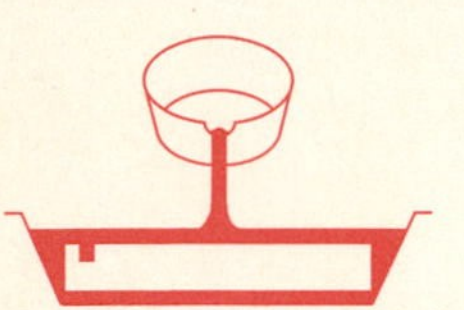

Staining time increases markedly for thicker gels. Band intensity and contrast increase dramatically if the gel is destained for 15–30 minutes in tap water. More simply, rinse and drain the gels, stack the staining trays, cover the top gel, and let them set overnight at room temperature.

1. Flood the gel with ethidium bromide solution (1 µg/ml), and allow it to stain for 5–10 minutes.
2. View the gel under an ultraviolet transilluminator or other UV source.

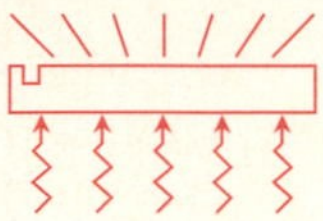

CAUTION

Ultraviolet light can damage your eyes. Never look directly at an unshielded UV light source without eye protection. View only through a filter or safety glasses that absorb the harmful wavelengths.

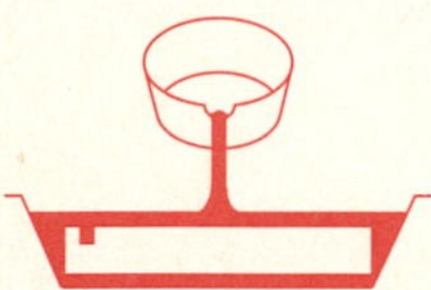

IVB. Stain the Gel with Methylene Blue and View

(30+ minutes)

Destaining time can be decreased by agitating and rinsing the gel in warm water. For best results, continue to destain overnight in a *small volume* of water. Cover the staining tray to retard evaporation.

1. Flood the gel with 0.025% methylene blue, and allow it to stain for 20–30 minutes.
2. Rinse the gel in running tap water. Let it soak for several minutes in several changes of fresh water. The DNA bands will become increasingly distinct as the gel destains.
3. View the gel over a light box; cover the surface with plastic wrap to prevent staining.

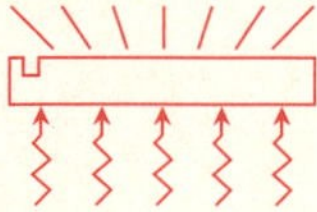

V. Photograph the Gel

(5 minutes)

1. *For ultraviolet (UV) photography of ethidium-bromide-stained gels:* Set the camera aperture to f/8 and the shutter speed to 1. Depress the shutter once or twice for a time exposure of 1–2 seconds.

   For white-light photography of methylene-blue-stained gels: Set the camera aperture to f/8 and the shutter speed to 1/125 of a second.

CAUTION

Avoid getting caustic developing jelly on your skin or clothing. If the jelly does get on your skin, wash immediately with plenty of soap and water.

2. Take the time for proper cleanup.
   a. Wipe down the camera, the light source, and the staining area.
   b. If you are using ethidium bromide, decontaminate the gel and any staining solution that will not be reused.
   c. Wash your hands before leaving the laboratory.

RESULTS AND DISCUSSION

Interpreting gels containing plasmid DNA is not a straightforward process and is further complicated by the impurities in miniprep DNA. Examine the gel, and determine which lanes contain cut and uncut control pAMP and which contain miniprep DNA. Even if you have not followed the prescribed loading order, the miniprep lanes can be distinguished by the following characteristics:

- A background "smear" of degraded and partially digested chromosomal DNA, plasmid DNA, and RNA will typically run much of the length of the miniprep lanes. The smear is composed of faint bands representing virtually every base-pair length. A heavy background smear, along with DNA of high molecular weight at the top of the undigested lane, indicates that the miniprep is contaminated with large amounts of chromosomal DNA.
- Undissolved material and DNA of high molecular weight are frequently trapped at the front edge of the sample well. These anomalies are not seen in commercial preparations, where plasmid DNA is separated from degraded nucleic acids by ultracentrifugation through a cesium chloride gradient.
- A "cloud" of RNA of low molecular weight is often seen in both the cut and uncut miniprep lanes, at a position corresponding to 100–200 bp. Again, various sizes of molecules are represented—the remnants of larger molecules that have been partially digested by the RNase. However, the majority of RNA is usually digested into fragments either too small to intercalate the ethidium bromide dye or else electrophorese off of the end of the gel.
- Only two bands (784 bp and 3,755 bp) are expected to be seen in the cut miniprep lane. However, it is common to see one or more faint bands higher up on the gel that comigrate with the uncut plasmid forms described below. Incomplete digestion is usually due to contaminants in the preparation, which inhibit restriction enzyme activity, or else it can occur when the miniprep solution contains a very high concentration of plasmid DNA.

It is especially confusing to see several bands in a lane known to contain only uncut plasmid. This occurs because the migration of plasmid DNA in an agarose gel depends on its molecular conformation and its molecular weight (base-pair size). Plasmid DNA can exist in any one of three major conformations:

- *Form I, supercoiled:* Although a plasmid is usually pictured as an open circle, within the *E. coli* cell *(in vivo)* the DNA strand is wound around histone-like proteins to form a compact structure. Adding these coils to the coiled DNA helix produces a *supercoiled* molecule. (Supercoiling can be demonstrated by adding twists to one end of a loop of rope.) The enzyme DNA gyrase (topoisomerase II) introduces the supercoils around the protein supports by passing a section of the double helix through a break in another area of the DNA. The extraction procedure strips proteins from plasmid, causing the molecule to coil about itself. A compact molecular shape allows supercoiled plasmid to be the fastest-moving form under most gel conditions. Therefore, the fastest-moving band of uncut plasmid is assumed to be supercoiled.

- *Form II, relaxed or nicked circle:* During DNA replication, the enzyme topoisomerase I introduces a nick into one strand of the DNA helix and rotates the strand to release the torsional strain that holds the molecule in a supercoil. The relaxed section of the DNA uncoils, allowing access to the enzymes involved in replication. Physical shearing and enzymatic cleavage during plasmid isolation introduce nicks in the supercoiled plasmid to produce the familiar, open circular structure. Thus, the percentage of supercoiled plasmid is an indicator of the care with which the DNA is extracted from the *E. coli* cell. The relaxed circle is the slowest-migrating form of plasmid; its "floppy" molecular shape impedes movement through the agarose matrix.
- *Form III, linear:* Linear DNA is produced when a restriction enzyme cuts the plasmid at a single recognition site or when damage results in strand nicks directly opposite each other on the DNA helix. Under most gel conditions, linear plasmid DNA migrates at a rate intermediate between supercoiled and relaxed circle DNA. The presence of linear DNA in a plasmid preparation is a sign of either nuclease contamination or sloppy lab procedure (for example, overmixing or mismeasuring SDS/NaOH and KOAc in steps 5–8 of the miniprep procedure).

MM294 and other strains of *E. coli*, termed *recA*$^+$, have an enzyme system that recombines plasmids to form large concatemers of two or more plasmid units. Homologous recombination, a general mechanism for shuffling DNA strands, occurs when identical regions of nucleotides are exchanged between two DNA molecules. In *recA*$^+$ hosts, homologous recombination occurs frequently between multiple, identical copies of a plasmid. The *recA* enzyme binds to single-stranded regions of nicked plasmids, promoting crossover and rejoining of homologous sequences. This results in multimeric ("super") plasmids that appear as a series of slowly migrating bands near the top of the gel. Since the concatemers form head to tail, they produce the identical restriction fragments as a monomer (single plasmid) when they are cut with a given restriction enzyme. To confuse matters further, multimers can exist in any of the three forms mentioned above. Supercoiled multimers can appear further down on the gel than relaxed or linear monomers with fewer base pairs.

1. Examine the photograph of your stained gel (or view it on a light box or overhead projector). Compare your gel with the ideal gel shown on the next page. Label the sizes of the fragments in each lane of your gel.

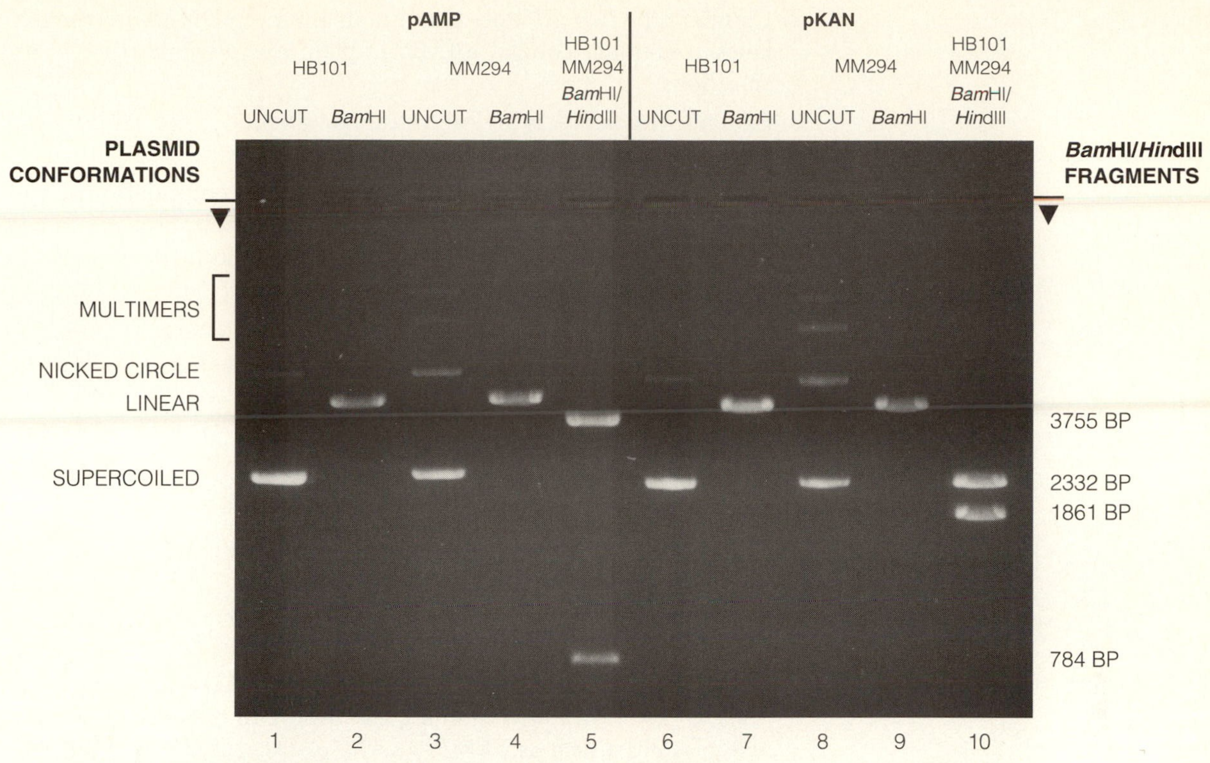

Comparison of Plasmid DNA Isolated from a *recA*– Strain (HB101) and *recA*+ Strain (MM294)

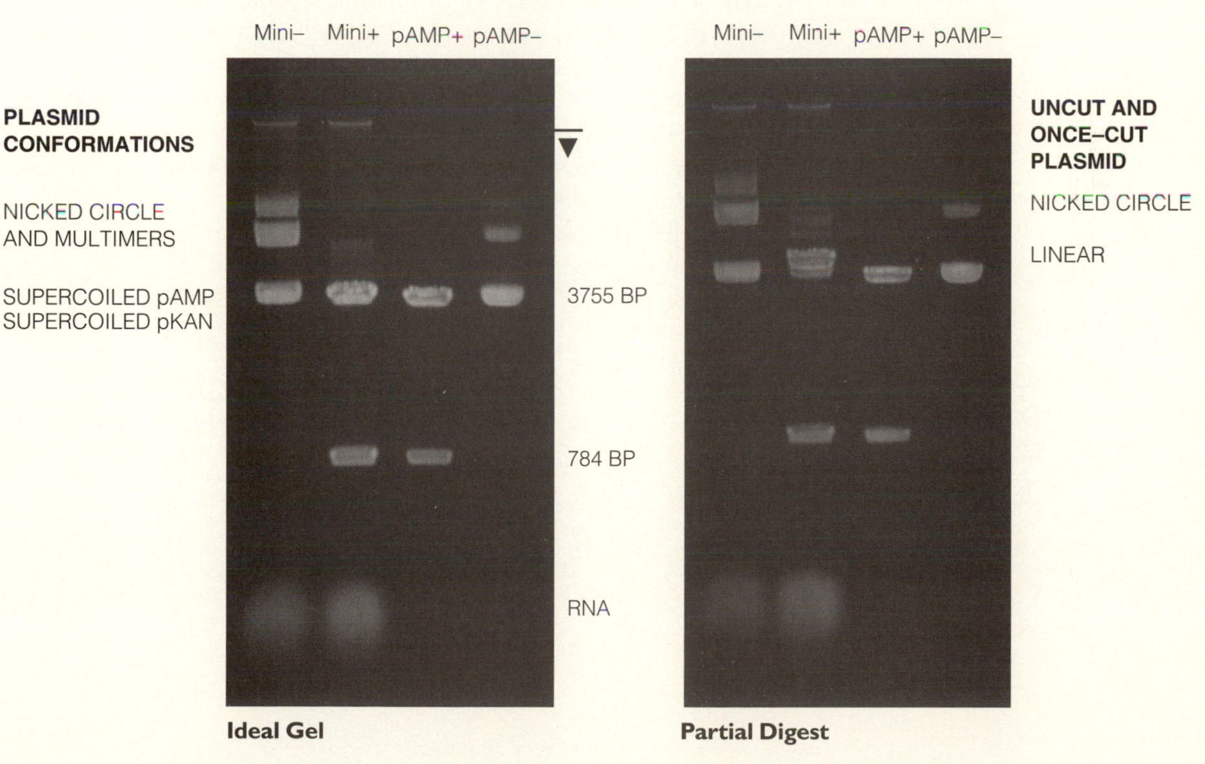

Ideal Gel

Partial Digest

2. Compare the two gel lanes containing miniprep DNA with the two lanes containing control pAMP. Explain the possible reasons for variations.
3. Explain why EDTA is an important component of the TE buffer in which the miniprep DNA is dissolved.
4. Assume that a plasmid preparation of pAMP is composed entirely of dimeric molecules (pAMP/pAMP). The two molecules are joined *head-to-head* at a "hot spot" for recombination located 655 bp from the *Hin*dIII site, near the origin of replication.
   a. Draw a map of the dimeric plasmid described above.
   b. Draw a map of the dimeric pAMP that actually forms by *head-to-tail* recombination at the site described above.
   c. Now draw the gel-banding patterns that would result from double digestion of each of these plasmids with *Bam*HI and *Hin*dIII. Label the base-pair size of the fragment in each band.

FOR FURTHER RESEARCH

1. Isolate and characterize an unknown plasmid. Make overnight cultures of *E. coli* strains containing any of several commercially available plasmids (such as pAMP, pKAN, pBLU, pUC19, or pBR322). Digest miniprep and control samples of each plasmid with *Bam*HI/*Hin*dIII, and electrophorese them in an agarose gel.
2. Transform pAMP or other plasmids into a *recA*$^+$ strain (MM294) and a *recA*$^-$ strain (HB101). Do minipreps from overnight cultures of each strain, and incubate samples of each with no enzyme, *Hin*dIII, and *Bam*HI+*Hin*dIII. Electrophorese these samples as long as possible in an agarose gel. Compare the banding patterns of the two strains, especially in the uncut lanes.
3. Research the potential use of homologous recombination in targeted gene therapy.

Plasmid Recombination and Identification

LABORATORY 7

Recombination of Antibiotic-Resistance Genes

This laboratory begins an experimental unit to construct and analyze a recombinant DNA molecule. The starting reactants are the plasmids pAMP and pKAN, each of which carries a single antibiotic-resistance gene—ampicillin in pAMP and kanamycin in pKAN. The goal is to construct a recombinant plasmid that contains both ampicillin- and kanamycin-resistance genes.

In Part A, "Restriction Digest of Plasmids pAMP and pKAN," samples of both plasmids are digested in separate restriction reactions with *Bam*HI and *Hin*dIII. Following incubation at 37°C, samples of digested pAMP and pKAN are electrophoresed in an agarose gel to confirm proper cutting. Each plasmid contains a single recognition site for each enzyme, so each plasmid yields only two restriction fragments. Cleavage of pAMP yields fragments of 784 bp and 3,755 bp; pKAN yields fragments of 1,861 bp and 2,332 bp.

In Part B, "Ligation of pAMP and pKAN Restriction Fragments," the restriction digests of pAMP and pKAN are heated to destroy *Bam*HI and *Hin*dIII activity. Samples of the restricted plasmids are mixed together and incubated at room temperature with DNA ligase, ATP, and Mg^{++}. The "sticky ends" of *Bam*HI and *Hin*dIII hydrogen-bond with their complementary sequences to align the restriction fragments. Ligase catalyzes the formation of covalent phosphodiester bonds that link complementary ends into stable recombinant-DNA molecules.

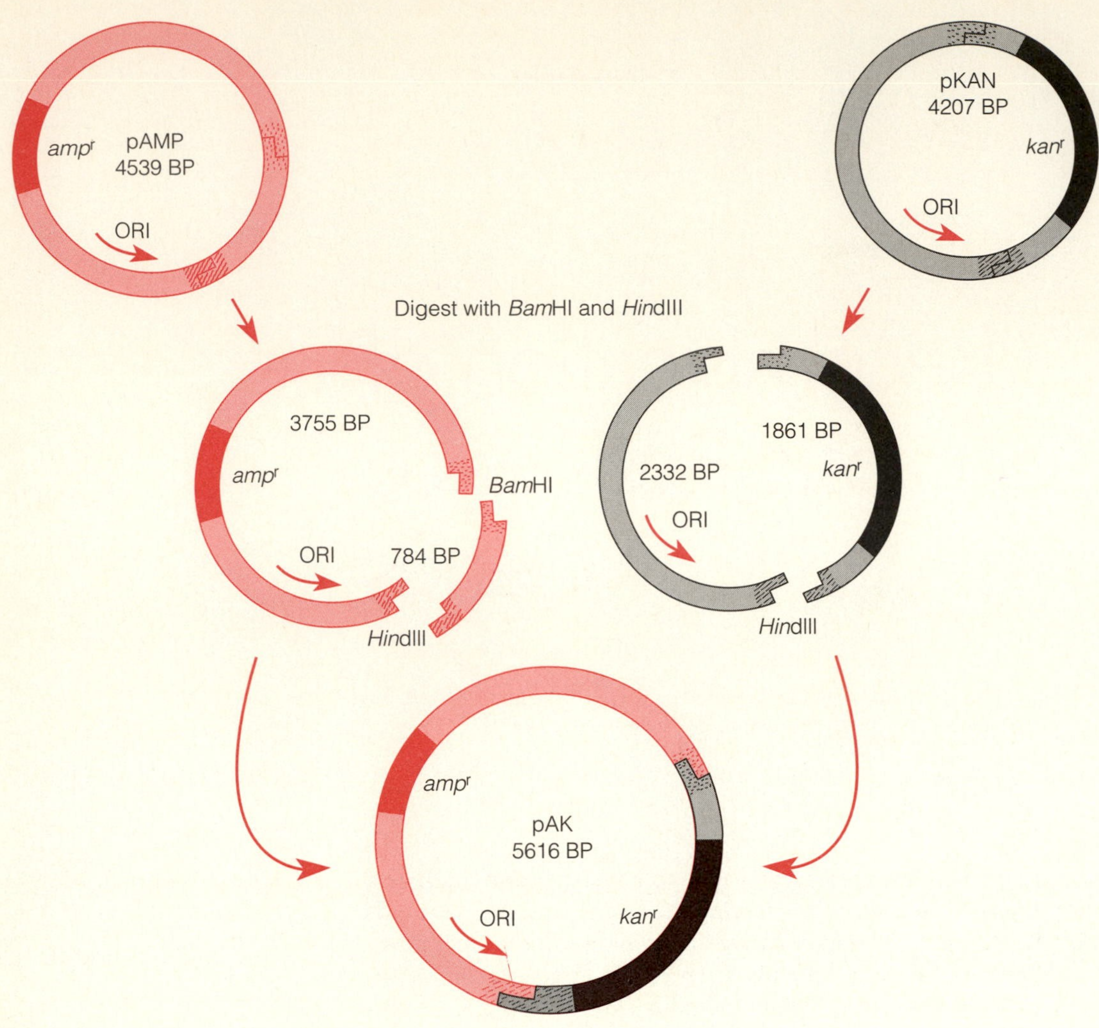

Formation of the "Simple Recombinant" pAK

I. Set Up Restriction Digests

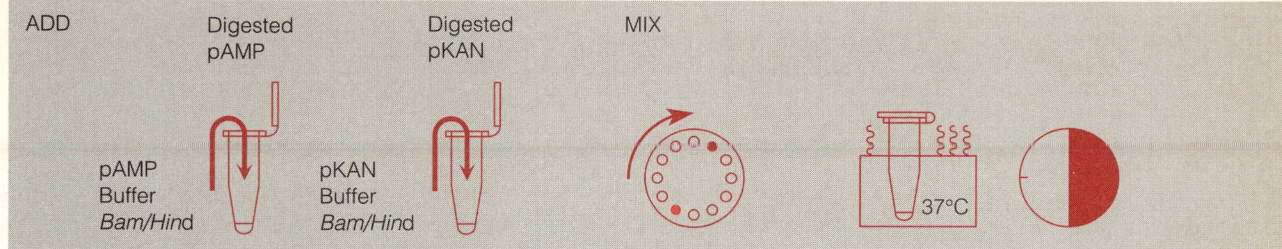

II. Cast 0.8% Agarose Gel

III. Load Gel and Electrophorese

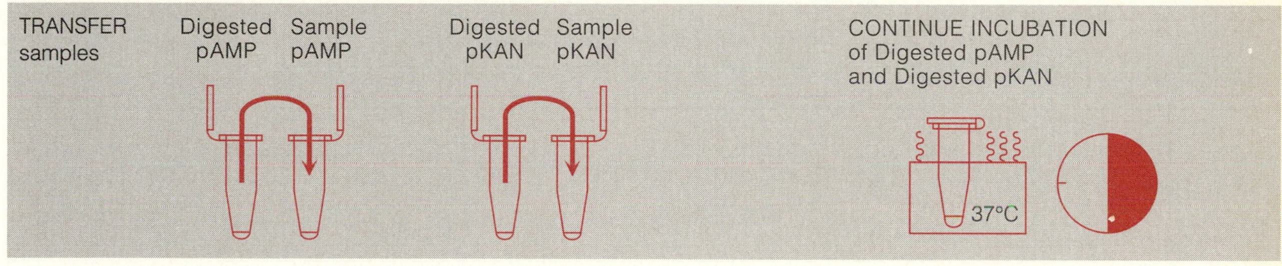

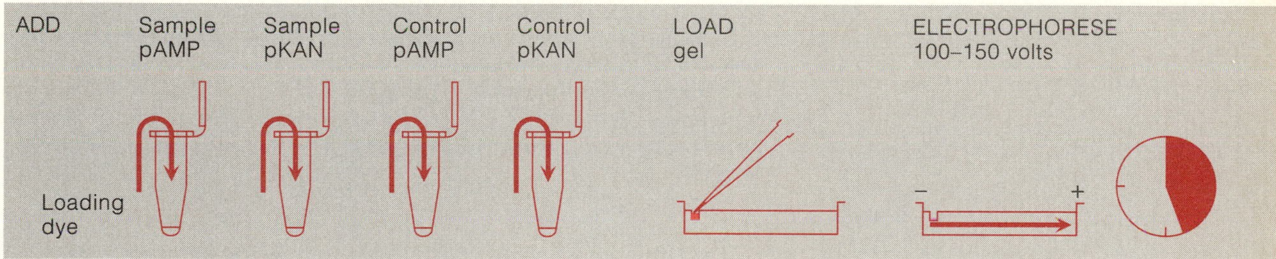

IV. Stain Gel, View, and Photograph

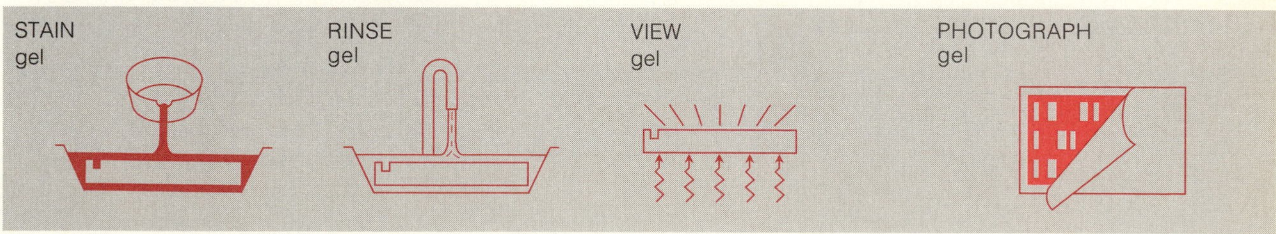

PRELAB NOTES

Review the Prelab Notes in Laboratory 3, "DNA Restriction and Electrophoresis."

The Prudent Control

In Section III, samples of the restriction digests are electrophoresed, before ligation, to confirm complete cutting by the endonucleases. This prudent control is standard experimental procedure. If you are pressed for time, you can opt to omit electrophoresis and ligate DNA directly following the restriction digest. However, be sure to pretest the activity of your *Bam*HI and *Hin*dIII to determine the incubation time needed for complete digestion.

It is fairly impractical to use methylene blue staining for this step, because it demands a rapid and sensitive assay to check for complete digestion of the plasmid DNAs. Methylene blue destaining requires *at least* 30 minutes, and it could fail to detect a small but possibly significant amount of uncut DNA. However, if you prefer to use methylene blue staining for this lab, refer to the staining procedure in Section IVB of Laboratory 3.

Saving DNA

To minimize expense, exact volumes of pAMP and pKAN are aliquoted, and restriction reactants are added directly to the pre-aliquoted DNA in Section I.

For Further Information

The protocol presented here is based on the following published methods:

Sharp, P. A., B. Sugden, and J. Sambrook. 1973. Detection of two restriction endonuclease activities in *Hemophilus parainfluenzae* using analytical agarose-ethidium bromide electrophoresis. *Biochemistry* 12: 3055.

Helling, R. B., H. M. Goodman, and H. W. Boyer. 1974. Analysis of *Eco*RI fragments of DNA from lambdoid bacteriophages and other viruses by agarose-gel electrophoresis. *Journal of Virology* 14: 1235.

Messing, J. and J. Vieira. 1982. The pUC plasmids, an M13mp7-derived system for insertion mutagenesis and sequencing with synthetic universal primers. *Gene* 19: 259.

RESTRICTION DIGESTION OF PLASMIDS pAMP AND pKAN

Reagents	Supplies and Equipment
For digest:	0.5–10-µl micropipettor + tips
0.20 µg/µl pAMP	1.5-ml tubes
0.20 µg/µl pKAN	aluminum foil
*Bam*HI/*Hin*dIII	beaker for agarose
2× restriction buffer	beaker for waste/used tips
	camera and film (optional)
For electrophoresis:	disposable gloves
0.1 µg/µl pAMP	electrophoresis box
0.1 µg/µl pKAN	masking tape
loading dye	microfuge (optional)
0.8% agarose solution	Parafilm or waxed paper (optional)
1× Tris/Borate/EDTA (TBE) buffer	permanent marker
1 µg/ml ethidium bromide	plastic wrap (optional)
	power supply
For decontamination:	test tube rack
0.05 M $KMnO_4$	transilluminator
0.25 N HCl	37°C water bath
0.25 N NaOH	60°C water bath (for agarose)

I. Set Up Restriction Digests

(10 minutes; then 30+ minutes of incubation)

Refer to Laboratory 3, "DNA Restriction and Electrophoresis," for more detailed instructions on setting up reactions.

1. Obtain labeled reaction tubes containing pAMP and pKAN. Use a permanent marker to add the label "Digested" to each tube. *Add other reagents directly to these tubes.*
2. Use the matrix below as a checklist while adding reagents to each reaction. Read down the columns, adding the same reagent to all appropriate tubes. *Use a fresh pipet tip for each reagent.*

Tube	pAMP 0.2 µg/µl	pKAN 0.2 µg/µl	2x Buffer	*Bam*HI/ *Hin*dIII
Digested pAMP	5.5 µl (preadded)	—	7.5 µl	2 µl
Digested pKAN	—	5.5 µl (preadded)	7.5 µl	2 µl

3. After adding all of the above reagents, close the tops of the tubes. Pool and mix reagents by pulsing them in a microfuge or by sharply tapping the bottom of the tube on the lab bench.
4. Place the reaction tubes in a 37°C water bath, and incubate for 30 minutes or longer.

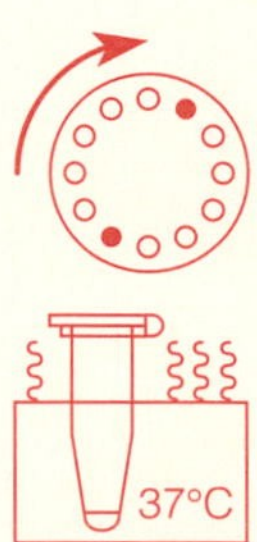

Stop Point

After a full 30-minute incubation (or longer), freeze the reactions at –20°C until you are ready to continue. Thaw the reactions before proceeding to Section III, step 1.

II. Cast 0.8% Agarose Gel

(15 minutes)

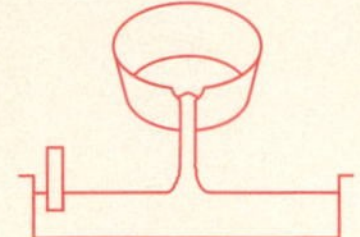

1. Carefully pour agarose solution into the casting tray to fill it to a depth of about 5 mm. The gel should cover only about one-third of the height of the comb teeth.
2. After the agarose solidifies, place the casting tray into the electrophoresis box, and set up for electrophoresis.

Stop Point

Cover the electrophoresis box, and save the gel until you are ready to continue. The gel will remain in good condition for several days, if it is completely submerged in buffer.

Refer to Laboratory 3, "DNA Restriction and Electrophoresis," for more detailed instructions on casting and loading the gel.

III. Load Gel and Electrophorese

(10 minutes; then 20+ minutes of electrophoresis)

Only a fraction of the *Bam*HI/ *Hin*dIII digests of pAMP and pKAN are electrophoresed to check that plasmids are completely cut. These restricted samples are electrophoresed along with uncut pAMP and pKAN as controls.

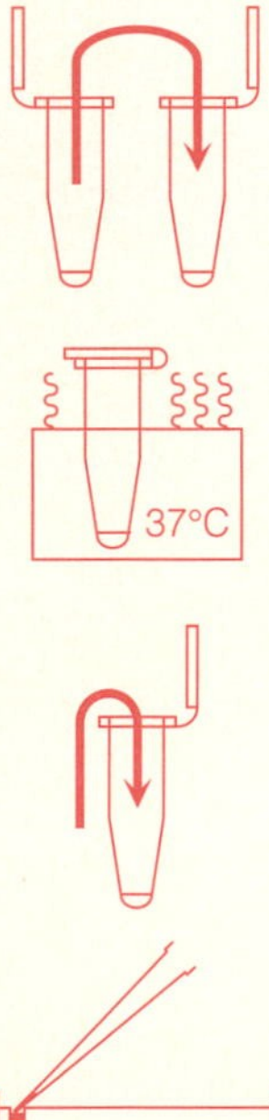

1. Use a permanent marker to label two clean 1.5-ml tubes: "Sample pAMP" and "Sample pKAN."
2. Remove the "Digested pAMP" and "Digested pKAN" tubes from the 37°C water bath. Transfer 5 µl of plasmid from the "Digested pAMP" tube into the "Sample pAMP" tube. Transfer 5 µl of plasmid from the "Digested pKAN" tube into the "Sample pKAN" tube.
3. *Immediately return the "Digested pAMP" and "Digested pKAN" tubes to the water bath, and continue incubating at 37°C during electrophoresis.*
4. Obtain 1.5-ml tubes containing 0.1 µg/µl pAMP and pKAN to serve as controls. Use a permanent marker to add the label "Control" to each tube.
5. Add 1 µl of loading dye to the digested "Sample" and undigested "Control" of pAMP and pKAN. Close the tops of the tubes, and mix the solution by either tapping the bottom of the tube on the lab bench, pipetting in and out, or pulsing in a microfuge.
6. Load the entire contents of each sample tube into a separate well in the gel, as shown in the diagram below. *Use a fresh pipet tip for each sample.*

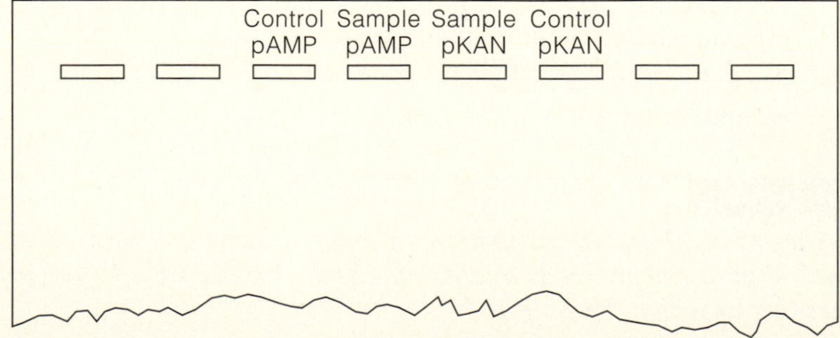

7. Electrophorese at 100–150 volts, until the bromophenol blue bands have moved 30–40 mm from the wells.
8. Turn off the power supply, remove the casting tray from the electrophoresis box, and transfer the gel to a disposable weigh boat (or other shallow tray) for staining.

Stop Point

The gel can be stored overnight, covered with a small volume of buffer in a staining tray or in a zip-lock plastic bag, for viewing or photographing the next day. However, over a longer time, the DNA will diffuse through the gel, and the bands will either become indistinct or disappear entirely.

IV. Stain the Gel with Ethidium Bromide, View, and Photograph

(10–15 minutes)

Staining can be performed by an instructor in a controlled area, when students are not present.

CAUTION

Review "Ethidium Bromide Staining and Responsible Handling" in Laboratory 3. Wear disposable gloves when staining, viewing, and photographing gels and during clean up. Confine all staining to a restricted sink area.

1. Flood the gel with ethidium bromide solution (1 μg/ml), and allow it to stain for 5–10 minutes.
2. View the gel under an ultraviolet transilluminator or other UV source.

Ethidium bromide solution can be reused to stain 15 gels or more.

CAUTION

Ultraviolet light can damage your eyes. Never look directly at an unshielded UV light source without eye protection. View only through a filter or safety glasses that absorb the harmful wavelengths.

3. Make an exposure of 1–2 seconds with the camera set at f/8. Develop the print for the recommended time (approximately 45 seconds at room temperature).
4. Take the time for proper cleanup.
   a. Wipe down the camera, the transilluminator, and the staining area.
   b. Decontaminate the gel and any staining solution that is not to be reused.
   c. Wash your hands before leaving the laboratory.

RESULTS AND DISCUSSION

Compare your stained gel with the ideal gel shown below, and check to ensure that both plasmids have been completely digested by *Bam*HI and *Hin*dIII.

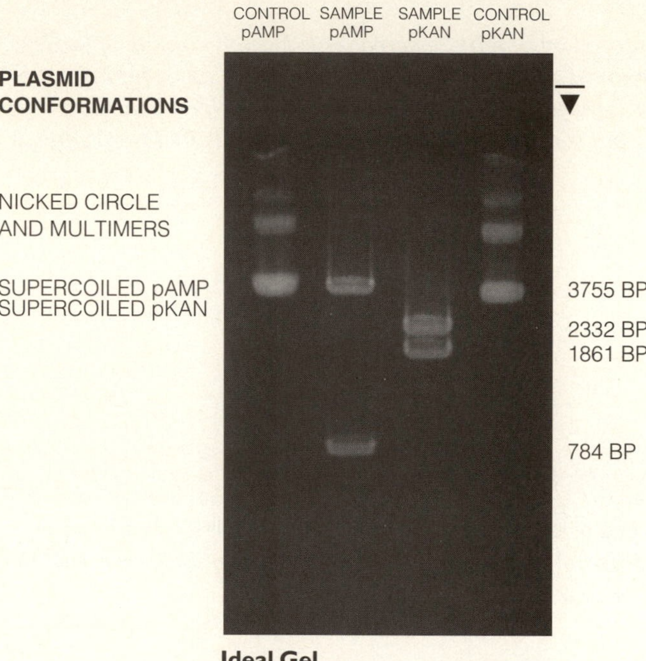

Ideal Gel

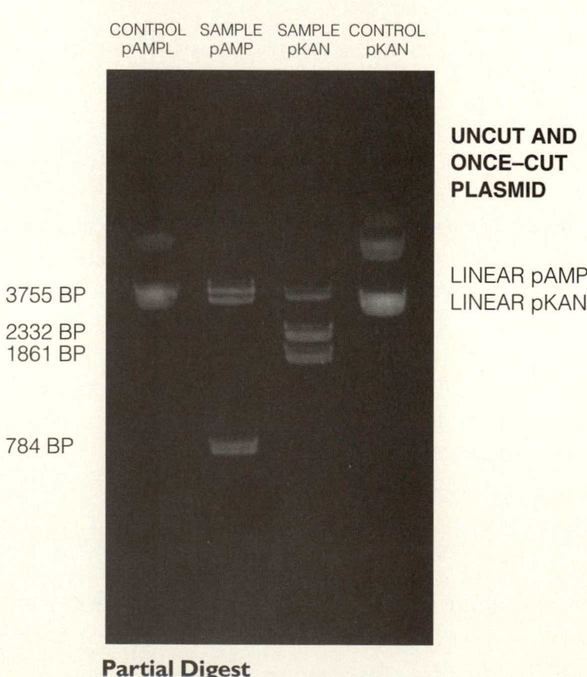

Partial Digest

1. The "Sample pAMP" lane should show two distinct bands of 784 bp and 3,755 bp.
2. The "Sample pKAN" lane should show two distinct bands of 1,861 bp and 2,332 bp.
3. The presence of any additional bands in the "Sample" lanes that comigrate with bands in the "Control" lanes indicates some degree of incomplete digestion.
4. If both digests look complete or nearly complete, continue on to Part B, "Ligation of pAMP and pKAN Restriction Fragments." The restriction reactions will be completed with the additional incubation during electrophoresis.
5. If either digest looks very incomplete, add another 1 µl of the *Bam/Hind* solution, and incubate it for 20 minutes longer. Then, continue on to Part B.

Ligate pAMP and pKAN

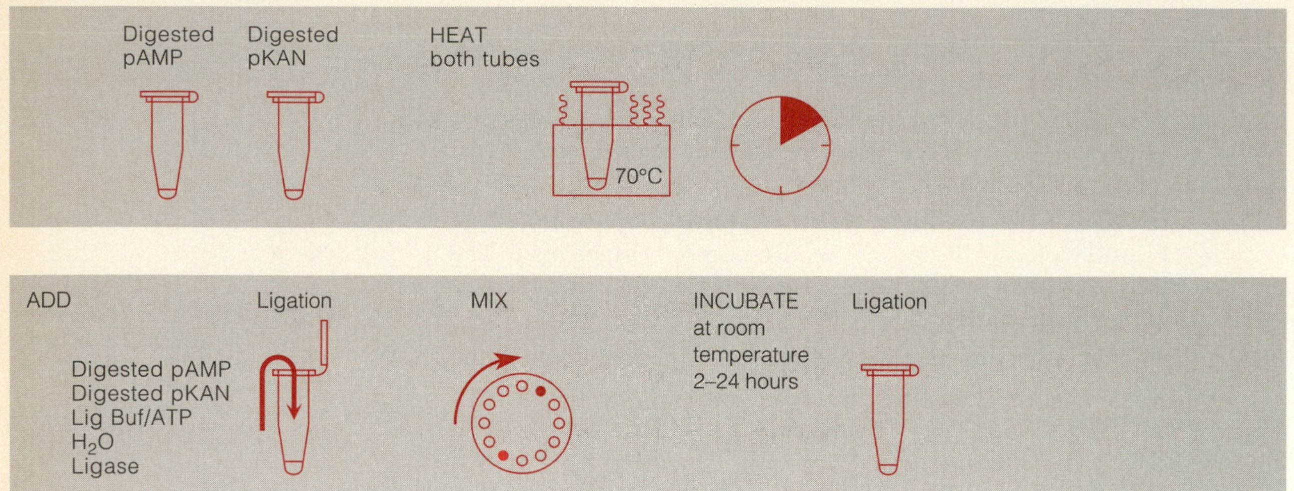

Laboratory 7/Part B
Ligation of pAMP and pKAN Restriction Fragments

PRELAB NOTES

DNA Ligase

Use only T4 DNA ligase. *E. coli* DNA ligase requires different reaction conditions and cannot be substituted in this experiment. Two different units are used to calibrate ligase activity:

- *Weiss Unit (Pyrophosphate-Exchange Unit):* One unit of enzyme catalyzes the exchange of 1 nmole of ^{32}P from pyrophosphate into [γ,β-^{32}P]ATP in 20 minutes at 37°C. This unit is used by Bethesda Research Laboratories (BRL), International Biotechnologies, Inc. (IBI), and Boehringer Mannheim.
- *Cohesive-End Unit:* One unit of enzyme ligates 50% of *Hind*III fragments of *Lambda* DNA (6 µg in 20 µl) in 30 minutes at 16°C. This unit is used by New England Biolabs (NEB) and the Carolina Biological Supply Company (CBS).

  1 Weiss unit = 67 cohesive end units

Researchers typically incubate a ligation reaction overnight at room temperature. *For brief ligations of a minimum of 2 hours, it is essential to choose a high-concentration T4 DNA ligase with at least 5 Weiss units/µl or 300–500 cohesive end ligation units/µl.*

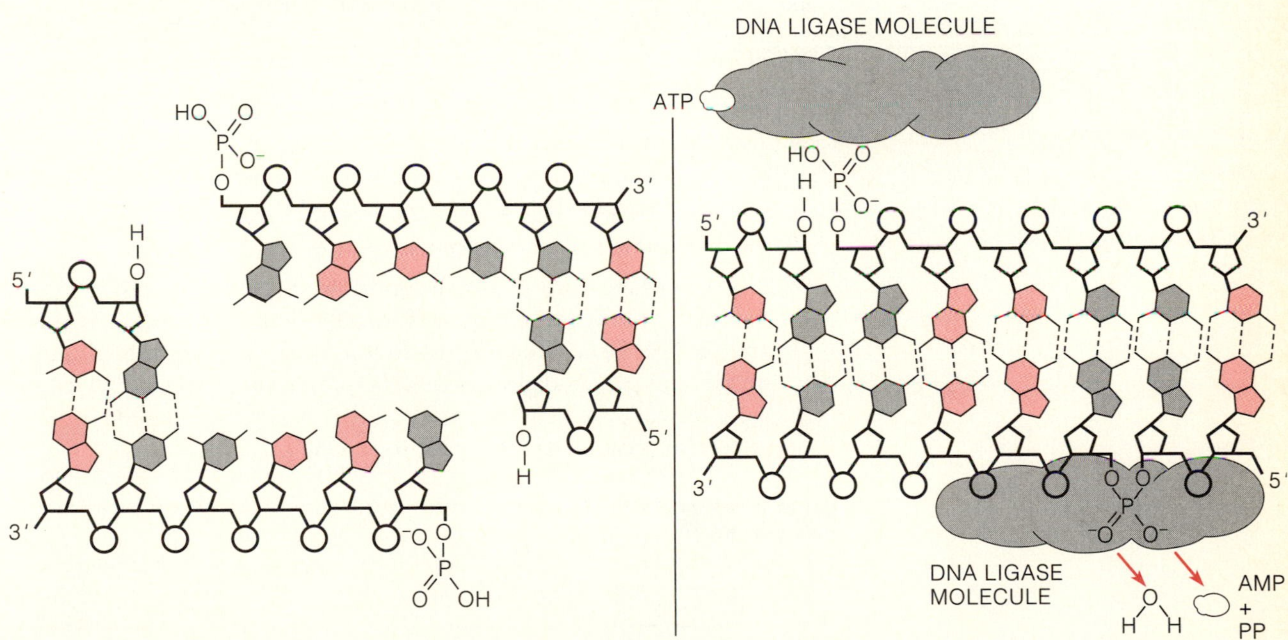

Hydrogen bonding between complementary nucleotides aligns *Bam*HI fragments while ligase reforms phosphodiester bonds on each side of the DNA molecule.

For Further Information

The protocol presented here is based on the following published method:

Cohen, S. N., A. C. Y. Chang, and H. W. Boyer. 1973. Construction of biologically functional bacteria plasmids in vitro. *Proceedings of the National Academy of Sciences USA* 70: 3240.

LIGATION OF pAMP AND pKAN RESTRICTION FRAGMENTS

Reagents	Supplies and Equipment
digested pAMP	0.5–10-µl micropipettor + tips
digested pKAN	1.5-ml tube
10× ligation buffer/ATP	beaker for used tips
T4 DNA ligase	microfuge (optional)
distilled water	test tube rack
	70°C water bath

Ligate pAMP and pKAN

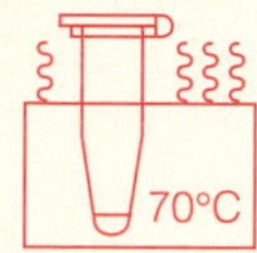

(30 minutes; then 2+ hours of incubation)

Step 1 is critical: Heat denaturation inactivates the restriction enzymes. For brief ligations of 2–4 hours, be sure to use high-concentration ligase with at least 5 Weiss units/µl or 300–500 cohesive-end units/µl.

1. Incubate the "Digested pAMP" and "Digested pKAN" tubes in a 70°C water bath for 10 minutes.
2. Label a clean 1.5-ml tube "Ligation."
3. Use the matrix below as a checklist while adding reagents to the ligation reaction. *Use a fresh pipet tip for each reagent.*

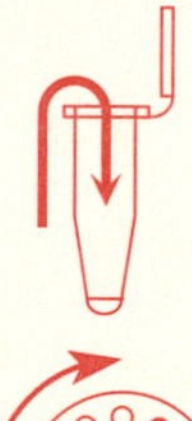

Tube	Digested pAMP	Digested pKAN	10x Ligation Buffer/ATP	Water	DNA Ligase
Ligation	3 µl	3 µl	2 µl	11 µl	1 µl

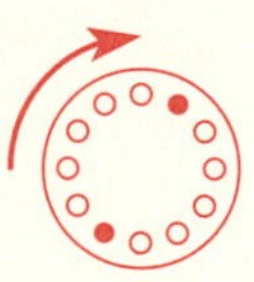

4. After adding all of the above reagents, close the top of the tube. Pool and mix the reagents by pulsing them in a microfuge or by sharply tapping the bottom of the tube on the lab bench.
5. Incubate the reaction at room temperature for 2–24 hours.
6. If time permits, ligation can be confirmed by electrophoresing 5 µl of the ligation reaction along with *Bam*HI/*Hin*dIII digests of pAMP and pKAN. Few, if any, of the low-molecular-weight bands in the pAMP and pKAN digests should be present in the ligated DNA gel lane. Instead, the ligated lane should show multiple bands of high-molecular-weight DNA near the top of the gel.

Stop Point

Freeze the ligation at –20°C until you are ready to continue. Thaw the reaction before proceeding to Laboratory 8.

RESULTS AND DISCUSSION

Ligation of the four *Bam*HI/*Hin*dIII restriction fragments of pAMP and pKAN produces many types of hybrid molecules, including plasmids composed of more than two fragments. However, only constructs possessing an origin of replication will be maintained and expressed. Three different replicating plasmids, with selectable antibiotic resistance, are created by ligating combinations of *two* *Bam*HI/*Hin*dIII fragments:

- Ligation of the 784-bp fragment to the 3,755-bp fragment regenerates pAMP.
- Ligation of the 1,861-bp fragment to the 2,332-bp fragment regenerates pKAN.
- Ligation of the 1,861-bp fragment to the 3,755-bp fragment produces the "simple recombinant" plasmid pAK, in which the kanamycin-resistance gene has been fused into the pAMP backbone.

1. Make a scale drawing of the simple recombinant molecule pAK described above. Include fragment sizes, the locations of *Bam*HI and *Hin*dIII restriction sites, the location of origins(s), and the location of antibiotic-resistance gene(s).
2. Make scale drawings of other theoretical two-fragment recombinant plasmids with the following properties.
   a. Three kinds of plasmids having two origins
   b. Three kinds of plasmids having no origin
3. Ligation of the 784-bp fragment, 3,755-bp fragment, 1,861-bp fragment, and 2,332-bp fragment produces a "super plasmid" pAMP/KAN. Make a scale drawing of the super plasmid pAMP/KAN.
4. Make scale drawings of several recombinant plasmids composed of any three of the four *Bam*HI/*Hin*dIII fragments of pAMP and pKAN. Include fragment sizes, the locations of *Bam*HI and *Hin*dIII restriction sites, the location of origin(s), and the location of antibiotic-resistance genes(s). What rules govern the construction of plasmids from the three kinds of restriction fragments?
5. What kind of antibiotic selection would identify *E. coli* cells that have been transformed with each of the plasmids you have drawn in questions 1–4?
6. Explain what is meant by "sticky ends." Why are they so useful for creating recombinant DNA molecules?
7. Why is ATP essential for the ligation reaction?

Transformation of *E. coli* with Recombinant DNA

In Part A, "Classic Procedure for Making Competent Cells," *E. coli* cells are rendered competent to uptake plasmid DNA, using a method essentially unchanged since its publication in 1970 by M. Mandel and A. Higa. The procedure begins with vigorous *E. coli* cells grown in suspension culture. The cells are harvested in mid-log phase by centrifugation and incubated on ice with two successive changes of calcium chloride solution. Although the classic procedure is more involved than the rapid colony protocol introduced in Laboratory 5, this procedure typically achieves transformation efficiencies of 10^5 to 10^6 colonies per microgram of plasmid—a 5–200-fold increase over the colony procedure. The enhanced efficiency helps to compensate for two difficulties inherent in this laboratory:

- Ligated plasmid DNA—composed of relaxed and linear molecules—transforms less efficiently than supercoiled monomeric plasmid.
- Dual selection with kanamycin and ampicillin is unforgiving and kills a percentage of transformants containing both resistance genes.

In Part B, "Transformation of *E. coli* with Recombinant pAMP/pKAN Plasmids," the competent *E. coli* cells are transformed with the ligation products from Laboratory 7, "Recombination of Antibiotic-Resistance Genes." Ligated plasmid DNA is added to one sample of competent cells, and purified pAMP and pKAN plasmids are added as controls to two other samples. The competent cells are incubated on ice with the plasmid DNAs, heat-shocked at 42°C, and recovered in LB broth for 40–60 minutes at 37°C. Unlike ampicillin selection in Laboratory 5, "Rapid Colony Transformation of *E. coli* with Plasmid DNA," the recovery step is essential for the kanamycin selection in this lab. Samples of transformed cells are plated onto three types of LB agar: with ampicillin (LB/amp), kanamycin (LB/kan), and both ampicillin and kanamycin (LB/amp+kan).

The ligation reaction produces many kinds of recombinant molecules composed of *Bam*HI/*Hin*dIII fragments, including the re-ligated parental plasmids pAMP and pKAN. The object is to select for transformed cells with dual antibiotic resistance, which must contain the 3,755-bp fragment from pAMP with the ampicillin-resistance gene (plus the origin of replication) and the 1,861-bp fragment from pKAN containing the kanamycin-resistance gene. Bacteria transformed with a single plasmid containing these fragments, or those doubly transformed with both pAMP and pKAN plasmids, form colonies on the LB/amp+kan plate.

Prepare Competent Cells

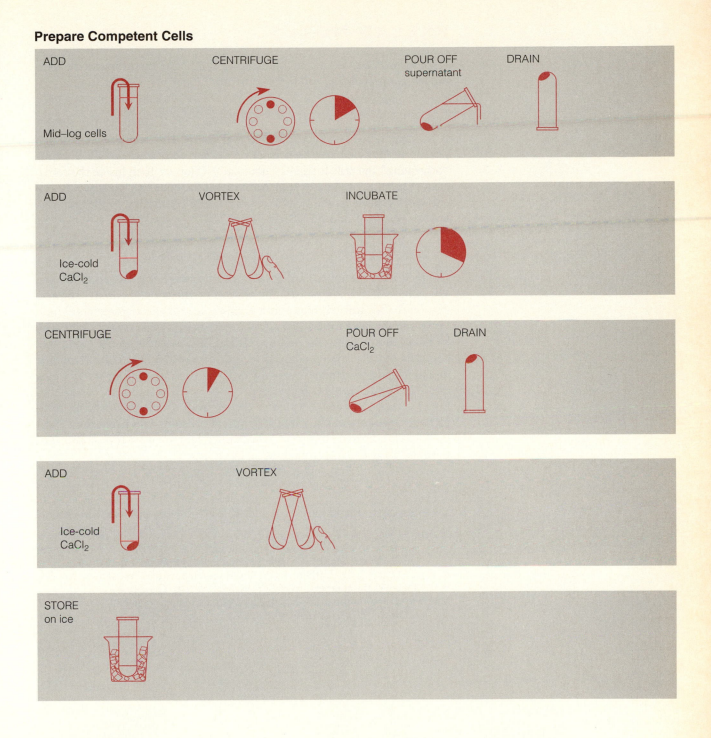

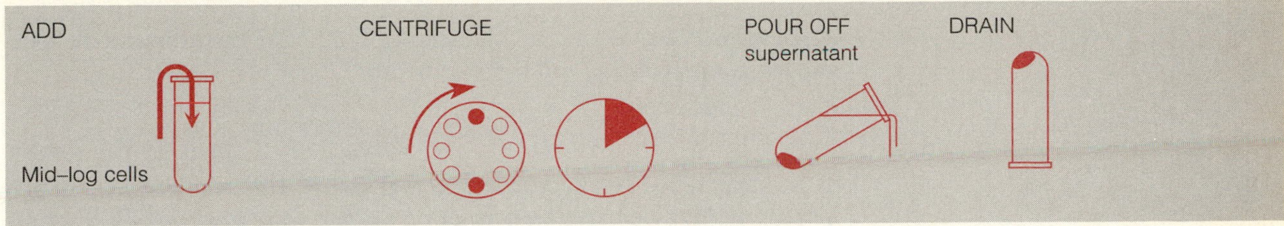

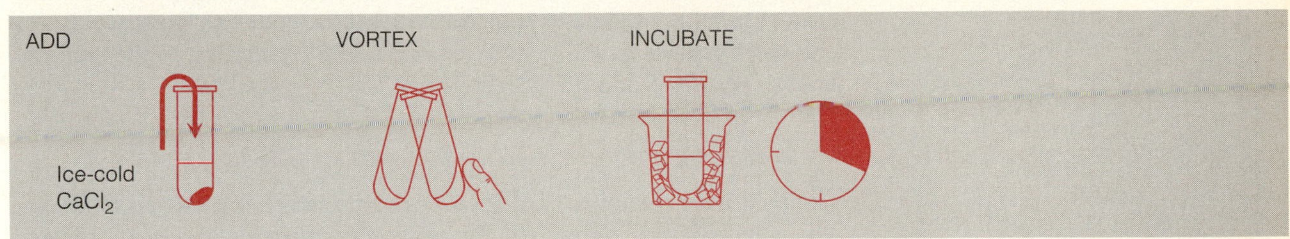

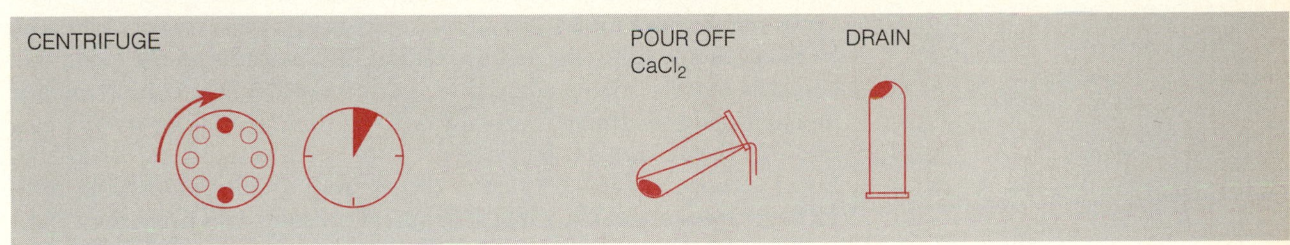

Laboratory 8/Part A
Classic Procedure for Making Competent Cells

PRELAB NOTES

Review the Prelab Notes in Laboratories 1, 2, and 5 regarding sterile techniques, *E. coli* culture, and transformation.

Seasoning Cells for Transformation

If possible, schedule experiments so that competent cells (Part A) are prepared one day before transformation with recombinant DNA (Part B). "Seasoning" the cells for 12–24 hours at 0°C (an ice bath inside the refrigerator) generally increases transformation efficiency five- to tenfold. This enhanced efficiency will help to assure successful cloning of the recombinant molecules produced in Laboratory 7.

Substituting Rapid Colony Transformation

Coupled with the inherent difficulties of transforming ligated DNA and selecting with two antibiotics, the relative inefficiency of colony transformation (Laboratory 5) makes it a poor choice for this experimental unit. However, we have found that small numbers of dual-resistant clones can be recovered by using the colony method. It is essential to use healthy *E. coli* cells that have been assayed for performance by the colony method and to include an artificially lengthy recovery period of 2.5 hours or longer at 37°C. Even overnight recovery does not overly amplify the number of transformants, possibly because of the presence of calcium chloride in the recovery medium, along with LB broth. We recommend using the colony method with ligated DNA only when either equipment or time constraints preclude using the classic method.

For Further Information

The protocol presented here is based on the following published methods:

Cohen, S. N., A. C. Y. Chang, and L. Hsu. 1972. Nonchromosomal antibiotic resistance in bacteria: Genetic transformation of *Escherichia coli* by R-factor DNA. *Proceedings of the National Academy of Sciences USA* 69: 2110.

Dagert, M. and S. D. Ehrlich. 1979. Prolonged incubation in calcium chloride improves the competence of *Escherichia coli* cells. *Gene* 6: 23.

Mandel, M. and A. Higa. 1970. Calcium-dependent bacteriophage DNA infection. *Journal of Molecular Biology* 53: 159.

CLASSIC PROCEDURE FOR MAKING COMPETENT CELLS

Culture and Reagents	Supplies and Equipment
10 ml of mid-log MM294 cells	5- or 10-ml pipets, sterile
50 mM of $CaCl_2$	pipet aid or bulb
	100–1,000-µl micropipettor + tips (or 1-ml pipet)
	beaker of crushed or cracked ice
	beaker for waste
	10%-bleach solution or disinfectant
	burner flame
	clean paper towels
	clinical centrifuge (500–1,000 × *g*)
	test tube rack

Prepare Competent Cells

(40–50 minutes)

This entire experiment must be performed under sterile conditions, using sterile tubes and pipet tips. Review the sterile techniques given in Laboratory 1, "Measurements, Micropipetting, and Sterile Techniques."

1. Place a sterile tube containing a 50-mM $CaCl_2$ solution on ice.
2. Obtain a 15-ml tube with 10 ml of mid-log cells, and label it with your name and the date.
3. *Securely close the cap,* and place the tube of cells in a *balanced* configuration with the other classmates' tubes in the rotor of a clinical centrifuge. Centrifuge at 500–1,000 × *g* for 10 minutes to pellet the cells on the bottom of the culture tube.
4. Sterilely pour off the supernatant from the tube into a waste beaker for later disinfection. *Take care not to disturb the cell pellet.*
   a. Remove the cap from the culture tube, and briefly flame the mouth. *Do not place the cap on the lab bench.*
   b. Carefully pour off the supernatant into a waste beaker for later disinfection. Invert the culture tube, and touch the mouth to a clean paper towel to wick off as much as possible of the remaining supernatant.
   c. Reflame the mouth of the culture tube, and replace the cap.
5. Sharply tap the bottom of the tube several times on the lab bench to loosen the cell pellet.

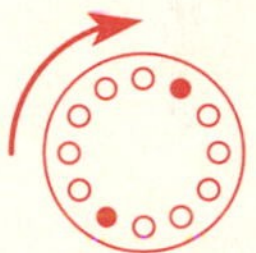

"Blank" tubes filled with 10 ml of water can be used for balance, if needed. 500–1,000 × *g* corresponds to 2,000–3,000 rpm for most tabletop clinical centrifuges. A tight-cell pellet should be visible at the bottom of the tube. If the pellet appears to be very loose or unconsolidated, centrifuge for another 5 minutes.

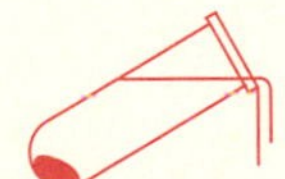

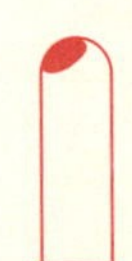

Organize materials on the lab bench, and plan out sterile manipulations. If you are working with a partner, one person can handle the pipet, and the other can remove the cap and flame the mouth of the tube.

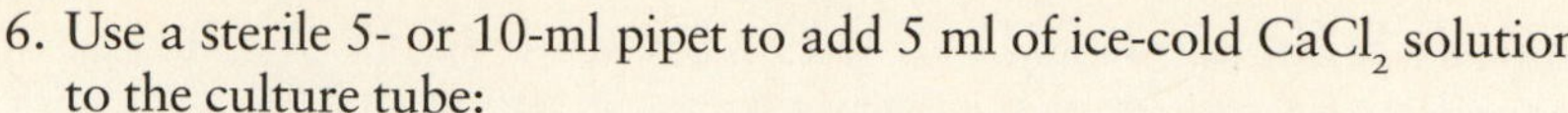

6. Use a sterile 5- or 10-ml pipet to add 5 ml of ice-cold $CaCl_2$ solution to the culture tube:
   a. Briefly flame the cylinder of the pipet.
   b. Remove the cap from the $CaCl_2$ tube, and flame the mouth of the tube. *Do not place the cap on the lab bench.*
   c. Withdraw 5 ml of $CaCl_2$ solution. Reflame the mouth of the tube, and replace the cap.
   d. Remove the cap of the culture tube, and flame the mouth. *Do not place the cap on the lab bench.*
   e. Expel $CaCl_2$ into the culture tube. Reflame the mouth of the tube, and replace the cap.
7. *Close the cap of the culture tube tightly.* Immediately vortex to *completely* resuspend the pelleted cells. Periodically, during vortexing, hold the tube up to the light to check for complete suspension of cells. *The finished cell suspension should be homogeneous, with no visible clumps of cells.* If a mechanical vortexer is unavailable, "finger vortex" as follows:

   Hold the upper part of the tube securely between your thumb and index finger. Vigorously hit the bottom of the tube with the index finger of the opposite hand to create a vortex that will lift the cell pellet off of the bottom of the tube.

The cell pellet will become increasingly difficult to resuspend the longer it sits in the calcium chloride. Complete cell suspension is probably the most important variable for obtaining competent cells.

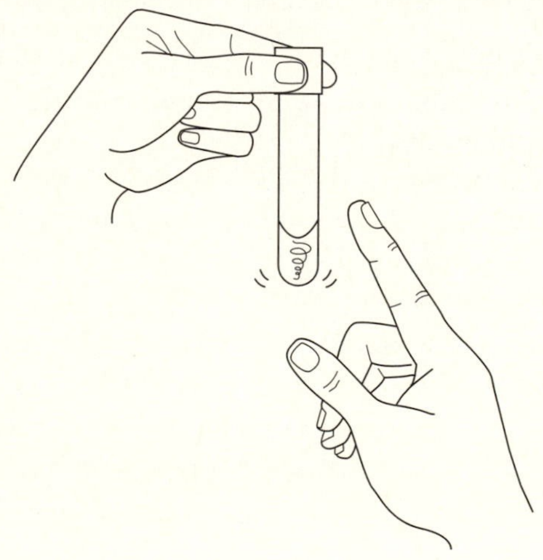

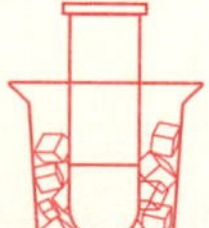

8. Return the tube to ice, and incubate it for 20 minutes.
9. Following incubation, respin the cells in a clinical centrifuge for 5 minutes at 500–1,000 × *g*.
10. *Be especially careful not to disturb the diffuse cell pellet.* Sterilely pour off the supernatant into a waste beaker for later disinfection, as in step 4. Invert the culture tube, and touch the mouth of the tube to a clean paper towel to wick off as much as possible of the remaining supernatant. Flame the mouth of the culture tube after removing the top and, again, before replacing the top.

The calcium chloride treatment alters the adhering properties of the *E. coli* cell membranes. At this point, the cell pellet typically appears diffuse or ring-shaped and should disperse more easily.

11. Use a 100–1,000-µl micropipettor (or a 1-ml pipet) to sterilely add 1,000 µl (1 ml) of fresh, ice-cold calcium chloride to the tube. Flame the mouth of the culture tube after removing the top and, again, before replacing the top.

Do not flame the tip of the micropipet.

12. *Close the cap of the culture tube tightly.* Vortex to *completely* resuspend the pelleted cells, as in step 7. *The finished cell suspension should be homogeneous, with no visible clumps of cells.*

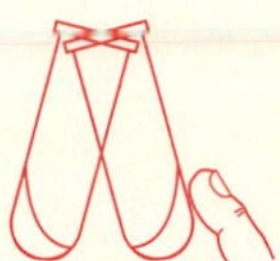

The diffuse cell pellet should resuspend very easily; do not vortex it too much.

Stop Point

Store the cells in a beaker of ice in the refrigerator (at approximately 0°C) until you are ready to use them. "Seasoning" at 0°C for as long as 24 hours increases the competency of cells five- to tenfold.

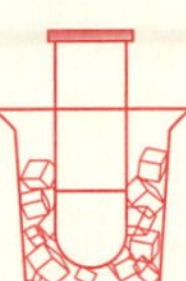

13. Take the time for proper cleanup:
    a. Segregate for proper disposal culture plates and tubes, pipets, and micropipet tips that have come into contact with *E. coli*.
    b. Disinfect the mid-log culture, tips, and supernatant from steps 4 and 10 with a 10%-bleach solution or disinfectant (such as Lysol).
    c. Wipe down the lab bench with soapy water, a 10%-bleach solution, or disinfectant.
    d. Wash your hands before leaving the laboratory.

Transform *E. coli* with Recombinant pAMP/pKAN Plasmids

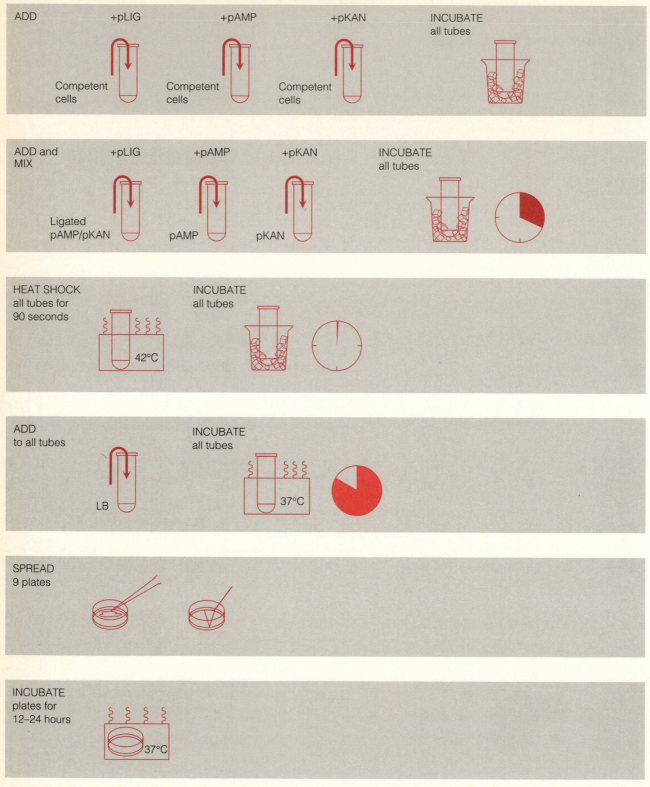

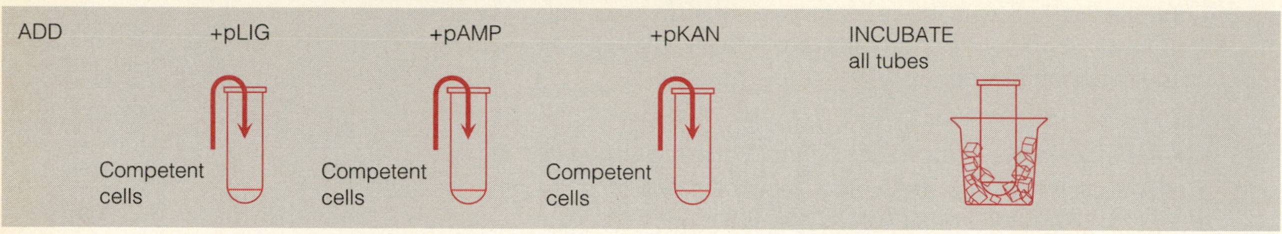

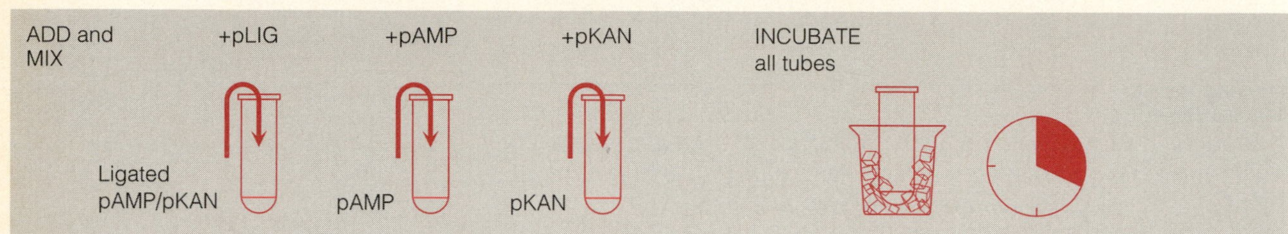

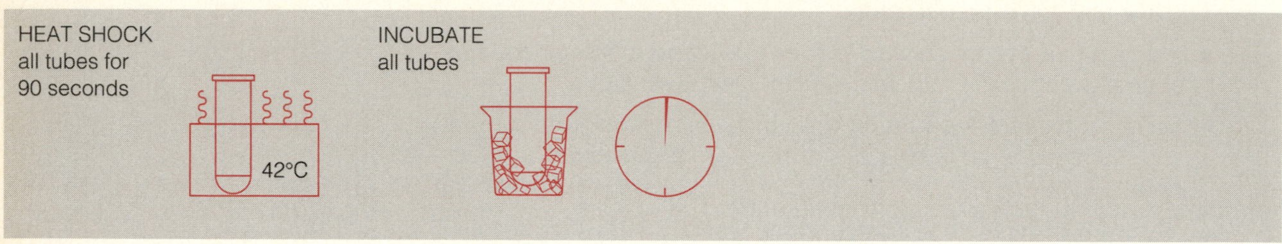

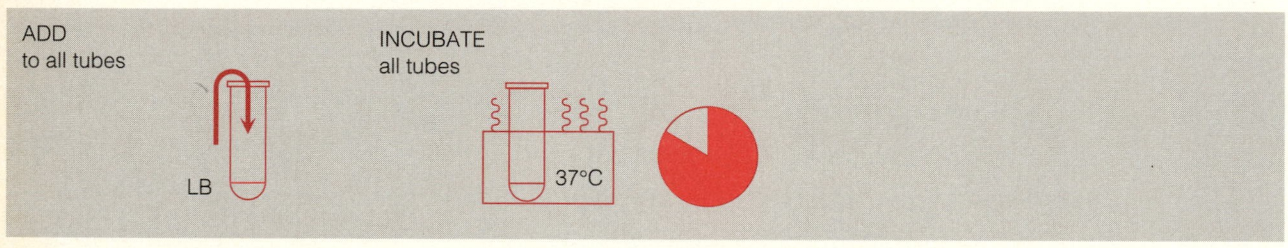

Laboratory 8/Part B

Transformation of *E. coli* with Recombinant pAMP/pKAN Plasmids

PRELAB NOTES

Review the Prelab Notes in Laboratory 5, "Rapid Colony Transformation of *E. coli* with Plasmid DNA."

Recovery Period

A postincubation recovery of 40–60 minutes at 37°C, with shaking, is essential before plating the transformed cells on kanamycin, which acts quickly to kill any cells that are not actively expressing the resistance protein.

For Further Information

The protocol presented here is based on the following published method:

Cohen, S. N., A. C. Y. Chang, and H. W. Boyer. 1973. Construction of biologically functional bacteria plasmids in vitro. *Proceedings of the National Academy of Sciences USA* 70: 3240.

TRANSFORMATION OF *E. COLI* WITH RECOMBINANT pAMP/pKAN PLASMIDS

Culture, Media, and Reagents

competent *E. coli* cells (from Part A)
ligation tube (from Laboratory 7)
0.005 µg/µl of pAMP
0.005 µg/µl of pKAN
LB broth
3 LB/amp plates
3 LB/kan plates
3 LB/amp+kan plates

Supplies and Equipment

0.5–10-µl micropipettor + tips
100–1,000-µl micropipettor + tips
3 15-ml culture tubes
beaker of crushed or cracked ice
beaker of 95% ethanol
beaker for waste/used tips
"bio-bag" or heavy-duty trash bag
10%-bleach solution or disinfectant
burner flame
cell spreader
permanent marker
test tube rack
37°C incubator
37°C shaking water bath (optional)
42°C water bath

Transform *E. coli* with Recombinant pAMP/pKAN Plasmids

(70–90 minutes)

This entire experiment must be performed under sterile conditions, using sterile tubes and pipet tips. Review the sterile techniques given in Laboratory 1, "Measurements, Micropipetting, and Sterile Techniques."

To save plates, experimenters can omit *either* pAMP or pKAN control.

1. Use a permanent marker to label three sterile 15-ml culture tubes:
   +pLIG = ligated DNA
   +pAMP = pAMP control
   +pKAN = pKAN control

Optimally, the mouth of the culture tube should be flamed after you remove the top and, again, before you replace the top.

2. Use a 100–1000-µl micropipettor with a *sterile tip* to add 200 µl of competent cells to each tube.
3. Place all three tubes on ice.

Store the remainder of the ligated DNA at 4°C. Electrophorese with cut pAMP and pKAN controls in Laboratory 10 to observe the products of ligation reaction.

4. Use a 0.5–10-µl micropipettor to add 10 µl of ligated pAMP/pKAN solution *directly into the cell suspension* in the +pLIG tube.
5. Use a *fresh tip* to add 10 µl of a 0.005-µg/µl pAMP solution *directly into the cell suspension* in the +pAMP tube.
6. Use a *fresh tip* to add 10 µl of 0.005-µg/µl pKAN solution *directly into the cell suspension* in the +pKAN tube.
7. Close the caps, and tap the tubes lightly with your finger to mix the solution. Avoid making bubbles in the suspension or splashing it up on the sides of the tubes.

8. Return all three tubes to ice for 20 minutes.

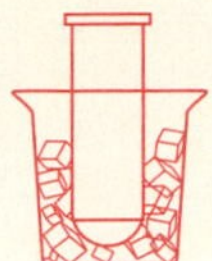

9. While the cells are incubating on ice, use a permanent marker to label all nine LB agar plates with your name and the date. Divide the plates into three sets, containing one of each plate type, and mark the plates as follows:

   L: LB/amp, L: LB/kan, and L: LB/amp+kan

   A: LB/amp, A: LB/kan, and A: LB/amp+kan

   K: LB/amp, K: LB/kan, and K: LB/amp+kan

To save plates, experimenters can omit *either* set A or set K.

10. Following the 20-minute incubation, heat shock the cells in all three tubes. *Cells must receive a sharp and distinct shock.*

    a. Carry the ice beaker to a water bath. Remove the tubes from the ice, and *immediately* immerse them in a 42°C water bath for 90 seconds.

    b. Immediately return all three tubes to ice, and let them stand on ice for at least 1 more minute.

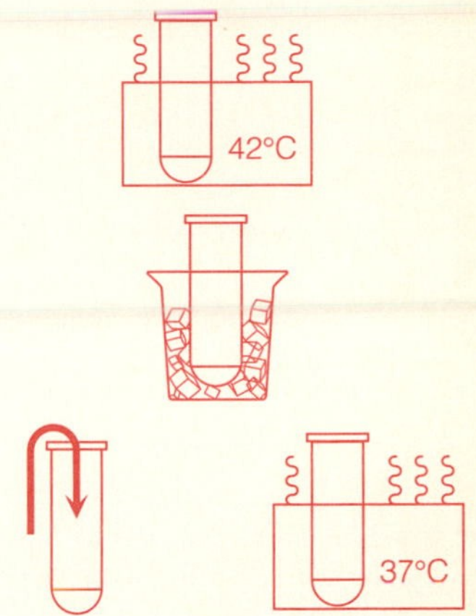

11. Use a 100–1,000-µl micropipettor with a sterile tip to add 800 µl of LB broth to each tube. Tap the tubes lightly with your finger to mix the solution.

12. Allow the cells to recover by incubating all three tubes at 37°C in a water bath (with moderate agitation) for 40–60 minutes.

If a shaking water bath is not available, warm the cells for several minutes in a 37°C water bath, then transfer the cells to a dry shaker inside a 37°C incubator. Alternatively, occasionally swirl the tubes by hand in a nonshaking 37°C water bath.

Stop Point

Cells can be allowed to recover for several hours. A longer recovery period will ensure the growth of as many kanamycin-resistant recombinants as possible and can help to compensate for either a poor ligation or cells of low competence.

13. Use the matrix below as a checklist as the +pLIG, +pAMP, and +pKAN cells are spread on each type of antibiotic plate in the following steps.

	Ligated DNA	**pAMP control**	**pKAN control**
	L	**A**	**K**
LB/amp	100 µl	100 µl	100 µl
LB/kan	100 µl	100 µl	100 µl
LB/amp+kan	100 µl	100 µl	100 µl

14. Use a 100–1,000-µl micropipettor with a sterile tip to add 100 µl of cell suspension from the +pLIG tube onto three plates marked "L." *Do not allow the suspensions to sit on the plates too long before proceeding to step 15.*

Do not allow cells to sit too long on the plates before spreading them or else they will become absorbed into the plates as concentrated spots. The object is to evenly distribute and separate the transformed cells on the plate surface so that each will give rise to a distinct colony of clones.

15. Sterilize the cell spreader, and spread the cells over the surface of each L plate in succession.
    a. Dip the spreader into a beaker of ethanol, and briefly pass it through a burner flame to ignite the alcohol. Allow the alcohol to burn off *away from the burner flame.*

The spreader, submerged in alcohol, is already sterile. The only purpose of using the flame is to burn off the alcohol before spreading the cells. The spreader will overheat if it is held directly in the burner flame for several seconds, and this can kill *E. coli* cells on the plate.

CAUTION

Be extremely careful not to ignite the ethanol in the beaker.

Take care not to gouge the agar.

    b. Lift the plate lid, like a clam shell, only wide enough to allow spreading.
    c. Cool the spreader by touching it to the agar, away from the cells, or to condensed water on the plate lid.
    d. Touch the spreader to the cell suspension, and gently drag the spreader back and forth several times across the agar surface. Rotate the plate one-quarter turn, and then repeat the spreading motion.
    e. Replace the plate lid. Return the cell spreader to the ethanol without flaming.

16. Use a 100–1,000-µl micropipettor with a fresh, sterile tip to add 100 µl of cell suspension from the +pAMP tube onto three A plates. Repeat steps 15a–e to sterilize the cell spreader, and spread the cells over the surface of each A plate in succession.
17. Use a 100–1,000-µl micropipetor with a fresh, sterile tip to add 100 µl of cell suspension from the +pKAN tube onto three K plates. Repeat steps 15a–e to sterilize the cell spreader and spread the cells over the surface of each K plate in succession.
18. Reflame the spreader once more before setting it down.

19. Let the plates set for several minutes to allow the cell suspensions to be absorbed.
20. Wrap the plates together with tape, place them upside down in a 37°C incubator, and incubate them for 12–24 hours.
21. After the initial incubation, store the plates at 4°C to arrest *E. coli* growth and to slow the growth of any contaminating microbes.
22. Take the time for proper cleanup:
    a. Segregate for proper disposal culture plates and tubes, pipets, and micropipettor tips that have come into contact with *E. coli.*
    b. Disinfect the overnight cell suspensions, tubes, and tips with a 10%-bleach solution or disinfectant (such as Lysol).
    c. Wipe down the lab bench with soapy water, a 10%-bleach solution, or disinfectant.
    d. Wash your hands before leaving the laboratory.

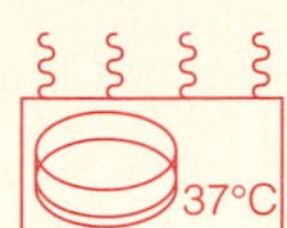

Save the "L" LB/amp and "L" LB/kan plates if you are planning to complete Laboratory 9, "Replica Plating to Identify Mixed *E. coli* Populations." Save the "L" LB/amp+kan plate, as a source of colonies to begin overnight suspension cultures, if you are planning to do Laboratory 10, "Purification and Identification of Recombinant Plasmid DNA."

RESULTS AND DISCUSSION

Observe the cell colonies through the bottom of the culture plates, using a permanent marker to mark visible colonies as they are counted. A total of 10–100 colonies should be observed on the "L" LB/amp+kan experimental plate, 500–5,000 colonies on the "A" LB/amp control plate, and 200–2,000 colonies on the "K" LB/kan control plate. Approximately tenfold fewer colonies should be observed on the corresponding "L" LB/amp and "L" LB/kan plates. An extended recovery period would elevate these numbers, however. (Question 3 explains how to compute transformation efficiency.) If plates have been overincubated or left at room temperature for several days, "satellite" colonies may be observed on the LB/amp plates. Satellite colonies are never observed on the LB/kan or LB/amp+kan plates.

1. Record your observation of each plate on the matrix below. If the cell growth is too dense for you to count individual colonies, record "lawn." Were the results as you expected? Explain the possible reasons for variations from the expected results.

	Ligated DNA L	**pAMP control A**	**pKAN control K**
LB/amp			
LB/kan			
LB/amp+kan			

2. Compare and contrast the growth on each of the following pairs of plates. What does each pair of results tell you about transformation or antibiotic selection?

   "L" LB/amp and "A" LB/amp

   "L" LB/kan and "K" LB/kan

   "A" LB/amp and "K" LB/kan

   "L" LB/amp and "L" LB/kan

   "L" LB/amp and "L" LB/amp+kan

   "L" LB/kan and "L" LB/amp+kan

3. Calculate transformation efficiencies of "A" LB/amp and "K" LB/kan positive controls. Remember that transformation efficiency is expressed as the number of antibiotic resistant colonies per microgram of intact plasmid DNA. The object is to determine the mass of pAMP or pKAN that was spread on each plate and was therefore responsible for the transfomants observed.

   a. Determine the *total mass (in μg) of pAMP* used in step 5 and of pKAN used in step 6: concentration × volume = mass.

   b. Determine the *fraction of cell suspension spread* onto "A" LB/amp plate (step 16) and "K" LB/kan (step 17): volume suspension spread/total volume suspension (steps 2 and 11) = fraction spread

   c. Determine the *mass of plasmid pAMP and pKAN in the cell suspension spread* onto "A" LB/amp and "K" LB/kan plate: total mass plasmid (a) × fraction spread (b) = mass plasmid spread.

   d. Determine the *number of colonies per μg of pAMP and pKAN.* Express your answer in scientific notation: colonies observed/mass plasmid spread (c) = transformation efficiency.

4. Calculate the transformation efficiencies using ligated DNA on the "L" LB/amp, "L" LB/kan, and "L" LB/amp+kan plates.

   a. Determine *the mass (in μg) of pAMP and pKAN* used in the initial restriction digests = the concentration of plasmid × the volume of plasmid (Laboratory 7A, Section I.2).

   b. *The total mass of plasmid used* in the experiment = the mass of pAMP + the mass of pKAN (a).

   c. Determine *the fraction of pAMP and pKAN digests* used in the ligation reaction = the volume of digest in the ligation (7B.3)/the total volume of each digest (7A.I.2).

   d. Determine *the fraction of ligation reaction* used for transformation = the volume of ligation reaction in transformation (8B.4)/ the total volume of ligation reaction (7B.3).

   e. Determine *the fraction of cell suspension spread* onto L plates = the volume of cell suspension spread (8B.14)/the total volume of cell suspension (8B.2 and 11).

   f. Determine *the mass of plasmid in cell suspension spread* onto L plates = the total mass of plasmid (b) × fraction c × fraction d × fraction e.

   g. Express *the transformation efficiency* in scientific notation as the number of colonies per μg of plasmid = the total number of colonies on the "L" LB/amp, "L" LB/kan, or "L" LB/amp+kan plate/the mass of plasmid in the cell suspension spread (f).

5. Compare the transformation efficiencies you calculated for the "A" LB/amp plate in this laboratory and the +AMP plate in Laboratory 5. By what factor is the classic procedure more or less efficient than colony transformation? What differences in the protocols contribute to the increase in efficiency?
6. Compare the transformation efficiencies you calculated for control pAMP and pKAN *versus* the ligated pAMP/pKAN. How can you account for the differences in efficiency? (Take into account the formal definition of transformation efficiency.)

FOR FURTHER RESEARCH

Interpretable experimental results can be obtained only when the classic transformation protocol can be repeated with reproducible results: 500–5,000 colonies on an LB/amp plate when transforming with intact pAMP plasmid.

1. Attempt to isolate pAMP/pKAN recombinants, using the colony transformation protocol in Laboratory 5. What would increase the likelihood of retrieving ampicillin/kanamycin-resistant colonies?
2. Design an experiment to compare the transformation efficiency of ligated versus intact circular plasmid DNA. Keep the molecular weight constant. What percentage of ligated DNA appears to be nontransforming?
3. Design an experiment to compare the transformation efficiencies of linear versus circular plasmid DNAs. Keep the molecular weight constant.
4. Design a series of experiments to test the relative importance of each of the four major steps of most transformation protocols: preincubation, incubation, heat shock, and recovery. Which steps are absolutely necessary?
5. Design a series of experiments to compare the transforming effectiveness of calcium chloride *versus* salts of other monovalent (+), divalent (++), and trivalent (+++) cations.
   a. Make up 50-mM solutions of each salt.
   b. Check the pH of each solution, and buffer them to approximately pH 7, when necessary.
   c. Is calcium chloride unique in its ability to facilitate transformation?
   d. Is there a consistent difference in the transforming effectiveness of monovalent *versus* divalent *versus* trivalent cations?
6. Design a series of experiments to determine saturating conditions for transformation.
   a. Transform *E. coli*, using the following DNA concentrations:

      0.00001 µg/µl
      0.00005 µg/µl
      0.0001 µg/µl
      0.0005 µg/µl
      0.001 µg/µl
      0.005 µg/µl

0.01 μg/μl

0.05 μg/μl

0.1 μg/μl

b. Plot a graph of DNA mass *versus* colonies per plate.

c. Plot a graph of DNA mass *versus* transformation efficiency.

d. At what mass does the reaction appear to become saturated?

e. Repeat the experiment with concentrations clustered on either side of the presumed saturation point to produce a fine saturation curve.

7. Repeat experiment 6 above, but transform using a 1:1 mixture of pAMP and pKAN at each concentration. Plate the transformants on LB/amp, LB/kan, and LB/amp+kan plates. *Be sure to include a recovery of 40–60 minutes, with shaking.*

a. Calculate the percentage of double transformations at each mass.

$$\frac{\text{colonies LB/amp+kan plate}}{\text{colonies LB/amp plate + colonies LB/kan plate}}$$

b. Plot a graph of DNA mass *versus* colonies per plate.

c. Plot a graph of DNA mass *versus* percentage of double transformations. Under saturating conditions, what percentage of bacteria are doubly transformed?

8. Plot a recovery curve for *E. coli* transformed with pKAN. Allow the cells to recover for 0–120 minutes at 20-minute intervals.

a. Plot a graph of recovery time *versus* colonies per plate.

b. At what point is antibiotic expression maximized?

c. Can you discern a point where the cells began to replicate?

L A B O R A T O R Y 9

Replica Plating to Identify Mixed *E. coli* Populations

Ligation of *Bam*HI/*Hin*dIII fragments of pAMP and pKAN in Laboratory 7 created constructs containing various combinations of an ampicillin-resistance gene (*amp*^{r}), a kanamycin-resistance gene (*kan*^{r}), or both genes together (*amp*^{r}/*kan*^{r}). Following transformation of the ligated DNA into competent *E. coli* cells in Laboratory 8, colonies with the *amp*^{r}/*kan*^{r} genotype can be clearly identified by their growth on the "L" LB/amp+kan plate. It is impossible, however, to be certain of the *amp*^{r}/*kan*^{r} genotypes of bacteria growing on "L" LB/amp or "L" LB/kan plates. Of course, colonies growing on an "L" LB/amp plate must possess *at least* an *amp*^{r} gene, and colonies growing on an "L" LB/kan must possess *at least* a *kan*^{r} gene. In the absence of a challenge by the second antibiotic, though, it cannot be determined whether a particular colony harbors both resistance genes.

In this laboratory, replica plating provides a rapid means of distinguishing between single- and dual-resistant colonies growing on "L" LB/amp and "L" LB/kan plates. Cells from 12 colonies on the "L" LB/amp plate and from 12 colonies on the "L" LB/kan plate are transferred onto one fresh LB/amp and one LB/kan plate to which numbered grids have been attached. An "L" colony is scraped with a sterile toothpick (or inoculating loop), and a sample of cells is streaked successively into the same numbered square of the fresh LB/amp and LB/kan plates. Following overnight incubation at 37°C, colonies that grow in the same square of both the LB/amp and LB/kan plates are identified as having the *amp*^{r}/*kan*^{r} genotype.

PRELAB NOTES

Replica plating provides a rapid means of screening "L" LB/amp and "L" LB/kan plates for dual-resistant colonies that potentially contain pAMP/pKAN recombinant plasmids. If no colonies were obtained on the "L" LB/amp+kan plate, replicate plating provides another chance to identify dual-resistant colonies from which to isolate recombinant plasmids for Laboratory 10, "Purification and Identification of Recombinant Plasmid DNA."

For Further Information

The protocol presented here is based on the following published method:

Lederberg, J. and E. M. Lederberg. 1952. Replica plating and indirect selection of bacterial mutants. *Journal of Bacteriology* 63: 399.

Replica Plate *E. coli*

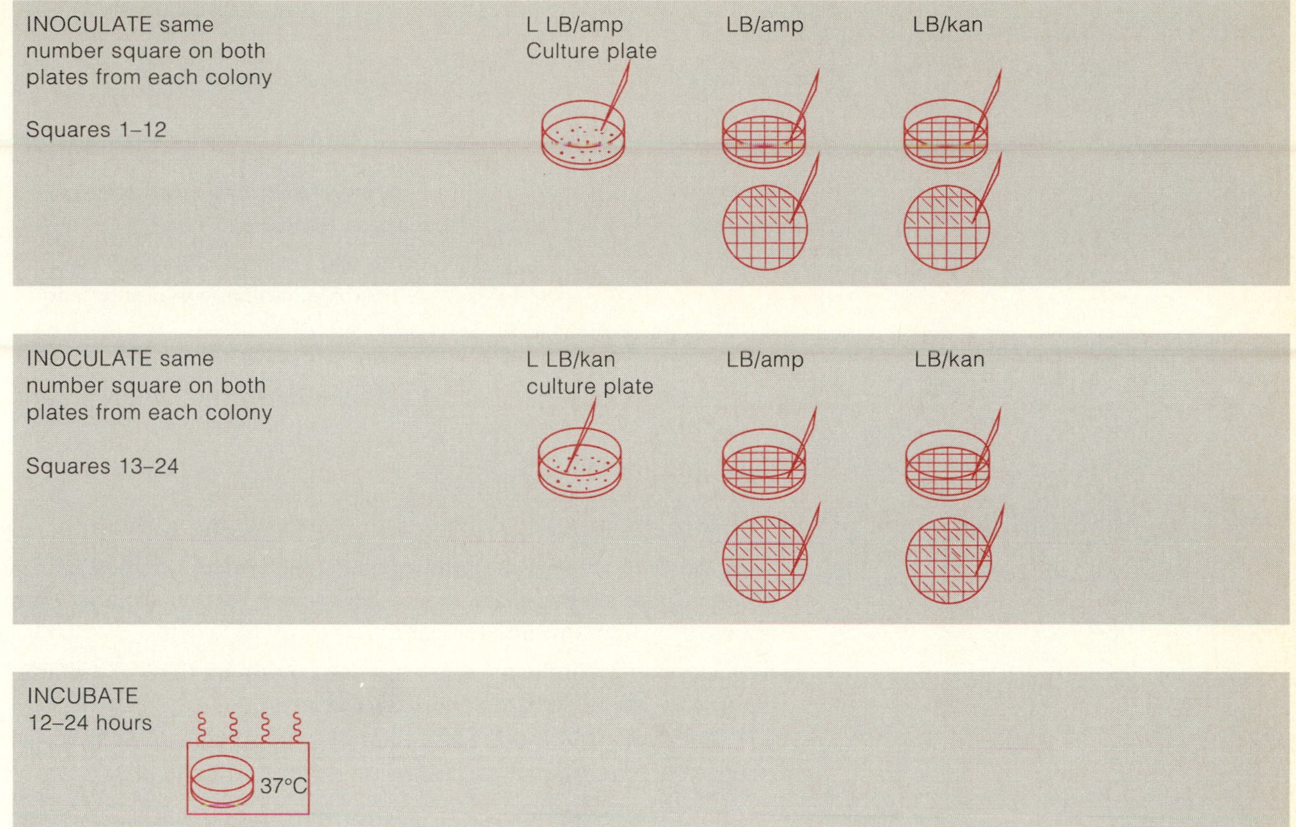
INOCULATE same number square on both plates from each colony
Squares 1–12
L LB/amp Culture plate
LB/amp
LB/kan
INOCULATE same number square on both plates from each colony
Squares 13–24
L LB/kan culture plate
LB/amp
LB/kan
INCUBATE 12–24 hours
37°C

REPLICA PLATING TO IDENTIFY MIXED *E. COLI* POPULATIONS

Cultures and Media	Supplies and Equipment
"L" LB/amp plate w/colonies	2 replica-plating grids
"L" LB/kan plate w/colonies (from Laboratory 8)	sterile toothpicks (or inoculating loop + burner flame)
1 LB/amp plate	beaker for waste
1 LB/kan plate	"bio bag" or heavy-duty trash bag
	10%-bleach solution or disinfectant (such as Lysol)
	permanent marker
	37°C incubator

Replica Plate *E. coli*

(10 minutes)

Alternatively, a 24-square grid can be drawn on the bottom of the plate with a permanent marker.

1. Use tape to attach replica-plating grids to the *bottoms* of the LB/amp plate and the LB/kan plate. Use a permanent marker to label each plate with your name and the date.
2. Replica plate a sample of cells from one colony on the "L" LB/amp plate onto gridded LB/amp and LB/kan plates.
   a. Use a sterile toothpick (or inoculating loop) to scrape up a cell mass from a well-defined colony on the "L" LB/amp plate.
   b. Immediately drag the *same* toothpick (or loop) gently across the agar surface to make a short diagonal (/) streak *within square 1* of the LB/amp plate.
   c. Immediately use the *same* toothpick (or loop) to make a diagonal (/) streak within square 1 of the LB/kan plate.
   d. Discard the toothpick in a waste beaker (or reflame and cool the inoculating loop).

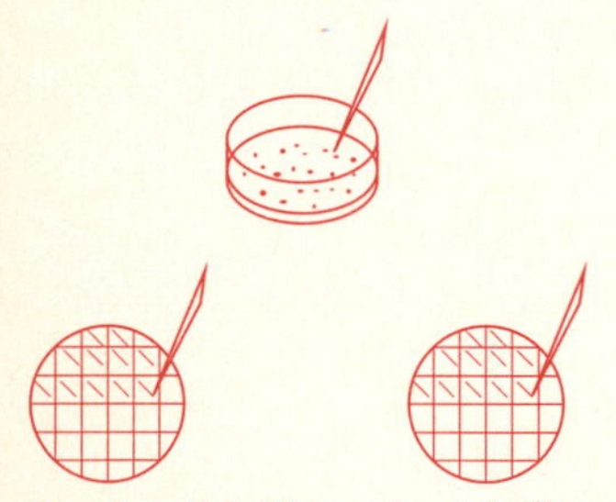

Lift the plate lids like a clam shell, only wide enough to select a colony and streak. Do not place the lids on the lab bench.

If you have fewer than 12 colonies on either plate, obtain a plate from another experimenter.

3. Repeat step 2a–d *with fresh toothpicks* (or a *flamed* and *cooled* inoculating loop) to streak cells from 11 *different* "L" LB/amp colonies onto squares 2–12 of both LB/amp and LB/kan plates.
4. Repeat step 2a–d *with fresh toothpicks* (or a *flamed* and *cooled* inoculating loop) to streak cells from 12 *different* "L" LB/kan colonies onto squares 13–24 of both LB/amp and LB/kan plates.
5. Place the plates upside down in a 37°C incubator, and incubate them for 12–24 hours.

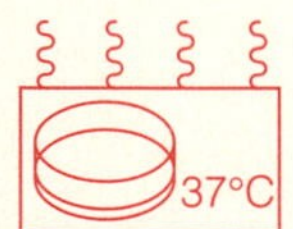

6. After the initial incubation, store the plates at 4°C to arrest *E. coli* growth and to slow the growth of any contaminating microbes.

If your "L" LB/amp+kan plate has less than two colonies, save the replica plates as a source of colonies from which to isolate plasmid DNA in Laboratory 10.

7. Take the time for proper cleanup:
   a. Segregate bacterial cultures *and* used toothpicks for proper disposal.
   b. Wipe down the lab bench with soapy water, a 10%-bleach solution, or disinfectant (such as Lysol).
   c. Wash your hands before leaving the laboratory.

RESULTS AND DISCUSSION

Typically, 30–70% of replica-plated colonies have dual resistance, with roughly equal numbers from "L" LB/amp and "L" LB/kan plates. In general, the results of replica plating indicate the success of the ligation in Laboratory 7 and parallel the results observed on the "L" LB/amp+kan plate from Laboratory 8. Thus, if there were a large number of colonies on the "L" LB/amp+kan plate, you will probably have a high percentage of dual-resistant colonies that grow on both the LB/amp and LB/kan replica plates.

1. Observe the LB/amp and LB/kan replica plates. Use the matrix below to record "+" in squares where new bacterial growth *has expanded beyond the width of the initial streak*. Record "–" in squares where no new growth has expanded the initial streak. Remember that non-resistant cells can survive, separated from the antibiotic, on top of a heavy initial streak; however, no *new growth* will be observed.

Colony Source ("L" LB/amp)	**REPLICA PLATES**		**Colony Source ("L" LB/kan)**	**REPLICA PLATES**	
	(LB/amp)	**(LB/kan)**		**(LB/amp)**	**(LB/kan)**
1			13		
2			14		
3			15		
4			16		
5			17		
6			18		
7			19		
8			20		
9			21		
10			22		
11			23		
12			24		

2. On the basis of your observations:

   a. Calculate the percentage of dual-resistant colonies taken from the "L" LB/amp plate (squares 1–12).

   b. Calculate the percentage of dual-resistant colonies taken from the "L" LB/kan plate (squares 13–24).

3. Draw restriction maps for different plasmid molecules that could be responsible for the dual resistance phenotype.

FOR FURTHER RESEARCH

1. We have discussed the phenomenon of satellite colonies on LB/amp plates: small nonresistant colonies that grow in a halo around a large resistant colony. Confirm the true phenotype of satellite colonies by replica plating them carefully onto fresh LB and LB/amp plates.

2. The *amp*^{r} protein, β-lactamase, is not actively secreted into the medium but is believed to leak through the cell envelope of *E. coli.* Satellite colonies do not form on kanamycin plates, because the antibiotic kills all nonresistant cells outright. The following experiment tests whether resistance protein escapes from ampicillin- and kanamycin-resistant cells.

   a. Grow separate 5-ml overnight cultures of an *amp*^{r} colony from the "A" LB/amp plate and a *kan*^{r} colony from the "K" LB/kan plate, according to the protocol in Laboratory 2B, "Overnight Suspension Culture."

   b. Following incubation, pass each overnight culture through a 0.22- or 0.45-μm filter, and collect the filtrate in a clean, sterile 15-ml tube. Filtering removes all *E. coli* cells.

   c. Mark a line on the bottom of one LB/amp plate and one LB/kan plate to divide each plate in half. Mark one half "+." Sterilely spread 100 μl of the "A" filtrate onto the " + " *half only* of the LB/amp plate. Sterilely spread 100 μl of the "K" filtrate onto the " + " *half only* of the LB/kan plate. Allow the filtrates to soak into the plates for 10–15 minutes.

   d. Sterilely streak wild-type (nontransformed) *E. coli* cells on each filtrate-treated plate, taking care to streak back and forth across the center dividing line.

   e. Incubate the plates at 37°C for 12–24 hours. Compare growth on the treated *versus* untreated sides of each plate.

3. The *amp*^{r} protein is believed to leak primarily from stationary-stage cells. The following experiment tests the hypothesis that leakage of β-lactamase is dependent on the growth phase.

   a. Grow an overnight culture of an ampicillin-resistant colony from an "A" LB/amp plate. Inoculate 1 ml of *plain* LB broth, according to the protocol in Laboratory 2B, "Overnight Suspension Culture."

   b. Use this overnight culture to inoculate 100 ml of fresh LB broth, and grow it according to the protocol in Laboratory 2C, "Mid-Log Suspension Culture."

c. Sterilely withdraw 10-ml aliquots from the culture after 1, 2, and 4 hours. Hold the aliquots on ice during the rest of the experiment.

d. Take the OD_{550} of each aliquot.

   Note: The objective is to test resistance-protein "leakage" as a function of culture age, *not as a function of cell number*. Because cell number increases over time, the 2- and 4-hour samples must be diluted with sterile LB to equal the *E. coli* concentration of the 1-hour sample. Because the OD_{550} values are proportional to cell number, they can be used to compute the dilution factor for the 2- and 4-hour samples.

e. Pass each of the three equalized samples through a 0.22- or 0.45-µm filter to remove *E. coli* cells. Collect the filtrates into sterile 15-ml tubes.

f. Prepare a ten- and hundredfold dilution of each equalized filtrate, using sterile LB broth.

g. Mark a line on the bottom of six LB/amp plates to divide each plate in half.

h. Sterilely spread 100 µl of undiluted filtrate from the 1-hour sample over one-half of one plate. Spread 100 µl of the equalized 2- and 4-hour filtrates over separate halves of the second plate. Label each half of each plate with the time and the dilution factor.

i. Repeat this spreading procedure for the ten- and hundredfold dilutions.

j. Allow all filtrates to soak into the plates for 10–15 minutes. Then, sterilely streak wild-type (nontransformed) *E. coli* cells on each half of each plate.

k. Incubate the plates at 37°C for 12–24 hours. Compare the growth for each time point across each dilution.

LABORATORY 10

Purification and Identification of Recombinant Plasmid DNA

The growth of *E. coli* colonies on the "L" LB/amp+kan plate in Laboratory 8 confirms that they have been transformed to a dual-resistance phenotype. This resistance is expressed by one or more replicating plasmids, which were assembled in Laboratory 7 by ligating four *Bam*HI/*Hin*dIII restriction fragments of the parental plasmids pAMP and pKAN:

- A 784-bp pAMP fragment
- A 3,755-bp pAMP fragment containing an origin of replication and an *amp*^{r} gene
- A 1,861-bp pKAN fragment containing a *kan*^{r} gene
- A 2,332-bp pKAN fragment containing an origin of replication

The goal of this laboratory is to determine the molecular genotypes of recombinant plasmids responsible for dual resistance. In Part A, "Plasmid Minipreparation of pAMP/pKAN Recombinants," plasmid DNA is isolated from overnight cultures of two different colonies from an "L" LB/amp+kan plate (Laboratory 8) or from replica plates (Laboratory 9). In Part B, "Restriction Analysis of Purified pAMP/pKAN Recombinants," samples of the plasmids isolated in Part A and a control sample of pAMP+pKAN are incubated with *Bam*HI and *Hin*dIII. The three digested samples and samples of uncut minipreps are electrophoresed together in an agarose gel, along with uncut pAMP and a *Hin*dIII digest of λ DNA. The comigration of *Bam*HI/*Hin*dIII fragments in the lanes of miniprep DNA and pAMP/pKAN controls, along with an evaluation of the relative sizes of uncut supercoiled DNAs, gives evidence of the structure, size, and number of plasmids present in each of the transformed clones.

Isolate Plasmid DNA

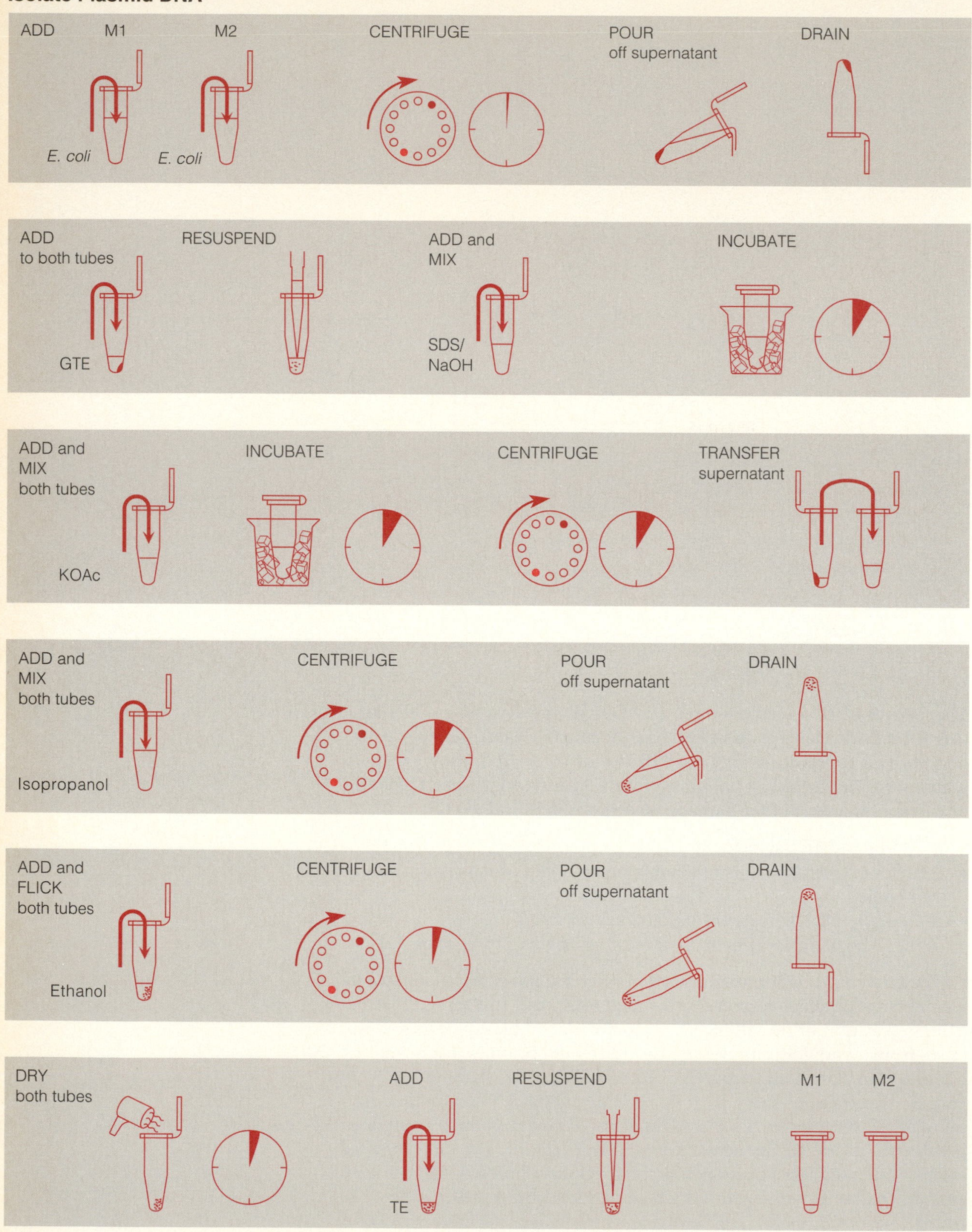

PRELAB NOTES

Review the prelab notes in Laboratory 6A, "Plasmid Minipreparation of pAMP."

Antibiotic Selection

It is safest to maintain antibiotic selection when making overnight cultures of *E. coli,* transformed with recombinant plasmid, although plain LB broth can be used. Single selection with LB/amp is sufficient.

Alternate Sources of Dual-Resistant Colonies

If colonies were not obtained on the "L" LB/amp+kan plate:

1. Use dual-resistant colonies identified by replica plating in Laboratory 9. These can be picked from either the LB/amp or the LB/kan replica plate.
2. If you do not have time to replica plate, prepare overnight cultures from at least five different colonies. The following selection strategies allow only dual-resistant colonies to grow; you should have at least two cloudy cultures in the morning.
   a. Inoculate LB/amp with colonies from the "L" LB/kan plate.
   b. Inoculate LB/kan with colonies from the "L" LB/amp plate.
   c. Inoculate LB/amp+kan with colonies from either the "L" LB/amp plate or the "L" LB/kan plate.

For Further Information

The protocol presented here is based on the following published methods:

Birnboim, H. C. and J. Doly. 1979. A rapid alkaline extraction method for screening recombinant plasmid DNA. *Nucleic Acids Research* 7: 1513.

Ish-Horowicz, D. and J. F. Burke. 1981. Rapid and efficient cosmid cloning. *Nucleic Acids Research* 9: 2989.

PLASMID MINIPREPARATION OF pAMP/pKAN RECOMBINANTS

Cultures and Reagents	Supplies and Equipment
2 *E. coli* pAMP/pKAN overnight cultures	100–1,000-μl micropipet + tips
glucose/Tris/EDTA (GTE)	0.5–10-μl micropipet + tips
SDS/sodium hydroxide (SDS/NaOH)	1.5-ml tubes
potassium acetate/acetic acid (KOAc)	beaker of crushed ice
isopropanol	beaker for waste/used tips
95% ethanol	10%-bleach solution or disinfectant
Tris/EDTA (TE)	paper towels
	hair dryer
	microfuge
	permanent marker
	test tube rack

Isolate Plasmid DNA

(50 minutes)

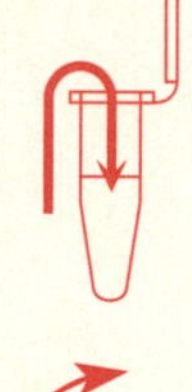

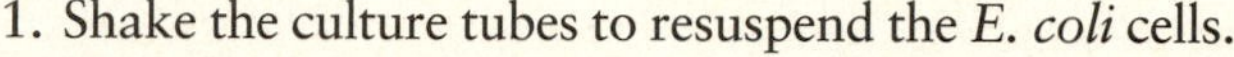

1. Shake the culture tubes to resuspend the *E. coli* cells.

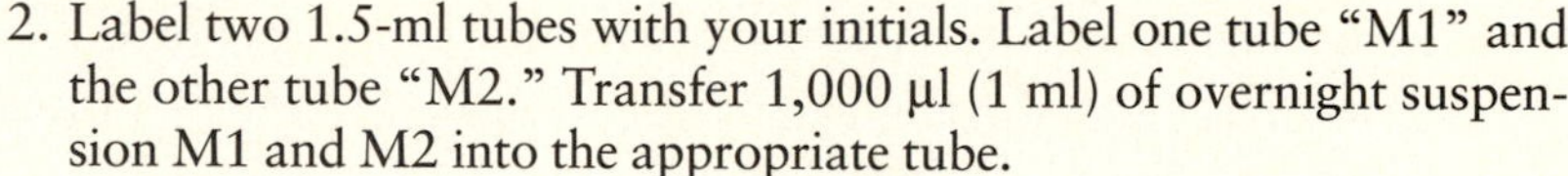

2. Label two 1.5-ml tubes with your initials. Label one tube "M1" and the other tube "M2." Transfer 1,000 μl (1 ml) of overnight suspension M1 and M2 into the appropriate tube.

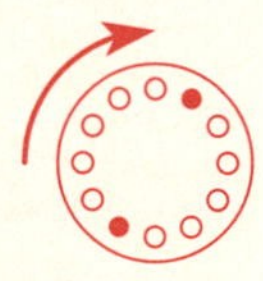

Double all centrifuge times when using a small microfuge that generates 2,000–3,000 × *g*.

3. Close the caps, and place the tubes in a *balanced* configuration in the microfuge rotor. Spin for 1 minute to pellet the cells.
4. Pour off the supernatant from both tubes into a waste beaker for later disinfection. *Take care not to disturb the cell pellets.* Invert the tubes, and touch the mouths of the tubes to a clean paper towel, to wick off as much as possible of the remaining supernatant.

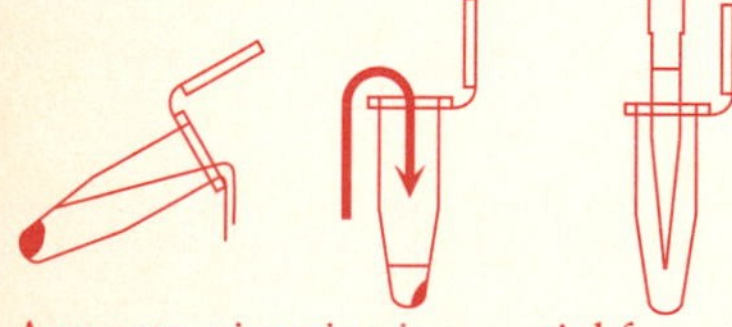

Accurate pipetting is essential for good plasmid yield. The volumes of reagents are precisely calibrated so that the sodium hydroxide added in step 6 will be neutralized by the acetic acid in step 8.

Do not overmix the solutions. Excessive agitation shears the single-stranded DNA and decreases plasmid yield.

5. Add 100 μl of *ice-cold* GTE solution to each tube. Resuspend the pellets by pipetting the solution in and out several times. Hold the tubes up to the light to check that the suspension is homogeneous and that no visible clumps of cells remain.
6. Add 200 μl of SDS/NaOH solution to each tube. Close the caps, and mix the solutions by rapidly inverting the tubes about five times.
7. Stand the tubes on ice for 5 minutes. The suspension will become relatively clear.

8. Add 150 μl of *ice-cold* KOAc solution to each tube. Close the caps, and mix the solutions by rapidly inverting the tubes about five times. A white precipitate will immediately appear.
9. Stand the tubes on ice for 5 minutes.
10. Place the tubes in a *balanced* configuration in a microfuge rotor, and spin for 5 minutes to pellet the precipitate.
11. Transfer 400 μl of supernatant from "M1" into a clean 1.5-ml tube labeled "M1." Transfer 400 μl of supernatant from "M2" into a clean 1.5-ml tube labeled "M2." *Avoid pipetting the precipitate*, and wipe off any precipitate clinging to the outside of the tip before expelling the supernatant. Discard old tubes containing precipitate.

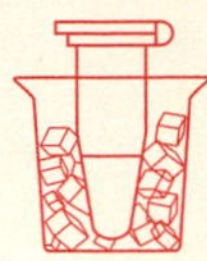

Precipitate typically collects along the side of the tube, rather than as a tight pellet.

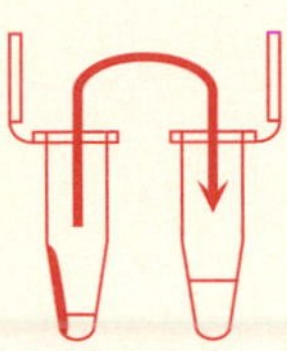

12. Add 400 μl of isopropanol to each tube of supernatant. Close the caps, and mix the solution by rapidly inverting the tubes about five times. *Stand the tubes at room temperature for only 2 minutes.*
13. Place the tubes in a *balanced* configuration in a microfuge rotor, and spin for 5 minutes to pellet the nucleic acids. Align the tubes in the rotor so that the cap hinges point outward. The nucleic acid residue, visible or not, will collect at the bottom of the tube under the hinge during centrifugation.
14. Pour off the supernatant from both tubes. *Take care not to disturb the nucleic acid pellets.* Wick off as much of the remaining alcohol as possible on a paper towel.

Do step 12 quickly, and make sure that the microfuge will be immediately available for step 13. The isopropanol preferentially precipitates nucleic acids rapidly; however, proteins and other cellular components remaining in the solution also begin to precipitate in time.

The cell pellet will likely appear as a tiny, teardrop-shaped smear or small particles on the bottom of each tube. However, pellet size is not a valid predictor of plasmid yield or quality. Large pellets are composed primarily of RNA, salt, and cellular debris carried over from the original precipitate; small pellets, or those impossible to see, often indicate a cleaner preparation.

15. Add 200 μl of 95% ethanol to each tube, and close the caps. Flick the tubes several times to wash the pellets.

Nucleic acid pellets are not soluble in ethanol and will not resuspend during washing.

Stop Point

Store the DNA in ethanol at –20°C until you are ready to continue.

16. Place the tubes in a *balanced* configuration in a microfuge rotor, and spin for 2–3 minutes to recollect the nucleic acid pellets.
17. Pour off the supernatant from both tubes. *Take care not to disturb the nucleic acid pellets.* Wick off as much of the remaining alcohol as possible on a paper towel.
18. Dry the nucleic acid pellets, using one of following methods:
    a. Direct a stream of warm air from a hair dryer across the open ends of the tubes for about 3 minutes. *Take care not to blow the pellets out of the tubes.*

    or

    b. Close the caps, and pulse the tubes in a microfuge to pool the remaining ethanol. *Carefully* use a micropipettor to draw off the

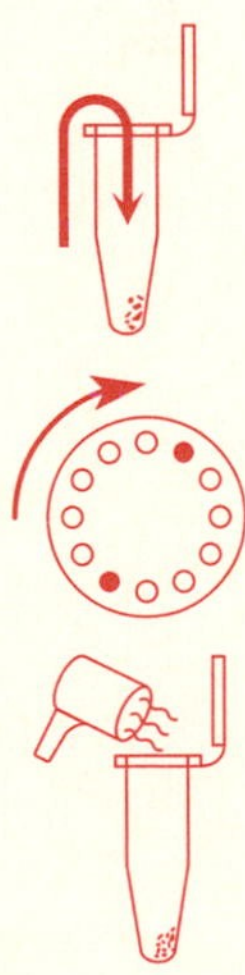

Air bubbles cast in the tube wall can be mistaken for ethanol droplets.

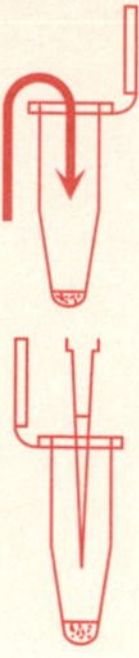

ethanol. Let the pellets air dry at room temperature for 10 minutes.

19. Be sure that the nucleic acid pellet is dry and that all ethanol has evaporated before proceeding to step 20. Hold each tube up to the light and confirm that no ethanol droplets remain and that the nucleic acid pellet, if visible, appears white and flaky. If the ethanol is still evaporating, an alcohol odor can be detected by sniffing the mouth of the tube.
20. Add 15 µl of TE to each tube. Resuspend the pellets by scraping the tubes with the tip of the pipet and pipetting in and out vigorously. Rinse down the side of the tube several times, concentrating on the area where the pellet should have formed during centrifugation (beneath the cap hinge). Make sure that all DNA is dissolved and that no particles remain in the tip or on the side of the tube.

If you are using a 0.5–10-µl micropipettor, set it to 7.5 µl, and pipet twice.

21. Keep the two DNA/TE solutions *separate*. Do *not* pool them into one tube!

Stop Point

Freeze the DNA/TE solutions at –20°C until you are ready to continue. Thaw before using.

22. Take the time for proper cleanup:
    a. Segregate for proper disposal the culture tubes and micropipettor tips that have come into contact with *E. coli*.
    b. Disinfect the overnight cultures, the tips, and the supernatant from step 4 with a 10%-bleach solution or disinfectant (such as Lysol).
    c. Wipe down the lab bench with soapy water, a 10%-bleach solution, or disinfectant.
    d. Wash your hands before leaving the laboratory.

RESULTS AND DISCUSSION

See the "Results and Discussion" section of Laboratory 6A, "Plasmid Minipreparation of pAMP," for a detailed discussion of the biochemistry of the alkaline lysis method for plasmid purification.

I. Set Up Restriction Digests

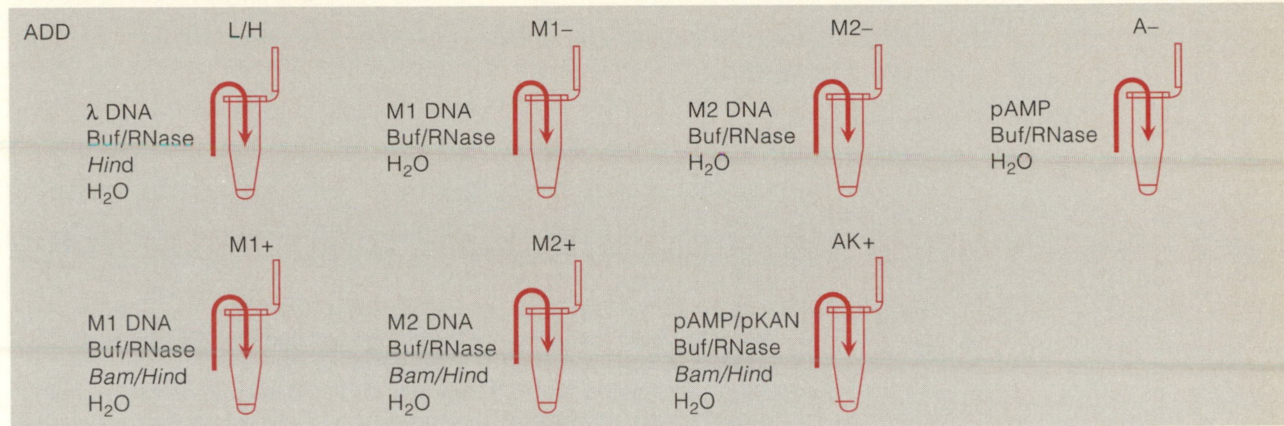

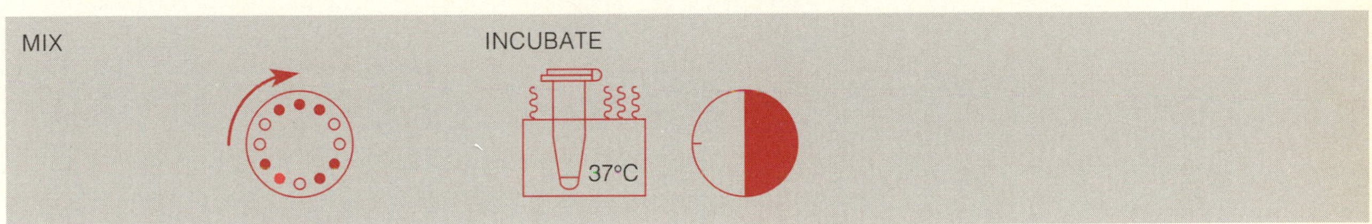

II. Cast 0.8% Agarose Gel

III. Load Gel and Electrophorese

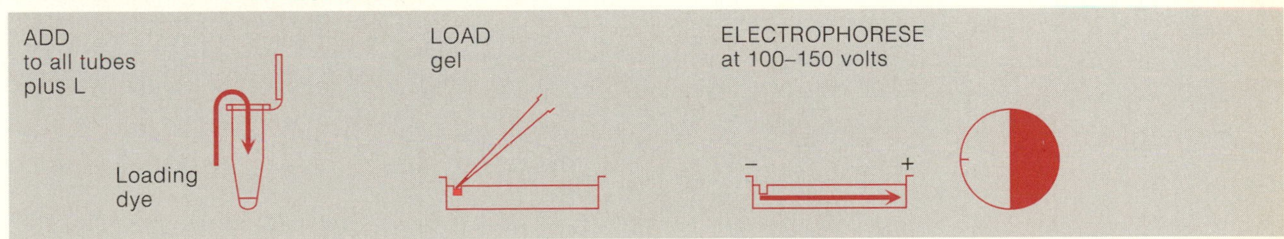

IV. Stain Gel and View

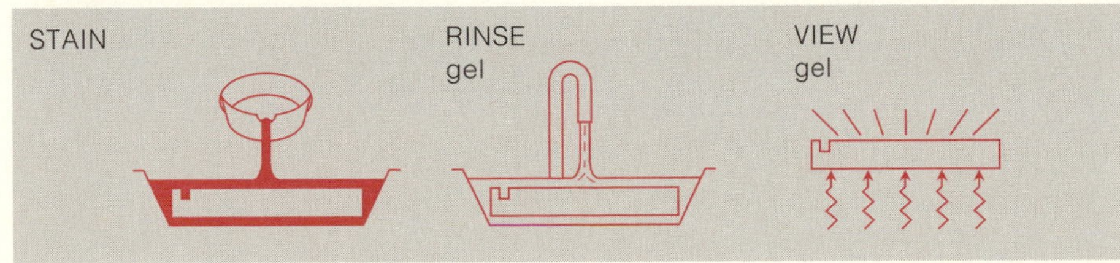

V. Photograph Gel

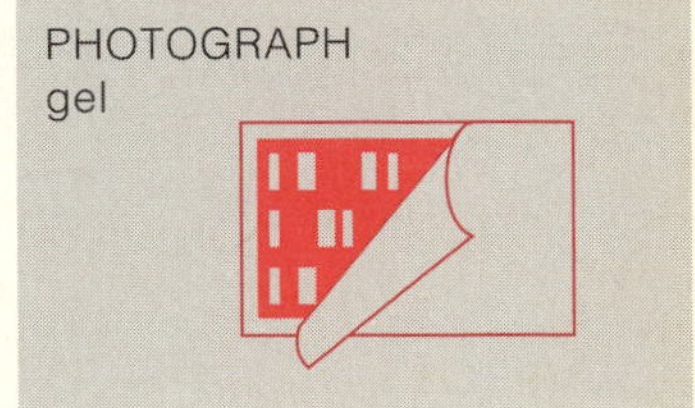

PRELAB NOTES

Review the prelab notes in Laboratory 6B, "Restriction Analysis of Purified pAMP."

For Further Information

The protocol presented here is based on the following published methods:

Cohen, S. N., A. C. Y. Chang, and H. W. Boyer. 1973. Construction of biologically functional bacteria plasmids in vitro. *Proceedings of the National Academy of Sciences USA* 70: 3240.

Helling, R. B., H. M. Goodman, and H. W. Boyer. 1974. Analysis of *Eco*RI fragments of DNA from lambdoid bacteriophages and other viruses by agarose-gel electrophoresis. *Journal of Virology* 14: 1235.

Sharp, P. A., B. Sugden, and J. Sambrook. 1973. Detection of two restriction endonuclease activities in *Haemophilus parainfluenzae* using analytical agarose-ethidium bromide electrophoresis. *Biochemistry* 12: 3055.

RESTRICTION ANALYSIS OF PURIFIED pAMP/pKAN RECOMBINANTS

Reagents

miniprep DNA/TE (M1, M2)
0.1 μg/μl pAMP
0.1 μg/μl pAMP/pKAN
0.1 μg/μl λ DNA
*Bam*HI/*Hin*dIII
5× restriction buffer/RNase
distilled water
loading dye
0.8% agarose
1× Tris/Borate/EDTA (TBE) buffer
1 μg/ml ethidium bromide (or 0.025% methylene blue)

for decontamination:
0.05 M $KMnO_4$
0.25 N HCl
0.25 N NaOH

Supplies and Equipment

0.5–10-μl micropipettor + tips
1.5-ml tubes
aluminum foil
beaker for agarose
beaker for waste/used tips
10%-bleach solution
camera (optional)
disposable gloves
electrophoresis box
microfuge (optional)
masking tape
Parafilm or waxed paper (optional)
permanent marker
plastic wrap (optional)
power supply
test tube rack
transilluminator (optional)
37°C water bath
60°C water bath (for agarose)

I. Set Up Restriction Digests

(10 minutes; then 30–40 minutes of incubation)

Refer to Laboratory 3, "DNA Restriction and Electrophoresis," for detailed instructions on setting up digests.

1. Use a permanent marker to label seven 1.5-ml tubes, as shown in the matrix below. Restriction reactions will be performed in these tubes.

Tube	λ DNA	M1	M2	pAMP	pKAN	Buffer/ RNase	*Bam*HI/ *Hin*dIII	*Hin*dIII	H_20
L/H	5μl	—	—	—	—	2 μl	—	1μl	2μl
M1–	—	5 μl	—	—	—	2 μl	—	—	3 μl
M2–	—	—	5 μl	—	—	2 μl	—	—	3 μl
A–	—	—	—	5 μl	—	2 μl	—	—	3 μl
M1+	—	5 μl	—	—	—	2 μl	2 μl	—	1 μl
M2+	—	—	5 μl	—	—	2 μl	2 μl	—	1 μl
AK+	—	—	—	3 μl	3 μl	2 μl	2 μl	—	—

2. Use the matrix as a checklist while adding reagents to each reaction. Read down each column, adding the same reagent to all appropriate tubes. *Use a fresh pipet tip for each reagent.*

Return unused miniprep DNA (M1 and M2) to the freezer at –20°C for possible use in further experiments, as suggested at the end of this laboratory.

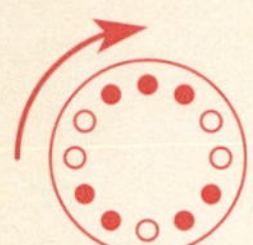
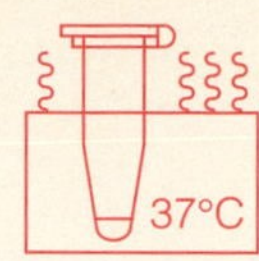

Do not overincubate. During a longer incubation, the DNases in the miniprep can degrade the plasmid DNA.

3. After adding all of the above reagents, close the tops of the tubes. Pool and mix the reagents by pulsing them in a microfuge or by sharply tapping the bottom of the tube on the lab bench.
4. Place the reaction tubes in a 37°C water bath, and incubate them for 30–40 minutes only.

Stop Point

Following incubation, freeze the reactions at –20°C until you are ready to continue. Thaw the reactions before continuing on to Section III, step 1.

II. Cast 0.8% Agarose Gel

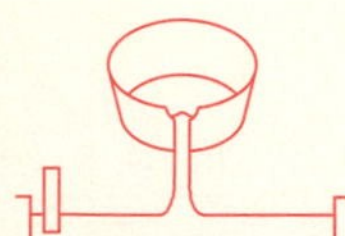

(15 minutes)

1. Carefully pour agarose solution into a casting tray to fill it to the depth of about 5 mm. The gel should cover only about one-third of the height of the comb teeth.
2. When the agarose has set, place the casting tray into the electrophoresis box, and set up for electrophoresis.

TBE buffer can be used several times; do not discard it. If you are using buffer remaining in the electrophoresis box from a previous experiment, rock the chamber back and forth to remix the ions that have accumulated at either end.

Stop Point

Cover the electrophoresis box and save the gel until you are ready to continue. The gel will remain in good condition for several days if it is completely submerged in buffer.

III. Load the Gel and Electrophorese

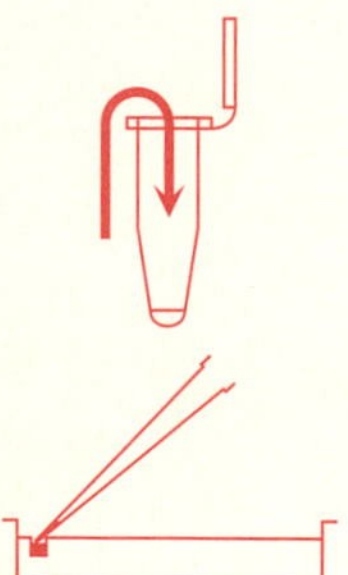

(10 minutes; then 20–40 minutes of electrophoresis)

1. Remove the restriction digests from the 37°C water bath.
2. Add 1 µl of loading dye to each reaction tube. Close the tube tops, and mix by tapping the tube on the lab bench or pulsing it in a microfuge.
3. Load 10 µl from each reaction tube into a separate well in the gel, as shown in the diagram below. Use a *fresh tip* for each reaction.

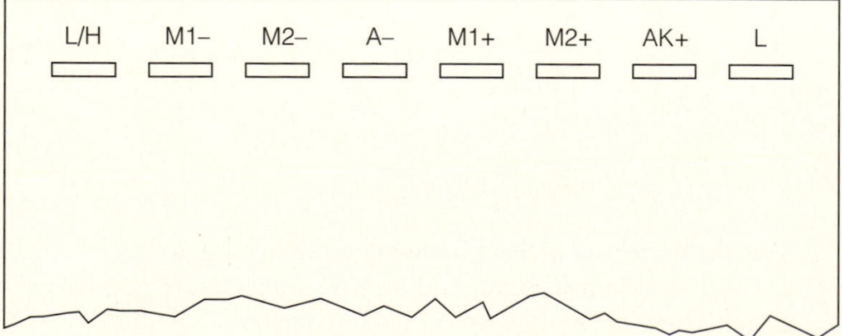

4. Add loading dye to the ligated DNA saved from Laboratory 7, "Recombination of Antibiotic-Resistance Genes." Load the entire contents of the "L" tube (5–10 µl) into well 8.
5. Electrophorese at 100–150 volts for 20–40 minutes. Good separation has occurred when the bromophenol blue band has moved 40–60 mm from the wells.
6. Turn off the power supply, remove the casting tray from the electrophoresis box, and transfer the gel to a disposable weigh boat or other shallow tray.

If time allows, electrophorese until the bromophenol blue markers near the end of the gel. This allows maximum separation of uncut DNA, which is important for differentiating a large "superplasmid" from a double transformation of two smaller plasmids. However, do not electrophorese bromophenol blue markers off the end of the gel; you can also lose the 784-bp *Bam*HI/*Hin*dIII fragment of pAMP that migrates just behind it.

Stop Point

The gel can be stored overnight, covered with a small volume of buffer in a staining tray or a zip-lock plastic bag, for viewing or photographing the next day. However, over a longer period of time, the DNA will diffuse through the gel, and the bands will either become indistinct or disappear entirely.

7. Stain and view the gel, using one of the methods described in Sections IVA and IVB.

IVA. Stain the Gel with Ethidium Bromide and View

(10–15 minutes)

CAUTION

Review "Ethidium Bromide Staining and Responsible Handling" in Laboratory 3. Wear disposable gloves when staining, viewing, and photographing gel, and during cleanup. Confine all staining to a restricted sink area.

Refer to Laboratory 3, "DNA Restriction and Electrophoresis," for detailed instructions on staining and photographing gel.

1. Flood the gel with ethidium bromide solution (1 µg/ml), and allow it to stain for 5–10 minutes. Destain, if desired.
2. View the gel under an ultraviolet transilluminator or other UV source.

The band intensity and contrast increase dramatically if the gel is destained for 15–30 minutes in tap water. More simply, rinse and drain the gels, stack the staining trays, cover the top gel, and let it set overnight at room temperature.

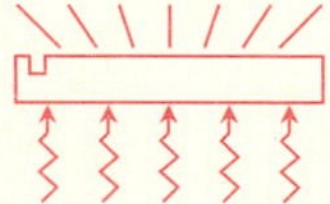

CAUTION

Ultraviolet light can damage your eyes. Never look directly at an unshielded UV light source without eye protection. View only through a filter or safety glasses that absorb the harmful wavelengths.

IVB. Stain the Gel with Methylene Blue and View

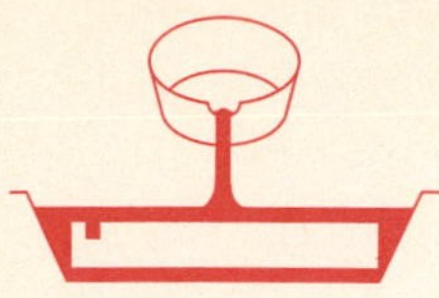

Destaining time is decreased by agitating and rinsing the gel in warm water. For best results, continue to destain overnight in a small volume of water. Cover the staining tray to retard evaporation.

(30+ minutes)

1. Flood the gel with 0.025% methylene blue, and allow it to stain for 20–30 minutes.
2. Rinse the gel in running tap water. Let the gel soak for several minutes in several changes of fresh water. The DNA bands will become increasingly distinct as the gel destains.
3. View the gel over a light box; cover the surface with plastic wrap to prevent staining.

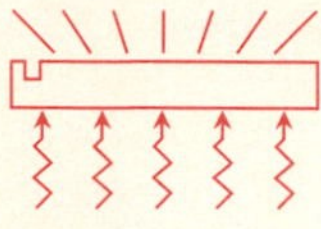

V. Photograph the Gel

(5 minutes)

1. *For ultraviolet (UV) photography of ethidium-bromide-stained gels:* Set the camera aperature to f/8 and the shutter speed to 1. Depress the shutter once or twice for a time exposure of 1–2 seconds.

   For white-light photography of methylene-blue-stained gels: Set the camera aperature to f/8 and the shutter speed to ⅟₁₂₅th of a second.
2. Make sure that the photograph shows the sample wells: Their positions are needed for measuring migration distances.

CAUTION

Avoid getting caustic developing jelly on your skin or clothing. If jelly does get on your skin, wash immediately with plenty of soap and water.

3. Take the time for proper cleanup:
   a. Wipe down the camera, the light source, and the staining area.
   b. If you are using ethidium bromide, decontaminate the gel and any staining solution that will not be reused.
   c. Wash your hands before leaving the laboratory.

RESULTS AND DISCUSSION

Refer to the "Results and Discussion" section of Laboratory 6B, "Restriction Analysis of Purified pAMP," for more details about interpreting miniprep gels and plasmid conformations.

Examine the photograph of your stained gel, and determine which lanes contain control pAMP/pKAN and which contain minipreps M1 and M2. Even if you have not followed the prescribed loading order, the miniprep lanes can be distinguished by the following characteristics:

- A background "smear" of degraded and partially digested chromosomal DNA, plasmid DNA, and RNA
- Undissolved material and high-molecular-weight DNA "trapped" at the front edge of the well

- A "cloud" of low-molecular-weight RNA at a position corresponding to 100–200 bp
- The presence of high-molecular-weight bands of uncut plasmid in lanes of digested miniprep DNA

Remember these three facts when considering possible constructions of the ligated plasmids in M1 and M2:

- Every replicating plasmid must have an origin of replication. Recombinant plasmids with more than one origin also replicate normally; however, only one origin is active.
- Each adjacent restriction fragment can only ligate at a like restriction site: *Bam*HI to *Bam*HI and *Hin*dIII to *Hin*dIII. Thus, intact plasmid must be constructed from an *even* number of fragments.
- Repeated copies of a restriction fragment cannot exist adjacent to one another; that is, they must alternate with other fragments. Adjacent duplicate fragments form "inverted repeats"; the sequences, one on either side of the restriction site, are complementary along the entire length of the duplicated fragment. Molecules with such inverted repeats cannot replicate properly. As the plasmid opens up to allow access to DNA polymerase, the single-strand regions on either side of the restriction site base pair to one another to form a large "hairpin loop," which fouls replication.

Follow questions 1–6 to interpret each pair of miniprep results (M1+/– and M2+/–).

1. Examine the photograph of your stained gel. Compare your gel with the ideal gels shown below and on the following page. Label the size of the fragments in each lane of your gel.

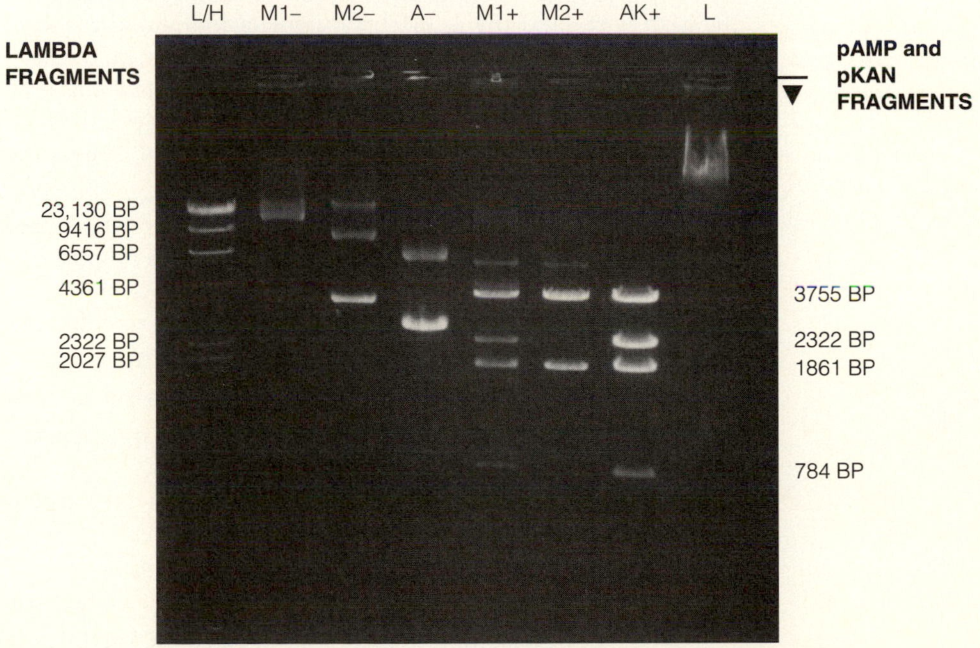

Restriction Analysis of Two pAMP/pKAN Recombinants (M1,M2)

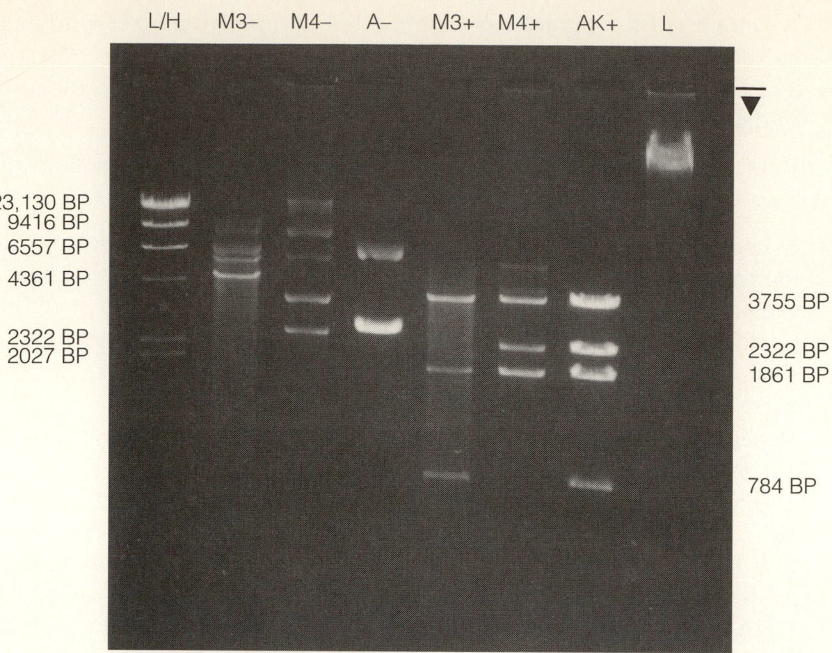

Restriction Analysis of Two pAMP/pKAN Recombinants (M3, M4)

2. Label the fragment sizes of the four bands in the AK+ lane (cut control pAMP and pKAN) from the bottom of the gel to the top: 784 bp, 1,861 bp, 2,332 bp, and 3,755 bp.

3. *The cut miniprep lanes (M+) give information about the number and types of restriction fragments in the construct.*

   a. Every miniprep must include the 3,755-bp fragment containing the *amp*^{r} gene and the 1,861-bp fragment containing the *kan*^{r} gene. Locate these bands by comparing the M+ lane with the AK+ lane (cut control). Bear in mind that the 3,755-bp band runs only slightly ahead of uncut plasmid and may therefore appear as the leading edge of a smear of high-molecular-weight DNA near the top of the gel.

   b. Now, look for evidence of other bands in the M+ lane. Compare the M+ lane with the AK+ lane. Either the 2,332-bp fragment or the 784-bp fragment may be present. If neither of these two additional bands is present, the molecule is the simple recombinant, pAK. (This is M2 of the ideal gel on the preceding page.)

   c. If a third band of 784-bp is present, the molecule may be:

      - a "superplasmid," in which one of the three fragments is repeated. (This is M3 of the ideal gel above.)
      - a double transformation of a simple recombinant *and* a religated pAMP.

   d. If a third band of 2,332-bp is present, the molecule may be:

      - a superplasmid in which one of the three fragments is repeated.
      - a double transformation of pAK *and* religated pKAN. (This is M4 of the ideal gel above.)
      - a double transformation of pAK *and* ligated 3,755-bp + 2,332-bp fragments.
      - a double transformation of religated pKAN *and* ligated 3,755-bp + 2,332-bp fragments.

e. If all four bands are present, the molecule may be:
   - a superplasmid containing all four fragments. (This is M1 at the ideal gel on page 161.)
   - a double transformation of religated pAMP *and* religated pKAN.
   - a double transformation of a simple recombinant *and* ligated 2,332-bp + 784-bp fragments.

4. *The uncut miniprep lanes (M–) give information about the overall size of the construct.* Compare the M– lane (uncut miniprep) with the A– lane (uncut pAMP) and L/H lane (λ markers). Remember that uncut plasmid can assume several conformations, but the fastest moving form is supercoiled.

   a. Locate the band that has migrated furthest in the A– lane; this is the supercoiled form of pAMP.

   b. Now examine the band(s) furthest down the M– lane. If this and the pAMP band have comigrated similar distances, your miniprep is most likely a double transformation. The possible molecules present in a double transformation range in size from 3,116 bp to 6,087 bp and can thus appear noticeably lower or higher on the gel than supercoiled pAMP.

   c. If the fastest moving band of the uncut miniprep is very high on the gel, your molecule is most likely a superplasmid. Compute the possible sizes of superplasmids composed of three or four fragments.

5. When bacteria are transformed with two different plasmids having related origins of replication, one of the two plasmids can be preferentially replicated within the host cell. In this case, over generations, one of the two plasmids is eventually lost. Thus, in double transformations with four different fragments, one pair of fragments may be fainter than the other pair.

6. Based on your above evaluation, make scale restriction maps of your M1 and M2 plasmids.

7. If your ligation did not proceed to completion, how might this affect the percentage of doubly resistant clones due to transformation by the simple (2-band) recombinant plasmid?

8. *Bam*HI is more heat stable than *Hin*dIII. If the heat inactivation step fails to completely abolish *Bam*HI activity, how might this affect the genotype distribution of recombinant plasmids?

FOR FURTHER RESEARCH

Further research may reveal with certainty the structure of perplexing recombinant plasmids. To obtain additional plasmid DNA for further experimentation, do double minipreps from your master colonies (M1 and M2) in Laboratory 8 or from colonies of interest from replica plates in experiment 1 below.

1. Perform a series of experiments to distinguish between a superplasmid and a double transformation in questions 3c, d, and e in the "Results and Discussion" section.

a. Make a 1:10 dilution of your miniprep DNA.

b. Use the dilute miniprep DNA to transform competent *E. coli* cells, and plate them onto LB/amp and LB/kan.

c. Replica plate colonies from each master plate onto fresh LB/amp and LB/kan plates.

d. Examine the proportion of dual-resistant colonies.

- If all the restriction fragments are contained in a single superplasmid, all transformants will have dual resistance.
- If three or four restriction fragments are distributed among separate plasmids, the transformants will have mixed antibiotic resistance. Matching the observed pattern of antibiotic resistance with alternate two-gene recombinants will often reveal the structure of the two plasmids involved.

2. Digesting miniprep DNA with the restriction enzyme *Xho*I can elucidate some of the structures of superplasmids and plasmids in double transformations. This enzyme has a single recognition site within the 1,861-bp pKAN fragment and *no sites* within any of the other three *Bam*HI/*Hin*dIII fragments. Electrophorese *Xho*I digests of miniprep DNA with samples of uncut pAMP and λ/*Hin*dIII size markers.

a. Results of a *Xho*I digest can distinguish between superplasmids and a double transformation in minipreps showing three fragments, including the 784-bp fragment (question 3c in the "Results and Discussion" section):

- A single 7,184-bp fragment, produced by linearizing a superplasmid with a repeated 784-bp fragment
- A single 10,155-bp fragment, produced by linearizing a superplasmid with a repeated 3,755-bp fragment
- Two fragments—2,645 bp and 5,616 bp—produced by cutting a superplasmid with a repeated 1,861-bp fragment containing the *Xho*I site
- A 5,616-bp fragment *plus* evidence of uncut pAMP plasmid, produced by a double transformation in which pAK is linearized and religated pAMP is left uncut.

b. Results of a *Xho*I digest can distinguish between superplasmids and double transformations in minipreps showing three fragments, including the 2,332-bp fragment (question 3d in the "Results and Discussion" section):

- A single 10,280-bp fragment, produced by linearizing superplasmid with a repeated 2,332-bp fragment
- A single 11,703-bp fragment, produced by linearizing superplasmid with a repeated 3,755-bp fragment
- Two fragments—4,193 bp and 5,616 bp—produced by cutting superplasmid with a repeated 1,861-bp fragment containing the *Xho*I site
- A 5,616-bp fragment *plus* evidence of an uncut 6,087-bp plasmid, produced by a double transformation in which pAK is linearized and a 3,755-bp + 2,332-bp construct is left uncut
- A 4,193-bp fragment *plus* evidence of an uncut 6,087-bp plasmid, produced by a double transformation in which religated pKAN is linearized and a 3,755-bp + 2,332-bp construct is left uncut

c. Results of a *Xho*I digest can distinguish between a superplasmid and double transformations in minipreps showing all four fragments (question 3e in the "Results and Discussion" section):

- A single 8,732-bp fragment, produced by linearizing a four-piece superplasmid
- A 4,193-bp fragment *plus* evidence of uncut pAMP plasmid, produced by a double transformation in which religated pKAN is linearized and religated pAMP is left uncut
- A 5,616-bp fragment *plus* evidence of uncut 3,106-bp plasmid, produced by a double transformation in which pAK is linearized and a 2,332-bp + 784-bp construct is left uncut

3. Make a restriction map of the simple recombinant plasmid, pAK, using *Bam*HI, *Hin*dIII, and *Pvu*I.

   a. Do a double miniprep to obtain additional plasmid from a master colony known to contain pAK.

   b. Digest aliquots of the miniprep DNA with:

      *Pvu*I

      *Pvu*I + *Bam*HI

      *Pvu*I + *Hin*dIII

      *Bam*HI + *Hin*dIII

      *Bam*HI + *Hin*dIII+*Pvu*I

   c. Electrophorese the digested samples on a 1.2% agarose gel; then stain and photograph them.

   d. The expected number and sizes of the fragments are shown in the diagram below:

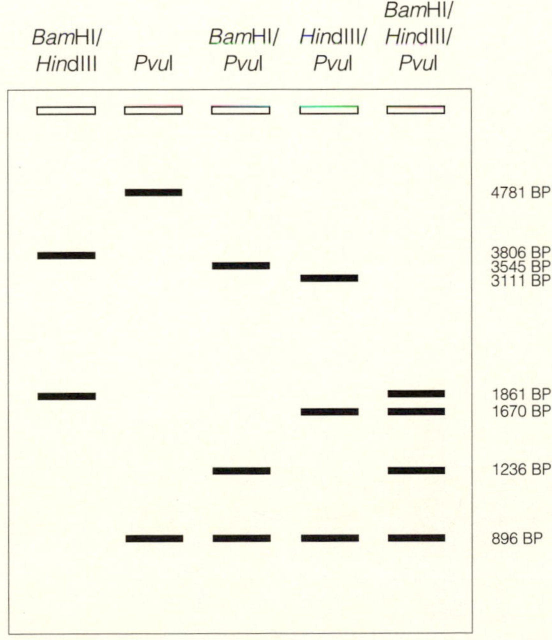

4. Using the data from experiment 5 above and by applying a little logic, the relative positions of the restriction sites can be positioned around a circle to produce a restriction map of the simple recombinant plasmid.

a. The *Bam*HI/*Hin*dIII digest reveals that the *Bam*HI and *Hin*dIII sites are separated by 1,861 bp.

b. The *Pvu*I digest reveals that the two *Pvu*I sites are separated by 896 bp.

c. The *Bam*HI/*Hin*dIII/*Pvu*I digest shows both 1,861-bp and 896-bp fragments. This means that the 1,615-bp and 1,244-bp fragments separate the 896-bp *Pvu*I fragment from the 1,861-bp *Bam*HI/*Hin*dIII fragment.

d. The *Pvu*I/*Bam*HI digest shows a 3,476-bp fragment that must be composed of the 1,861-bp fragment plus the 1,615-bp fragment.

e. The *Pvu*I/*Hin*dIII digest shows a 3,105-bp fragment that must be composed of the 1,861-bp fragment plus the 1,244-bp fragment.

f. Results from d and e indicate that the 1,244-bp fragment is adjacent to the *Bam*HI site and the 1,615-bp fragment is adjacent to the *Hin*dIII site.

g. Complete the restriction map showing all restriction sites and the distances between them.

Restriction Mapping of Linear and Circular Chromosomes

LABORATORY 11

Restriction Mapping of the λ Chromosome

Generally, the first step in exploring a cloned DNA fragment is to construct its restriction map. By digesting the DNA with various restriction enzymes, alone and in combination, we can determine the number and relative positions of cutting sites along the DNA for each restriction enzyme. By identifying overlaps between the pattern of restriction sites of mapped fragments, we build up large-scale restriction maps of chromosome regions. The resulting map can be used to determine the smallest restriction fragment containing an intact gene or to provide restriction fragments suitable for subcloning in preparation for DNA sequencing. With DNA sequence data, a complete restriction map showing exact cut sites can be produced.

This laboratory recalls experiments performed in the early 1970s, when λ was among the first viral chromosomes to be mapped using restriction enzymes and agarose gel electrophoresis. The object of this laboratory is to determine the relative cutting positions of the restriction enzymes *Apa*I, *Eco*O109I, and *Sna*BI on the linear λ chromosome. Samples of λ DNA are incubated at 37°C in single and double digests with these endonucleases. The digested samples are electrophoresed in an agarose gel, stained, and photographed.

A separate digest of λ DNA cut with *Hin*dIII generates a series of fragments of known sizes; these fragments serve as markers to help gauge the size of λ fragments from the *Apa*I, *Eco*O109I, and *Sna*BI digests. Data from the single digests indicate how far from the end of the genome each enzyme cuts; data from the double digests indicate whether the enzymes cut at the same or at opposite ends of the λ chromosome. *Apa*I and *Sna*BI each cut the λ chromosome at a single position. *Eco*O109I cuts at three sites; however, two of the cuts produce fragments that are either too small or too large to be resolved by gel electrophoresis, thus leaving only one *Eco*O109I site to be mapped.

I. Set Up Restriction Digests

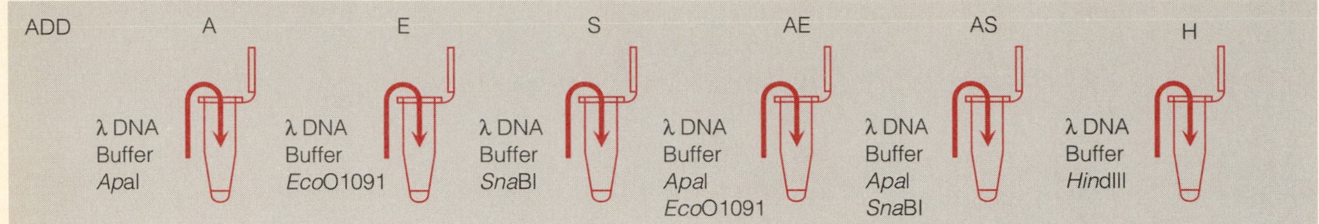

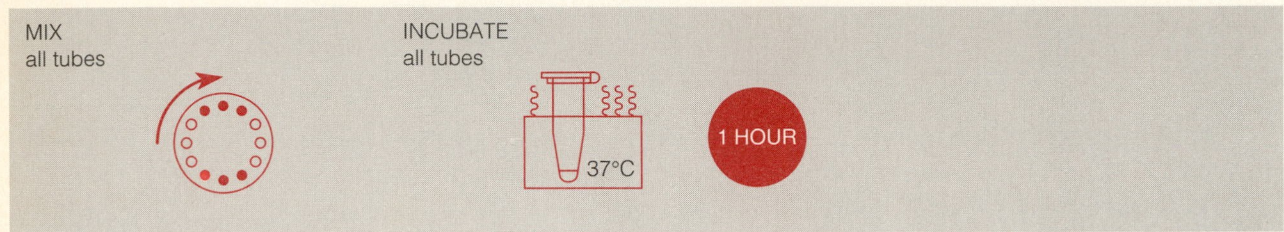

II. Cast 0.8% Agarose Gel

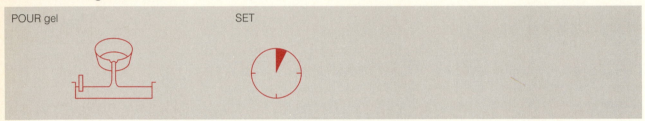

III. Disrupt COS Hydrogen Bonding, Load Gel, and Electrophorese

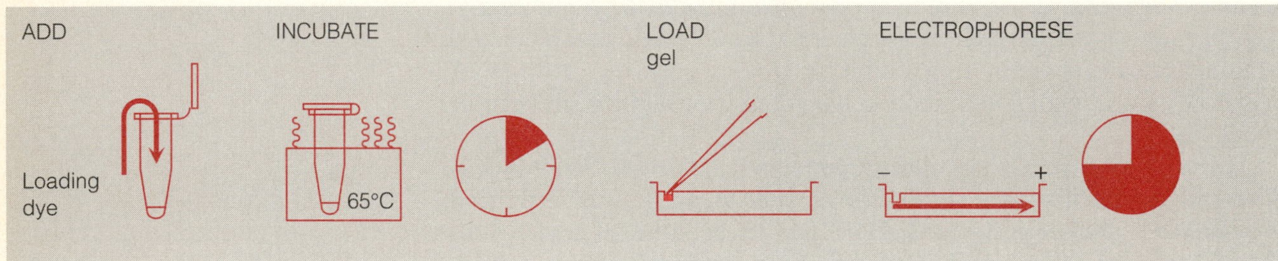

IV. Stain Gel, View, and Photograph

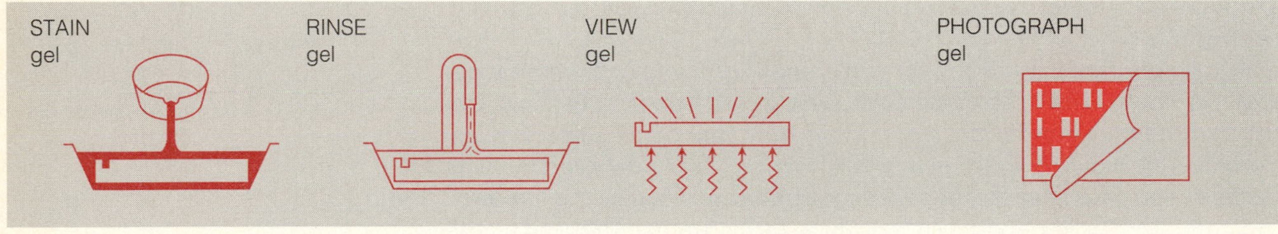

PRELAB NOTES

Review the prelab notes in Laboratory 3, "DNA Restriction and Electrophoresis."

λ DNA

The λ chromosome can exist as either a circular or linear molecule. At each end of the linear λ molecule is a single-stranded sequence of 12 nucleotides, called a COS site (for cohesive). The COS sites at each end are complementary to one another and can thus base pair to produce the circular form. Some circlets are linked only by hydrogen bonds, and others are covalently linked by the addition of phosphodiester bonds between 5′ and 3′ carbons at each end of the COS site.

Commercially available λ DNA is a mixture of linear and circular molecules. Restriction enzymes with a single recognition site cut linear λ DNA to produce two identifiable fragments but cut the circular molecule into a single linear fragment. Because the circular molecules do not provide useful map information in this experiment, it is important to maximize the amount of linear λ molecules. Circular λ molecules, whose COS sites are held together by hydrogen bonds, can be converted to the linear form by heating them at 65°C for 10 minutes, which disrupts the hydrogen bonds. This treatment will not affect circular molecules that are covalently closed.

Single digests with each restriction enzyme generate one small and one large restriction fragment from linear λ molecules. The larger fragments do not resolve well; their sizes can be estimated by subtracting the sizes of the smaller fragments from the size of the λ genome (48,502 base pairs). Although the smaller fragments provide key mapping information, their yield is reduced by the percentage of λ molecules that remain in the COS circlet conformation. The covalently linked circlets are unaffected by heating at 65°C, and some noncovalently bonded circlets reform hydrogen bonds immediately after heating. To compensate for this effect, a large mass of DNA (0.8 μg) is loaded in each gel lane.

Double the DNA mass to 1.6 μg per lane when you are staining with methylene blue.

10x Restriction Buffer

This experiment uses New England Biolabs (NEB) buffer #4, which optimizes the activity of the three enzymes to be mapped.

Semilog Graph Paper

Restriction fragments used in this analysis range in size from about 1,000 to 20,000, bp., so paper with two log cycles will best spread the data points.

For Further Information

The protocol presented here is based on the following published methods:

Danna, K. and D. Nathans. 1971. Specific cleavage of simian virus 40 DNA by restriction endonuclease of *Hemophilus influenzae*. *Proceedings of the National Academy of Sciences USA* 68: 2913.

Sharp, P. A., B. Sugden, and J. Sambrook. 1973. Detection of two restriction endonuclease activities in *Haemophilus parainfluenzae* using analytical agarose-ethidium bromide electrophoresis. *Biochemistry* 12: 3055.

Helling, R. B., H. M. Goodman, and H. W. Boyer. 1974. Analysis of *Eco*RI fragments of DNA from lambdoid bacteriophages and other viruses by agarose-gel electrophoresis. *Journal of Virology* 14: 1235.

RESTRICTION MAPPING OF THE λ CHROMOSOME

Reagents	**Equipment and Supplies**
0.1 µg/µl / DNA	0.5–10-µl micropipettor + tips
restriction enzymes:	10–100-µl micropipettor + tips
*Apa*I	1.5-ml tubes
*Eco*O109I	aluminum foil
*Hind*III	beaker for agarose
*Sna*BI	beaker for waste/used tips
	disposable gloves
10× NEB #4 restriction buffer	camera and film (optional)
distilled water	electrophoresis box
loading dye	masking tape
0.8% agarose	metric ruler
1× Tris/Borate/EDTA (TBE buffer	microfuge (optional)
1 µg/ml ethidium bromide (or 0.025% methylene blue)	permanent marker
	power supply
	semilog graph paper
for decontamination:	test tube rack
0.05 M $KMnO_4$	transilluminator
0.25 N HCl	37°C water bath
0.25 N NaOH	65°C water bath

I. Set Up Restriction Digests

(70 minutes, including incubation)

Refer to Laboratory 3, "DNA Restriction and Electrophoresis," for detailed instructions on setting up digests.

1. Use a permanent marker to label six 1.5-ml tubes as in the matrix below. Restriction reactions will be performed in these test tubes.
2. Use the matrix below as a checklist while adding reagents to each reaction. Read down each column, adding the same reagent to all appropriate tubes. Refer to the detailed instructions that follow.

Tube	λ DNA	10x Buffer	*Apa*I	*Eco*O109I	*Sna*BI	*Hind*III	H_2O
A	8 µl	2 µl	1.5 µl	—	—	—	8.5 µl
E	8 µl	2 µl	—	1.5 µl	—	—	8.5 µl
S	8 µl	2 µl	—	—	1.5 µl	—	8.5 µl
AE	8 µl	2 µl	1.5 µl	1.5 µl	—	—	7.0 µl
AS	8 µl	2 µl	1.5 µl	—	1.5 µl	—	7.0 µl
H	8 µl	2 µl	—	—	—	1.5 µl	8.5 µl

3. Add 8 µl of DNA to each reaction tube.
4. Use a *fresh tip* to add 2 µl of 10× buffer to a clean spot on each reaction tube.
5. Use a *fresh tip* to add an appropriate volume of distilled water to each reaction tube.
6. Use *fresh tips* to add restriction enzymes to the appropriate reaction tubes.

It is important to change pipet tips after each enzyme addition, because even a small amount of carryover into an inappropriate tube will produce unexpected restriction fragments.

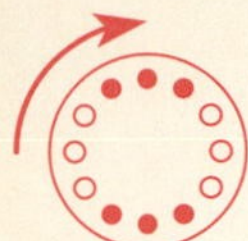

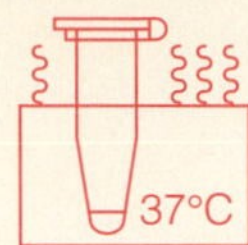

7. After adding all of the above reagents, close the tops of the tubes and mix.
8. Place the reactions in a 37°C water bath, and incubate for a minimum of 60 minutes.

Stop Point

Following incubation, freeze the reactions at –20°C until you are ready to continue. Thaw the reactions before continuing on to Section III, step 1.

II. Cast 0.8% Agarose Gel

Refer to Laboratory 3, "DNA Restriction and Electrophoresis," for detailed instructions on casting and loading gel.

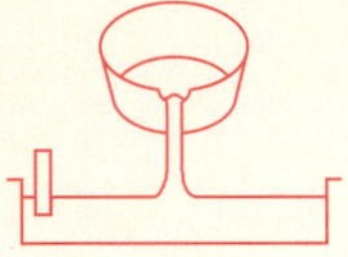

(15 minutes)

1. Carefully pour agarose solution into the casting tray to fill it to a depth of about 6 mm. The gel should cover about half of the height of the comb teeth.
2. After agarose solidifies, place the casting tray into the elctrophoresis box and set up for electrophoresis.

Stop Point

Cover the electrophoresis box, and save the gel until you are ready to continue. The gel will remain in good condition for several days if it is completely submerged in buffer.

III. Disrupt COS Hydrogen Bonding, Load Gel, and Electrophorese

Heating disrupts hydrogen bonding at COS sites, converting circular / molecules to the linear form. Heating does not affect covalently linked circles.

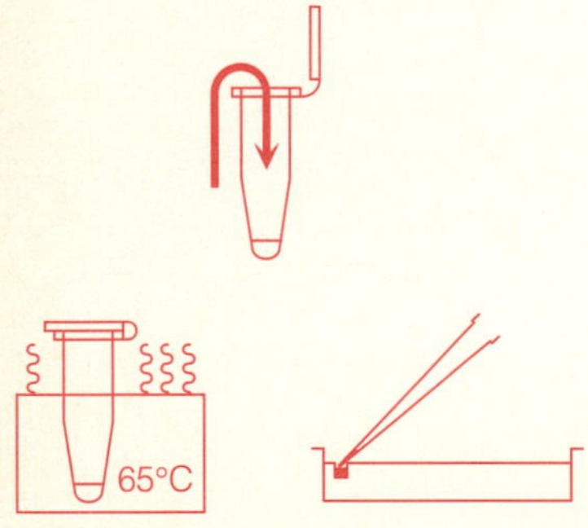

(50–70 minutes)

1. Remove the restriction digests from the 37°C water bath.
2. Add 2 µl of loading dye to each reaction tube. Close the tops of the tubes and mix.
3. Incubate the reaction tubes in a 65°C water bath for 10 minutes.
4. Load 20 µl from reaction tubes A, E, S, AE, and AS into separate sample wells, as shown in the diagram below. Use a fresh tip for each sample. Expel any air in the tip before loading, and take care not to punch the tip of the pipet through the bottom of the gel.

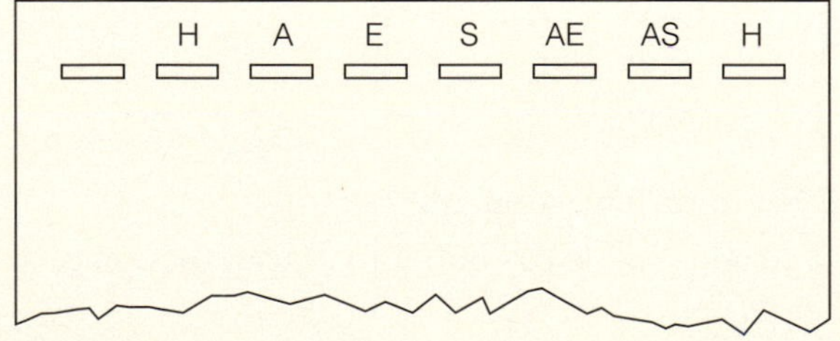

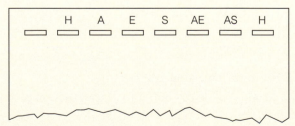

5. Use a fresh tip to load 10 μl from the H tube into the sample wells on each side of the restriction digests.
6. Electrophorese at 50–125 volts until the xylene cyanol bands have migrated 65–75 mm from the wells.
7. Turn off the power supply, remove the casting tray from the electrophoresis box, and transfer the gel to a disposable weigh boat or other shallow tray for staining.

Xylene cyanol is the aqua-colored, slower migrating of the two bands in the loading dye. A longer electrophoresis run is essential to adequately resolve / restriction fragments for the analysis.

Stop Point

The gel can be stored overnight, covered with small volume of buffer, in a staining tray or a zip-lock plastic bag, for viewing or photographing the next day. However, over a longer period of time, the DNA will diffuse through the gel, and the bands will either become indistinct or disappear entirely.

IV. Stain Gel with Ethidium Bromide, View, and Photograph

(10–15 minutes)

CAUTION

Review "Ethidium Bromide Staining and Responsible Handling" in Laboratory 3. Wear disposable gloves when staining, viewing, and photographing gel, and during cleanup. Confine all staining to a restricted sink area.

Refer to Laboratory 3, "DNA Restriction and Electrophoresis," for detailed instructions on staining and photographing gel.

1. Flood the gel with ethidium bromide solution (1 μg/ml), and allow it to stain for 5–10 minutes. Destain, if desired.
2. View the gel under an ultraviolet transilluminator or other UV light source.

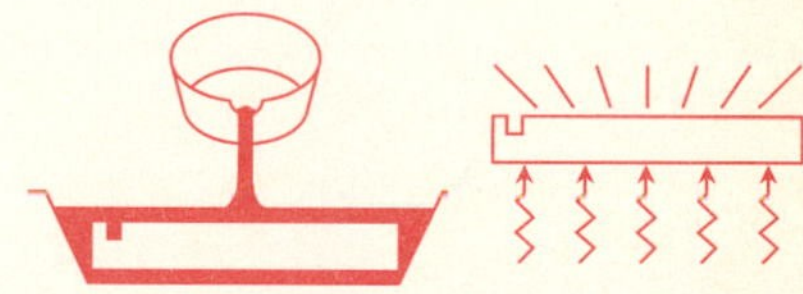

Band intensity and contrast increase dramatically if the gel is destained for 15–30 minutes in tap water. More simply, rinse and drain the gels, stack the staining trays, cover the top gel, and let it set overnight at room temperature. Alternatively, stain the gels with methylene blue for 20–30 minutes and then destain.

CAUTION

Ultraviolet light can damage your eyes. Never look directly at an unshielded UV light source without eye protection. View only through a filter or safety glasses that absorb the harmful wavelengths.

3. Make an exposure of 1–2 seconds with your camera set at f/8. Develop the print for the recommended time (approximately 45 seconds at room temperature).

Make sure that the photograph shows the sample wells; their positions are needed to measure migration distances.

4. Take the time for proper cleanup:
   a. Wipe down the camera, the transilluminator, and the staining area.
   b. Decontaminate the gels and any staining solution that will not be reused.
   c. Wash your hands before leaving the laboratory.

RESULTS AND DISCUSSION

1. Compare your stained gel with the ideal gel (below), and confirm that your gel has the same number and arrangement of bands in each lane. If your gel is missing bands, you may want to use the ideal gel shown below to answer the following questions.

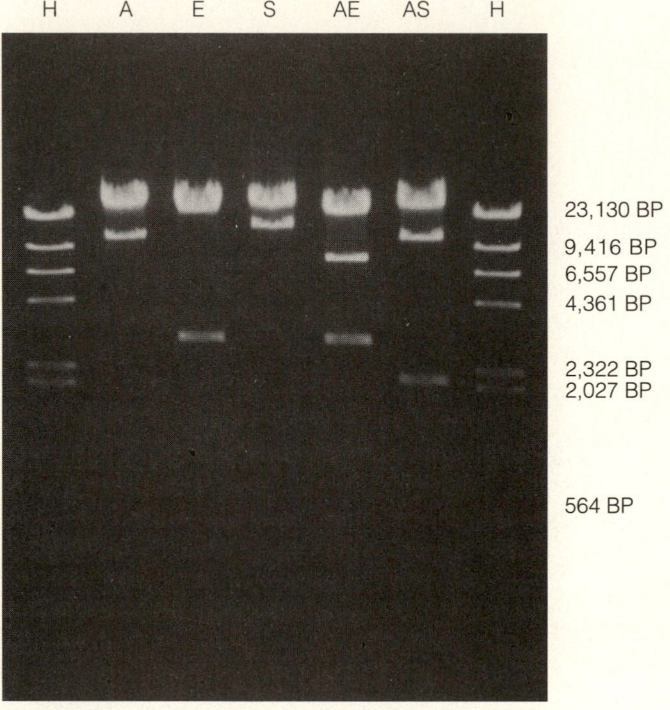

Ideal Gel

2. Linear DNA fragments migrate at rates inversely proportional to the $\log_{10}$ of their molecular weight. For simplicity's sake, base-pair length is substituted here for molecular weight. Use the photograph of your stained gel and a ruler to construct a graph that relates the base-pair size of each restriction fragment to the distance migrated through the gel. The matrix on the next page gives the size in base pairs (BP) of λ DNA fragments generated by the *Hin*dIII digest:

*Hind*III		*Apa*I		*Eco*O109I		*Sna*BI		*Apa*I/*Eco*O109I		*Apa*I/*Sna*BI	
D	BP	D	BP	D	BP	D	BP	D	BP	D	BP
	23,130										
	9,416										
	6,557										
	4,361										
	2,322										
	2,027										
	564										
	125*										

*usually runs off the end of the mini-gel during electrophoresis

a. Orient your gel photo with the wells at the top, and locate one of the two H lanes. Working from top to bottom, match the base-pair sizes of the *Hind*III fragments with the bands that appear in the H lane. Carefully measure the distance (in mm) that each *Hind*III fragment migrated from the sample well. Measure from the front edge of the well to the leading edge of each band. Enter the distance migrated by each fragment into the D column.

b. Set up semilog graph paper using the distance migrated as the x (arithmetic) axis and the base-pair length as the y (logarithmic) axis. (The first cycle of the y axis should begin with 1,000 bp and end with 10,000 bp—counting in 1,000-bp increments. The second cycle begins with 10,000 bp and ends with 100,000 bp—counting in 10,000 bp increments.) Then, plot the distance migrated *versus* base-pair length for each *Hind*III fragment. Finally, connect the data points.

c. Measure the distance migrated by the *smallest Apa*I restriction fragment, and locate this distance on the x axis of your graph. Then, use a ruler to draw a vertical line from this point to its intersection with the *Hind*III data line. Now, extend a horizontal line from this point to the y axis. This gives the bp size of the smallest *Apa*I fragment. Enter the result into the matrix above. Do not calculate the length of the largest fragment at this time.

d. Repeat step c to calculate the bp sizes of the *smallest* restriction fragments in the *Eco*O109I and *Sna*BI single digests and the *two smallest* fragments in the double digests. Enter the data into the matrix above.

e. The gel does not provide an accurate determination of the sizes of the largest fragments in each of the digests, because they are outside the linear resolving range of the gel. Therefore, determine the length of the largest fragment in each digest by subtracting the length(s) of the smaller fragment(s) from the length of the λ genome (48,502 base pairs). Enter the data into the matrix above.

3. Construct a restriction map of the cutting sites of λ DNA, showing the cutting sites for *Apa*I, *Eco*O109I, and *Sna*BI.

a. The data from the single digests indicates how far each enzyme cuts from the end of the λ chromosome.

b. Examine the double digest to determine whether the enzymes in each pair cut at the same or at opposite ends of the λ chromosome:

If both enzymes of a pair cut at *opposite ends* of the chromosome, then both fragments in the single digests will also be observed in the double digest.

If the enzymes of a pair cut at *the same end* of the chromosome, the fragment produced by the enzyme nearest the end will be observed in the double digest. However, a new fragment equal to the distance between the two cutting sites will also be observed.

c. To test your tentative map, check to see whether the map distances between cutting sites correspond to the gel bands in the single and double digests.

4. What is the function of a restriction map?
5. Explain the significance of COS sites on λ DNA.
6. How does the fact that λ molecules can exist as linear or circular molecules affect your restriction map?
7. Limitations inherent in restriction analysis preclude obtaining an accurate restriction map from a single experiment. Explain how the following problems would affect your results and how you might overcome them.
   a. Restriction digestion produces large fragments (over 15 kilobases).
   b. Restriction fragments comigrate and do not resolve.
   c. A restriction enzyme has two cut sites very close together.

For Further Research

1. Repeat this laboratory with other restriction enzymes to add additional cutting sites to your λ restriction map.
2. Repeat this laboratory with λ gt10 or λ gt11, modified λ phages that are used as cloning vectors. By comparing the restriction maps that you obtain, attempt to determine points in the λ genome that have been altered by deletion or insertion in the vector genome. Remember to take linear and circular forms into account.

LABORATORY 12

Restriction Mapping of the Plasmid pBR322

pBR322 was the first artificially produced plasmid to be used routinely in recombinant-DNA experiments. To be useful as a vector, the coordinates of cutting sites of various restriction enzymes had to be accurately determined, especially in relation to functional sequences, such as the origin of replication and antibiotic-resistance genes. Restriction mapping is also the first step in characterizing a DNA fragment that has been ligated into the vector, providing necessary data for applications such as DNA sequencing, gene localization, and site-directed mutagenesis.

This laboratory recalls early mapping experiments with pBR322 and other circular plasmids and contrasts with the mapping of the linear λ chromosome in Laboratory 11. The goal is to determine the number and relative positions of cut sites for three restriction enzymes—*Eco*RI, *Hinc*II, and *Pvu*II on the circular pBR322 plasmid. Samples of pBR322 are incubated at 37°C in single, double, and triple digests with the three enzymes. Each sample is loaded into a separate well of an agarose gel, together with a *Hin*dIII digest of λ DNA. Following electrophoresis, the gel is stained and photographed.

The λ/*Hin*dIII digest generates a series of fragments of known size; these fragments serve as markers to help gauge the size of pBR322 fragments from the *Eco*RI, *Hinc*II, and *Pvu*II digests. Data from the single digests indicate the number of cutting sites for each enzyme; the double and triple digest allow the cutting sites to be mapped relative to one another.

I. Set Up Restriction Digests

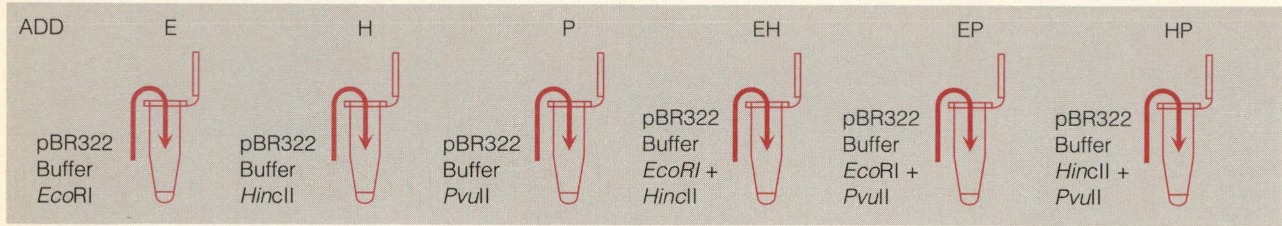

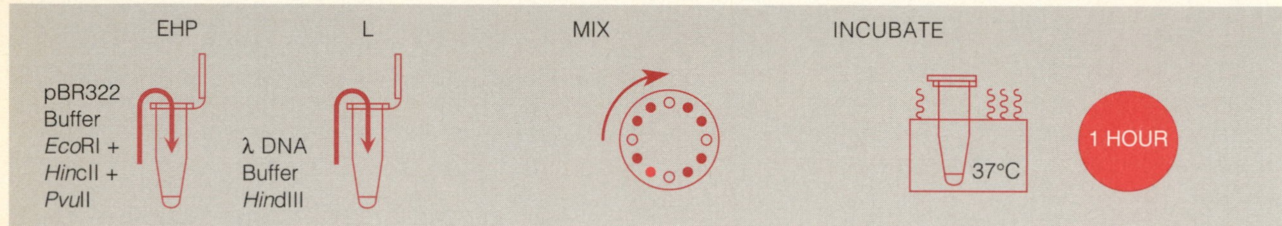

II. Cast 1.0% Agarose Gel

III. Load Gel and Electrophorese

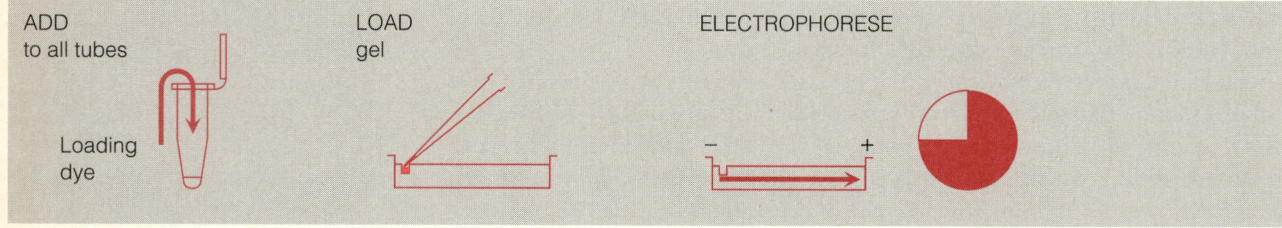

IV. Stain Gel, View, and Photograph

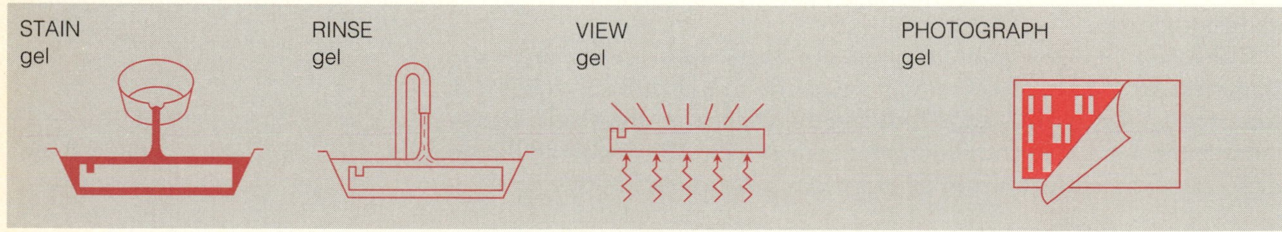

Laboratory 12
Restriction Mapping of the Plasmid pBR322

PRELAB NOTES

Review the prelab notes in Laboratory 3, "DNA Restriction and Electrophoresis."

Plasmid pBR322

Because each lane of the experiment contains a maximum of four restriction fragments, 0.3 μg of pBR322 per lane is adequate for visualization by ethidium bromide staining. Double the DNA mass to 0.6 μg per lane when you are staining with methylene blue.

Restriction Enzymes

Restriction enzymes are supplied in a buffer containing 50% glycerol, which can influence enzyme activity and specificity. When you are performing double and triple digests, it is advisable to keep the volume contributed by the enzymes to a minimum. If nonspecific cutting (star activity) is observed at the recommended volumes, reduce the enzyme volume to reduce glycerol concentration. It may be necessary to incubate digests for a longer period of time, to compensate for using less enzyme.

10x Restriction Buffer

The three enzymes used in this experiment have different salt and pH requirements for optimal activity. Since the *Hinc*II enzyme cuts the least efficiently, we use a 10× buffer that works well for it: New England Biolabs (NEB) buffer #1.

Semilog Graph Paper

λ restriction fragments used as size markers range in size from about 500 bp to 10,000 bp. Thus, paper with two log cycles will best spread the data points.

For Further Information

The protocol presented here is based on the following published methods:

Bolivar, F., R. L. Rodriguez, P. J. Greene, M. C. Betlach, H. L. Heyneker, H. W. Boyer, J. H. Crosa, and S. Falkow. 1977. Construction and characterization of new cloning vehicles. II. A multipurpose cloning system. *Gene* 2: 95.

Danna, K. and D. Nathans. 1971. Specific cleavage of Simian Virus 40 DNA by restriction endonuclease of *Hemophilus influenzae. Proceedings of the National Academy of Sciences USA* 68: 2913.

Sharp, P. A., B. Sugden, and J. Sambrook. 1973. Detection of two restriction endonuclease activities in *Haemophilus parainfluenzae* using analytical agarose-ethidium bromide electrophoresis. *Biochemistry* 12: 3055.

Helling, R. B., H. M. Goodman, and H. W. Boyer. 1974. Analysis of *Eco*RI fragments of DNA from lambdoid bacteriophages and other viruses by agarose-gel electrophoresis. *Journal of Virology* 14: 1235.

RESTRICTION MAPPING OF THE PLASMID pBR322

Reagents	Equipment and Supplies
0.1 μg/μl pBR322 DNA	0.5–10-μl micropipettor + tips
0.1 μg/μl λ DNA	10–100-μl micropipettor + tips (optional)
restriction enzymes:	1.5-ml tubes
*Eco*RI	aluminum foil
*Hinc*II	beaker for agarose
*Hind*III	beaker for waste/used tips
*Pvu*II	camera and film (optional)
10× NEB #1 restriction buffer	disposable gloves
distilled water	electrophoresis box
loading dye	masking tape
1.0% agarose	microfuge (optional)
1× TBE buffer (Tris/Borate/EDTA)	metric ruler
1 μg/ml ethidium bromide (or 0.025% methylene blue)	permanent marker
	power supply
	semilog graph paper
for decontamination:	test tube rack
0.05 M $KMnO_4$	transilluminator
0.25 N HCl	37°C water bath
0.25 N NaOH	60°C water bath (for agarose)

I. Set Up Restriction Digests

Refer to Laboratory 3, "DNA Restriction and Electrophoresis," for detailed instructions on setting up digests.

(75 minutes, including incubation)

1. Use a permanent marker to label eight 1.5-ml tubes, as shown in the matrix below. Restriction reactions will be performed in these tubes.
2. Use the matrix below as a checklist while adding reagents to each reaction. Read down each column, adding the same reagent to all appropriate tubes. Refer to the detailed instructions that follow.

Tube	pBR322	λ DNA	10x Buffer	*Eco*RI	*Hinc*II	*Pvu*II	*Hind*III	H_2O
E	3 μl	—	2 μl	1 μl	—	—	—	14 μl
H	3 μl	—	2 μl	—	1 μl	—	—	14 μl
P	3 μl	—	2 μl	—	—	1 μl	—	14 μl
EH	3 μl	—	2 μl	1 μl	1 μl	—	—	13 μl
EP	3 μl	—	2 μl	1 μl	—	1 μl	—	13 μl
HP	3 μl	—	2 μl	—	1 μl	1 μl	—	13 μl
EHP	3 μl	—	2 μl	1 μl	1 μl	1 μl	—	12 μl
L	—	6 μl	2 μl	—	—	—	1 μl	11 μl

3. Add 3 μl of pBR322 to the appropriate reaction tubes.
4. Use a *fresh tip to* add 6 μl of λ DNA to the "L" reaction tube.
5. Use a *fresh tip* to add 2 μl of 10× buffer to a clean spot on each reaction tube.
6. Use a *fresh tip* to add an appropriate volume of distilled water to each reaction tube.
7. Use fresh tips to add restriction enzymes to the appropriate reaction tubes.
8. After adding all of the above reagents, close the tops of the tubes and mix.
9. Place the reactions in a 37°C water bath, and incubate for a minimum of 60 minutes.

It is important to change pipet tips after each enzyme addition, since even a small amount of carryover into an inappropriate tube will produce unexpected restriction fragments.

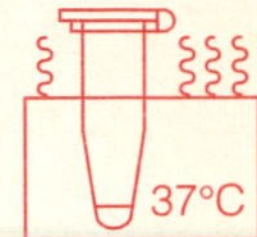

Stop Point

Following incubation, freeze the reactions at –20°C until you are ready to continue. Thaw the reactions before continuing on to Section III, step 1.

II. Cast 1.0% Agarose Gel

(15 minutes)

1. Carefully pour agarose solution into the casting tray to fill it to a depth of about 6 mm. The gel should cover about half the height of the comb teeth.
2. When the agarose has solidified, place the casting tray into the electrophoresis box and set up for electrophoresis.

Stop Point

Cover the electrophoresis box and save the gel until you are ready to continue. The gel will remain in good condition for several days if it is completely submerged in buffer.

III. Load Gel and Electrophorese

(10 minutes; then 40+ minutes of electrophoresis)

Refer to Laboratory 3, "DNA Restriction and Electrophoresis," for detailed instructions on casting and loading gel.

1. Remove the restriction digests from the 37°C water bath.
2. Add 2 μl of loading dye to each reaction tube. Close the tops of the tubes, and mix by either tapping the tube bottoms on the lab bench, pipetting in and out, or pulsing in a microfuge.

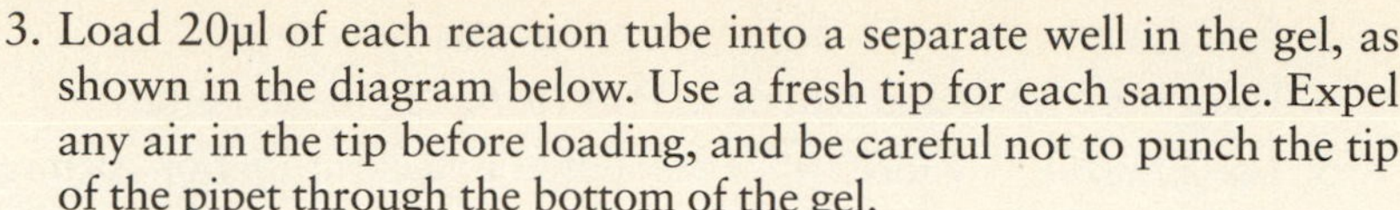

3. Load 20µl of each reaction tube into a separate well in the gel, as shown in the diagram below. Use a fresh tip for each sample. Expel any air in the tip before loading, and be careful not to punch the tip of the pipet through the bottom of the gel.

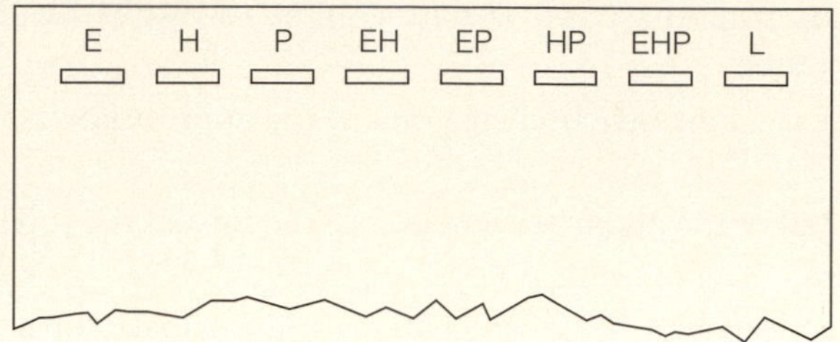

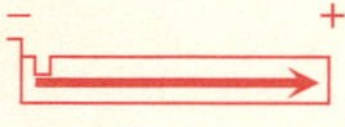

4. Electrophorese at 50–125 volts until the bromophenol blue bands have migrated 65–75 mm from the wells.
5. Turn off the power supply, remove the casting tray from the electrophoresis box, and transfer the gel to a disposable weigh boat (or other shallow tray) for staining.

Refer to Laboratory 3, "DNA Restriction and Electrophoresis," for detailed instructions on staining and photographing the gel.

IV. Stain Gel with Ethidium Bromide, View, and Photograph

(15 minutes)

CAUTION

Review the section on "Ethidium Bromide Staining and Responsible Handling" in Laboratory 3. Wear disposable gloves when staining, viewing, and photographing the gel, and during cleanup. Confine all staining to a restricted sink area.

Band intensity and contrast increase dramatically if the gel is destained for 15–30 minutes in tap water. More simply, rinse and drain the gels, stack the staining trays, cover the top gel, and let it set overnight at room temperature. Alternatively, stain the gels with methylene blue for 20–30 minutes and then destain.

1. Flood the gel with ethidium bromide solution (1 µg/ml), and allow it to stain for 5–10 minutes.
2. View under an ultraviolet transilluminator or other UV source.

CAUTION

Ultraviolet light can damage your eyes. Never look directly at an unshielded UV light source without eye protection. View only through a filter or safety glasses that absorb the harmful wavelengths.

3. Make an exposure of 1–2 seconds with a camera set at f/8. Develop the print for the recommended time (approximately 45 seconds at room temperature).
4. Take the time for proper cleanup:
   a. Wipe down the camera, the transilluminator, and the staining area.
   b. Decontaminate the gels and any staining solution that will not be reused.
   c. Wash your hands before leaving the laboratory.

Make sure that the photograph shows the sample wells; their positions are needed for measuring migration distances.

RESULTS AND DISCUSSION

1. Compare your stained gel with the ideal gel (below), and confirm that your gel has the same number and arrangement of bands in each lane. If your gel is missing bands, you may want to use the ideal gel to answer the following questions.

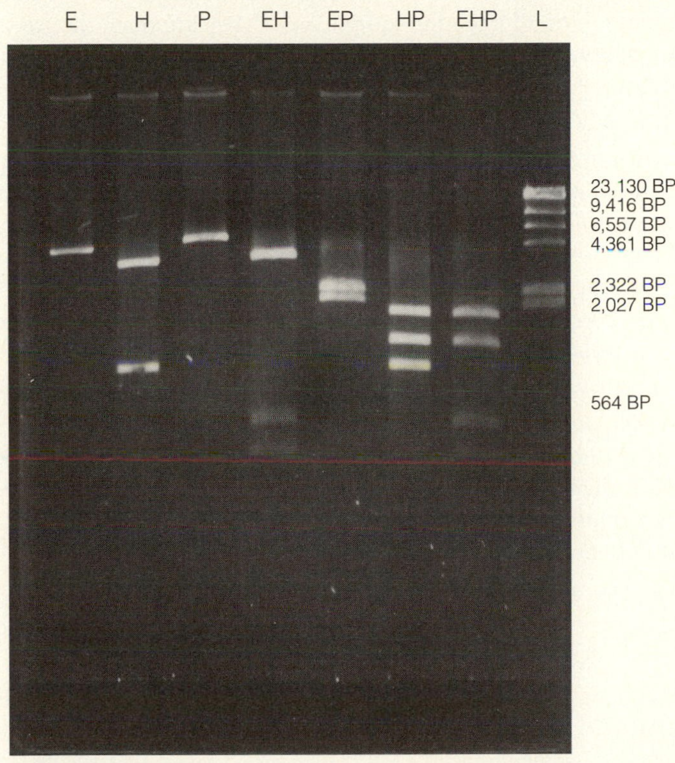

Ideal Gel

2. Linear DNA fragments migrate at rates that are inversely proportional to the $\log_{10}$ of their molecular weight. For simplicity's sake, base-pair length is substituted here for molecular weight. Use the photograph of your stained gel and a ruler to construct a graph that relates the base-pair size of each restriction fragment to the distance migrated through the gel. The matrix on the next page gives the size in base pairs (BP) of λ DNA fragments generated by a *Hin*dIII digest:

λ HindIII		EcoRI		HincII		PvuII		EcoRI/ PvuII		HincII/ PvuII		EcoRI/ HincII		EcoRI/ HincII/PvuII	
D	**BP**	**D**	**BP**	**D**	**BP**	**D**	**BP**	**D**	**BP**	**D**	**BP**	**D**	**BP**	**D**	**BP**
	23,130														
	9,416														
	6,557														
	4,361														
	2,322														
	2,027														
	564														
	125*														

*usually runs off the end of the mini-gel during electrophoresis

a. Orient your gel photo with the wells at the top and locate the L lane. Working from top to bottom, match the base-pair sizes of the *Hin*dIII fragments with the bands that appear in the L lane. Carefully measure the distance (in mm) that each *Hin*dIII fragment migrated from the sample well. Measure from the front edge of the well to the leading edge of each band. Enter the distance migrated by each fragment into the D column.

b. Set up semilog graph paper using the distance migrated as the *x* (arithmetic) axis and base-pair length as the *y* (logarithmic) axis. (The first cycle of the *y* axis should begin with 100 bp and end with 1,000 bp, counting in 100-bp increments. The second cycle begins with 1,000 bp and ends with 10,000 bp, counting in 1,000-bp increments.) Then, plot the distance migrated *versus* base-pair length for each *Hin*dIII fragment. Finally, connect the data points. Do not plot the 23,130 bp fragment.

c. Measure the distance migrated by the fragment in the *Eco*RI digest, and locate this distance on the x axis of your graph. Then, use a ruler to draw a vertical line from this point to its intersection with the λ/*Hin*dIII data line. Now, extend a horizontal line from this point to the y axis. This gives the base-pair size of the *Eco*RI fragment. Enter the result into the above matrix.

d. Repeat step c, measuring the distances migrated by the fragment(s) in the *Eco*RI, *Hin*cII, and *Pvu*II single, double, and triple digests. Enter the data into the above matrix.

3. Construct a restriction map of pBR322, showing the cutting sites of *Eco*RI, *Hin*cII, and *Pvu*II.

a. The data from the single digests indicates the number of cutting sites for each enzyme on the pBR322 molecule. Because the *Eco*RI digest produces a single fragment, there must be only one *Eco*RI site on the plasmid, and the fragment size must correspond to the plasmid size.

b. First compare the results of each single digest with the double digest using the same enzyme. Check to determine which bands in the single digest are missing in the double digest. The sum of the bp sizes of two fragments in the double digest should equal the bp size of the missing fragment—thereby indicating the relative locations of the cutting sites of the two enzymes.

c. Bear in mind that the sizes of the restriction fragments determined by the graph are only approximate. Although the sum of all fragments in each digest must equal the bp size of the pBR322 plasmid, the calculated sums will differ somewhat. Therefore, use logic to help you decide the relative sizes of the fragments.

d. Typically, you must construct alternative maps that place a cutting site for an enzyme in one of two possible orientations relative to the sites for another enzyme. One orientation should yield restriction fragment sizes that more closely match your results than the other.

e. To test your tentative map, check to see whether the distances between cutting sites on the map correspond to the gel bands in the single, double, and triple digests.

4. This type of analysis produces size estimates for restriction fragments. How can their sizes be determined precisely?
5. How can a restriction map help you to subclone a gene?

For Further Research

1. Repeat this laboratory with other restriction enzymes to add additional cutting sites to your pBR322 restriction map.
2. Repeat this laboratory with other plasmids, such as pAMP, pKAN, pBLU, or pUC 18-19. By comparing the restriction maps you obtain, attempt to determine which regions are shared by the vectors and which regions have been altered by deletion or insertion.

Southern Hybridization

LABORATORY 13

Southern Hybridization of λ DNA

Southern hybridization allows one to visualize a specific DNA fragment against the background of a complex genome. Genomic DNA is digested with one or more restriction enzymes, and the fragments are separated by gel electrophoresis. The DNA fragments are transferred from the gel to a filter membrane and then exposed to a solution containing a DNA probe.

This technique relies on the hydrogen bonds that hold together complementary adenine-thymine and guanine-cytosine base pairs in double-stranded DNA. At temperatures above 90°C, or a pH greater than 10.5, these hydrogen bonds are disrupted, causing the complementary strands to denature into single strands. Under proper conditions of salt content, temperature, and pH, complementary single-stranded molecules can renature to restore the original duplex DNA molecule. This process of complementary single-stranded molecules aligning and forming double-stranded molecules is known as *hybridization.*

A single-stranded DNA probe—derived from either a restriction fragment, a plasmid containing a cloned sequence of interest, or a synthetic oligonucleotide—is employed to recognize and hybridize to its complementary sequence in a sample of genomic (or other target) DNA. A radioactive, colorimetric, or chemiluminescent group attached to the probe allows areas of probe hybridization to be visualized. Under reaction conditions of "high stringency," stable DNA duplexes form only when complementary base pairing is essentially perfect along the entire length of the hybridizing probe and target DNA sequence. Under conditions of low stringency, stable DNA duplexes can form in regions of partial hybridization, where the probe and the target DNA strand have occasional mismatches.

In this laboratory, Southern hybridization is used to locate a sequence in the λ genome. In Part A, "Restriction and Electrophoresis of λ DNA," samples of λ DNA are digested with *Bam*HI, *Eco*RI, and *Hin*dIII (both singly and in combination). The resulting restriction fragments are separated, using agarose gel electrophoresis, and stained with ethidium bromide. The gel is photographed next to a metric ruler, so that accurate measurements of the migration of restriction fragments can be obtained.

In Part B, "Southern Transfer of λ Restriction Fragments," the gel is incubated in a sodium hydroxide solution to denature the restriction fragments into single strands. Following neutralization, capillary action is used to transfer the single-stranded restriction fragments from the gel onto the surface of a nylon membrane, using a technique called blotting. The membrane is washed, air dried, and baked at 70°C to bond the DNA to the membrane.

In Part C, "Hybridization of the λ Probe," the membrane is prehybridized with a protein-blocking reagent, which prevents the probe from binding nonspecifically to the membrane. Next, the membrane is incubated with the hybridization probe, a 25-nucleotide oligomer labeled with the plant steroid digoxigenin. The single-stranded probe will base pair to its complementary sequence at position 22,915–22,939 of the λ chromosome.

In Part D, "Nonradioactive Probe Detection," the membrane is washed and incubated with an antidigoxigenin antibody, which binds to the digoxigenin label on the probe. The resulting immune complex is detected by color-forming reactions initiated by the enzyme alkaline phosphatase, which is attached to the antibody. A purplish-brown precipitate indicates areas on the membrane where the probe has bound to restriction fragments containing its complementary DNA sequence.

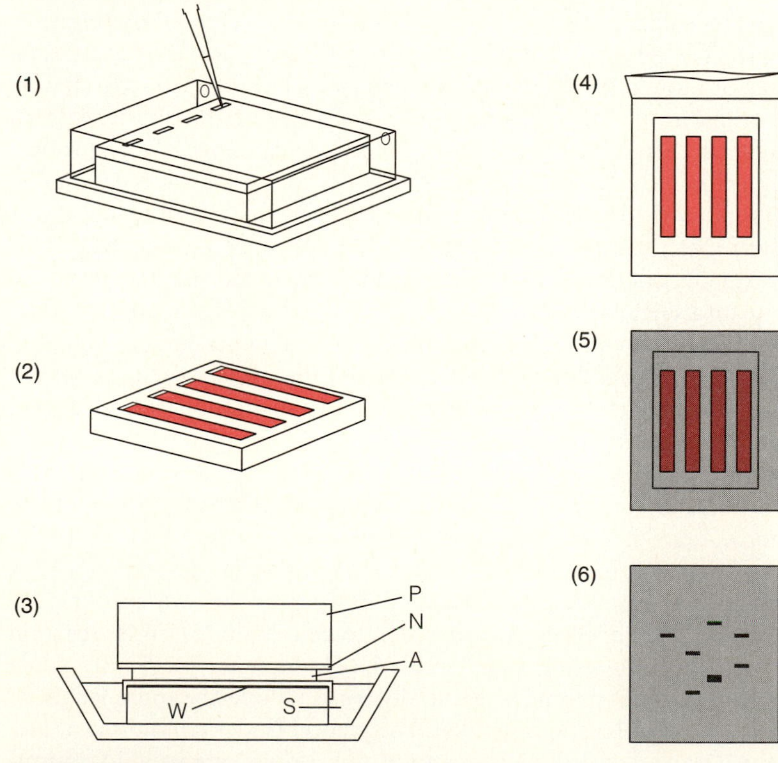

Southern Blotting
(1) Load digested DNA into agarose gel and electrophorese.
(2) Visualize DNA and denature in gel.
(3) Transfer DNA to nylon membrane by capillary action:
P = paper towels A = agarose gel S = high salt solution
N = nylon membrane W = wick
(4) Hybridize probe to membrane, which, if stained, would be a mirror image replica of gel.
(5) Wash membrane, and develop bands using colorimetric detection system.
(6) Developed membrane reveals location of bands to which probe binds.

I. Set Up Restriction Digests

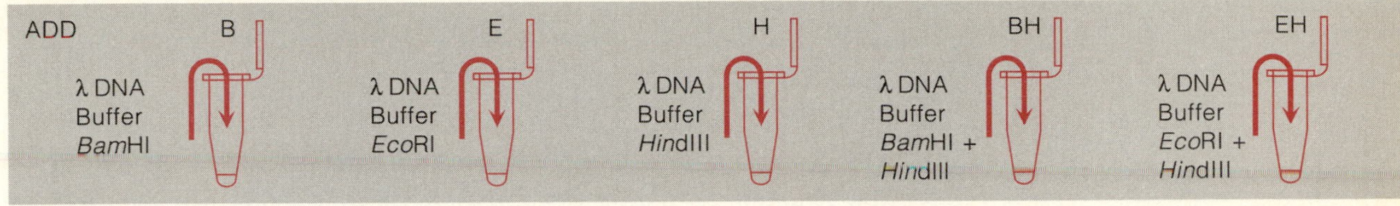

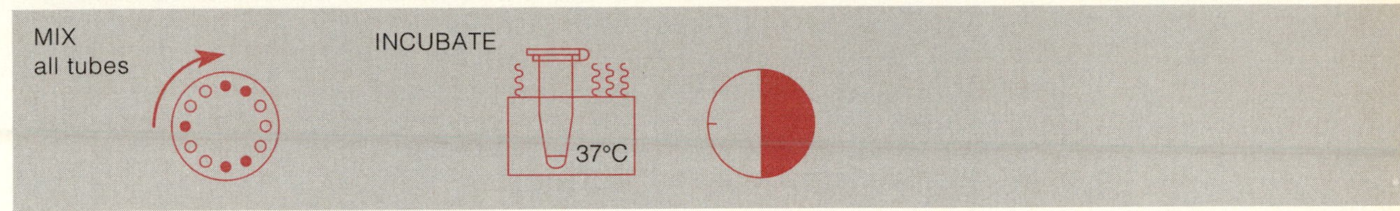

II. Cast 0.8% Agarose Gel

III. Load Gel and Electrophorese

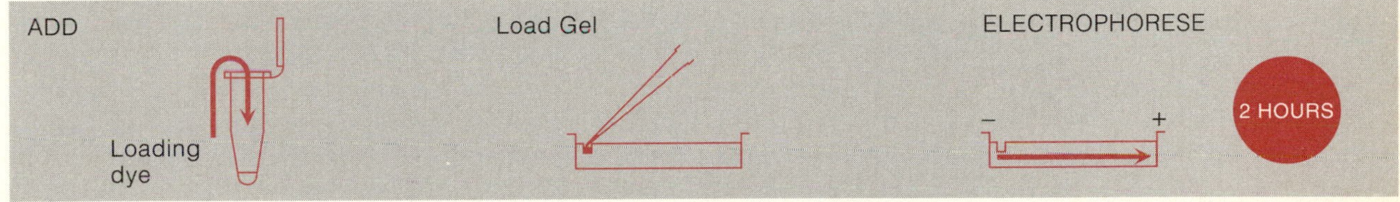

IV. Stain Gel, View, and Photograph

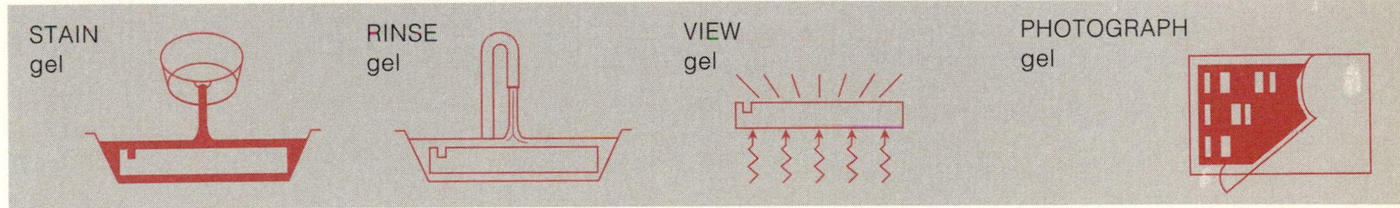

Laboratory 13/Part A
Restriction and Electrophoresis of λ DNA

PRELAB NOTES

Review the prelab notes in Laboratory 3, "DNA Restriction and Electrophoresis."

λ *versus* Complex Genomes

At the beginning of a typical Southern hybridization experiment, 10 μg of genomic DNA is cut with one or more restriction enzymes. An endonuclease with a 6-base-pair recognition sequence, such as those used in this laboratory, statistically cuts once every 4,096 basepairs (4^6)—producing many thousands of restriction fragments in genomes of higher eukaryotes:

Organism	bp in Genome (haploid, monoploid)	Restriction Fragments
λ	48,502	12
E. coli	4,000,000	977
yeast	14,000,000	3,418
Arabidopsis	100,000,000	24,414
C. elegans	100,000,000	24,414
Drosophila	165,000,000	40,283
mouse	3,000,000,000	732,422
human	3,500,000,000	854,492

Following electrophoresis and ethidium bromide staining, digests of eukaryotic DNA produce a smear composed of many thousands of restriction fragments in which individual fragments cannot be resolved. However, Southern hybridization is sensitive enough to detect a given sequence within any of the thousands of restriction fragments composing a complex genome. In this experiment, the restriction digests will display distinct banding patterns, rather than smears. This is because of the small size of the λ genome (48,502 base pairs) and the relatively large mass of DNA loaded into each sample well (0.6 μg). Use of a small genome and a relatively large mass of DNA helps to ensure reproducible results and shorter hybridization times. Pushed to the limit, Southern hybridization is an extremely sensitive technique, capable of detecting a restriction fragment that is present only once in the human genome.

A scientist using a DNA probe to search a complex genome for a particular sequence cannot visualize a hybridizing restriction fragment in the stained gel. In contrast, following restriction and electrophoresis, the tiny λ genome is displayed as a collection of discrete restriction fragments. Examination of the stained gel allows fragments that are predicted to hybridize with the probe to be identified. These predictions will be confirmed or refuted at the end of the laboratory.

For Further Information

The protocol presented here is based upon the following published methods:

Helling, R. B., H. M. Goodman, and H. W. Boyer. 1974. Analysis of *Eco*RI fragments of DNA from lambdoid bacteriophages and other viruses by agarose-gel electrophoresis. *Journal of Virology* 14: 1235.

Sharp, P. A., B. Sugden, and J. Sambrook. 1973. Detection of two restriction endonuclease activities in *Haemophilus parainfluenzae* using analytical agarose-ethidium bromide electrophoresis. *Biochemistry* 12: 3055.

RESTRICTION AND GEL ELECTROPHORESIS OF λ DNA

Reagents

0.3 μg/μl λ DNA
restriction enzymes
- *Bam*H
- *Eco*RI
- *Hin*dIII

2× restriction buffer
distilled water
loading dye
0.8% agarose
1× Tris/Borate/EDTA (TBE)
1 μg/ml ethidium bromide

for decontamination:
- 0.05 M $KMnO_4$
- 0.25 N HCl
- 0.25 N NaOH

Supplies and Equipment

0.5–10-μl micropipettor + tips
1.5-ml tubes
aluminum foil
beaker for agarose
beaker for waste/used tips
camera and film
disposable gloves
electrophoresis box
masking tape
microfuge (optional)
Parafilm or waxed paper (optional)
permanent marker
plastic wrap (optional)
power supply
semilog graph paper (3-cycle)
test tube rack
transilluminator
transparent ruler
37°C water bath
60°C water bath (for agarose)

I. Set Up Restriction Digests

(10 minutes; then 30+ minutes of incubation)

Refer to Laboratory 3, "DNA Restriction and Electrophoresis," for detailed instructions on setting up reactions.

1. Use a permanent marker to label five 1.5-ml tubes, as shown in the matrix below. Restriction reactions will be performed in these tubes.
2. Use the matrix below as a checklist while adding reagents to each reaction. Read down each column, adding the same reagent to all appropriate tubes. Refer to the detailed instructions that follow.

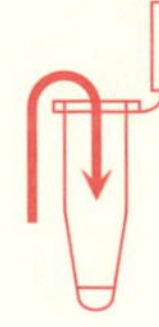

Tube	λ DNA	2x Buffer	*Bam*HI	*Eco*RI	*Hin*dIII	Water
B	2 μl	5 μl	1 μl	—	—	2 μl
E	2 μl	5 μl	—	1 μl	—	2 μl
H	2 μl	5 μl	—	—	1 μl	2 μl
BH	2 μl	5 μl	1 μl	—	1 μl	1 μl
EH	2 μl	5 μl	—	1 μl	1 μl	1 μl

3. Add 2 μl of DNA to each reaction tube.
4. Use a *fresh tip* to add 5 μl of 2× buffer to a clean spot on each reaction tube.
5. Use a *fresh tip* to add an appropriate volume of distilled water to each reaction tube.
6. Use *fresh tips* to add restriction enzymes to appropriate reaction tubes.

You must change pipet tips after each enzyme addition, because even a small amount of carryover into an inappropriate tube will produce unexpected restriction fragments.

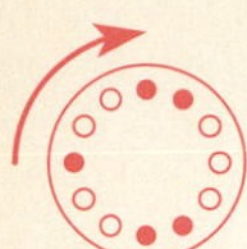

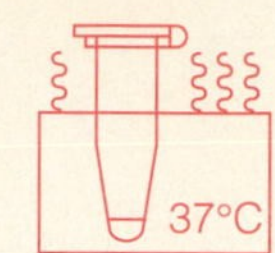

7. After adding all of the above reagents, close the tops of the tubes and mix.
8. Place the reactions in a 37°C water bath and incubate for a minimum of 30 minutes.

Stop Point

Following incubation, freeze the reactions at –20°C until you are ready to continue. Thaw the reactions before continuing on to Section III, step 1.

II. Cast 0.8% Agarose Gel

(15 minutes)

1. Carefully pour agarose solution into the casting tray to fill it to a depth of about 5 mm. The gel should cover at least one-third of the height of the comb teeth.
2. After the agarose solidifies, place the casting tray into the electrophoresis box and set up for electrophoresis.

Stop Point

Cover the electrophoresis box and save the gel until you are ready to continue. The gel will remain in good condition for several days if it is completely submerged in buffer.

III. Load Gel and Electrophorese

Refer to Laboratory 3, "DNA Restriction and Electrophoresis," for detailed instructions on casting and loading gel.

(10 minutes; then 120+ minutes of electrophoresis)

1. Remove the restriction digests from the 37°C water bath.
2. Add 1 μl of loading dye to each reaction tube. Close the tops of the tubes and mix.
3. Load 10 μl from reaction tubes B, E, H, BH, and EH into separate wells, as shown in the diagram below. Use a *fresh tip* for each sample. Expel any air in the tip before loading, and take care not to punch the tip of the pipet through the bottom of the gel.

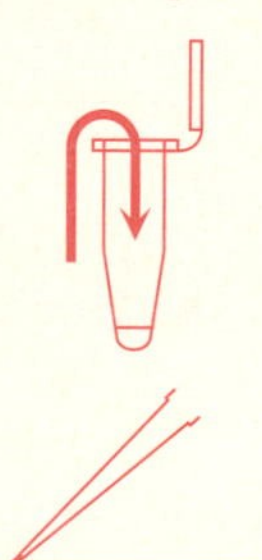

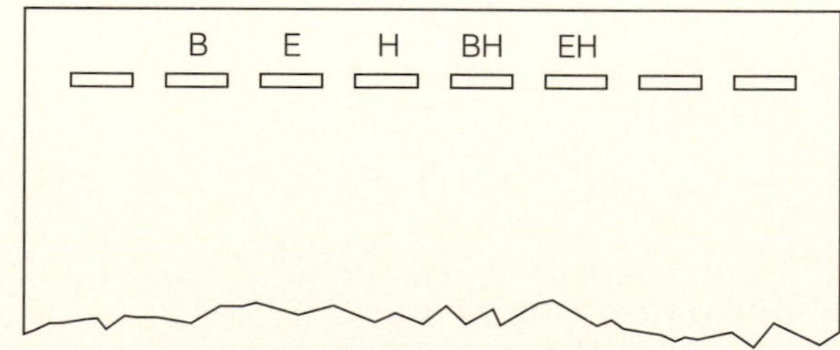

4. Electrophorese *at 60 volts* until the bromophenol blue band is about 80 mm from the wells (about 2 hours).

Electrophoresis conducted at a relatively low voltage produces more tightly focused bands in the gel, which aids in the analysis.

5. Turn off the power supply, remove the casting tray from the electrophoresis box, and transfer the gel to a disposable weigh boat or other shallow tray for staining.

IV. Stain Gel with Ethidium Bromide, View, and Photograph

(15 minutes)

CAUTION

Review the section on "Ethidium Bromide Staining and Responsible Handling" in Laboratory 3. Wear disposable gloves when staining, viewing, and photographing gel, and during cleanup. Confine all staining to a restricted sink area.

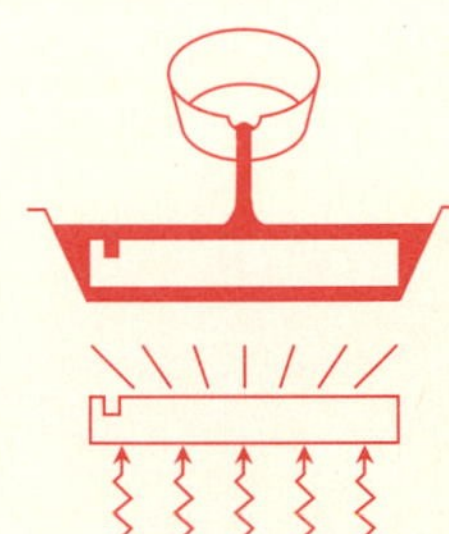

1. Flood the gel with ethidium bromide solution (1 µg/ml), and allow it to stain for 5–10 minutes.
2. View the gel under an ultraviolet transilluminator or other UV light source.

CAUTION

Ultraviolet light can damage your eyes. Never look directly at an unshielded UV light source without eye protection. View only through a filter or safety glasses that absorb the harmful wavelengths.

Make sure that the photograph shows the sample wells; their positions are needed for measuring migration distances. Hybridization signals on the membrane must be compared with the restriction fragments on the gel photograph. If the camera does not produce a 1:1 subject-to-image ratio, then be sure to include a transparent ruler alongside the gel in the photograph.

3. Make an exposure of 1–2 seconds with the camera aperture set at f/8. Develop the print for the recommended time (approximately 45 seconds at room temperature).
4. Take the time for proper cleanup:
   a. Wipe down the camera, the transilluminator, and the staining area.
   b. Decontaminate any staining solution that will not be reused.
   c. Wash your hands before leaving the laboratory.

RESULTS AND DISCUSSION

1. Examine the photograph of your stained gel. Compare your stained gel with the ideal gel (below), and try to account for the fragments of λ DNA in each lane. How can you account for the differences in separation and band intensity between your gel and the ideal gel? If your gel is missing bands, you may want to use the photo of the ideal gel to answer the following questions.

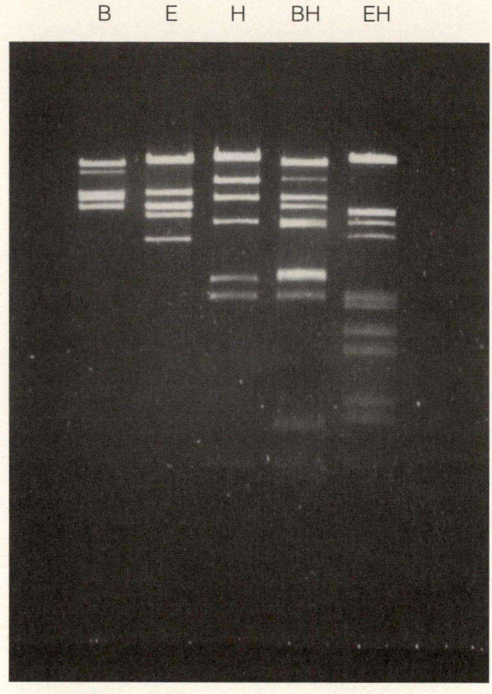

Ideal Gel

2. Linear DNA fragments migrate at rates that are inversely proportional to the $\log_{10}$ of their molecular weight. For simplicity's sake, base-pair length is substituted here for molecular weight. Use the photograph of your stained gel and a ruler to construct a graph that relates the base-pair size of each restriction fragment to the distance migrated through the gel. The matix below gives the size in base pairs (bp) of λ DNA fragments generated by a *Hin*dIII digest:

***Hin*dIII**		***Bam*HI**		***Eco*RI**		***Bam*HI/*Hin*dIII**		***Eco*RI/*Hin*dIII**	
D	**BP**	**D**	**BP**	**D**	**BP**	**D**	**BP**	**D**	**BP**
	23,130								
	9,416								
	6,557								
	4,361								
	2,322								
	2,027								
	564								
	125*								

* not usually visible

a. Orient your gel photo with the wells at the top, and locate the H lane. Working from top to bottom, match the base-pair sizes of the *Hin*dIII fragments with the bands that appear in the H lane. Carefully measure the distance (in mm) that each *Hin*dIII fragment migrated from the sample well. Measure from the front edge of the well to the leading edge of each band. Enter the distance migrated by each fragment into the D column.

b. Set up semilog graph paper using the distance migrated as the *x* (arithmetic) axis and base-pair length as the *y* (logarithmic) axis. (The first cycle of the *y* axis should begin with 100 bp and end with 1,000 bp, counting in 100-bp increments. The second cycle begins with 1,000 bp and ends with 10,000 bp, counting in 1,000-bp increments. The third cycle begins with 10,000 bp and ends with 100,000 bp, counting in 10,000-bp increments.) Then, plot the distance migrated *versus* the base-pair length for each *Hin*dIII fragment. Finally, connect the data points.

c. Locate the sequence of the hybridization probe (nucleotides 22,915–22,939) on the *Hin*dIII, *Eco*RI, and *Bam*HI restriction maps of the λ genome (page 50). Determine the base-pair size of the restriction fragment that you predict will hybridize with the probe for each of the restriction digests: *Hin*dIII, *Bam*HI, *Eco*RI, *Bam*HI/*Hin*dIII, and *Eco*RI/*Hin*dIII.

d. Continue by one of the two following methods:

As in step a, measure the distance migrated by each restriction fragment in each lane of your gel, and record the results in column D of the appropriate restriction digest. Then, determine the bp size of each fragment: (1) Locate the distance on the *x* axis of your graph. (2) Draw a vertical line from this point to its intersection with the *Hin*dIII data line. (3) Extend a horizontal line from this point to the *y* axis, and record the result in column BP. Finally, identify the restriction fragment in each digest to which the probe should bind, according to the map information in step c.

or

For each restriction digest in step c: (1) Locate on the *y* axis the base-pair length of the fragment expected to hybridize with the probe. (2) Draw a horizontal line from this point to its intersection with the *Hin*dIII data line. (3) Extend a vertical line from this point to the *x* axis, to determine the expected migration distance for this fragment. (4) Measuring from the well, identify the fragment in the digest that best matches the migration distance predicted from the graph.

When you are attempting to correlate bands on the gel photograph with fragments from the restriction map, it is helpful to begin with the small fragments at the bottom of the gel lane, because they are well separated and easier to identify.

e. Use a highlighter to mark the bands in your photograph that you predict will hybridize with the probe.

3. What mass of DNA is contained in the band representing the smallest restriction fragment predicted to hybridize with the probe in question 2?

4. Southern hybridization has enough sensitivity to detect a 1,000-bp sequence that occurs at the single-copy level. Assuming that 10 μg of mammalian DNA (haploid genome = 3.5 billion base pairs) is loaded in the gel lane, how many picograms of DNA are contained in the 1,000-bp sequence?

I. Denature and Neutralize DNA

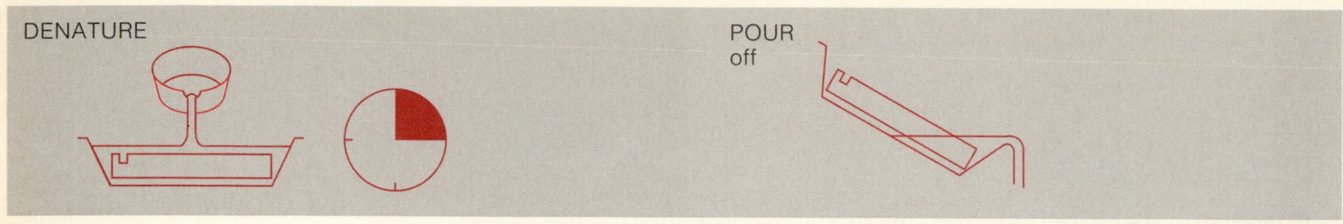

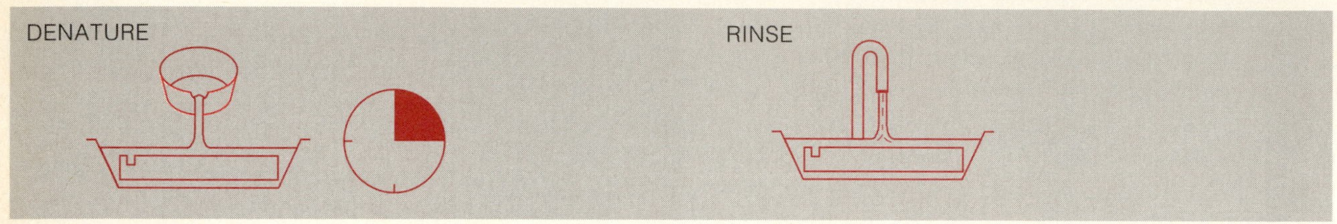

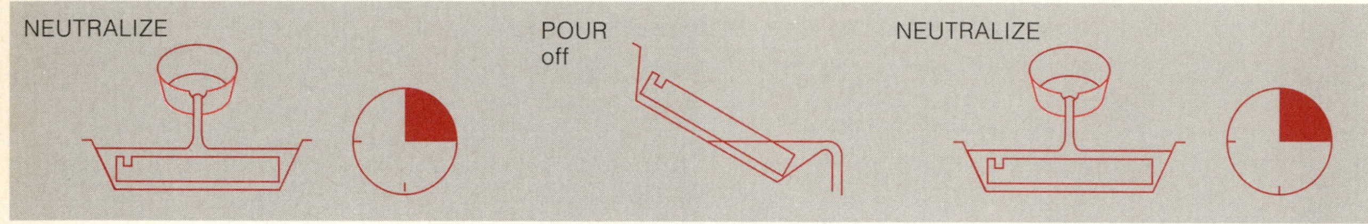

II. Set Up Blotting Apparatus and Transfer to Nylon Membrane

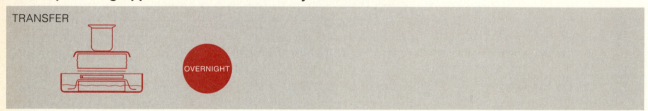

III. Wash and Bake Nylon Membrane

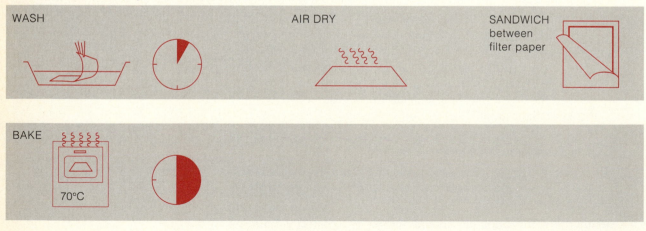

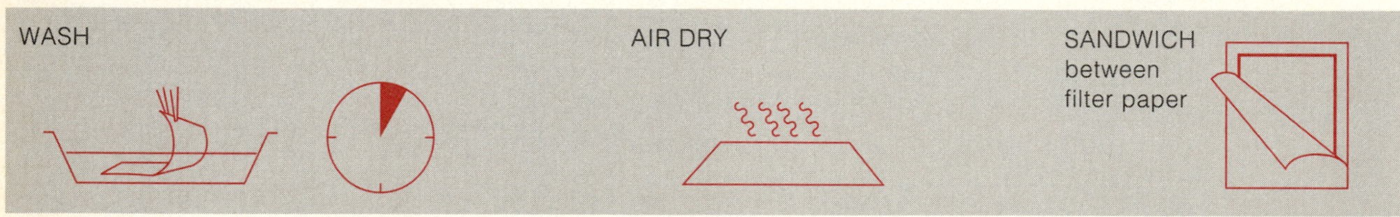

PRELAB NOTES

Blotting Methods

Three methods are commonly used to transfer nucleic acids from gels to nitrocellulose or nylon membranes: capillary, electrophoretic, and vacuum transfer. Although each method has certain advantages, we recommend the capillary method because it requires no special equipment.

1. The capillary method was introduced by Edwin Southern in 1975 and uses the flow of a transfer buffer to deposit the DNA fragments on the membrane. A stack of dry paper towels acts as a wick to draw the buffer up through the gel/membrane sandwich. Small DNA fragments are efficiently transferred in about 1 hour; fragments longer than 15,000 base pairs require an overnight transfer and, even then, may not be completely transferred.
2. Electrophoretic transfer uses current to propel DNA fragments out of the gel and onto the membrane. This method became popular after the development of nylon membranes that can bind DNA fragments using buffers of low ionic strength. Transfers can be completed in 2–3 hours, but cooled buffer must be continuously cycled through the apparatus to dissipate the ohmic heat generated during electrophoresis. This is the method of choice when DNA fragments are transferred out of polyacrylamide gels.
3. Vacuum transfer is the fastest, most efficient method for depositing DNA fragments on membranes. Any of several commercially available devices place the gel/membrane sandwich on a porous support over a vacuum chamber. Buffer is then drawn from an upper reservoir, carrying the DNA fragments out of the gel and onto the membrane.

Nylon *versus* Nitrocellulose Membranes

When choosing a membrane for hybridization experiments, you should consider the type of label to be used. Radioactive and nonradioactive labels perform differently depending on the type of membrane used. Manufacturers generally recommend the membranes found to work best with their products.

For years, nitrocellulose was the only membrane available for nucleic acid hybridization, despite its several drawbacks. Nucleic acids are attracted to nitrocellulose via hydrophobic interactions; because binding is reversible, DNA is gradually released from the membrane during hybridization. Furthermore, nitrocellulose is fragile and does not stand up well to repeated cycles of washing and hybridization. Finally, nitrocellulose is flammable, which means that it must be baked in a vacuum oven.

Nylon membranes bind nucleic acids irreversibly under the low ionic strengths needed for electrophoretic transfer. They are relatively durable, nonflammable, and can undergo repeated hybridizations. Nylon membranes are available either unmodified or in a positively charged form that binds larger quantities of DNA and is preferred for Southern hybridizations. The sole disadvantage of nylon is that it sometimes gives higher levels of background, especially with RNA probes.

Filter Paper

During transfer, baking, and drying steps, the gel or membrane is placed in contact with sheets of relatively thick filter paper of medium porosity. All references to filter paper in this laboratory are to Whatman 3mm filter paper, the type most commonly used for Southern hybridization.

Plastic Containers

Inexpensive plastic containers, such as those sold by Rubbermaid, are ideal for hybridization. These are available in several convenient sizes, notably 15 × 15 × 4.5 cm (470 ml) and 17 × 28 × 6.5 cm (1.7 liter).

For Further Information

The protocol presented here is based upon the following published method:

Southern, E. M., 1975. Detection of specific sequences among DNA fragments separated by gel electrophoresis. *Journal of Molecular Biology* 98: 503.

SOUTHERN TRANSFER OF λ RESTRICTION FRAGMENTS

Reagents

distilled water
denaturing buffer
neutralizing buffer
10× SSC buffer
2× SSC buffer

Equipment and Supplies

ballpoint pen
beaker, 400 ml
disposable gloves
flat forceps
2 filter paper wicks (9.5 × 25 cm)
2 pieces of filter paper (9.5 × 8 cm)
3 pieces of filter paper (12 × 14 cm)
gel-casting tray
masking tape
nylon membrane (9.5 × 8 cm)
paper towels (9.5 × 8 cm)
4 Parafilm strips (12 × 2 cm)
small plastic container (15 × 15 × 4.5 cm)
large plastic container (17 × 28 × 6.5 cm)
scissors
weigh boat
70° C oven

I. Denature and Neutralize DNA

(60 minutes)

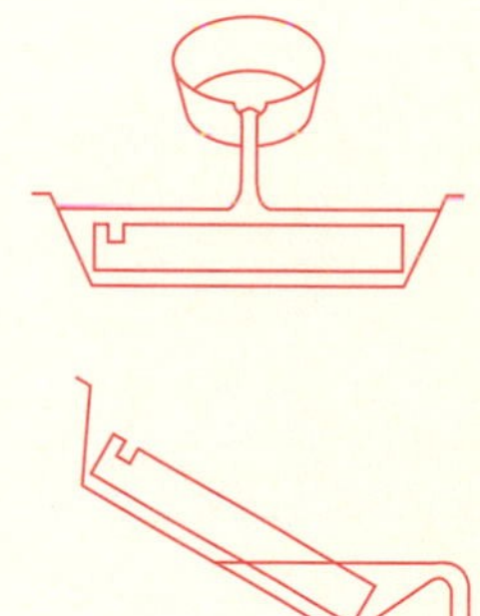

1. Using gloved hands, transfer the agarose gel to small plastic container. Flood the gel with 100 ml of denaturing buffer (0.5 N NaOH, 1.5 M NaCl). Soak the gel, with occasional agitation, for 15 minutes.
2. Pour off the denaturing buffer, and replace it with 100 ml of fresh denaturing buffer. Soak the gel, with occasional agitation, for another 15 minutes.
3. Pour off the denaturing buffer again, and rinse the gel briefly with 100 ml of distilled water.
4. Pour off the distilled water, and add 100 ml of neutralizing buffer (0.5 M Tris pH 7.5, 1.5 M NaCl). Soak the gel, with occasional agitation, for 15 minutes.
5. Pour off the neutralizing buffer, and replace it with 100 ml of fresh neutralizing buffer. Soak the gel, with occasional agitation, for another 15 minutes.

II. Set Up Blotting Apparatus and Transfer DNA to Nylon Membrane

(20 minutes; then overnight transfer)

1. Obtain a nylon membrane, using forceps or disposable gloves. Label the bottom corner of the membrane with your initials, using a ball-

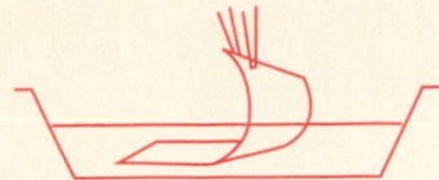

Do not label the membrane with a felt-tip pen, because the ink will bleed when it becomes wet.

Never touch the nylon membrane with your bare hands. Oils from your skin may be deposited on the membrane and can prevent the DNA fragments from binding.

The gel is inverted to facilitate transfer; this brings the DNA fragments, which migrate near the bottom of the gel, closer to the membrane.

The mask prevents the wick from directly contacting the paper towels to be added in step 8 and channels the transfer buffer up through the central part of the gel containing the DNA fragments. If the paper towels are cut exactly to the size of the gel, then the precaution of using a mask will be unnecessary.

Aligning the membrane with the bottom of the sample wells facilitates measuring the migration distance of the hybridized bands.

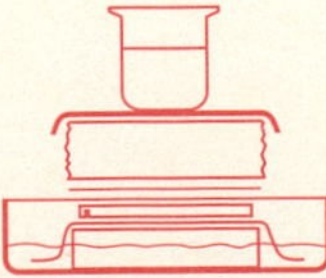

point pen. Wet the membrane by floating it and then submerging it in a clean container of distilled water.

2. Place the gel-casting tray (or other gel-sized support), upside down in the bottom of a large plastic container. Add about 500 ml of 10× SSC buffer to the container; *the buffer level must be below the top of the casting tray (or support).*
3. Place one filter paper wick (9.5 × 25 cm) atop the other to form a wick of double thickness, and wet the wick by submerging in 10× SSC buffer. Center the wet wick on top of the casting tray (or support), with the overlapping ends submerged in buffer.
4. Remove the agarose gel from the neutralization solution in Section I. Turn the gel over (with the sample wells facing downward), and center it on top of the saturated wick.
5. Place the wetted nylon membrane on top of the gel. Carefully align the top edge of the membrane flush with the bottom edge of the sample wells of the gel. *Use a gloved finger to gently push out any air bubbles trapped between the gel and the membrane.*
6. Mask the membrane with four strips of Parafilm. Each Parafilm strip should cover only about 2 mm of one edge of the membrane. *The mask must not interfere with the central area of the membrane containing the DNA fragments.*

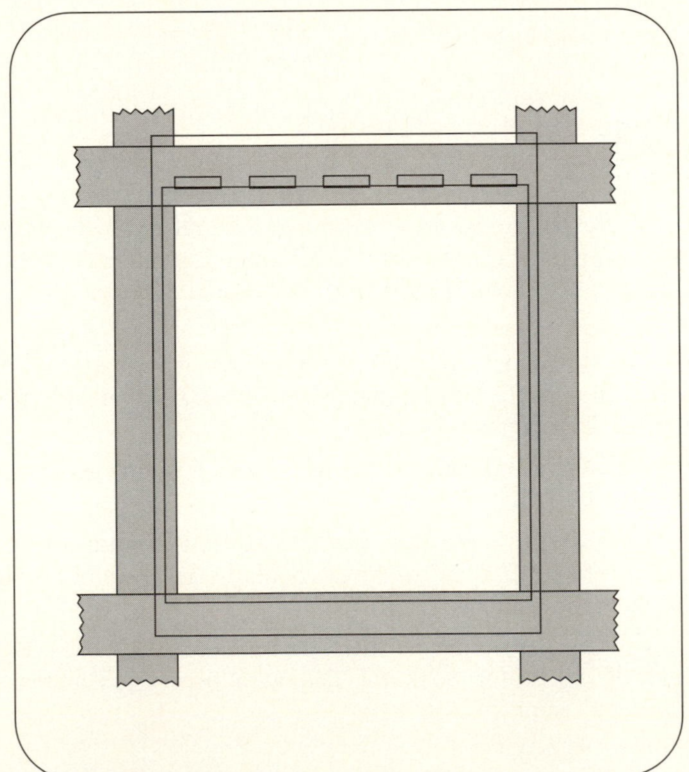

Parafilm Mask, Top View

7. Center two pieces of gel-sized filter paper (9.5 × 8 cm) on top of the nylon membrane.
8. Center a 5-cm stack of gel-sized paper towels (9.5 × 8 cm) on top of the filter paper.

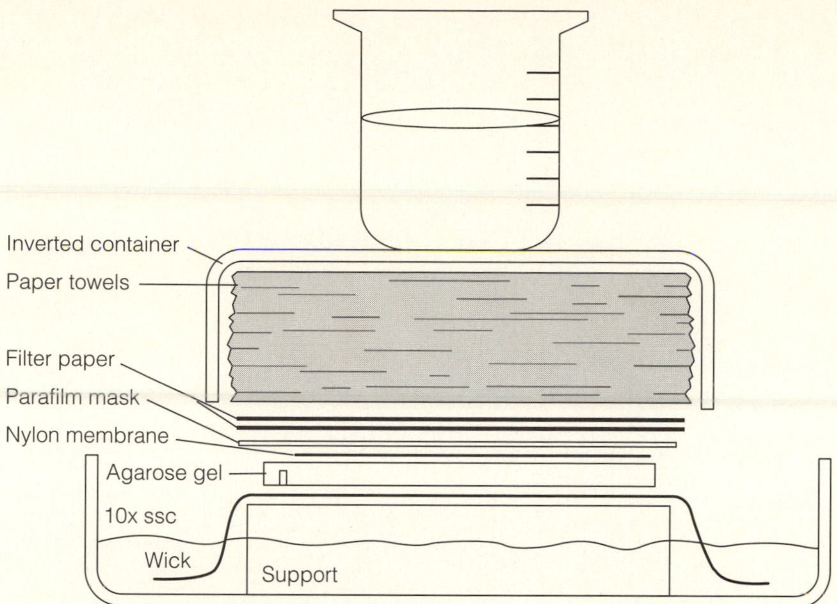

Blotting Set-Up for Southern Transfer

9. Invert the small plastic container, and place it over the stack of paper towels.
10. Carefully place a 400-ml beaker filled with water on top of the plastic container.

The weighted beaker presses the gel/membrane sandwich, the filter paper, and the paper towels into close contact, thus encouraging capillary flow through the blotting apparatus.

Stop Point

Transfer takes place overnight.

III. Wash and Bake Nylon Membrane

(60 minutes)

1. Remove the saturated paper towels, the filter paper, and the Parafilm mask from the blotting apparatus.
2. With the nylon membrane still in contact with the agarose gel, use scissors to cut off the bottom right corner of the gel/membrane sandwich.
3. Use forceps to peel the nylon membrane from the gel. Transfer the membrane into 100 ml of 2× SSC buffer in a small plastic container. Agitate for 1 minute.
4. Transfer the membrane to a clean piece of filter paper (12 × 14 cm), and allow it to air dry for 5 minutes.
5. Sandwich the membrane between two fresh pieces of filter paper (12 × 14 cm). Tape the four sides, and label the filter paper with your name and the date.
6. Bake the membrane at 70°C for 30 minutes or longer.

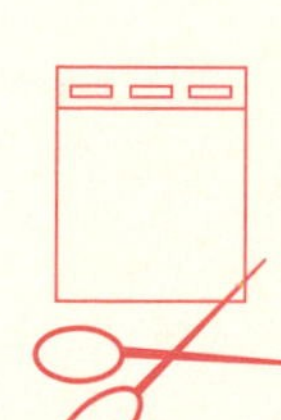

Cutting one corner provides registration to orient the hybridized bands on the membrane with the bands in the stained gel.

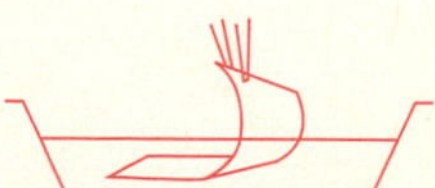

At this point, you can restain the agarose gel with ethidium bromide, to assess the efficiency of DNA transfer. Typically, the larger fragments will remain visible in the restained gel; this indicates that they have not been completely transferred onto the nylon membrane. This is expected and will not affect the results.

Stop Point

After it is baked, the membrane can be stored indefinitely at room temperature.

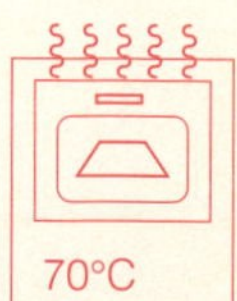

I. Prehybridize Nylon Membrane

II. Hybridize Nylon Membrane

Laboratory 13/Part C
Hybridization of the λ Probe

PRELAB NOTES

Hybridization Conditions

Hybridizations to eukaryotic genomes are usually carried out at 68°C in an aqueous solution or at 42°C in 50% formamide; both methods give comparable results. Hybridization at lower temperatures in formamide is less damaging to the fragile nitrocellulose membranes but takes two or three times longer than at higher temperature in an aqueous solution. Furthermore, formamide is a hazardous chemical. Use of a nylon membrane, in combination with the less-demanding conditions of hybridization to the small λ genome, will allow efficient hybridization at a lower temperature (50°C) *without* using formamide.

Minimizing Buffer Volumes

The rate of DNA reassociation increases as the volume decreases, thus reducing the amount of probe required and shortening the hybridization times. Use only just enough buffer to cover the membranes. This is often accomplished by either heat-sealing the membrane in a plastic bag containing about 15 ml of hybridization solution or by using a "batch" procedure (if you are hybridizing several membranes to the same probe), placing the membranes into a single container. In batch hybridization, be sure to use continuous agitation to prevent the membranes from adhering to one another.

Blocking Reagents

All protocols employ a reagent to block nonspecific attachment of the probe to the membrane. The blocking reagent usually contains a protein, such as nonfat dried milk or bovine serum albumin, often in combination with denatured, fragmented salmon sperm DNA and a detergent, such as SDS. Most protocols make use of a blocking reagent both during prehybridization and hybridization; when using nylon membranes, however, the blocking agent is sometimes omitted from the hybridization solution, because high concentrations of protein can interfere with probe binding.

The Advantages of Colorimetric Probe Detection

Hybridization probes can be restriction fragments, plasmids containing a sequence of interest, or synthetic oligonucleotides. Traditionally, researchers have used probes labeled with radioactive ^{32}P and detected by autoradiography—a technique where the hybridized membrane is sandwiched against X-ray film that becomes exposed in positions of probe binding. Radio-labeled probes, although extremely sensitive, have several serious drawbacks. Since the half-life of ^{32}P is only 14 days, the usefulness of a radio probe is limited to several weeks, depending upon the specific activity of the probe. Furthermore, use of radioactivity requires licensing and entails significant safety and disposal concerns.

Therefore, researchers are increasingly switching to nonradioactive probes. Colorimetric and chemiluminescent detection systems require no special licensing and do not present the same health and environmental hazards as radioactivity. Colorimetric systems typically employ digoxigenin or biotin and are well suited for teaching purposes. Results of colorimetric detection can be directly observed, and color development can be monitored. Chemiluminescent systems, although safe and more sensitive, entail exposing and developing film, which requires darkroom facilities, and they are also more difficult to monitor.

Probe Options

The probe for this experiment is a 25-nucleotide sequence (oligonucleotide), GAGTAGATGCTTGCTTTTCTGAGCC, that corresponds to base pairs 22,915–22,939 of the λ chromosome. For convenience, your instructor may have purchased a ready-to-use digoxigenin-labeled oligonucleotide probe from Carolina Biological Supply Company (CBS). However, several options are available for experimenters who would like to prepare or label the probe:

1. The unlabeled oligonucleotide can be obtained from Carolina Biological Supply Company and labeled using the Genius 6 Oligonucleotide Tailing Kit (Boehringer Mannheim catalog number 1417 231).
2. The oligonucleotide probe sequence can be custom-synthesized at a university or commercial facility and labeled using the Genius tailing kit. *Following synthesis, oligonucleotides must be further purified by HPLC or butanol extraction before being labeled.* Some custom synthesis facilities will also attach the digoxigenin label, for an additional charge.
3. The 784 base-pair *Bam*HI/*Hin*dIII fragment of plasmid pAMP contains the probe sequence. This fragment can be gel-purified and labeled with the Genius tailing kit. When using this restriction fragment as the probe, you should increase the stringency of the hybridization (65°C and 5× SSC buffer).

For Further Information

The protocol presented here is based upon the following published methods:

Genius System User's Guide, Boehringer Mannheim Corporation, Biochemical Products, 9115 Hague Road, P.O. Box 50414, Indianapolis, IN 46250-0414.

Lathe, J. Synthetic oligonucleotide probes deduced from amino acid sequence data: Theoretical and practical considerations. 1985. *Journal of Molecular Biology* 183: 1.

Martin, R., C. Hoover, S. Grimme, C. Grogan, J. Holtke, and C. Kessler. 1990. A highly sensitive, nonradioactive DNA labeling and detection system. *BioTechniques* 9(6): 762–768.

Southern, E. M., 1975. Detection of specific sequences among DNA fragments separated by gel electrophoresis. *Journal of Molecular Biology* 98: 503.

HYBRIDIZATION OF THE λ PROBE

Reagents	Equipment and Supplies
prehybridization buffer	disposable gloves
hybridization buffer (prehybridization buffer + labeled probe)	flat forceps
	small plastic container (15 × 15 × 4.5 cm)
	50°C shaking water bath

I. Prehybridize Nylon Membrane

(5 minutes; then 60+ minutes of incubation)

1. Using gloved hands, remove the nylon membrane from the filter paper sandwich of Part B, and transfer it to a container with prehybridization buffer. *If the membrane sticks to the filter paper, do not try to separate them; place both in the prehybridization buffer, and they will separate when wet.*
2. Incubate, with gentle shaking, at 50°C for 60 minutes or overnight. Cover the container to prevent evaporation.

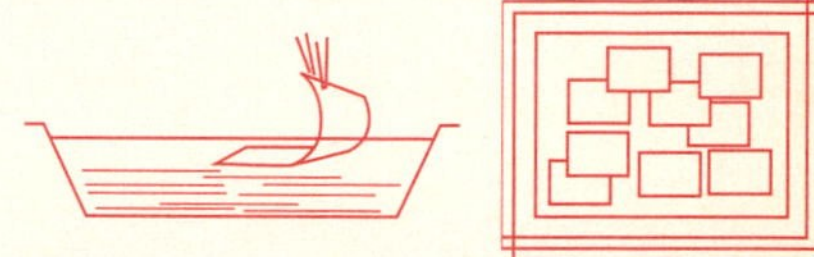

Never touch the nylon membrane with your bare hands. Oils from your skin may be deposited on the membrane and can prevent the probe from binding.

II. Hybridize Nylon Membrane

(5 minutes; then 120+ minutes of incubation)

1. Using gloved hands, transfer the nylon membrane to a container with hybridization buffer.
2. Incubate, with gentle shaking, at 50°C for two hours or overnight. Cover the container to prevent evaporation.

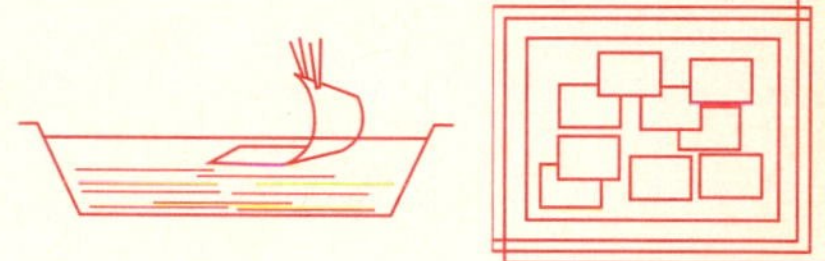

Wash and Develop Color

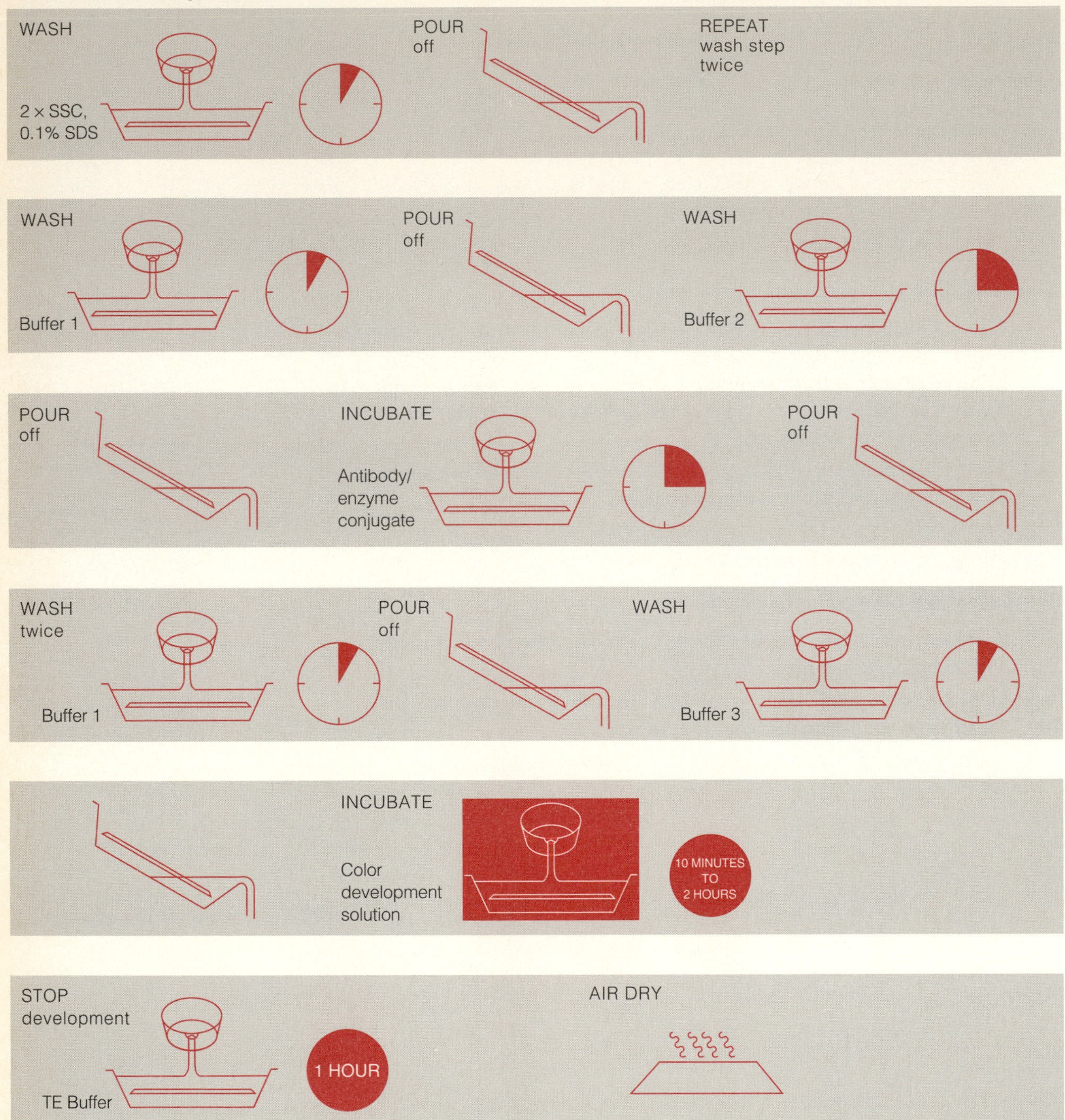

Laboratory 13/Part D
Nonradioactive Probe Detection

PRELAB NOTES

Review the instructions pertaining to color development that accompany the Genius 3 Nucleic Acid Detection Kit. The numbered buffers in this laboratory correspond to those in the Genius kit.

Antibody Preparation and Colorimetric Detection

Digoxigenin, which is used to label the probe in this laboratory, is widely used for immune detection. This plant steroid molecule is highly antigenic. Because it is found exclusively among foxglove and digitalis plants, antibodies directed against it will not cross-react with antigens from other organisms.

To prepare antidigoxigenin antibodies, sheep are immunized with digoxigenin, and the IgG fraction is purified from serum by ion exchange chromatography. Digoxigenin-specific antibodies are isolated by immunosorption: The IgG fraction is passed through a column packed with resin cross-linked to digoxigenin. The purified polyclonal antidigoxigenin antibodies are then covalently linked to the enzyme alkaline phosphatase.

The antibody/enzyme conjugate is used to bind the digoxigenin label on the probe, producing an immune complex. In a separate step, the antibody/digoxigenin complex is detected by two color-forming reactions. First, alkaline phosphatase catalyzes the removal of a phosphate group from 5-bromo-4-chloro-3-indoylphosphate (X-phosphate); the resulting oxidation product dimerizes to form an indigo precipitate. Hydride ions released during dimerization reduce a second compound, nitro blue tetrazolium (NBT), to form a purple precipitate. The amount of the two precipitates deposited on the membrane is proportional to the mass of target DNA present.

The undiluted antibody/enzyme conjugate is stable at 4°C. *Do not freeze it.* Diluted antibody/enzyme conjugate will remain stable for about 12 hours at 4°C. The color development substrates X-phosphate and NBT are stable at –20°C.

Blocking Reagent

During the washing steps, some blocking reagent may be stripped from the nylon membrane. Because the antibody can also bind nonspecifically to the membrane, you must reblock the membrane before adding the antibody/enzyme conjugate.

Reuse the Hybridization Solution

Remember that the hybridization solution (from Part C) is reusable and can be stored frozen for as long as 1 year.

Washing

It is best to wash membranes under very stringent conditions (high temperature and low salt), typically at the hybridization temperature and with 2× SSC or even 0.5× SSC.

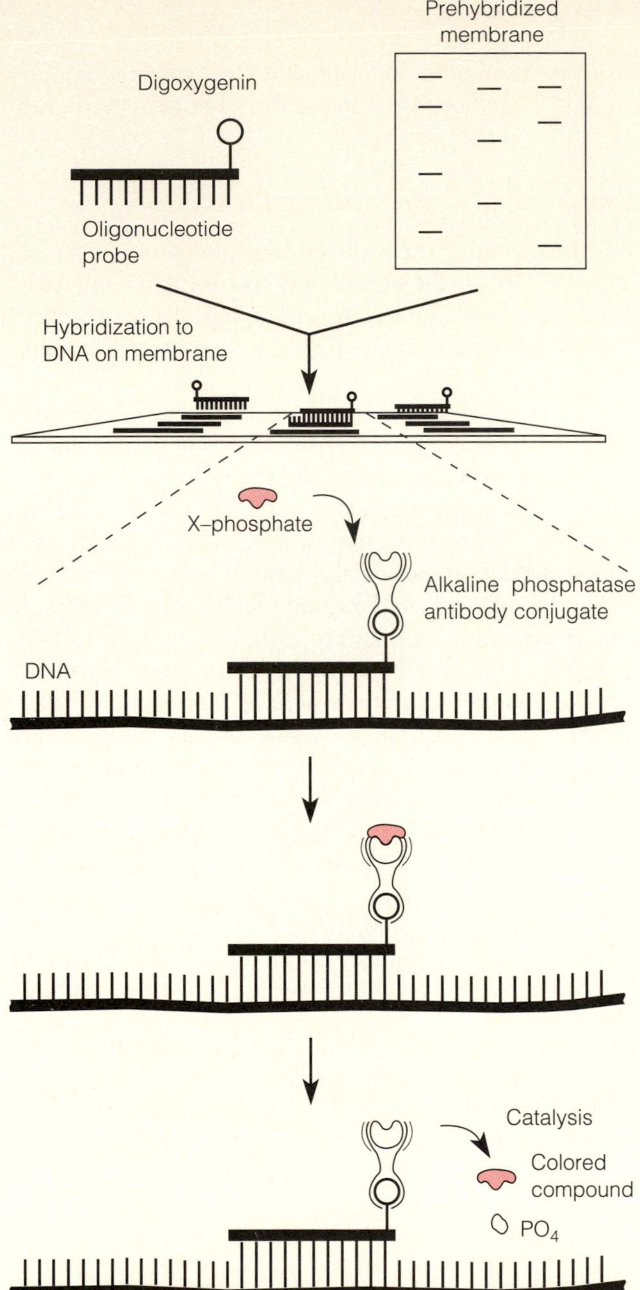

For Further Information

The protocol presented here is based upon the following published methods:

Genius System User's Guide, Boehringer Mannheim Corporation, Biochemical Products, 9115 Hague Road, P.O. Box 50414, Indianapolis, IN 46250-0414.

Martin, R., C. Hoover, S. Grimme, C. Grogan, J. Holtke, and C. Kessler. 1990. A highly sensitive, nonradioactive DNA labeling and detection system. *BioTechniques* 9(6): 762.

NONRADIOACTIVE PROBE DETECTION

Reagents	Equipment and Supplies
wash buffer (2× SSC, 0.1% SDS)	disposable gloves
buffer 1 (0.1*M* Tris pH 7.5, 0.15*M* NaCl)	filter paper
buffer 2 (buffer 1 + 1% w/v blocking reagent)	flat forceps
diluted antibody/enzyme solution	petri dish (150-mm)
buffer 3 (0.1*M* Tris pH 9.5, 0.1*M* NaCl, 0.05*M* $MgCl_2$)	small plastic container (15 × 15 × 4.5 cm)
color development solution	50°C water bath
TE buffer	

Wash and Develop Color

(70 minutes; then 10+ minutes of development)

Gentle agitation by hand should accompany each wash (incubation), *except for the color development in step 11.*

Do not discard the hybridization buffer; collect it for reuse!

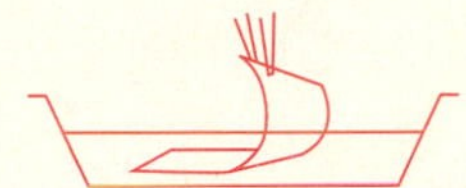

Never touch the nylon membrane with your bare hands. Oils from your skin may be deposited on the membrane and can prevent the antibody/enzyme conjugate from binding.

1. Using gloved hands and forceps, remove the membrane from the hybridization solution (Part C), and transfer it to a small plastic container containing 100 ml of wash buffer. Incubate for 5 minutes at room temperature.
2. Pour off the first wash, and replace it with 100 ml of fresh wash buffer. Incubate, with agitation, for 5 minutes at room temperature.
3. Pour off the second wash, and replace it with 100 ml of fresh wash buffer. Incubate, with agitation, at *50°C* for 5 minutes.
4. Pour off the third wash, and add 100 ml of buffer 1. Incubate, with agitation, for 5 minutes at room temperature.

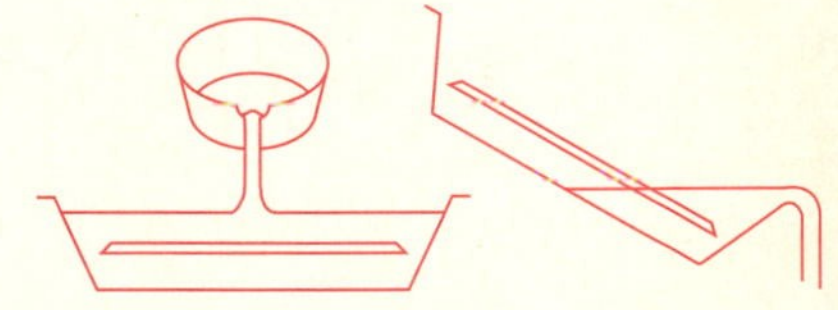

5. Pour off buffer 1, and add 100 ml of buffer 2. Incubate, with agitation, for 15 minutes at room temperature.
6. Pour off buffer 2, and add 40 ml of the antibody/enzyme solution. Incubate, with agitation, for 15 minutes at room temperature.
7. Pour off the antibody/enzyme solution, and add 100 ml of buffer 1. Incubate, with agitation, for 5 minutes at room temperature.
8. Pour off buffer 1, and replace with 100 ml of *fresh* buffer 1. Incubate, with agitation, for 5 minutes at room temperature.
9. Pour off buffer 1, and add 100 ml of buffer 3. Incubate, with agitation, for 5 minutes at room temperature.
10. Transfer the membrane to a clean 150-mm petri dish or other suitable container.
11. Add 20 ml of freshly diluted color development solution to the membrane, and incubate it at room temperature *in the dark* until bands become clearly visible (from 10 minutes to 1 hour). *Make sure that the membrane is positioned with the DNA side up. Do not agitate!*

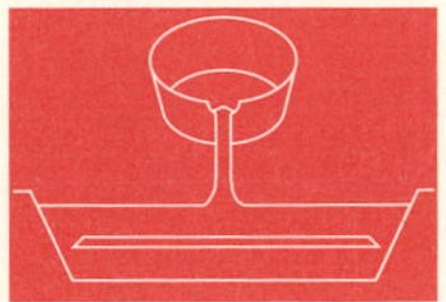

If the hybridization signal is faint, color development can be extended overnight. However, overincubation will cause the entire membrane to begin turning brown.

The final wash in TE buffer stops the color development and prevents further darkening of the background.

12. When the hybridization signals reach an acceptable level, pour off the development solution, and add 20 ml of TE buffer. Incubate, with agitation, at room temperature for at least 30 minutes.
13. Air dry the membrane on a piece of clean filter paper, and store it in the dark.

RESULTS AND DISCUSSION

Use the cut corner of your hybridized membrane to orient it with the photo of your stained gel from Part A. Does the pattern of hybridized bands on the membrane agree with your predicted result? Compare your result to the ideal membrane pattern (below), showing the sizes of all hybridizing fragments. Account for any differences in separation or band intensity between your result and the ideal membrane pattern.

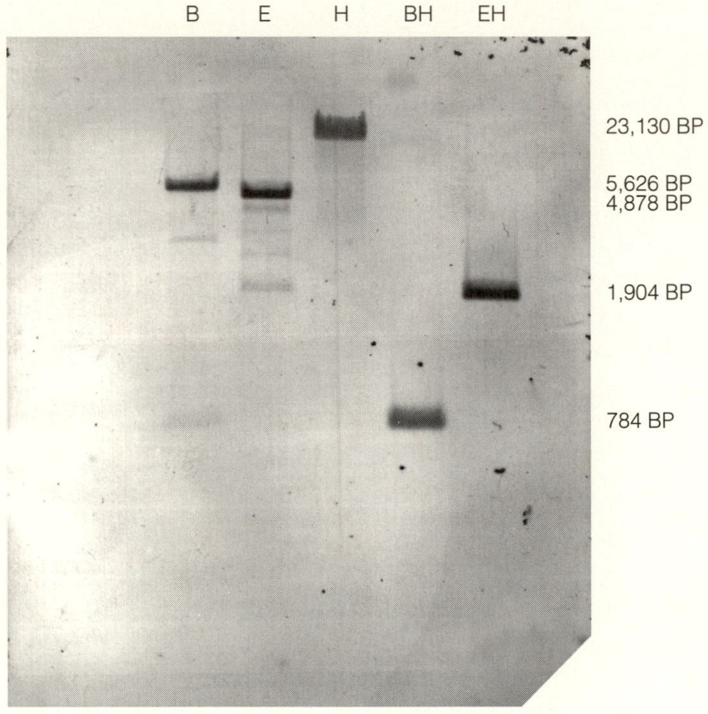

Ideal Membrane

1. Based on your results, how many copies of the probe sequence are present in the λ genome?
2. How can you account for the faint bands that occur on the membrane in unexpected positions?
3. How can you distinguish between signals produced by nonspecific hybridization and those produced by partial digestion of the target DNA?
4. What factors influence the efficiency of transferring DNA restriction fragments to the membrane?
5. Why must the DNA fragments be exposed to an alkaline solution before hybridization?

6. Why is it unnecessary to denature the oligonucleotide probe?
7. What is the purpose of the prehybridization step?
8. What factors influence hybridization of the probe to the target sequence?
9. What are the advantages and disadvantages of using a radioactively labeled probe?

For Further Research

1. Research the application of Southern hybridization in:
   a. screening for genetic disease.
   b. forensic and paternity testing.
   c. molecular evolution.
2. Design and execute an experiment to test the sensitivity of Southern hybridization.
3. Design and execute a series of experiments to study the effects of hybridization stringency.
4. Design and execute an experiment that uses pBLU to probe the genomes of different *E. coli* strains to detect the presence of the *lacZ* gene.
5. Prepare restriction digests of λ DNA, and make various dilutions with salmon sperm or other eukaryotic DNA. Test the ability of the probe to detect the λ sequence against the background of the eukaryotic genome.
6. Gel-purify a 500–1,000-bp restriction fragment from λ, and label it by using the Genius 6 Oligonucleotide Tailing Kit (Boehringer Mannheim catalog number 1417 231). Hybridize your probe to restricted DNAs from other bacteriophages or modified λ vectors, such as λ gt10 and λ gt11. Compare and contrast the location of the probe sequence across the various bacteriophages.

Genomic Library Construction and Analysis

LABORATORY 14

Construction of a Genomic Library of λ DNA

A genomic library is the collection of an organism's total genetic complement. Chromosomal DNA is typically extracted and digested with one or more restriction enzymes. The resulting restriction fragments are ligated into a vector—usually a modified bacteriophage or yeast chromosome—and transformed into an appropriate host cell line. The genomic library then consists of thousands of plaques or host cells of transformants, each of which potentially contains a different fragment of chromosomal DNA. The library can be repeatedly screened by hybridization to identify those host cells that have taken up a vector containing a gene or DNA sequence of interest.

The term λ *library* usually refers to a genomic library in which eukaryotic DNA is cloned into a modified λ vector. In this laboratory, however, the λ genome is cloned into a plasmid vector. In Part A, "Restriction Digest of λ and Plasmid pBLU," samples of genomic and vector DNA are each double-digested with *Bam*HI and *Hin*dIII. Following incubation at 37°C, samples of the DNA are electrophoresed in an agarose gel to confirm proper cutting. The double digest of λ DNA generates 13 fragments, ranging in size from 125 to 16,841 base pairs. Plasmid pBLU has a single recognition site for each enzyme, located only a few base pairs apart, within a cloning site called a "polylinker." The double digest opens the plasmid by removing the short sequence between the *Bam*HI and *Hin*dIII sites.

In Part B, "Ligation of λ and pBLU Restriction Fragments," the restriction digests of λ and pBLU are first heated to destroy restriction enzyme activity. Samples from each reaction are mixed together and incubated at room temperature in a buffer containing DNA ligase, ATP, and magnesium. The *Bam*HI and *Hin*dIII "sticky ends" hydrogen-bond with their complementary sequences to align restriction fragments. Ligase catalyzes the formation of covalent phosphodiester bonds that link complementary ends into stable recombinant-DNA molecules.

I. Set Up Restriction Digests

II. Cast 0.8% Agarose Gel

III. Load Gel and Electrophorese

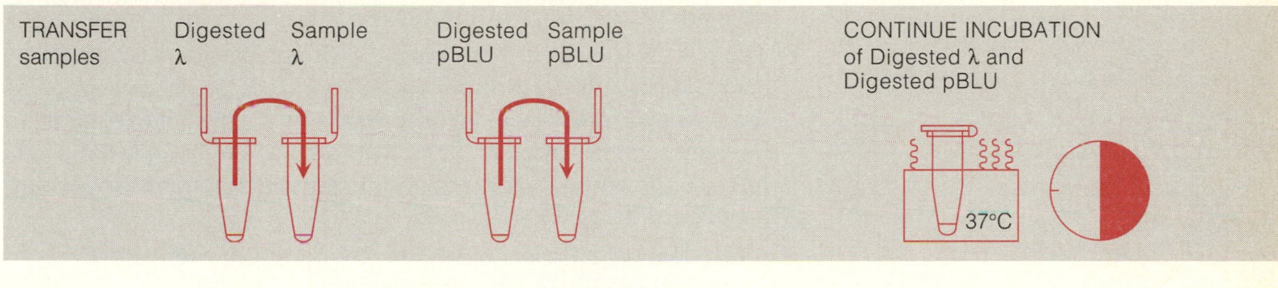

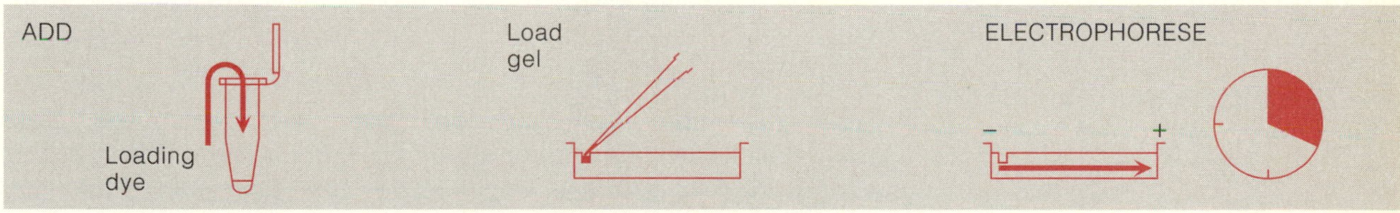

IV. Stain Gel with Ethidium Bromide, View, and Photograph

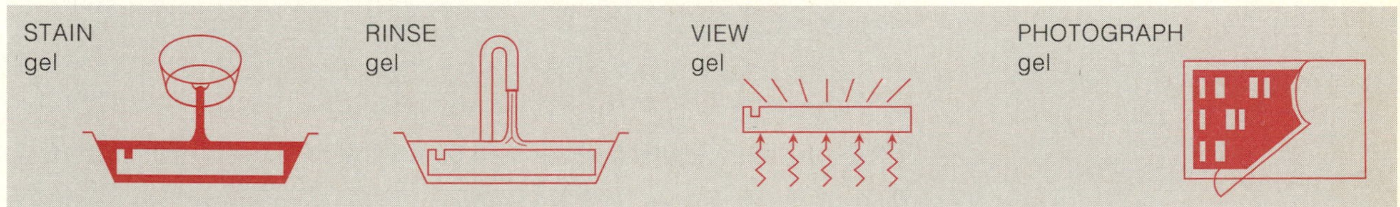

Laboratory 14/Part A
Restriction Digest of λ and Plasmid pBLU

PRELAB NOTES

Review the prelab notes in Laboratory 3, "DNA Restriction and Electrophoresis."

Plasmid pBLU

The process of constructing and analyzing recombinant-DNA molecules in Laboratories 14–17 is not trivial. However, good results can be expected, if the directions are followed carefully. Almost any plasmid containing a selectable antibiotic-resistance marker can be used to demonstrate the cloning of foreign DNA sequences into *E. coli*. This experimental unit has been optimized for the plasmid pBLU, and the extensive analysis of results is based entirely on recombinant molecules derived from this parent plasmid. It is unlikely that your instructor will substitute pBLU with other plasmids, because pBLU was constructed specifically as a teaching plasmid. It offers the following advantages:

1. pBLU is derived from a pUC expression vector that replicates to a high number of copies per cell. Yields from plasmid preparations are significantly greater than those obtained with pBR322 and other plasmids with a lower copy number.
2. The pUC plasmids encode a partial *lacZ* gene product that must be complemented by a corresponding partial *lacZ* product encoded by the host cell. pBLU contains the *entire lacZ gene* and requires no host cell complementation. Therefore, any *lacZ* minus a host (a host lacking a functional *lacZ* gene) can be used; recommended strains include JM101 and JM107.
3. pBLU does not have *lacI* repressor activity, which eliminates the need for the inducer IPTG in the growth medium.

The Prudent Control

In Section III, samples of the restriction digests are electrophoresed, before ligation, to confirm complete cutting by the endonucleases. This prudent control is standard experimental procedure. If you are pressed for time, your instructor may advise you to omit electrophoresis and ligate DNA directly following the restriction digest. However, be sure to pretest the activity of the *Bam*HI and *Hin*dIII enzymes, and incubate long enough for complete digestion.

For Further Information

The protocol presented here is based on the following published methods:

Sharp, P. A., B. Sugden, and J. Sambrook. 1973. Detection of two restriction endonuclease activities in *Hemophilus parainfluenzae* using analytical agarose-ethidium bromide electrophoresis. *Biochemistry* 12: 3055.

Helling, R. B., H. M. Goodman, and H. W. Boyer. 1974. Analysis of *Eco*RI fragments of DNA from lambdoid bacteriophages and other viruses by agarose-gel electrophoresis. *Journal of Virology* 14: 1235.

Messing, J. and J. Vieira. 1982. The pUC plasmids, an M13mp7-derived system for insertion mutagenesis and sequencing with synthetic universal primers. *Gene* 19: 259.

RESTRICTION DIGEST OF λ AND pBLU DNAS

Reagents	**Supplies and Equipment**
For digest:	0.5–10-µl micropipettor + tips
0.30 µg/µl / DNA	10–100-µl micropipettor + tips
0.15 µg/µl pBLU	1.5-ml tubes
*Bam*HI/ *Hin*dIII	aluminum foil
2× restriction buffer	beaker for waste/used tips
	camera and film (optional)
For electrophoresis:	disposable gloves
0.10 µg/µl / DNA	electrophoresis box
0.10 µg/µl pBLU	masking tape
loading dye	microfuge (optional)
1.0% agarose	Parafilm or waxed paper (optional)
1× Tris/Borate/EDTA (TBE) buffer	permanent marker
1 µg/ml ethidium bromide	plastic wrap (optional)
	power supply
For decontamination:	test tube rack
0.05 M $KMnO_4$	transilluminator
0.25 N HCl	37°C water bath
0.25 N NaOH	60°C water bath (for agarose)

I. Set Up Restriction Digests

(10 minutes; then 30+ minutes of incubation)

Refer to Laboratory 3, "DNA Restriction and Electrophoresis," for detailed instructions on setting up reactions.

1. Obtain labeled reaction tubes containing λ and pBLU. Use a permanent marker to add the label "digested" to each tube. *Add other reagents directly to these tubes.*
2. Use the matrix below as a checklist while adding reagents to each reaction. *Use a fresh pipet tip for each reagent.*

Tube	λ DNA	pBLU	2x Buffer	*Bam*HI/ *Hin*dIII
Digested λ	12 µl (preadded)	—	15 µl	3 µl
Digested pBLU	—	12 µl (preadded)	15 µl	3 µl

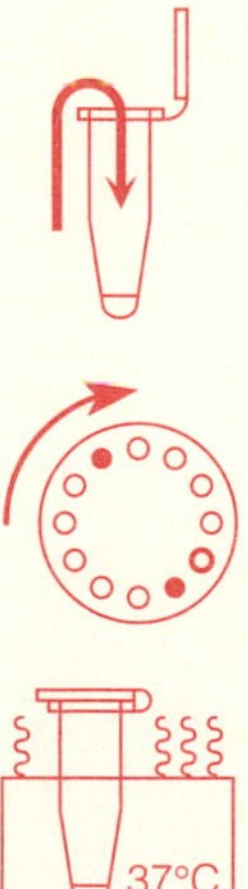

3. After adding all of the above reagents, close the tops of the tubes. Pool and mix the reagents by pulsing them in a microfuge or by sharply tapping the bottom of the tube on the lab bench.
4. Place the reaction tubes in a 37°C water bath, and incubate them for 30 minutes or longer.

Stop Point

After a full 30-minute incubation (or longer), freeze the reactions at −20°C until you are ready to continue. Thaw the reactions before proceeding on to Section III, step 1.

II. Cast 1.0% Agarose Gel

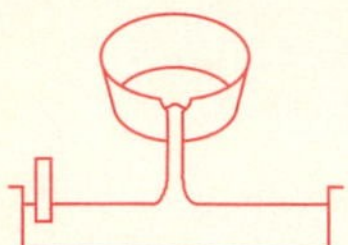

(15 minutes)

1. Carefully pour agarose solution into the casting tray to fill it to a depth of about 5 mm. The gel should cover only about one-third of the height of the comb teeth.
2. After the agarose solidifies, place the gel-casting tray into the electrophoresis box, and set up for electrophoresis.

Stop Point

Cover the electrophoresis box, and save the gel until you are ready to continue. The gel will remain in good condition for several days if it is completely submerged in buffer.

III. Load Gel and Electrophorese

Refer to Laboratory 3, "DNA Restriction and Electrophoresis," for detailed instructions on casting and loading gel.

(10 minutes; then 20+ minutes of electrophoresis)

Only a fraction of the *Bam*HI/*Hin*dIII digests of λ and pBLU are electrophoresed to determine whether the DNAs are completely cut. These restricted samples are electrophoresed along with uncut λ and pBLU as controls.

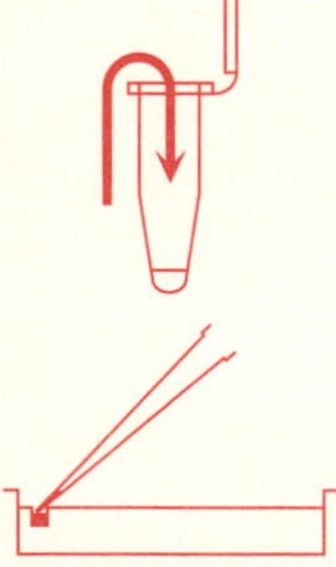

1. Use a permanent marker to label two clean 1.5-ml tubes "sample λ" and "sample pBLU." Remove the "digested λ" and "digested pBLU" tubes from the 37°C water bath. Transfer 5 µl of DNA from the "digested λ" tube into the "sample λ" tube. Transfer 5 µl of plasmid from the "digested pBLU" tube into the "sample pBLU" tube.
2. *Immediately return the "digested λ" and "digested pBLU" tubes to the water bath, and continue incubating at 37°C during electrophoresis.*
3. Obtain 1.5-ml tubes containing control λ and pBLU DNAs. Use a permanent marker to add the label "uncut" to each tube.
4. Add 1 µl of loading dye to the digested "samples" and undigested "controls" of λ and pBLU. Close the tops of the tubes and mix by either tapping the bottom of the tube on the lab bench, pipetting in and out, or pulsing in a microfuge.
5. Load the entire contents of each sample tube into a separate well in the gel, as shown in the diagram below. *Use a fresh pipet tip for each sample.*

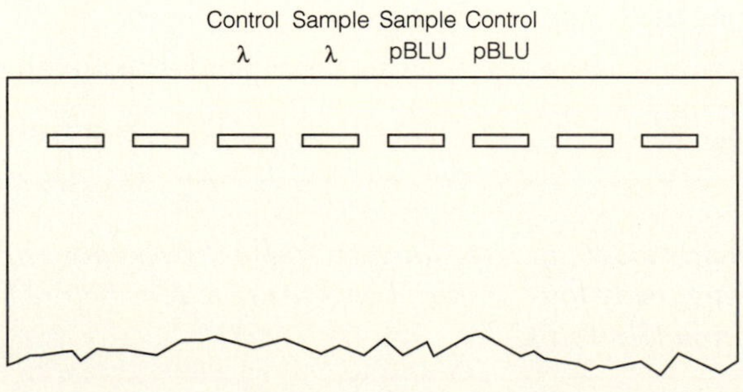

6. Electrophorese at 100–150 volts, until the bromophenol blue bands have moved 30–40 mm from the wells.

7. Turn off the power supply, remove the casting tray from the electrophoresis box, and transfer the gel to a disposable weigh boat (or other shallow tray) for staining.

IV. Stain Gel with Ethidium Bromide, View, and Photograph

(10–15 minutes)

CAUTION

Review the section on "Ethidium Bromide Staining and Responsible Handling" in Laboratory 3. Wear disposable gloves when staining, viewing, and photographing gel, and during cleanup. Confine all staining to a restricted sink area.

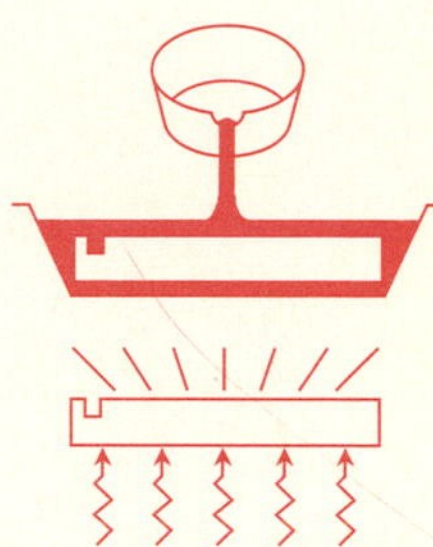

1. Flood the gel with an ethidium bromide solution (1 μg/ml), and allow it to stain for 5–10 minutes. Destain, if desired.
2. View the gel under an ultraviolet transilluminator or other UV light source.

CAUTION

Ultraviolet light can damage your eyes. Never look directly at an unshielded UV light source without eye protection. View only through a filter or safety glasses that absorb the harmful wavelengths.

3. Make an exposure of 1–2 seconds with the camera aperature set at f/8. Develop the print for the recommended time (approximately 45 seconds at room temperature).
4. Take the time for proper cleanup:
   a. Wipe down the camera, the transilluminator, and the staining area.
   b. Decontaminate the gel and any staining solution that will not be reused.
   c. Wash your hands before leaving the laboratory.

RESULTS AND DISCUSSION

Compare your stained gel with the ideal gel (below), and check to determine whether the pBLU and λ DNAs have been completely digested by *Bam*HI and *Hin*dIII. Also refer to the restriction maps in Appendix 3 and page 50.

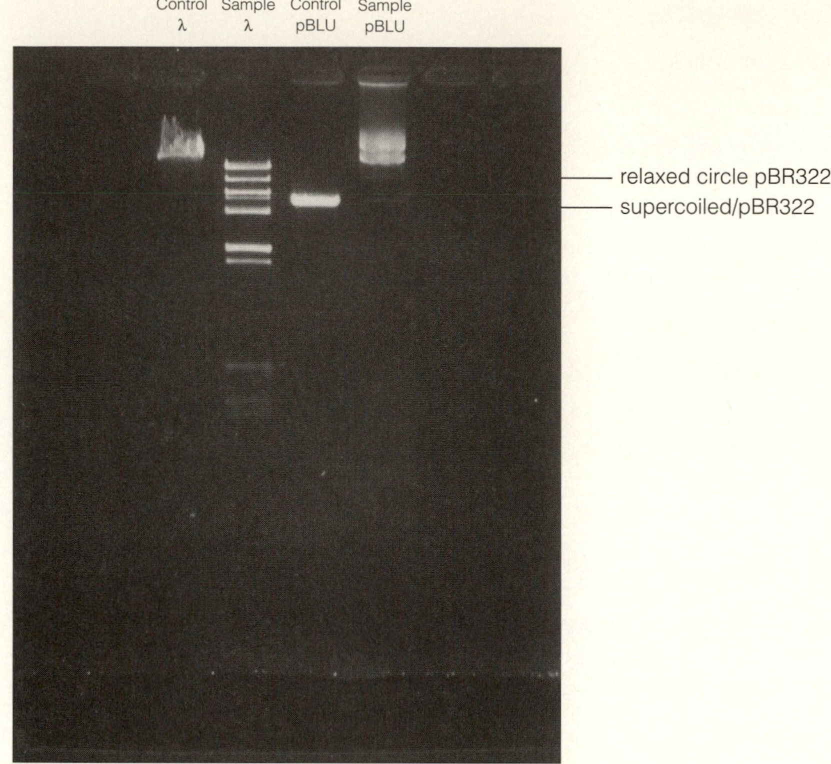

Ideal Gel

1. The sample pBLU lane should show a *single band* of 5,400 bp, usually in a position near the supercoiled form in the control pBLU lane. The presence of additional bands in the sample pBLU lane indicates some degree of incomplete digestion. (Recall that the digest actually produces an additional 30-bp fragment, which quickly runs off of the end of the gel during electrophoresis.)
2. The sample λ lane should show 9–10 distinct fragments, ranging in size from 493 to 16,841 bp; however, there is also a 125-bp fragment that is often too faint to be seen or else it runs off of the end of the gel during electrophoresis. The digest actually produces a total of 13 fragments, but several fragments are similar in size and do not resolve well from one another (eg. 2,409, 2,396, and 2,322 bp). Commercially available λ DNA is a mixture of both linear and circular molecules, and this adds further confusion. Recall that the single-stranded COS sites at each end of the λ genome are complementary and can hydrogen-bond to one another to form a closed circle. A percentage of circular molecules are covalently linked by phosphodiester bonds. A cut at the left-most restriction site in the linear λ chro-

mosome (by *Bam*HI) produces a 5,505-bp fragment; the right-most site (by *Hin*dIII) produces a 4,361-bp fragment. However, *Bam*HI/*Hin*dIII digestion of the COS circlet produces a 9,866-bp fragment, which is the sum of the left- and right-most restriction fragments. Thus, the 5,505-bp and 4,361-bp bands may appear fainter than expected, depending on the percentage of circular molecules in the λ sample. (Heating λ DNA for 10 minutes at 65°C immediately before electrophoresis disrupts COS circles held only by hydrogen bonds and increases the yield of the end fragments; however, heating has no effect on covalently closed circles.)

3. If both digests look complete or nearly complete, continue on to Part B, "Ligation of λ and pBLU Restriction Fragments." The reaction will most likely have gone to completion with the additional incubation during electrophoresis.
4. If either digest looks very incomplete, add another 1 μl of *Bam*HI/*Hin*dIII solution, and incubate for an additional 20 minutes. Then, continue on to Part B.

Ligate λ and pBLU

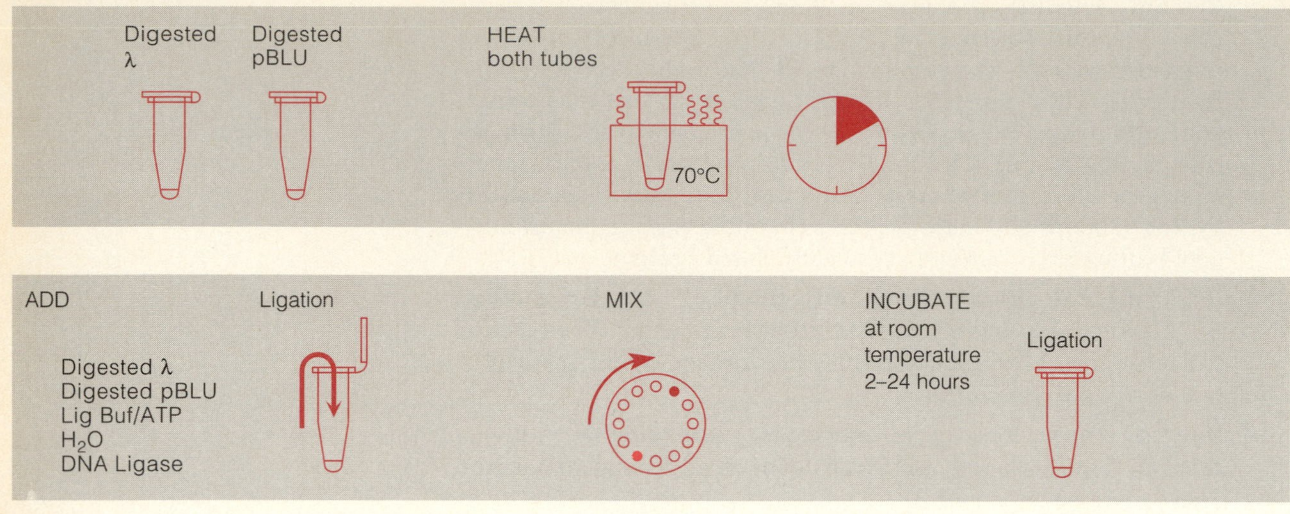

Laboratory 14/Part B
Ligation of λ and pBLU Restriction Fragments

PRELAB NOTES

Review the prelab notes in Laboratory 7B, "Ligation of pAMP and pKAN Restriction Fragments."

For Further Information

The protocol presented here is based on the following published method:

Cohen, S. N., A. C. Y. Chang, and H. W. Boyer. 1973. Construction of biologically functional bacterial plasmids in vitro. *Proceedings of the National Academy of Sciences USA* 70: 3240.

LIGATION OF λ AND pBLU RESTRICTION FRAGMENTS

Reagents	Supplies and Equipment
digested / (from Part A)	0.5–10-μl micropipettor + tips
digested pBLU (from Part A)	1.5-ml tube
10× ligation buffer/ATP	beaker for waste/used tips
T4 DNA ligase	disposable gloves
distilled water	microfuge (optional)
	test tube rack
	70°C water bath

Ligate λ and pBLU

(30 minutes; then 2+ hours of ligation)

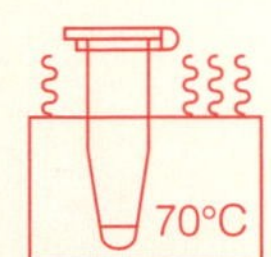

Step 1 is critical: Heat denaturation inactivates the restriction enzymes.

1. Incubate the "digested λ" and "digested pBLU" tubes in a 70°C water bath for 10 minutes.
2. Label a clean 1.5-ml tube "ligation."
3. Use the matrix below as a checklist while adding reagents to the reaction. *Use a fresh pipet tip for each reagent.*

Tube	Digested λ	Digested pBLU	10x Ligation Buffer/ATP	Water	DNA Ligase
Ligation	4 μl	4 μl	2 μl	8 μl	2 μl

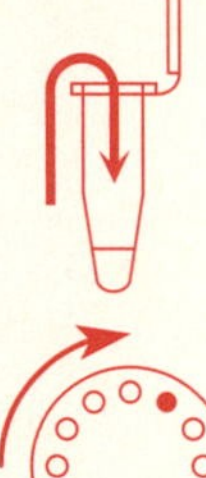

4. After adding all of the above reagents, close the top of the tube. Pool and mix reagents by pulsing them in a microfuge or by sharply tapping the bottom of the tube on the lab bench.
5. Incubate the reaction at room temperature for 2–24 hours.
6. If time permits, the ligation can be confirmed by electrophoresing 5 μl of the ligation reaction, along with *Bam*HI/*Hin*dIII digests of λ and pBLU. Few, if any, of the low-molecular-weight bands in the λ and pBLU digests should be present in the ligated DNA gel lane. Instead, the ligated lane should show multiple bands of high-molecular-weight DNA near the top of the gel.

For brief ligations of 2–4 hours, be sure to use a high-concentration ligase with at least 5 Weiss units/μl or 400 cohesive-end units/μl.

Stop Point

Freeze the ligation at –20°C until you are ready to continue. Thaw the reactions before proceeding on to Laboratory 15.

RESULTS AND DISCUSSION

Ligation of the multiple λ fragments and pBLU produces a number of different hybrid molecules, although not as many as you might expect. Each clonable λ fragment must have one *Bam*HI and one *Hin*dIII end, to combine with the *Bam*HI and *Hin*dIII ends of the pBLU backbone and form a circular recombinant plasmid. Only 5 of the 13 *Bam*HI/*Hin*dIII fragments from the linear λ genome satisfy this requirement: 493 bp, 784 bp, 2,396 bp, 2,409 bp, and 4,148 bp. A sixth clonable fragment of 9,866 bp is derived from the COS circle. Multiple copies of a single fragment or combinations of two different λ fragments can also be ligated into a single pBLU backbone. Three fragment plasmids can be constructed by combining the pBLU backbone with one of the five fragments having one *Bam*HI end and one *Hin*dIII end and with one of the eight fragments having two *Bam*HI or two *Hin*dIII ends. Constructs with more than one unit of pBLU are also formed. Only constructs with at least one unit of pBLU backbone contain an origin necessary for replication and the ampicillin-resistance gene required for growth on selective media.

1. Make scale drawings of all potential two-fragment (vector + insert) recombinant plasmids. Include the total bp size, fragment bp sizes, and the locations of *Bam*HI sites and *Hin*dIII sites, the origin, the ampicillin-resistance gene, and the *lacZ* gene.
2. Many different recombinant plasmids are produced in this experiment. Draw several examples of recombinant plasmids that contain three fragments including an origin of replication.
3. How many three-fragment recombinant plasmids containing the 784-bp fragment can be constructed?
4. Describe how four fragments can recombine to form a recombinant plasmid.

LABORATORY 15

Transformation of *E. coli* with λ Library

In Part A, "Classic Procedure for Preparing Competent Cells," *E. coli* cells are made competent to take up plasmid DNA by being incubated with cold calcium chloride. In Part B, "Transformation of *E. coli* with Recombinant λ/pBLU Plasmids," competent cells are transformed with the ligation products from Laboratory 14. To prepare for replica plating in the subsequent laboratory, nylon membranes are placed atop LB plates containing ampicillin and the lactose analog X-gal. Samples of transformed cells are spread on the plates and incubated overnight at 37°C. The ampicillin in the medium selects for cells that have taken up plasmids and that express the ampicillin-resistance gene of pBLU. Cells that take up a religated pBLU plasmid also express the *lacZ* gene product, β-galactosidase, which hydrolyzes X-gal to form a blue product.

The *Bam*HI/*Hin*dIII digest of pBLU in Laboratory 14 removes a 30-base-pair fragment from the *lacZ* gene but leaves the origin of replication and the ampicillin-resistance gene intact. Thus, all pBLU recombinants replicate and express ampicillin resistance. However, only recombinant plasmids containing the 30-bp fragment regenerate an intact *lacZ* gene; transformants of this molecule metabolize X-gal and give rise to blue colonies. Insertion of a λ restriction fragment into the pBLU backbone disrupts the *lacZ* gene; transformants of hybrid pBLU/λ constructs do not metabolize X-gal and therefore give rise to white (uncolored) colonies.

The remaining laboratories in this experimental unit analyze the λ inserts found in plasmids from these white colonies.

Prepare Competent Cells

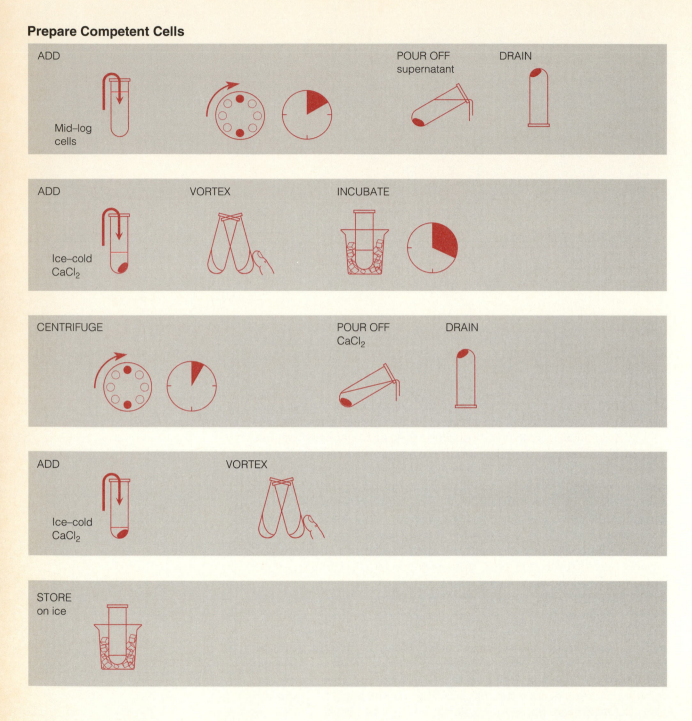

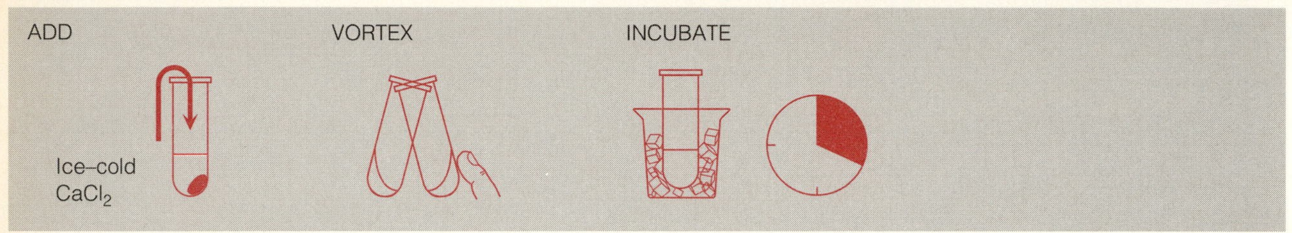

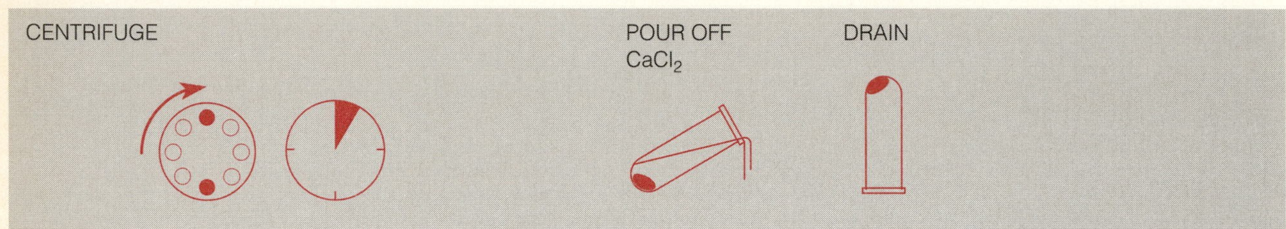

Laboratory 15/Part A
Classic Procedure for Preparing Competent Cells

PRELAB NOTES

Review the prelab notes in Laboratories 2A–C, 5, and 8A regarding *E. coli* culture and transformation.

E. coli Strains

It is essential to use a *lacZ* minus *E. coli* strain that is unable to metabolize X-gal. The protocols in this unit have been tested and optimized with the *lacZ* minus strain JM101. Other *lacZ* minus strains commonly used for molecular biological studies should give comparable results. However, growth properties of other *E. coli* strains in suspension culture may differ significantly—such as in the time needed to reach the mid-log phase and in the cell number represented by optical densities.

Competent Cell Yield

If competent cells are being prepared in quantity for group use, remember that 100 ml of mid-log culture yields 10 ml of competent cells, enough for 50 transformations (200 µl each).

For Further Information

The protocol presented here is based on the following published methods:

Hanahan, D. 1983. Studies on transformation of *Escherichia coli* with plasmids. *Journal of Molecular Biology* 166: 557.

Hanahan, D. Techniques for transformation of *E. coli*. 1987. In D. M. Glover (ed.), *DNA Cloning: A Practical Approach*. vol. 1. Oxford: IRL Press.

Mandel, M. and A. Higa. 1970. Calcium-dependent bacteriophage DNA infection. *Journal of Molecular Biology* 53: 159.

CLASSIC PROCEDURE FOR PREPARING COMPETENT CELLS

Culture and Reagents

10 ml mid-log JM101 culture
50 mM calcium chloride ($CaCl_2$)

Supplies and Equipment

5- or 10-ml sterile pipets
pipet aid or bulb
100–1,000-μl micropipettor + tips (or 1-ml pipet)
beaker of crushed or cracked ice
beaker for waste/used tips
10%-bleach solution or disinfectant
burner flame
clean paper towels
clinical centrifuge
test tube rack

Prepare Competent Cells

(40–50 minutes)

This entire experiment must be performed under sterile conditions, using sterile tubes and pipet tips.

"Blank" tubes with 10 ml of water can be used for balance, if needed. 500–1,000 × *g* corresponds to 2,000–3,000 rpm for most tabletop clinical centrifuges. A tight cell pellet should be visible at the bottom of the tube. If the pellet appears very loose or unconsolidated, centrifuge for another 5 minutes.

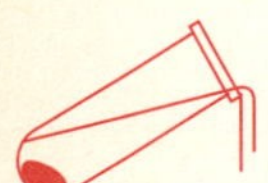

Organize the materials on the lab bench and plan out your sterile manipulations. If you are working as a team, one person can handle the pipet, and the other can remove the cap and flame the mouth of the tube.

1. Place a sterile tube of 50 mM $CaCl_2$ solution on ice.
2. Obtain a 15-ml tube with 10 ml of mid-log cells, and label it with your name and the date.
3. *Securely close the cap,* and place the tube of cells in a *balanced* configuration with other classmates' tubes in the rotor of a clinical centrifuge. Centrifuge at 500–1,000 × *g* for 10 minutes to pellet the cells on the bottom of the culture tube.
4. *Take care not to disturb the cell pellet.* Sterilely pour off the supernatant into a waste beaker for later disinfection. Invert the culture tube, and touch the mouth of the tube to a clean paper towel, to wick off as much as possible of the remaining supernatant. Flame the mouth of the culture tube after removing the top and again before replacing the top.
5. Sharply tap the bottom of the tube several times on the lab bench to loosen the cell pellet.

6. Use a sterile 5- or 10-ml pipet to add 5 ml of ice-cold calcium chloride to the culture tube. Briefly flame the pipet cylinder before using it. Flame the mouths of the calcium chloride supply tube and the culture tube, after removing the tops and again before replacing the tops.

7. *Close the cap of the culture tube tightly.* Immediately vortex the tube to *completely* resuspend the pelleted cells. During vortexing, periodically hold the tube up to the light to check on its progress. *The finished cell suspension should be homogeneous, with no visible clumps of cells.* If a mechanical vortexer is unavailable, "finger vortex" as follows:

   Hold the upper part of the tube securely between your thumb and index finger. Vigorously hit the bottom of the tube with the index finger of the opposite hand, to create a vortex that lifts the cell pellet off of the bottom of the tube.

The cell pellet will become increasingly difficult to resuspend the longer it sits in the calcium chloride. A complete cell suspension is probably the most important variable for obtaining competent cells.

8. Return the tube to ice, and incubate for 20 minutes.

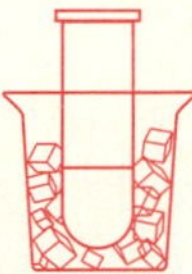

9. Following incubation, respin the cells in a clinical centrifuge for 5 minutes at 500–1,000 × *g*.
10. *Take special care not to disturb the diffuse cell pellet.* Sterilely pour off the supernatant into a waste beaker for later disinfection. Invert the culture tube, and touch the mouth of the tube to a clean paper towel, to wick off as much as possible of the remaining supernatant. Flame the mouth of the culture tube, after removing the top and again before replacing the top.

The calcium chloride treatment alters the adhering properties of the *E. coli* cell membranes. At this point, the cell pellet typically appears diffuse or ring-shaped and should resuspend more easily.

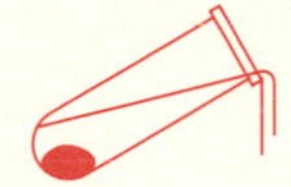

11. Use a 100–1,000-µl micropipettor (or 1-ml pipet) to sterilely add 1,000 µl (1 ml) of fresh, ice-cold $CaCl_2$ to the tube. Flame the mouth of the culture tube, after removing the top and again before replacing the top.
12. *Close the cap of the culture tube tightly.* Vortex to *completely* resuspend the pelleted cells. *The finished cell suspension should be homogeneous, with no visible clumps of cells.*

The diffuse cell pellet should resuspend very easily; do not overvortex it.

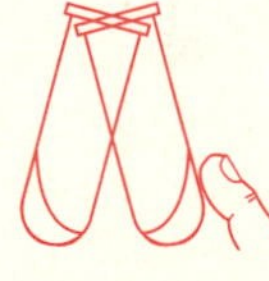

Stop Point

Store the cells in a beaker of ice in the refrigerator (at approximately 0°C) until you are ready to use them. "Seasoning" at 0°C for as long as 24 hours increases the competency of cells five- to tenfold.

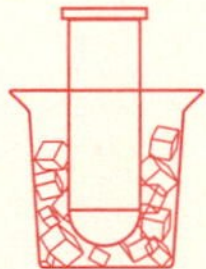

13. Take the time for proper cleanup:
    a. Segregate for proper disposal culture plates and tubes, pipets, and micropipet tips that have come into contact with *E. coli*.
    b. Disinfect the mid-log culture, the tips, and the supernatant from steps 4 and 10 with a 10%-bleach solution or a disinfectant (such as Lysol).
    c. Wipe down the lab bench with soapy water, a 10%-bleach solution, or disinfectant.
    d. Wash your hands before leaving the laboratory.

Transform *E. coli* with λ/pBLU Recombinants

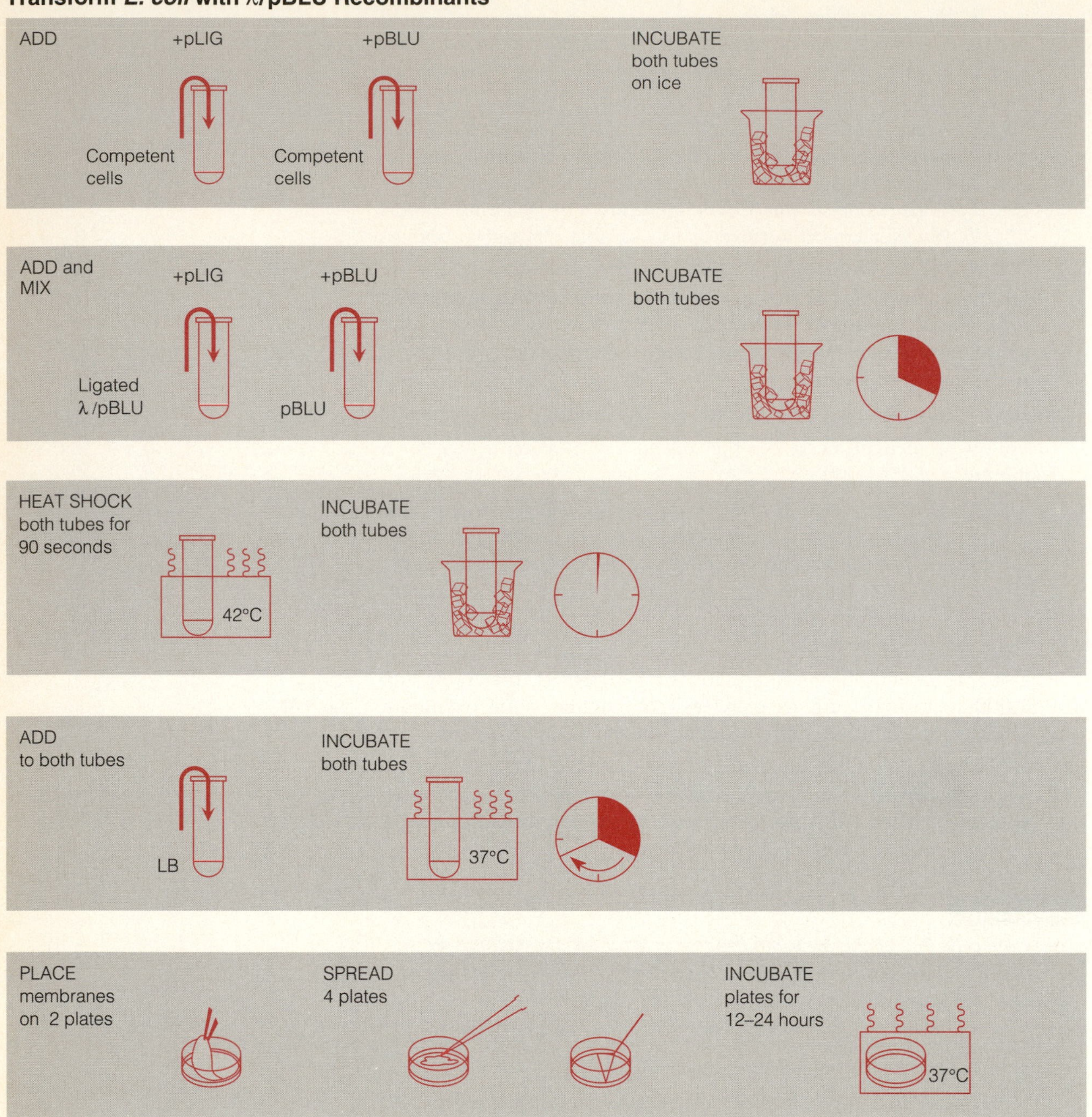

Laboratory 15/Part B
Transformation of *E. coli* with Recombinant λ/pBLU Plasmids

PRELAB NOTES

Review the prelab notes in Laboratory 5, "Rapid Colony Transformation of *E. coli* with Plasmid DNA."

X-Gal

The chromogenic substrate X-gal (5-bromo-4-chloro-3-indoyl-—D-galactoside) is chemically related to X-phosphate, one of the substrates for color development in Laboratory 13, "Southern Hybridization of λ DNA." Like antibiotics, X-gal is broken down by heat. When preparing agar plates containing X-gal, make sure to cool the agar solution until the container can be held comfortably in the hand (approximately 60°C) before adding the reagent. X-gal is expensive; one economical alternative to adding X-gal to an entire batch of culture medium is to sterilely spread 50 μl of a 20-mg/ml stock solution over the surface of an LB ampicillin plate. If time permits, allow the X-gal solution to diffuse into the medium overnight or incubate it at 37°C for several hours before use.

Nylon Membranes for Replica Plating

In preparation for Laboratory 16A, "Replica Transfer of λ Clones," transformed cells are spread onto a sterile nylon membrane, resting on the surface of an LB agar plate containing ampicillin and X-gal. Technically, nylon membranes used for replica plating should be sterile. However, the fresh clean membranes provided by the manufacturer are rarely contaminated and can be used without sterilization. This is another application of Max Delbrück's principle of "limited sloppiness" in molecular biology: It usually does not harm results, and antibiotic selection helps compensate for it. If you wish to be scrupulous, sandwich the membranes individually between sheets of filter paper, and autoclave them.

In typical protocols, a dry membrane is placed directly on the surface of a fresh agar plate. However, this simple method can lead to uneven wetting and crumbling of the membrane, especially when relatively dry plates are being used. We recommend thoroughly prewetting each nylon membrane by submerging it in a culture plate containing 10 ml of sterile LB media in a petri dish. Since nylon membranes are somewhat expensive, they are not placed on control plates that will not undergo replica transfer.

For Further Information

The protocol presented here is based on the following published methods:

Cohen, S. N., A. C. Y. Chang, L. Hsu. 1972. Nonchromosomal antibiotic resistance in bacteria: Genetic transformation of *Escherichia coli* by r-factor DNA. *Proceedings of the National Academy of Sciences* 69: 2110.

Grunstein, M. and D. S. Hogness. 1975. Colony hybridization: A method for the isolation of cloned DNAs that contain a specific gene. *Proceedings of the National Academy of Sciences USA* 72: 3961.

Lederberg, J. and E. M. Lederberg. 1952. Replica plating and indirect selection of bacterial mutants. *Journal of Bacteriology* 63: 399.

Messing, J. and J. Vieira. 1982. The pUC plasmids, an M13mp7-derived system for insertion mutagenesis and sequencing with synthetic universal primers. *Gene* 19: 259.

TRANSFORMATION OF *E. COLI* WITH λ/pBLU RECOMBINANTS

Culture, Media, and Reagents	Supplies and Equipment
competent *E. coli* cells (from Part A)	10–100-μl micropipettor + tips
ligated λ/pBLU (from Laboratory 14B)	100–1000-μl micropipettor + tips
0.005 μg/μl pBLU	0.5–10-μl micropipettor + tips
LB broth	2 15-ml culture tubes
four LB/amp + X-gal plates	beaker of crushed or cracked ice
two nylon membranes	beaker for waste/used tips
LB broth in culture plate	"biobag" or heavy-duty trash bag
	10%-bleach solution or disinfectant (such as Lysol)
	burner flame
	cell spreader
	disposable gloves
	forceps
	permanent marker
	test tube rack
	37°C incubator
	37°C shaking water bath (optional)
	42°C water bath

Transform *E. coli* with λ/pBLU Recombinants

(70–90 minutes)

This entire experiment must be performed under sterile conditions, using sterile tubes and pipet tips.

1. Use a permanent marker to label two sterile 15-ml culture tubes:
   +pLIG = ligated λ/pBLU
   +pBLU = pBLU control
2. Add 200 μl of competent cells to each tube. Return the remaining competent cells to ice, and store them at 0°C.
3. Place both tubes on ice.
4. Add 10 μl of ligated λ/pBLU *directly into the cell suspension* in the +pLIG tube.
5. Add 10 μl of 0.005 μg/μl pBLU *directly into the cell suspension* in the +pBLU tube.
6. Lightly tap each tube with your finger to mix the solution. Avoid making bubbles in the suspension or splashing the suspension up on the side of the tube.
7. Return both tubes to ice for 20 minutes.
8. While the cells are incubating on ice, use a permanent marker to label four LB/amp + X-gal plates with your name and the date. Divide the plates into pairs, and mark them as follows:
   +pLig: Mark one plate L20 and the other plate L100.
   +pBLU: Mark one plate B20 and the other plate B100.

Use a fresh tip to pipet each type of DNA. Optimally, flame the mouth of the culture tubes after removing the tops and again before replacing the tops.

Store the remainder of the ligated DNA at 4°C in preparation for electrophoretic analysis in Laboratory 17.

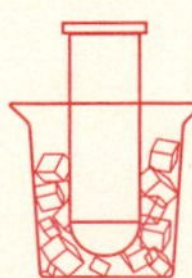

9. Using gloved hands or forceps, prewet a nylon membrane by submerging it in sterile LB broth contained within a culture plate. When the membrane has been completely wetted, carefully place it on the agar surface of the L20 plate. *Avoid trapping air bubbles between the membrane and the agar surface.*

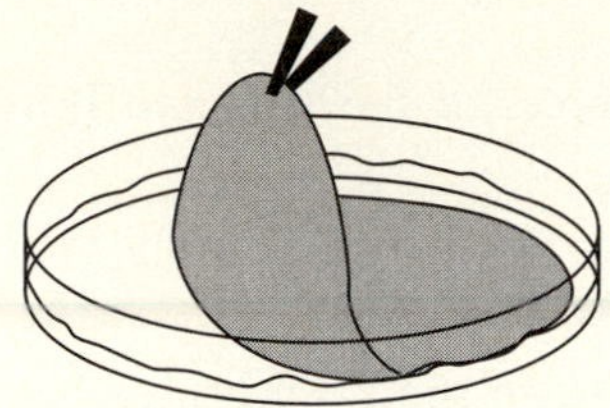

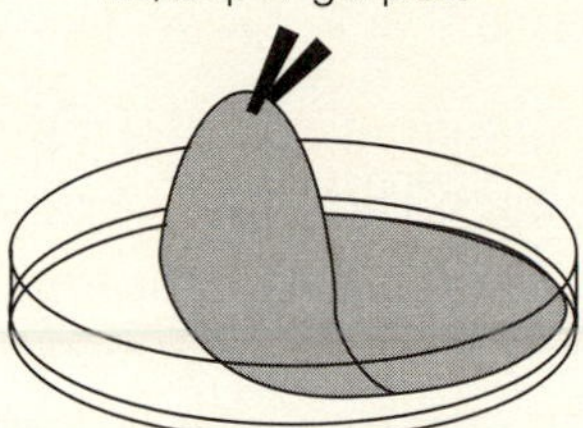

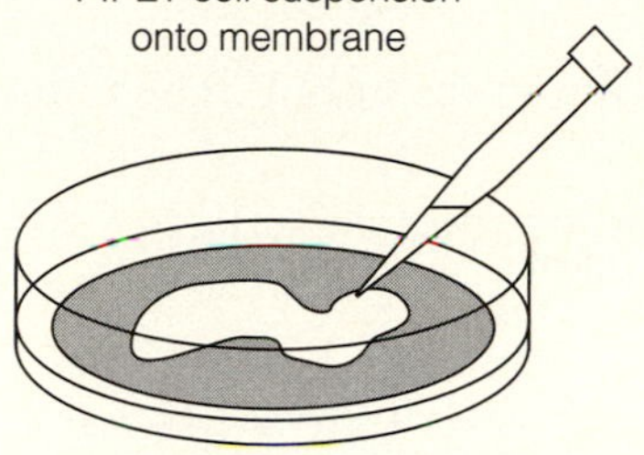

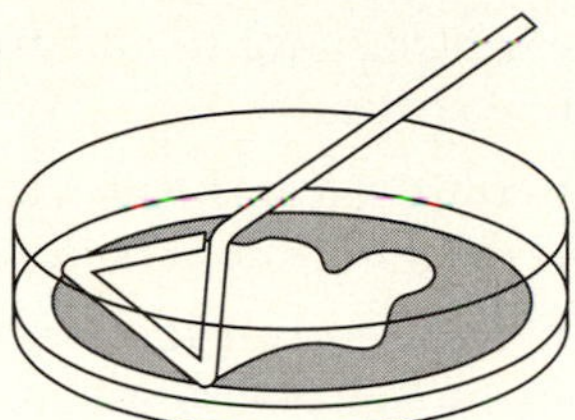

Colony Membrane Wetting

10. Wet a second membrane, as described above, and place it on the agar surface of the L100 plate.
11. Following the 20-minute incubation on ice, heat shock the cells in both tubes. *The cells must receive a sharp and distinct shock.*
    a. Carry the ice beaker to a 42°C water bath. Remove the tubes from the ice, and immediately immerse them in the water bath for 90 seconds.
    b. Immediately return both tubes to ice, and let them stand for at least 1 more minute.
12. Add 800 μl of LB broth to each tube. Lightly tap the tubes with your finger to mix the solution.

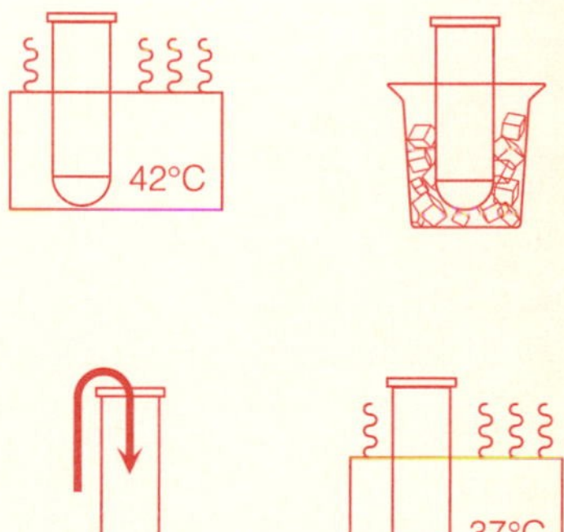

13. Allow the cells to recover by incubating both tubes at 37°C in a shaking water bath (with moderate agitation) for 20–40 minutes.

The recovery increases the number of transformants that express the resistance protein by two to fourfold, but it can be omitted if there are time constraints.

Stop Point

The cells can be allowed to recover for several hours. A longer recovery period can help to compensate for a poor ligation or for cells of low competence.

14. Use the matrix below as checklist as the +pLig and +pBLU cells are spread on plates in the following steps.

Plate	LB	+pLIG	+pBLU
L20	80 µl	20 µl	—
L100	—	100 µl	—
B20	80 µl	—	20 µl
B100	—	—	100 µl

The added LB broth increases the total volume of liquid on each plate to help to evenly spread the small volume (20 µl) of transformed cells on the L20 and B20 plates.

15. Add 80 µl of sterile LB broth to the centers of the L20 and B20 plates. Then *use fresh tips:*
    a. Add 20 µl of cells from the +pLig tube to the LB broth on the L20 plate.
    b. Add 20 µl of cells from the +pBLU tube to the LB broth on the B20 plate.
16. *Use a fresh tip* to add 100 µl of cells from the +pLig tube to the L100 plate.
17. *Use a fresh tip* to add 100 µl of cells from the +pBLU tube to the B100 plate.
18. Sterilize a cell spreader, and spread the cells over the surface of each of the four plates in succession.
    a. Dip the spreader into the ethanol beaker and briefly pass it through a burner flame to ignite the alcohol. Allow the alcohol to burn off *away from the burner flame.*

Do not allow the cells to sit too long on the plates before spreading them. The object is to evenly distribute and separate the transformed cells so that each gives rise to a distinct colony of clones.

The spreader, submerged in alcohol, is already sterile. Thus, the only purpose of the flame is to burn off the alcohol before spreading. The spreader will become too hot if it is held directly in the burner flame for several seconds, and this may kill *E. coli* cells on the plate.

CAUTION

Be extremely careful not to ignite the ethanol in the beaker.

    b. Lift the plate lid, like a clam shell, only enough to allow spreading.
    c. Cool the spreader by touching it to the agar, to the membrane surface away from the cells, or to the condensed water on the plate lid.
    d. Touch the spreader to the cell suspension, and gently drag it back and forth several times across the agar or membrane surface. Rotate the plate one-quarter turn, and repeat the spreading motion.
    e. Replace the plate lid. Return the cell spreader to the ethanol without flaming.

Take care not to gouge the agar or the membrane.

19. Reflame the spreader once more before setting it down.
20. Let the plates set for several minutes to allow the cell suspensions to be absorbed.

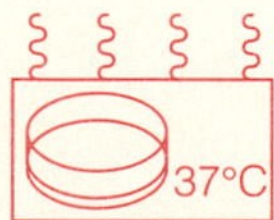

21. Wrap the plates together with tape, place them upside down in a 37°C incubator, and incubate them for 12–24 hours.
22. After the initial incubation, store the plates at 4°C to arrest *E. coli* growth and to slow the growth of any contaminating microbes. *Save either the L20 or the L100 plate to begin Laboratory 16.*

23. Take the time for proper cleanup:
    a. Segregate for proper disposal culture plates and tubes, pipets, and micropipettor tips that have come into contact with *E. coli.*
    b. Disinfect the cell suspensions, the tubes, and the tips overnight with a 10%-bleach solution or disinfectant (such as Lysol).
    c. Wipe down the lab bench with soapy water, a 10%-bleach solution, or disinfectant.
    d. Wash your hands before leaving the laboratory.

RESULTS AND DISCUSSION

Count the number of blue and white colonies growing on each plate. Use a permanent marker to mark the bottom of the culture plate as each colony is counted. A total of 50–200 colonies should be observed on the L20 experimental plate, with about equal numbers of blue and white colonies. Approximately five times as many colonies should be observed on the B20 control plate—all, or nearly all, blue. The corresponding L100 and B100 plates should contain approximately five times as many colonies. An extended recovery period (step 13) would most likely inflate these expected results.

Blue colonies may be pale in color and difficult to distinguish from white colonies. Refrigerating the plates for an hour or more can help to intensify the color. If the plates are overincubated at 37°C or left at room temperature for several days, small, white "satellite" colonies may be observed growing in a circle around large, resistant colonies. Satellite colonies are clones of the ampicillin-sensitive cells that begin to grow as the antibiotic is gradually reduced below the selection threshold.

1. Record your colony counts for each plate in the matrix below. Were the results as you expected? Explain the possible reasons for variations from the expected results.

	White Colonies	Blue Colonies
L20		
L100		
B20		
B100		

2. Compare and contrast the growth on each of the following pairs of plates. What does each pair of plates tell you about transformation, antibiotic selection, and cloning efficiency?
    a. Total colonies on the L20 versus the B20 *(and the L100 versus the B100)* plate
    b. The percentage of blue (or white) colonies on the L20 versus the B20 *(and the L100 versus the B100)* plate.
3. Explain why *white* satellite colonies appear around a *blue* colony.

4. Calculate the transformation efficiency of the control pBLU transformation (B20 or B100 plate): the number of antibiotic-resistant colonies per microgram of plasmid DNA. The object is to determine the mass of plasmid DNA that was spread on a plate and that was therefore responsible for the number of transformants observed.
   a. Determine *the total mass (in μg) of pBLU* used in the transformation = concentration of pBLU × volume of pBLU.
   b. Determine *the fraction of the cell suspension spread* onto the B20 or B100 plate = the volume of cell suspension spread/the total volume of the cell suspension.
   c. Determine *the mass of pBLU in the cell suspension spread* onto the B20 or B100 plate = the total mass of pBLU (a) × the fraction of the cell suspension spread (b).
   d. Express *transformation efficiency* in scientific notation as the number of colonies per μg of pBLU = the number of colonies on the B20 or B100 plate/the mass of pBLU in the cell suspension spread (c).
5. Calculate the transformation efficiency of the experimental transformation with ligated DNA (L20 or L100 plate).
   a. Determine *the total mass (in μg) of pBLU* used in the initial restriction digest (Laboratory 14A) = the concentration of pBLU × the volume of pBLU.
   b. Determine *the fraction of pBLU digest* used in the ligation reaction (Laboratory 14B) = the volume of pBLU digest in the ligation/the total volume of pBLU digest.
   c. Determine *the fraction of ligation reaction* used for transformation = the volume of the ligation reaction in the transformation/the total volume of the ligation reaction.
   d. Determine *the fraction of cell suspension spread* onto the L20 or L100 plate = the volume of cell suspension spread/the total volume of the cell suspension.
   e. Determine *the mass of pBLU in the cell suspension spread* onto the L20 or L100 plate = the total mass of pBLU (a) × fraction b × fraction c × fraction d.
   f. Express the *transformation efficiency* in scientific notation as the number of colonies per μg of pBLU = the total number of colonies on the L20 or L100 plate/the mass of pBLU in the cell suspension spread (e).
6. The "classic" transformation protocol used in this laboratory typically achieves efficiencies of 5×10^4 to 1×10^6 colonies per microgram. How do the efficiencies that you calculated in questions 4 and 5 compare to this range? Can you account for the differences?
7. How does the length of postincubation recovery influence transformation efficiency?
8. Why is high-efficiency transformation important for constructing a genomic library of eukaryotic DNA?

For Further Research

1. Research different research applications of the β-galactosidase colorimetric system.
2. Design a series of experiments to determine the optimum ratio of restricted λ to pBLU DNA for the ligation in Laboratory 14. The objective is to maximize the number of white colonies growing on an LB/amp X-gal plate. For each experiment, record the total number of colonies on each plate and the ratio of white to blue colonies. (Do not use nylon membranes; a hybridization protocol is unnecessary.)
3. Design an experiment to compare the transformation efficiencies of linear versus circular pBLU/λ constructs.
4. Design an experiment to determine the fraction of cells that took up plasmid DNA.

LABORATORY 16

Colony Hybridization of the λ Library

In this laboratory, a hybridization protocol is used to screen the genomic λ library to identify clones containing the 784-bp *Bam*HI/*Hin*dIII λ fragment. The starting point of the experiment is a nylon membrane containing colonies from either the L20 or the L100 plate from Laboratory 15B.

Because the hybridization procedure requires lysing the *E. coli* colonies on the membrane, it is essential to work with a replica membrane derived from the original. In Part A, "Replica Transfer of λ Clones," the membrane from either the L20 or the L100 plate from Laboratory 15B is sandwiched with a fresh membrane. When the membranes are pressed together, some of each colony is transferred to the fresh membrane, creating a mirror image of the transformants on the replica. The replica membrane is incubated at 37°C for several hours to allow the *E. coli* colonies to grow to a suitable size for hybridization. The original membrane is stored at 4°C as a source of clones for subsequent analysis.

Part B, "Hybridization of the Replica Membrane," begins a series of steps that are analogous to those in Laboratory 13, "Southern Hybridization of λ DNA." First, the membrane is exposed to a sodium hydroxide solution that lyses *E. coli* cells on the membrane and denatures the chromosomal and plasmid DNA. Following neutralization, the membrane is baked at 70°C to fix the DNA to the membrane. The membrane is prehybridized with a blocking agent that binds to all areas of the membrane not already occupied by bound DNA, thus preventing the probe from binding nonspecifically to the membrane. Then, the membrane is incubated with the digoxigenin-labeled probe, which hybridizes to only those colonies containing its complementary sequence, a 25-base oligonucleotide within the 784-bp *Bam*HI/*Hin*dIII λ fragment.

In Part C, "Nonradioactive Probe Detection," the hybridized membrane is washed to remove excess probe and incubated with the antidigoxigenin antibody/enzyme conjugate. The antibody binds to areas of the membrane where the digoxigenin-labeled probe has bound to a recombinant plasmid containing its complementary DNA sequence. Alkaline phosphatase is conjugated to the antibody. When this enzyme is incubated with a colorimetric substrate solution, it initiates a series of reactions that produce a purplish-brown precipitate, revealing the clones containing the DNA sequence of interest.

GROW colonies of
recombinant plasmids

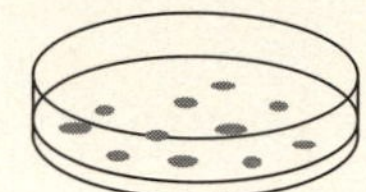

OVERLAY plate with
nylon membrane

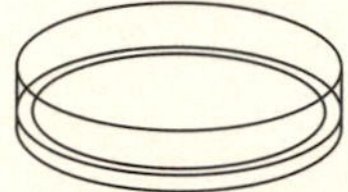

LIFT membrane containing
replica of colonies

LYSE cells; DENATURE DNA and
HYBRIDIZE to labeled probe in
sealed plastic bag

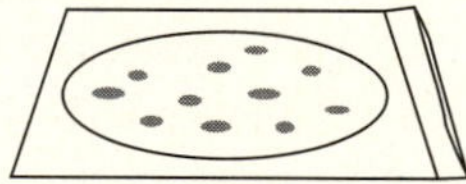

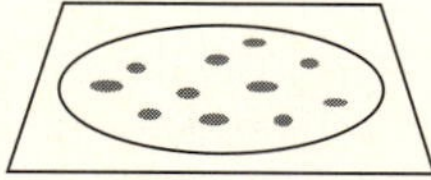

COLORIMETRIC DEVELOPMENT
reveals location of colonies to which
probe hybridized

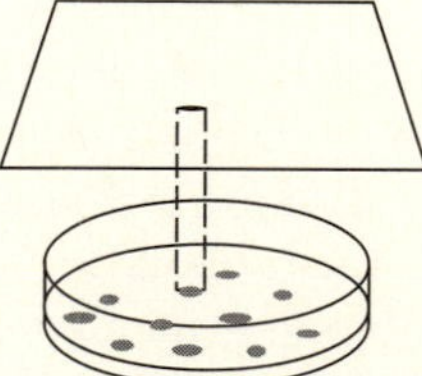

Screening a Genomic Library by Colony Hybridization

Prepare the Replica Membrane

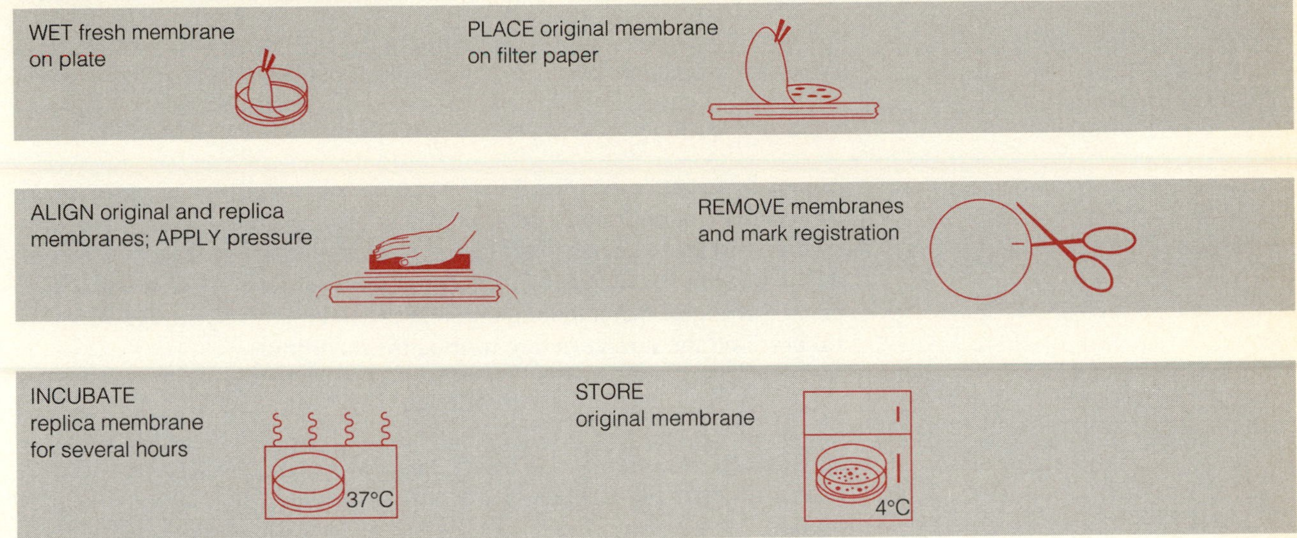

Laboratory 16/Part A
Replica Transfer of λ Clones

PRELAB NOTES

Review the prelab notes in Laboratory 13B, "Southern Transfer of λ Restriction Fragments."

Wetting the Nylon Membrane

The replica membrane must be wetted before being placed in contact with the original membrane. However, an overly wet membrane can cause colonies to run together, affecting the precision of the analysis. We recommend wetting the replica membrane by placing it on top of a moist agar plate. If the agar plates are old and relatively dry, sterilely spread 100 μl of LB broth on the surface before wetting the membrane.

Filter Paper

All mention of filter paper in Part A and the ensuing parts refers to Whatman 3mm filter paper, which is most commonly used for colony hybridization.

Felt

Felt (or a feltlike material) is used to help ensure uniform contact between the membranes. Filter paper can be substituted.

For Further Information

The protocol presented here is based on the following published methods:

Grunstein, M. and D. S. Hogness. 1975. Colony hybridization: A method for the isolation of cloned DNAs that contain a specific gene. *Proceedings of the National Academy of Sciences USA* 72: 3961.

Lederberg, J. and E. M. Lederberg. 1952. Replica plating and indirect selection of bacterial mutants. *Journal of Bacteriology* 63: 399.

REPLICA TRANSFER OF λ CLONES

Reagents	Equipment and Supplies
L20 or L100 plate (from Laboratory 15B)	ballpoint pen
LB/amp plate	disposable gloves
	felt (12 × 14 cm)
	filter paper (12 × 14 cm)
	flat forceps
	gel-casting tray or book
	nylon membrane (82 mm in diameter)
	sterile 18-gauge needle (or scissors)
	37°C incubator

Prepare the Replica Membrane

(15 minutes; then several hours of incubation)

1. Using flat forceps or disposable gloves, obtain a fresh nylon membrane: This is the replica membrane. Label the edge of the membrane with your initials using a ballpoint pen. Wet the membrane by carefully placing it on the surface of the LB/amp plate. Cover the plate, and leave the replica membrane in contact with the agar until you are ready to use it in step 4.

Never touch the nylon membrane with your bare hands. Oils from your skin may be deposited on the membrane and can prevent the DNA from binding following cell lysis. Do not label the membrane with a felt-tip pen, because the ink will bleed when it is wet.

2. Use the forceps to carefully peel the membrane off of the L20 or L100 plate from Laboratory 15B (whichever one has 30–300 white colonies on it): This is the L membrane.
3. Place the L membrane, colony side up, on top of six sheets of filter paper.
4. Use the forceps to remove the replica membrane from the LB/amp plate (step 1). Carefully align the edges of the wetted replica membrane with the edges of the L membrane. Then, carefully lower the replica membrane to contact the L membrane, so that they are superimposed as closely as possible.

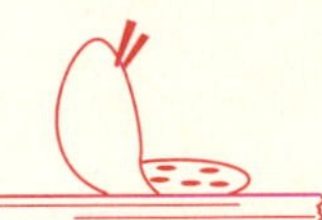

It is important to align the replica and L membranes as closely as possible. Be careful, though; once the membranes come into contact, they cannot be repositioned.

5. Place a square of felt (or filter paper) on top of the superimposed membranes.
6. Place a gel-casting tray (or book) on top of the filter paper/membrane/felt sandwich. Press firmly and evenly on the gel-casting tray (or book) for several seconds.

Uneven pressure can cause the membranes to slide against one another, smearing the colonies.

7. Remove the gel-casting tray (or book) and the felt. Use a sterile 18-gauge needle to poke three holes through the edge of the sandwiched membranes and into the filter papers below. The object is to make an asymmetric pattern of marks that will allow the two membranes to be reoriented after they are separated.

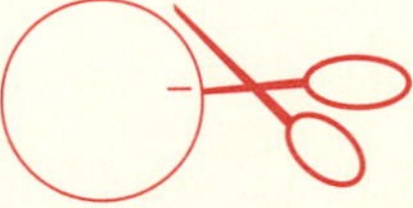

Alternately, use the scissors to cut several triangular notches into the edges of the membranes.

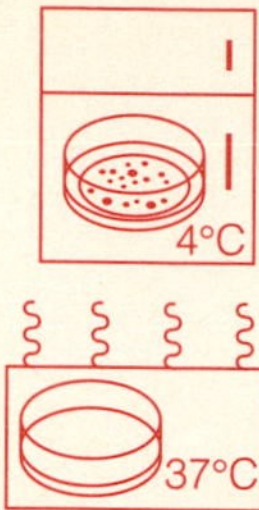

Incubating the replica membrane on an LB/amp plate without X-gal will cause previously blue colonies to appear white. This is desirable, because blue colonies can be mistaken for purple hybridization signals.

Incubate only long enough to obtain *E. coli* colonies of 1–2 mm in diameter. Do not overincubate: Satellite colonies may appear to complicate the comparison with the original membrane.

8. Using gloved hands and forceps, carefully peel the two membranes apart. Return the L membrane to its original plate (with the colonies facing up) and store it at 4°C.
9. Return the replica membrane to the LB/amp plate from step 1 (with the colonies facing up), invert the plate, and incubate at 37°C for several hours or overnight.

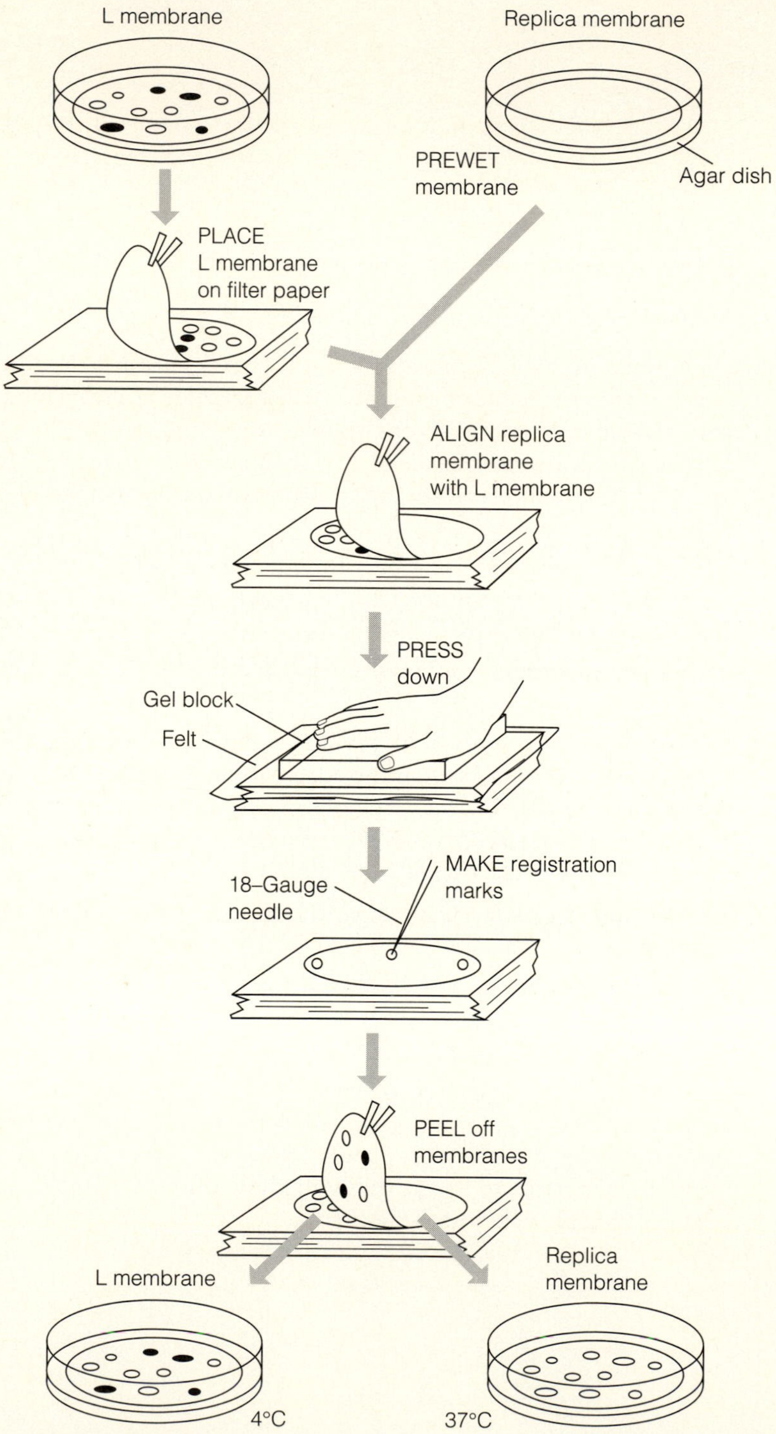

L membrane
Replica membrane
PREWET membrane
Agar dish
PLACE L membrane on filter paper
ALIGN replica membrane with L membrane
PRESS down
Gel block
Felt
18–Gauge needle
MAKE registration marks
PEEL off membranes
L membrane
Replica membrane
4°C
37°C

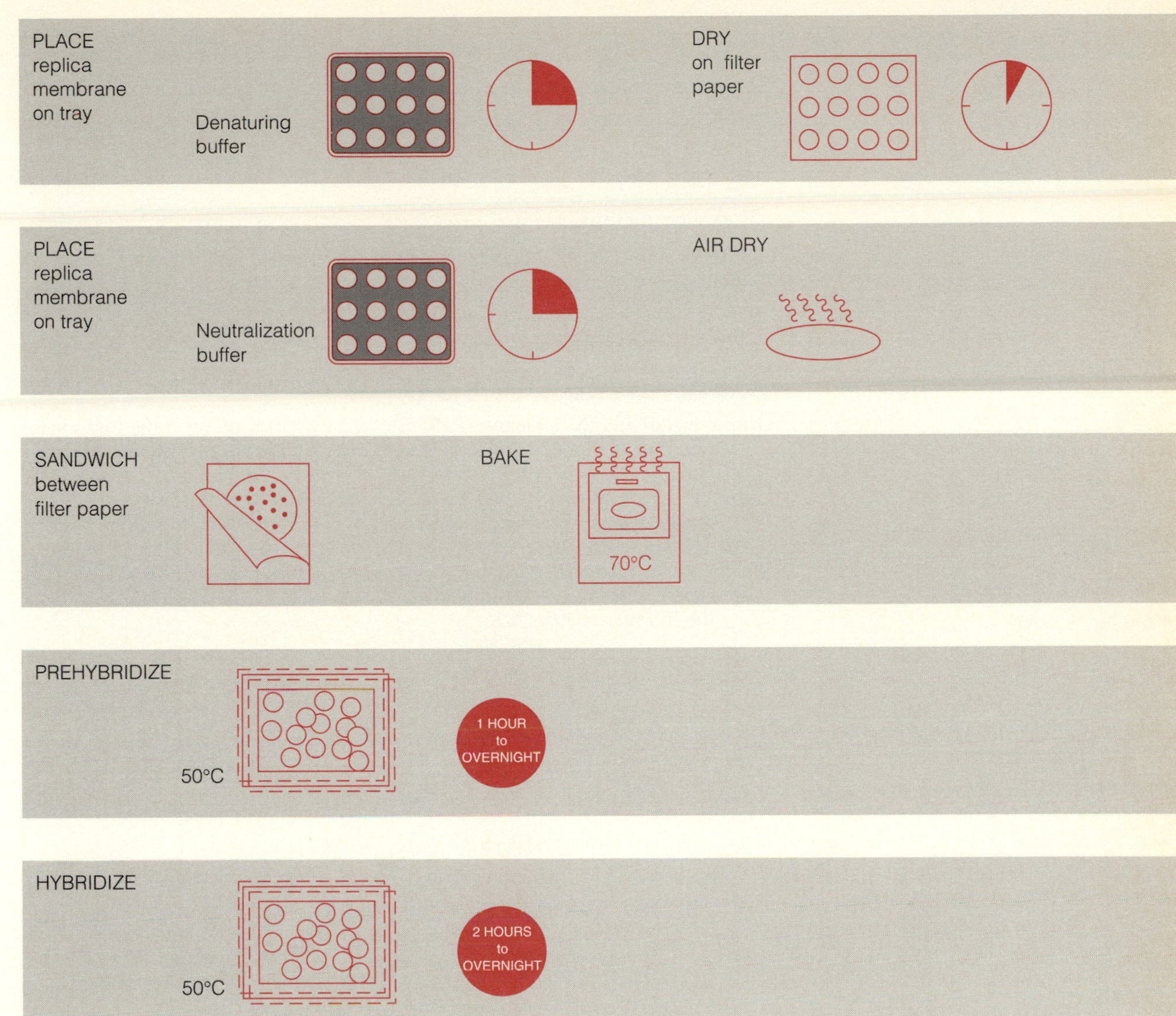

Laboratory 16/Part B
Hybridization of the Replica Membrane

PRELAB NOTES

Review the prelab notes in Laboratory 13C, "Hybridization of the λ Probe."

Batch Processing

All replica membranes are processed together in a "batch" procedure, requiring preparation of just one appropriately sized container of each reagent.

For Further Information

The protocol presented here is based on the following published methods:

Grunstein, M. and D. S. Hogness. 1975. Colony hybridization: A method for the isolation of cloned DNAs that contain a specific gene. *Proceedings of the National Academy of Sciences USA* 72: 3961.

Lathe, J. 1985. Synthetic oligonucleotide probes deduced from amino acid sequence data: Theoretical and practical considerations. *Journal of Molecular Biology* 183: 1.

Martin, R., C. Hoover, S. Grimme, C. Grogan, J. Holtke, and C. Kessler. 1990. A highly sensitive, nonradioactive DNA labeling and detection system. *BioTechniques* 9(6): 762.

HYBRIDIZATION OF THE REPLICA MEMBRANE

Reagents	**Equipment and Supplies**
denaturing buffer	ballpoint pen
neutralizing buffer	disposable gloves
prehybridization buffer	filter paper (12 × 14 cm)
hybridization buffer	flat forceps
	masking tape
	oven set at 70°C
	plastic container (15 × 15 × 4.5 cm)
	plastic container (17 × 28 × 6.5 cm)
	50°C shaking water bath
	tissue paper
	trays

I. Denature/Neutralize the Replica Membrane

(50 minutes; then 30+ minutes of baking)

1. Using forceps and gloved hands, place the replica membrane (colony side up) on a filter paper saturated with denaturing buffer. Let it stand for 15 minutes.
2. Transfer the membrane (colony side up) to dry filter paper, and allow the excess solution to drain off it for 5 minutes.

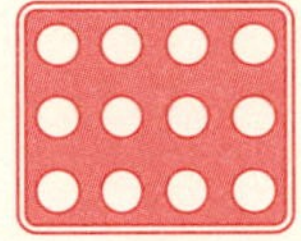
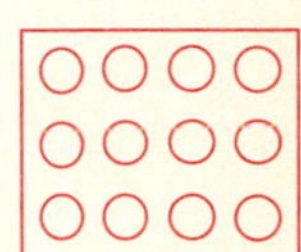

Never touch the nylon membrane with your bare hands. Oils from your skin may be deposited on the membrane and can prevent the DNA probe from binding.

3. Place the membrane (colony side up) on a filter paper saturated with neutralizing buffer. Let it stand for 15 minutes.
4. Transfer the membrane to a fresh piece of filter paper to drain excess solution; then air dry it for 10 minutes.
5. Sandwich the membrane between two fresh pieces of filter paper (12 × 14 cm). Tape the four sides, and label the filter paper with your name, using a soft lead pencil or ballpoint pen.

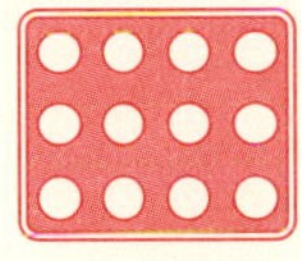

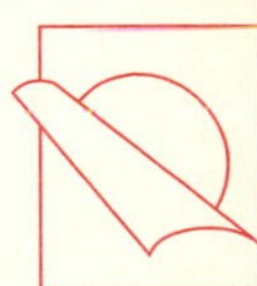

6. Bake the membrane at 70°C for at least 30 minutes.

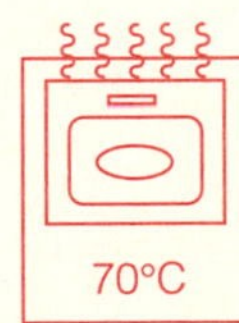

Stop Point

After baking, the membrane can be stored indefinitely at room temperature.

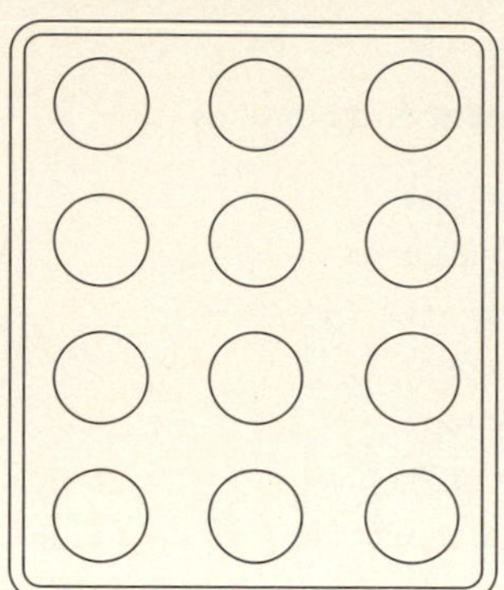

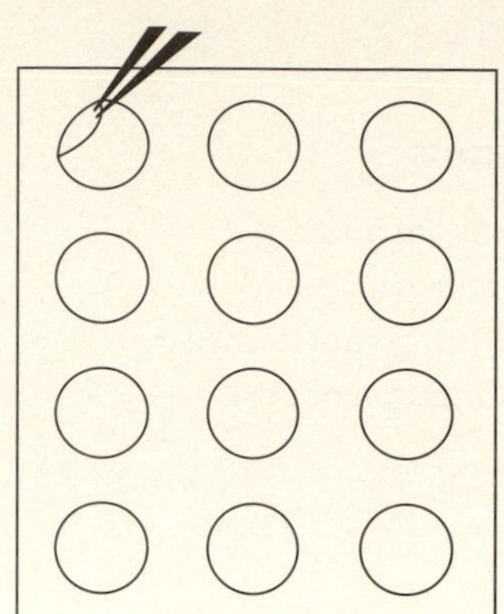

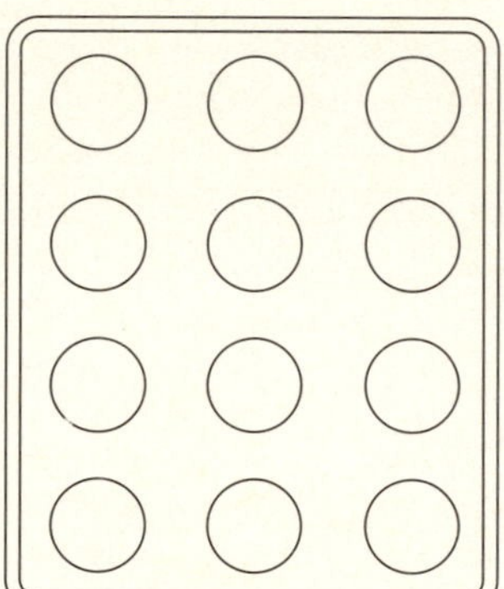

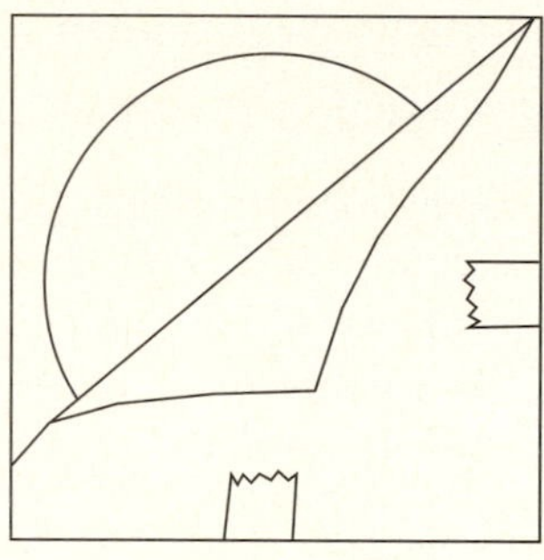

2. BLOT briefly on dry 3 mm paper

3. NEUTRALIZE on communal TRIS/NaCl surface

4. Place membrane between two sheets of 3 mm paper, TAPE sides, and BAKE

Colony Membrane Batch Treatment

II. Prehybridize the Replica Membrane

(5 minutes; then 60+ minutes of incubation)

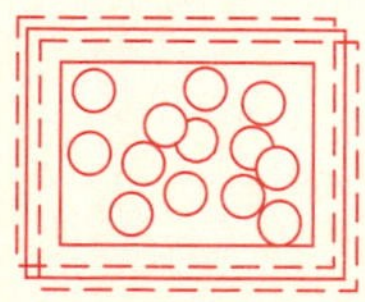

1. Using gloved hands, remove the nylon membrane from the filter paper sandwich from Section I, and transfer it to a common container with prehybridization buffer. *If the membrane sticks to the filter paper, do not try to separate them. Place both in the prehybridization buffer, and they will separate when wet.*
2. Incubate, with gentle shaking, at 50°C for 60 minutes to overnight. Cover the container to prevent evaporation.

III. Hybridize the Replica Membrane

(5 minutes; then 120+ minutes of incubation)

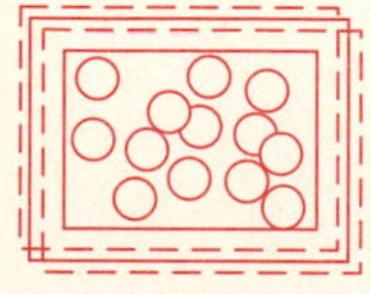

1. Using gloved hands, transfer the nylon membrane to the container with hybridization buffer.
2. Incubate, with gentle shaking, at 50°C for two hours to overnight. Cover the container to prevent evaporation.

Wash and Develop Color

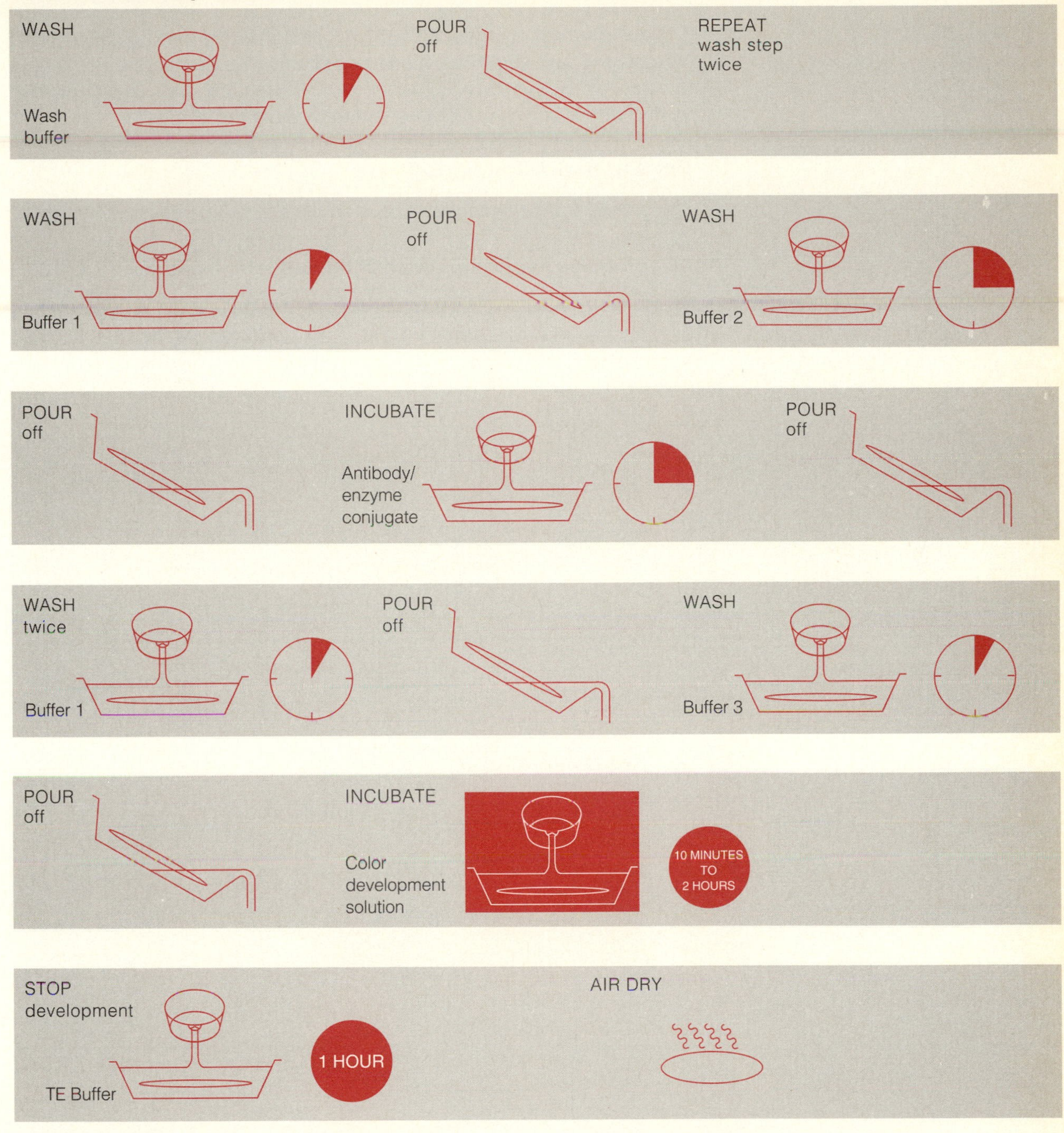

Laboratory 16/Part C
Nonradioactive Probe Detection

PRELAB NOTES

Review the prelab notes in Laboratory 13D, "Nonradioactive Probe Detection," and the instructions pertaining to color development that accompany the Genius 3 Nucleic Acid Detection Kit. The numbered buffers in this laboratory correspond to those in the Genius kit.

For Further Information

The protocol presented here is based on the following published methods:

Genius System User's Guide, Boehringer Mannheim Corporation, Biochemical Products, 9115 Hague Road, P.O. Box 50414, Indianapolis, IN 46250-0414.

Martin, R., C. Hoover, S. Grimme, C. Grogan, J. Holtke, and C. Kessler. 1990. A highly sensitive, nonradioactive DNA labeling and detection system. *BioTechniques* 9(6): 762.

NONRADIOACTIVE PROBE DETECTION

Reagents

wash buffer (2× SSC, 0.1% SDS)

buffer 1 (0.1*M* Tris pH 7.5, 0.15*M* NaCl)

buffer 2 (buffer 1 + 1% w/v blocking reagent)

diluted antibody/enzyme solution

buffer 3 (0.1*M* Tris pH 9.5, 0.1*M* NaCl, 0.05*M* $MgCl_2$)

color development solution

TE buffer

Equipment and Supplies

disposable gloves

filter paper

flat forceps

petri dish (150 mm)

small plastic container (15 × 15 × 4.5 cm)

50°C water bath

Wash and Develop Color

(70 minutes; then 10+ minutes of development)

Gentle agitation by hand should accompany each wash (incubation), *except for the color development in step 11.*

1. Using gloved hands and forceps, remove the membrane from the hybridization solution (Part B), and transfer it to a small plastic container with 100 ml of wash buffer. Incubate at room temperature for 5 minutes.

Never touch the nylon membrane with your bare hands. Oils from your skin may be deposited on the membrane and can prevent the antibody/enzyme conjugate from binding.

Do not discard the hybridization solution; collect it for reuse!

2. Pour off the first wash, and replace it with 100 ml of fresh wash buffer. Incubate, with agitation, at room temperature for 5 minutes.
3. Pour off the second wash, and replace it with 100 ml of fresh wash buffer. Incubate, with agitation, at *50°C* for 5 minutes.
4. Pour off the third wash, and add 100 ml of buffer 1. Incubate, with agitation, at room temperature for 5 minutes.
5. Pour off buffer 1, and add 100 ml of buffer 2. Incubate, with agitation, at room temperature for 15 minutes.

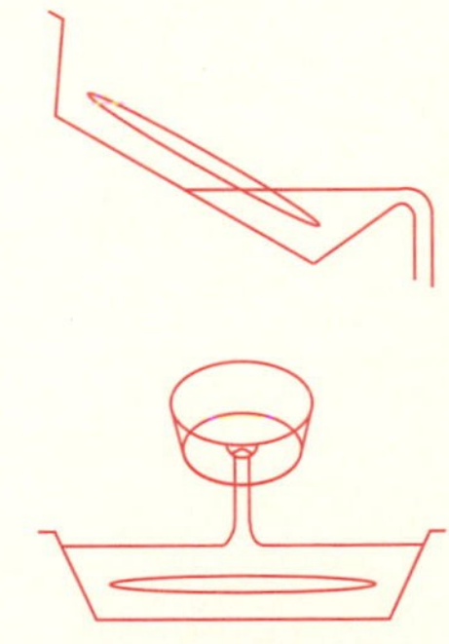

6. Pour off buffer 2, and add 40 ml of the antibody/enzyme solution. Incubate, with agitation, at room temperature for 15 minutes.
7. Pour off the antibody/enzyme solution, and add 100 ml of buffer 1. Incubate, with agitation, at room temperature for 5 minutes.
8. Pour off buffer 1, and replace it with 100 ml of fresh buffer 1. Incubate, with agitation, at room temperature for 5 minutes.
9. Pour off buffer 1, and add 100 ml of buffer 3. Incubate, with agitation, at room temperature for 5 minutes.
10. Transfer the membrane to a clean 150-mm petri dish or other suitable container.

11. Add 20 ml of freshly diluted color development solution to the membrane, and incubate at room temperature *in the dark* until the positive colonies become *clearly visible* (for 10 minutes to 1 hour). *Make sure that the membrane is positioned colony side up. Do not agitate!*

If the hybridization signal is faint, color development can be extended overnight. However, overincubation will cause the entire membrane to begin to turn brown.

The final wash in TE buffer stops the color development and prevents further darkening of the background.

12. When the signals reach an acceptable level, pour off the development solution, and add 20 ml of TE buffer. Incubate, with occasional agitation, at room temperature for a minimum of 30 minutes.
13. Air dry the membrane on a piece of clean filter paper, and store it in the dark.

RESULTS AND DISCUSSION

Examine your hybridized replica membrane. The dark purplish-brown signals result from the detection of the immune/digoxigenin complex and identify transformants containing the 784-bp *Bam*HI/*Hin*dIII λ fragment. Compare your result to the ideal membrane below.

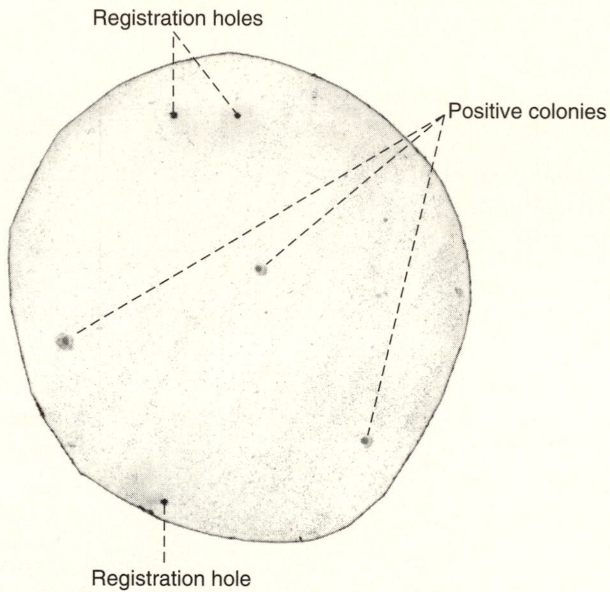

Ideal Membrane

Use the registration marks to orient the replica membrane with its original L20 or L100 plate. (Remember that these two membranes are mirror images of each other.) Note the locations of colonies on the original membrane plate that correspond to hybridization signals on the replica membrane. Often, positive-reacting clones can be easily identified by comparing the pattern of hybridization signals on the replica membrane to the pattern of transformants on the original L20 or L100 plate. If this proves difficult, use a felt-tip marker to transfer the pattern of hybridization signals (including registration marks) from the replica membrane to the lid of a petri dish. The lid is then placed over the original plate and the registration marks are aligned, thus simplifying the identification of positive clones.

Remember that all colonies containing plasmids with *any type* of λ insert are white. However, those with inserts other than 784 bp do not react with the probe and should not be visible on the hybridized membrane. Faint signals will become visible from negative white colonies when the color is overdeveloped, but these can be readily distinguished from the darker signals of positive colonies.

1. Fill in the matrix below with the following:
   a. The number of positive hybridization signals on your replica membrane (# Positive)
   b. The total number of white colonies on the master L20 or L100 plate (# White).
   c. The percentage of positive-reacting white colonies (% Positive).
   d. The number of clonable *Bam*HI/*Hin*dIII λ fragments (# Fragments).
   e. The predicted percentage of positive-reacting white colonies containing the 784-bp fragment, if each λ fragment is cloned with equal frequency (%784 bp).
   f. Enter data for items a–c from several other students, and the class means (averages).

# Positive	# White	% Positive	# Fragments	% 784 bp

2. What factors influence the cloning frequency of the 784-bp λ fragment versus other clonable fragments?
3. a. Calculate the mean (average) size of all restriction fragments generated by the initial *Bam*HI/*Hin*dIII digest of λ that was used to construct the genomic library in Laboratory 14.
   b. Using the average restriction fragment size from step a, estimate the number of fragments generated by a *Bam*HI/*Hin*dIII digest of the human genome (haploid = 3 billion bp).
   c. Calculate the total number of fragments created when the human genome is restricted into fragments averaging 50,000 bp and 500,000 bp.
   d. What are the advantages of using larger restriction fragments when constructing a genomic library of human DNA?
   e. What reagents and techniques are employed to generate large restriction fragments?
4. The human genome has a G + C content of 41%. Use this fact to estimate the cutting frequency of the restriction enzymes *Eco*RI (GAATTC) and *Sma*I (CCCGGG).
5. The number of recombinant clones needed (N) to detect a particular DNA sequence in a genomic library is calculated as follows

$$N = \frac{\ln(1-P)}{\ln(1-f)},$$

   where ln is the natural logarithm, P is the desired probability (level of confidence) that the library contains a specific recombinant of interest, and f is the average fraction (or percent) of the genome in a single recombinant.

Calculate the number of recombinant clones needed to have a 99% probability of identifying a single-copy sequence in a mammalian genome (3×10^9 bp), using the following vector systems:

a. plasmid — average insert size, 10,000 bp
b. λ phage — average insert size, 20,000 bp
c. cosmid — average insert size, 40,000 bp
d. YAC — average insert size, 1,000,000 bp

6. Identifying a suitable probe is a major challenge when screening a gene library. In some cases, the gene product can be isolated from tissue and amino acid sequence obtained from a portion of the purified protein. Using this data, it is possible to synthesize an oligonucleotide probe corresponding to the codons that specify the known sequence of amino acids. However, the degeneracy of the genetic code presents a problem, because most amino acids are specified by more than one codon. The *brute force solution* is to prepare a family of oligonucleotides that correspond to all possible DNA codons. A more streamlined approach is to select an amino acid sequence with the least redundancy—avoiding amino acids such as leucine, which has six codons.

   Referring to the codon table, select the amino acid sequence from the partial protein sequence below that will generate the smallest family of oligonucleotide probes, each 17 bases in length.

   Ala-Leu-Met-Phe-Asp-Glu-Ile-Tyr-Ala-Ser-Asn-Gly

   a. Calculate how many different oligonucleotides must be prepared to ensure that one will perfectly match the target gene sequence.
   b. Screening a genomic library with a family of oligonucleotide probes yields a number of candidate clones. How can you determine which positive-reacting clones contain the bona fide gene?

1st position (5′ end)	**2nd position**				**3rd position (3′ end)**
	U	**C**	**A**	**G**	
U	Phe	Ser	Tyr	Cys	U
	Phe	Ser	Tyr	Cys	C
	Leu	Ser	STOP	STOP	A
	Leu	Ser	STOP	Trp	G
C	Leu	Pro	His	Arg	U
	Leu	Pro	His	Arg	C
	Leu	Pro	Gln	Arg	A
	Leu	Pro	Gln	Arg	G
A	Ile	Thr	Asn	Ser	U
	Ile	Thr	Asn	Ser	C
	Ile	Thr	Lys	Arg	A
	Met	Thr	Lys	Arg	G
G	Val	Ala	Asp	Gly	U
	Val	Ala	Asp	Gly	C
	Val	Ala	Glu	Gly	A
	Val	Ala	Glu	Gly	G

LABORATORY 17

Purification and Identification of λ Clones

Laboratory 16, "Colony Hybridization of the λ Library," identifies colonies that have been transformed with recombinant plasmids containing the 784-bp λ fragment. However, hybridization alone cannot confirm the structure or genotype of the recombinant plasmid, which may well contain several λ restriction fragments. Furthermore, positive-reacting transformants can actually carry two or more different plasmids. The objective of this laboratory is to confirm the molecular genotypes of plasmid DNA from two of the positive-reacting clones from Laboratory 16.

In Part A, "Plasmid Minipreparation of λ/pBLU Recombinants," plasmid DNA is isolated from overnight cultures of two different positive-reacting clones from the original L20 or L100 plate (Laboratories 15–16). In Part B, "Restriction Analysis of Purified λ/pBLU Recombinants," samples of the plasmids isolated in Part A and control samples of λ and pBLU are incubated with *Bam*HI and *Hin*dIII. The digested samples and samples of uncut plasmids are coelectrophoresed in an agarose gel. Comigration of a 784-bp fragment in the digested minipreps and the digested λ control confirms the successful cloning of the sequence of interest. Comparison of miniprep fragments with λ restriction fragments—plus evaluation of the relative sizes of uncut, supercoiled plasmids—provides evidence of the structure, size, and number of plasmids present in each of the transformed clones.

Isolate Plasmid DNA

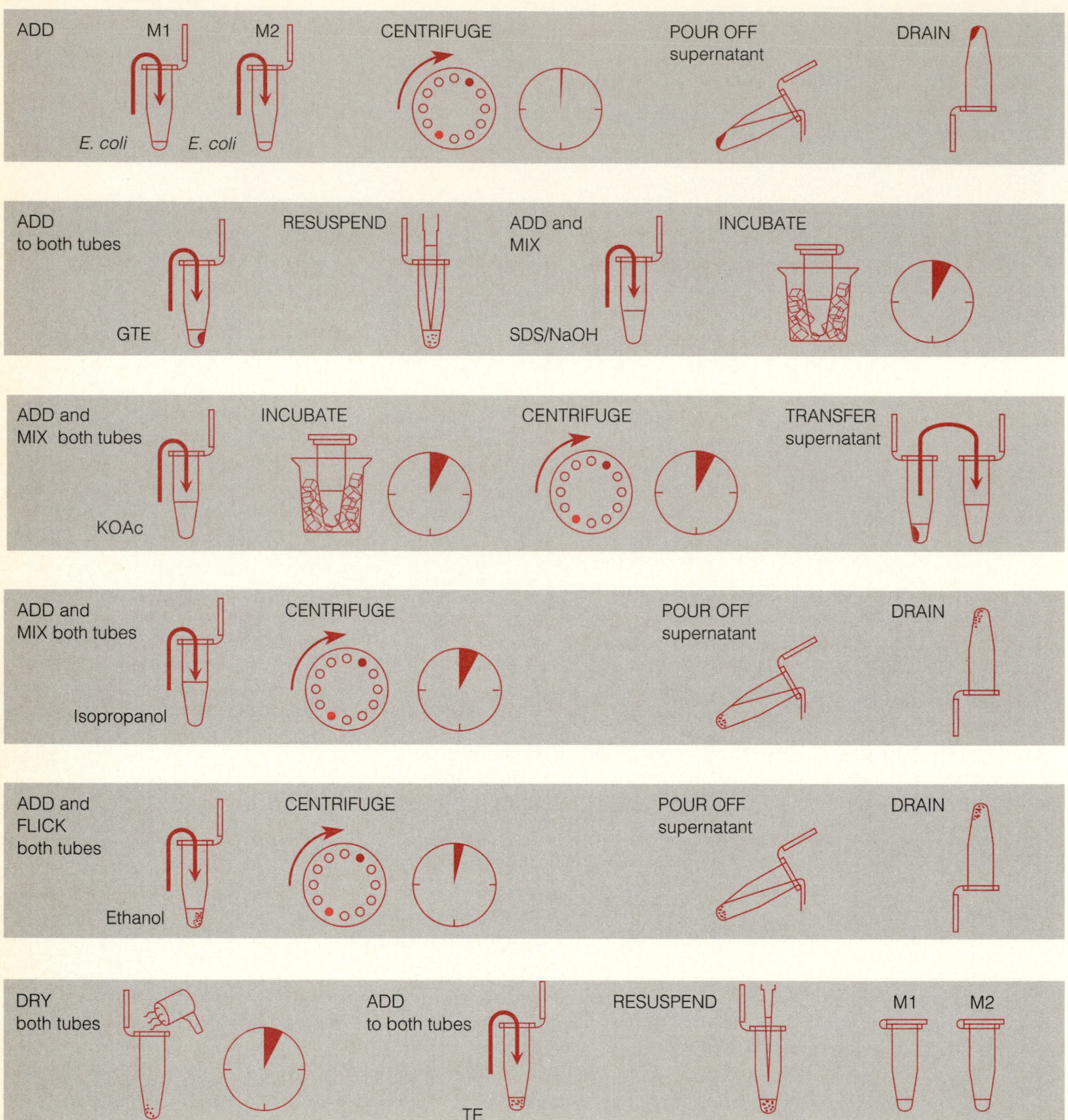

Laboratory 17/Part A
Plasmid Minipreparation of λ/pBLU Recombinants

PRELAB NOTES

Review the prelab notes in Laboratory 6A, "Plasmid Minipreparation of pAMP."

For Further Information

The protocol presented here is based on the following published methods:

Birnboim, H. C. and J. Doly. 1979. A rapid alkaline extraction method for screening recombinant plasmid DNA. *Nucleic Acids Research* 7: 1513.

Ish-Horowicz, D. and J. F. Burke. 1981. Rapid and efficient cosmid cloning. *Nucleic Acids Research* 9: 2989.

PLASMID MINIPREPARATION OF λ/pBLU RECOMBINANTS

Reagents

two *E. coli*/ λ/pBLU overnight cultures
glucose/Tris/EDTA (GTE)
SDS/sodium hydroxide (SDS/NaOH)
potassium acetate/acetic acid (KOAc)
isopropanol
95% ethanol
Tris/EDTA (TE)

Supplies and Equipment

100–1,000-μl micropipettor + tips
10–100-μl micropipettor + tips (optional)
0.5–10-μl micropipettor + tips
1.5-ml tubes
beaker of crushed ice
beaker for waste/used tips
10%-bleach solution or disinfectant (such as Lysol)
disposable gloves
hair dryer
microfuge
paper towels
permanent marker
test tube rack

Isolate Plasmid DNA

(50 minutes)

1. Shake the culture tubes to resuspend the *E. coli* cells.
2. Label two 1.5-ml tubes with your initials. Label one tube "M1" and the other tube "M2." Transfer 1,000 μl of each overnight culture into a separate 1.5-ml tube.
3. Close the caps, and place the tubes in a *balanced* configuration in a microfuge rotor. Spin for 1 minute to pellet the cells.

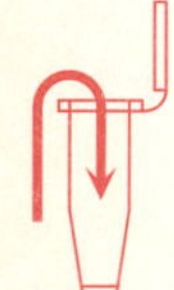
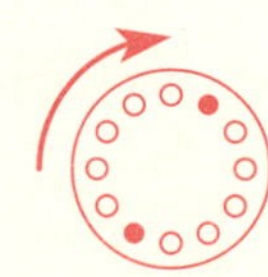

Double all centrifuge times if you are using a small microfuge that generates approximately 2,000 × *g*.

4. Pour off the supernatant from both tubes into a waste beaker for later disinfection. *Take care not to disturb the cell pellets.* Invert the tubes, and touch the mouths of the tubes to a clean paper towel, to wick off as much as possible of the remaining supernatant.

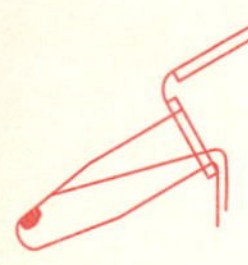
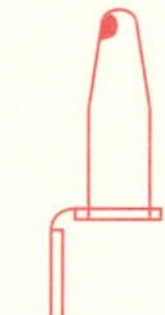

5. Add 100 μl of *ice-cold* GTE solution to each tube. Resuspend the pellets by pipetting the solution in and out several times. Hold the tubes up to the light to verify that the suspension is homogeneous and that no visible clumps of cells remain.

Accurate pipetting is essential to good plasmid yield. The volumes of reagents are precisely calibrated so that the sodium hydroxide added in step 6 will be neutralized by the acetic acid in step 8.

6. Add 200 μl of SDS/NaOH solution to each tube. Close the caps, and mix the solutions by rapidly inverting the tubes about five times.

Do not overmix. Excessive agitation shears the single-stranded DNA and decreases plasmid yield.

7. Stand the tubes on ice for 5 minutes. The suspension will become relatively clear.
8. Add 150 μl of *ice-cold* KOAc solution to each tube. Close the caps, and mix the solutions by rapidly inverting the tubes about five times. A white precipitate will immediately appear.
9. Stand the tubes on ice for 5 minutes.

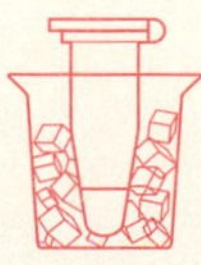

10. Place the tubes in a *balanced* configuration in a microfuge rotor, and spin for 5 minutes to pellet the precipitate along the side of the tube.

Precipitate typically collects along the side of the tube, rather than as a tight pellet.

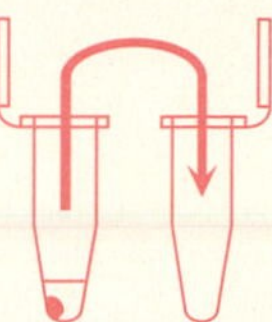

11. Transfer 400 μl of supernatant from the M1 tube into a clean 1.5-ml tube labeled "M1." Transfer 400 μl of supernatant from the M2 tube into a clean 1.5-ml tube labeled "M2." *Avoid pipetting the precipitate*, and wipe off any precipitate that clings to the outside of the tip, before expelling the supernatant. Discard old tubes containing precipitate.
12. Add 400 μl of isopropanol to each tube of supernatant. Close the caps, and mix the solutions by rapidly inverting the tubes about five times. *Stand the tubes at room temperature for only 2 minutes.*

Do step 12 quickly; make sure that the microfuge will be immediately available for step 13. The isopropanol preferentially precipitates nucleic acids rapidly; however, proteins and other cellular components remaining in solution will also begin to precipitate with time.

13. Place the tubes in a *balanced* configuration in a microfuge rotor, and spin for 5 minutes to pellet the nucleic acids. Align the tubes in the rotor so that the cap hinges point outward. The nucleic acid residue, visible or not, will collect under the hinge during centrifugation.

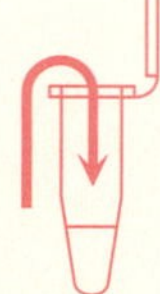

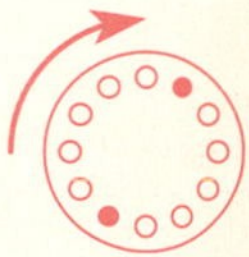

14. Pour off the supernatant from both tubes. *Take care not to disturb the nucleic acid pellets.* Wick off as much as possible of the remaining alcohol on a paper towel.
15. Add 200 μl of 95% ethanol to each tube, and close the caps. Flick the tubes several times to wash the nucleic acid pellets.

The pellet will likely appear either as a tiny, teardrop-shaped smear or small particles on the bottom of each tube. However, pellet size is not a valid predictor of plasmid yield or quality. A large pellet is composed primarily of RNA, salt, and cellular debris carried over from the original precipitate; a small pellet, or one difficult to see, often means a cleaner preparation.

Stop Point

Store the DNA in ethanol at –20°C until you are ready to continue.

16. Place the tubes in a *balanced* configuration in a microfuge rotor, and spin for 2–3 minutes to recollect the nucleic acid pellets.
17. Pour off the supernatant from both tubes. *Take care not to disturb the nucleic acid pellets.* Wick off as much as possible of the remaining alcohol on a paper towel.

Nucleic acid pellets are not soluble in ethanol and will not resuspend during washing.

18. Dry the nucleic acid pellets, using one of following methods:
    a. Direct a stream of warm air from a hair dryer over the openings of the tubes for about 3 minutes. *Be careful not to blow the pellets out of the tubes.*

    or

    b. Close the caps, and pulse the tubes in a microfuge to pool the remaining ethanol. *Carefully* use a micropipettor to draw off the ethanol. Let the pellets air dry at room temperature for 10 minutes.

Air bubbles cast in the tube wall can be mistaken for ethanol droplets.

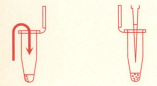

19. Be sure that the nucleic acid pellet is dry and that all ethanol has evaporated before proceeding on to step 20. Hold each tube up to the light to confirm that no ethanol droplets remain and that the nucleic acid pellet, if visible, appears white and flaky. If ethanol is still evaporating, you will detect an alcohol odor by sniffing the mouth of the tube.

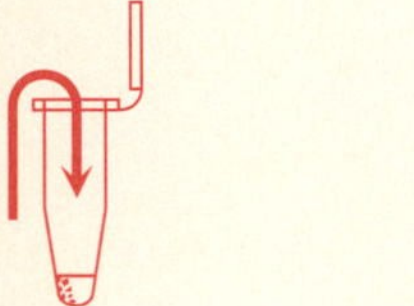

If you are using a 0.5–10-µl micropipettor, set it to 7.5 µl, and pipet twice.

20. Add 15 µl of TE to each tube. Resuspend the pellets by scraping with the pipet tip and vigorously pipetting in and out. Rinse down the side of the tube several times, concentrating on the area where the pellet should have formed during centrifugation (beneath the cap hinge). Make sure that all DNA has dissolved and that no particles remain in the tip or on the side of the tube.
21. Keep the two DNA/TE solutions *separate*. *Do not* pool them into one tube!

Stop Point

Freeze the DNA/TE solutions at –20°C until you are ready to continue. Thaw before using.

22. Take the time for proper cleanup:
    a. Segregate for proper disposal culture tubes and micropipettor tips that have come into contact with *E. coli.*
    b. Disinfect the overnight cultures, the tips, and the supernatant from step 4 with a 10%-bleach solution or disinfectant (such as Lysol).
    c. Wipe down the lab bench with soapy water, a 10%-bleach solution, or disinfectant.
    d. Wash your hands before leaving the laboratory.

RESULTS AND DISCUSSION

See the "Results and Discussion" section of Laboratory 6A, "Plasmid Minipreparation of pAMP," for a detailed discussion of the biochemistry of the alkaline lysis method for plasmid purification.

I. Set Up Restriction Digests

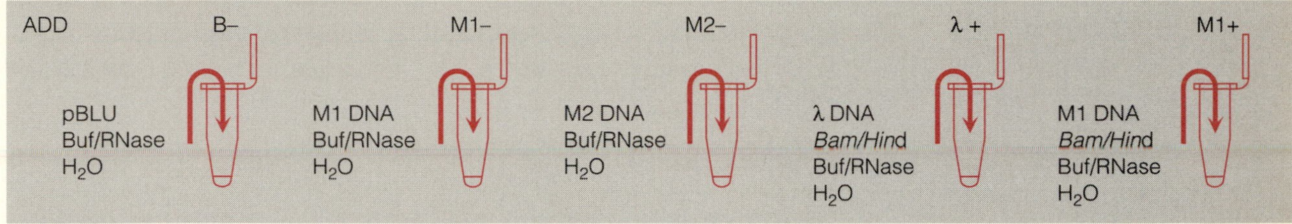

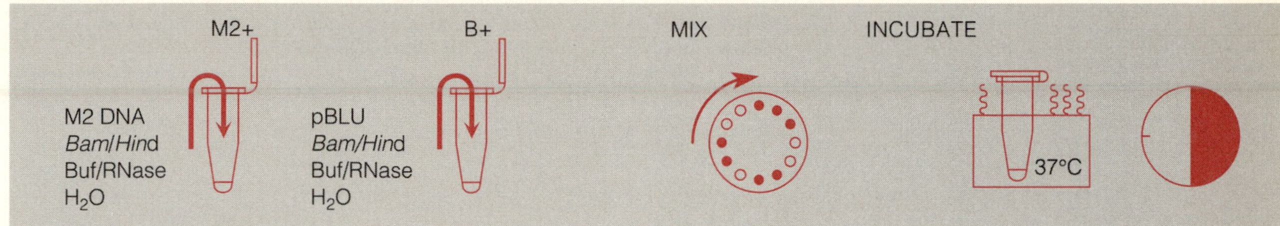

II. Cast 0.8% Agarose Gel

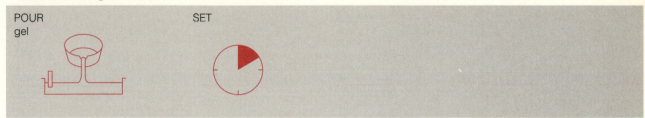

III. Load Gel and Electrophorese

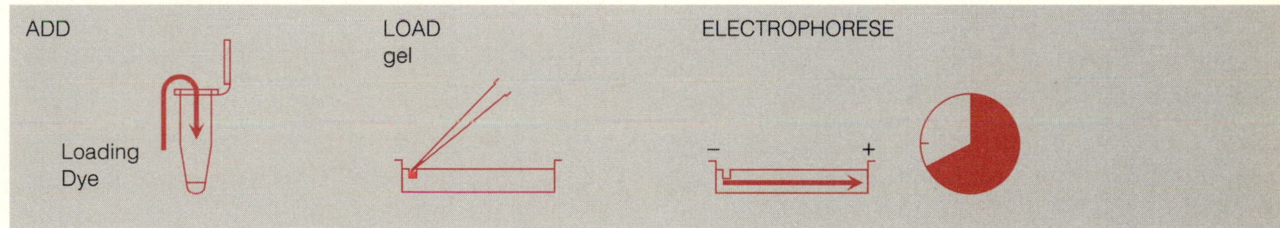

IV. Stain, View, and Photograph Gel

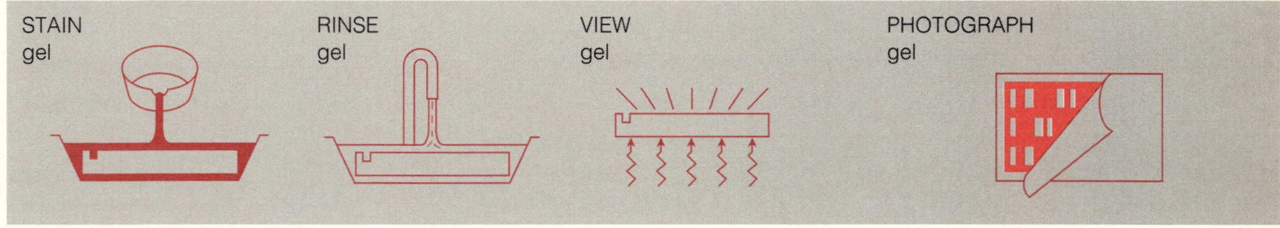

Laboratory 17/Part B
Restriction Analysis of Purifed λ/pBLU Recombinants

PRELAB NOTES

Review the prelab notes in Laboratory 3, "DNA Restriction and Electrophoresis," and Laboratory 6B, "Restriction Analysis of Purified pAMP."

For Further Information

The protocol presented here is based on the following published methods:

Helling, R. B., H. M. Goodman, and H. W. Boyer. 1974. Analysis of *Eco*RI fragments of DNA from lambdoid bacteriophages and other viruses by agarose-gel electrophoresis. *Journal of Virology* 14: 1235.

Sharp, P. A., B. Sugden, and J. Sambrook. 1973. Detection of two restriction endonuclease activities in *Haemophilus parainfluenzae* using analytical agarose–ethidium bromide electrophoresis. *Biochemistry* 12: 3055.

RESTRICTION ANALYSIS OF PURIFED λ/pBLU RECOMBINANTS

Reagents	Supplies and Equipment
M1 and M2 miniprep DNA/TE	0.5–10-μl micropipettor + tips
0.1 μg/μl pBLU	1.5-ml tubes
0.1 μg/μl λ DNA	aluminum foil
*Bam*HI/*Hin*dIII	beaker for agarose
5× restriction buffer/RNase	beaker for waste/used tips
distilled water	camera and film (optional)
loading dye	disposable gloves
0.8% agarose	electrophoresis box
1× Tris/Borate/EDTA (TBE) buffer	masking tape
1 μg/ml ethidium bromide	microfuge (optional)
	Parafilm or waxed paper (optional)
for decontamination:	permanent marker
0.05 M $KMnO_4$	plastic wrap (optional)
0.25 N HCl	power supply
0.25 N NaOH	test tube rack
	transilluminator
	37°C water bath
	60°C water bath (for agarose)

I. Set Up Restriction Digests

(10 minutes; then 30–40 minutes of incubation)

Refer to Laboratory 3, "DNA Restriction and Electrophoresis," for detailed instructions on setting up digests.

1. Use a permanent marker to label seven 1.5-ml tubes, as shown in the matrix below. Restriction reactions will be performed in these tubes.
2. Use the matrix below as a checklist while adding reagents to each reaction. Read down each column, adding the same reagent to all appropriate tubes. *Use a fresh pipet tip for each reagent.*

Tube	pBLU	M1	M2	λ *DNA*	Buffer/ RNase	*Bam*HI/ *Hin*dIII	H_2O
B–	5 μl	—	—	—	2 μl	—	3 μl
M1–	—	5 μl	—	—	2 μl	—	3 μl
M2–	—	—	5 μl	—	2 μl	—	3 μl
λ+	—	—	—	5 μl	2 μl	2 μl	1 μl
M1+	—	5 μl	—	—	2 μl	2 μl	1 μl
M2+	—	—	5 μl	—	2 μl	2 μl	1 μl
B+	5 μl	—	—	—	2 μl	2 μl	1 μl

3. After adding all of the above reagents, close the tops of the tubes and mix.

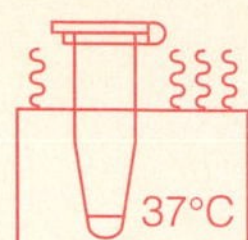

Do not overincubate: During a longer incubation, the DNases in the miniprep may degrade the plasmid DNA.

4. Place the reactions in a 37°C water bath, and incubate them for 30–40 minutes only.

Stop Point

Following incubation, freeze the reactions at –20°C until you are ready to continue. Thaw the reactions before continuing on to Section III, step 1.

II. Cast 0.8% Agarose Gel

(15 minutes)

1. Carefully pour agarose solution into the gel-casting tray to fill it to a depth of about 5 mm. The gel should cover about one-third the height of the comb teeth.
2. After the agarose solidifies, place the gel-casting tray into the electrophoresis box and set up for electrophoresis.

Stop Point

Cover the electrophoresis box, and save the gel until you are ready to continue. The gel will remain in good condition for several days if it is completely submerged in buffer.

Refer to Laboratory 3, "DNA Restriction and Electrophoresis," for detailed instructions on casting and loading gel.

III. Load Gel and Electrophorese

(10 minutes; then 30–45 minutes of electrophoresis)

1. Remove the restriction digests from the 37°C water bath.
2. Add 1 µl of loading dye to each reaction tube. Close the tops and mix.

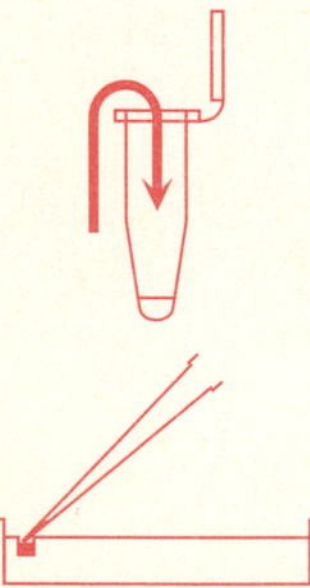

3. Load 10 µl from each reaction tube into a separate well in the gel, as shown in the diagram below. Use a *fresh tip* for each reaction.

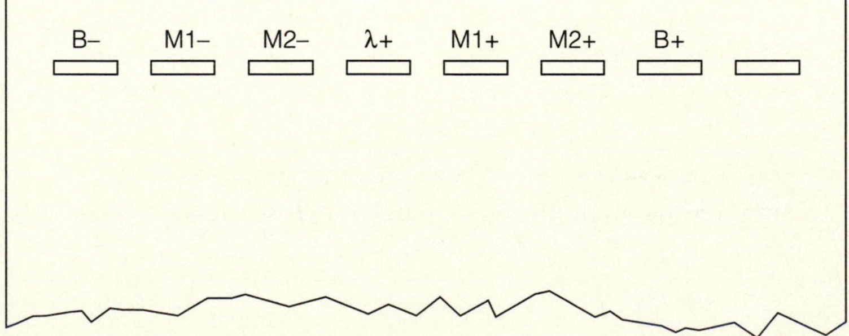

4. Electrophorese at 50–125 volts until the bromophenol blue bands have moved 50–60 mm from the wells.
5. Turn off the power supply, remove the casting tray from the electrophoresis box, and transfer the gel to a disposable weigh boat (or other shallow tray) for staining.

The 784-bp λ fragment migrates just behind the bromophenol blue marker. *Take care not to run it off of the end of the gel.* However, if time allows, electrophorese until the bromophenol blue bands *near* the end of the gel. This allows maximum separation of uncut plasmids, which is important for differentiating a large "superplasmid" from a double transformation of two smaller plasmids.

Stop Point

The gel can be stored overnight, covered with small volume of buffer, in a staining tray or in a zip-lock plastic bag, for viewing or photographing the next day. However, over a longer period of time, the DNA will diffuse through the gel, and the bands will either become indistinct or disappear entirely.

IV. Stain Gel with Ethidium Bromide, View, and Photograph

(15 minutes)

Staining may be performed by an instructor in a controlled area, when students are not present.

CAUTION

Review the section on "Ethidium Bromide Staining and Responsible Handling" in Laboratory 3. Wear disposable gloves when staining, viewing, and photographing gel and during cleanup. Confine all staining to a restricted sink area.

1. Flood the gel with ethidium bromide solution (1 µg/ml), and allow it to stain for 5–10 minutes.
2. View the gel under an ultraviolet transilluminator or other UV light source.

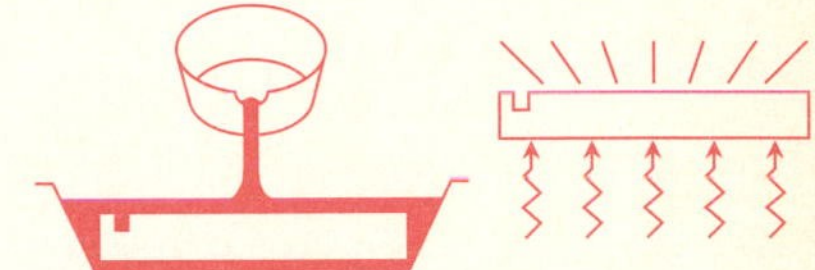

Band intensity and contrast increase dramatically if the gel is destained for 15–30 minutes in tap water. More simply, rinse and drain the gels, stack the staining trays, cover the top gel, and let it set overnight at room temperature.

CAUTION

Ultraviolet light can damage your eyes. Never look directly at an unshielded UV light source without eye protection. View only through a filter or safety glasses that absorb the harmful wavelengths.

3. Make an exposure of 1–2 seconds with the camera set at f/8. Develop the print for the recommended time (approximately 45 seconds at room temperature).
4. Take the time for proper cleanup:
   a. Wipe down the camera, the transilluminator, and the staining area.

b. Decontaminate the gel and any staining solution that will not be reused.

c. Wash your hands before leaving the laboratory.

RESULTS AND DISCUSSION

Refer to the "Results and Discussion" section of Laboratory 6B, "Restriction Analysis of Purified pAMP," for more about interpreting miniprep gels and plasmid conformation.

Examine the photograph of your stained gel, and determine which lanes contain control DNAs (pBLU and λ) and which contain minipreps M1 and M2. Even if you have not followed the prescribed loading order, the miniprep lanes can be distinguished by the following characteristics:

- A background "smear" of degraded and partially digested chromosomal and plasmid DNA
- Undissolved material and DNA of high molecular weight trapped at the front edge of the sample well
- A "cloud" of RNA of low molecular weight at a position corresponding to 100–200 bp
- The presence of high-molecular-weight bands of uncut plasmid in lanes of digested miniprep DNA

Remember this when you are considering the possible structures of the recombined plasmids in M1 and M2:

- Every replicating plasmid must have an origin of replication. Recombinant plasmids with more than one origin also replicate normally.
- Adjacent restriction fragments in a recombinant plasmid must be ligated at like restriction sites: *Bam*HI to *Bam*HI and *Hin*dIII to *Hin*dIII. Recall that, in Laboratory 10, "Purification and Identification of Recombinant Plasmid DNA," recombinant plasmids were ligated from restriction fragments with one *Bam*HI end and one *Hin*dIII end. Because of this, each recombinant plasmid had to be composed of an even number of fragments. The situation is more complicated in this laboratory, however. The *Bam*HI/*Hin*dIII digest of λ in Part A created numerous restriction fragments; some have two *Bam*HI ends or two *Hin*dIII ends. These can combine with *Bam*HI/*Hin*dIII restriction fragments to form plasmids composed of odd numbers of fragments.
- Repeated copies of the same restriction fragment cannot exist adjacent to one another; that is, they must alternate with other fragments. Adjacent duplicate fragments form "inverted repeats;" the sequences, one on either side of the restriction site, are complementary along the entire length of the duplicated fragment. Molecules with such inverted repeats cannot replicate properly. As the plasmid opens up to allow access to DNA polymerase, the single-stranded regions on either side of the restriction site base pair to one another, forming a large "hairpin loop," which fouls replication.

Use questions 1–10, to interpret each pair of miniprep results (M1+/– and M2+/–).

1. Examine the photograph of your stained gel. Compare your gel to the ideal gel on the next page, and label the size of the fragments in each lane.

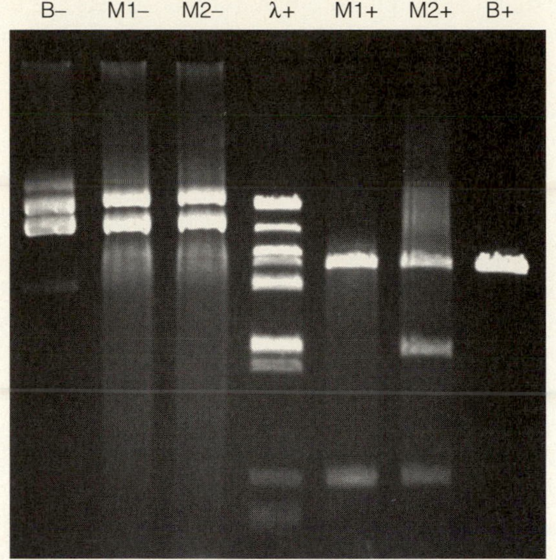

Ideal Gel

2. Label the band in the cut pBLU lane (B+) "5,400 bp." Then label the nine bands that appear in the cut λ lane (λ+). It is easiest to start with the smallest fragment at the bottom of the gel and then progress toward the sample well. Remember that you must account for 13 *Bam*HI/*Hin*dIII restriction fragments. Notice that several thick bands are actually composed of multiple fragments, and the smallest fragment (125 bp) either runs off of the end of the gel or is too faint to be seen. A 14th fragment of 9,866 bp arises from digesting circular λ molecules joined at the COS sites.
3. Label the type(s) of restriction sites at the ends of each λ fragment (*Bam*HI-*Hin*dIII, *Bam*HI-*Bam*HI, or *Hin*dIII-*Hin*dIII).
4. *The cut miniprep lane (M+) provides information about the number and types of restriction fragments in the construct.* Every digested miniprep must contain the 5,400-bp pBLU backbone with the amp^r gene and the 784-bp λ fragment that reacts with the hybridization probe.
   a. Locate these bands by comparing the cut miniprep lanes with the cut pBLU and λ lanes.
   b. It is possible to find minipreps that do *not* contain the 784-bp λ fragment. This results from beginning the overnight culture with a nonreacting colony from the L20 or L100 original plate. Inexact alignment with the hybridized membrane makes it easy to mistakenly select a colony located very close to a positive-reacting clone, especially if the master plate contains numerous, closely spaced colonies.
5. Now, look for evidence of other bands in the cut miniprep lanes. If no additional fragments are visible, the molecule is then termed a "simple recombinant" of 6,184 bp. (This is M1 of the ideal gel above.)

6. Other than the simple recombinant, constructs are typically seen to contain one additional λ fragment, which can be either
   a. a three-fragment "superplasmid" composed of the pBLU backbone, the 784-bp λ fragment, and an additional λ fragment with two *Bam*HI ends or two *Hin*dIII ends. (This is M2 of the ideal gel on the previous page.)
   b. a four-fragment superplasmid in which one of the three fragments above is repeated.
   c. a double transformation of a simple recombinant (pBLU backbone plus the 784-bp fragment) and another recombinant composed of pBLU plus the other visible fragment. *In this situation, the additional fragment must have one* Bam*HI end and one* Hind*III end.*
7. Use the sizes of the λ fragments below to compute the sizes of several types of constructs:

left end - *Bam*HI	5,505 bp
*Bam*HI - *Bam*HI	16,841 bp
*Bam*HI - *Hin*dIII	784 bp
*Hin*dIII - *Hin*dIII	2,027 bp
*Hin*dIII - *Hin*dIII	2,322 bp
*Hin*dIII - *Bam*HI	493 bp
*Bam*HI - *Bam*HI	6,527 bp
*Bam*HI - *Hin*dIII	2,396 bp
*Hin*dIII - *Hin*dIII	564 bp
*Hin*dIII - *Hin*dIII	125 bp
*Hin*dIII - *Bam*HI	4,148 bp
*Bam*HI - *Hin*dIII	2,409 bp
*Hin*dIII - right end	4,361 bp

   a. three-fragment superplasmids
   b. four-fragment superplasmids
   c. two-fragment plasmids, other than the simple recombinant
8. The uncut miniprep lanes (M–) provide information about the overall size of the construct. Compare the M– (uncut miniprep) lane with the B– (uncut pBLU) lane. Remember that the uncut plasmid can assume any of several conformations, but the fastest-moving form is supercoiled.
   a. Locate the band that has migrated farthest in the B– lane; this is the supercoiled form of pBLU.
   b. Now examine the band farthest down the M– lane. If this band is only slightly higher on the gel, than the molecule is most likely a simple recombinant containing the 784-bp fragment.
   c. The presence of supercoiled uncut miniprep DNA higher on the gel indicates a larger construct. Distinguishing between double transformations and superplasmids can be difficult, because some two-fragment constructs are larger than other three- or four-fragment constructs. Examining the type of ends on the additional

fragment(s) in the M+ lane can often narrow the range of possible constructs, however.

9. When bacteria are transformed using two different plasmids, one of the two may be preferentially replicated within the host cell. Over generations, the copy numbers of the two plasmids become increasingly different. Thus, in double transformations showing four different fragments, one pair of fragments can be fainter than the other pair.
10. Based on your above evaluation, make scale restriction maps of your M1 and M2 plasmids.

FOR FURTHER RESEARCH

Analyze several transformants to ascertain whether the five clonable λ fragments with one *Bam*HI end and one *Hin*dIII end are represented with equal frequency in the two-fragment constructs.

1. Start overnight cultures from several white colonies picked from an L20 or L100 plate.
2. Prepare plasmid DNA from each overnight culture, using the alkaline lysis method.
3. Restriction analyze the purified plasmids with *Bam*HI and *Hin*dIII, using appropriate controls.

Cloning by Polymerase Chain Reaction

LABORATORY 18

Amplification and Purification of a λ DNA Fragment

Like Southern hybridization, polymerase chain reaction (PCR) allows scientists to detect a specific DNA sequence against the background of a complex genome. PCR is a test tube system for DNA replication that employs the essential enzyme of cellular DNA replication, DNA polymerase, to selectively amplify a "target" DNA region. Also key to the system is a pair of oligonucleotide primers—single-stranded DNA sequences of 20–30 nucleotides—that serve as points of attachment for the polymerase. The primers bracket the region to be amplified: One primer is complementary to a sequence at the *beginning* of the target region, and the second is complementary to a sequence at the *end* of the target region on the antiparallel DNA strand.

To perform a PCR reaction, a small quantity of the target DNA is added to a buffered solution containing DNA polymerase, oligonucleotide primers, the four deoxynucleotide building blocks of DNA, and the cofactor Mg^{++}. The PCR mixture is taken through 20–30 cycles of replication; each cycle doubles the copy number of the target region. Each PCR cycle consists of

- 1 to several minutes at 94–96°C, during which the DNA is denatured into single strands
- 1 to several minutes at 50–65°C, during which the primers hybridize or "anneal" (by way of hydrogen bonds) to their complementary sequences on either side of the target region
- 1 to several minutes at 72°C, during which the polymerase binds and extends a complementary DNA strand from each primer

This experimental unit combines specific PCR amplification and cellular amplification (molecular cloning) for a two-stage enrichment of a target region. Besides speed, another advantage of this cloning strategy is that all recombinant plasmids should contain the desired fragment, thus eliminating the need for a complicated hybridization screening protocol such as that used in Laboratory 16.

In Part A, "PCR Amplification of a λ DNA Fragment," PCR is used to selectively amplify a 1,106-base-pair region from the λ genome. The "upstream primer" has the sequence 5′ TTCTGAACTCGGTCCGTTAC 3′ and corresponds to the λ map positions 22,233-22,252. The "downstream primer" has the sequence 5′ TCGCCAACATCATTCGACTC 3′, and corresponds to nucleotides 23,339-23,320 (counting backward on the antiparallel strand).

In Part B, "Purification of a λ PCR Product," the amplified λ fragment is treated either by phenol/chloroform extraction or by column chromatography to remove the DNA polymerase and deoxynucleotides that can interfere with the subsequent ligation reaction. Intact DNA polymerase can use the nucleotides remaining in the sample to fill in the 5′ single-stranded (sticky) ends generated by the restriction digestion, thus blocking activity of the ligase.

In Part C, "Electrophoresis of a λ PCR Product," amplified and unamplified samples of λ DNA are coelectrophoresed to confirm that amplification has occurred and that the amplified DNA was not lost during phenol/chloroform extraction or chromatography. The expected 1.1 kb amplification product is verified by comparison to a *Bam*HI/*Hin*dIII digest of λ. A *Bam*HI site is located near the left-hand side of the amplified region (λ map position 22,346), and a *Hin*dIII site is located toward the right-hand side at position 23,130. These restriction sites define the 784-bp λ fragment cloned in Laboratories 14–17 that will again be cloned in this unit.

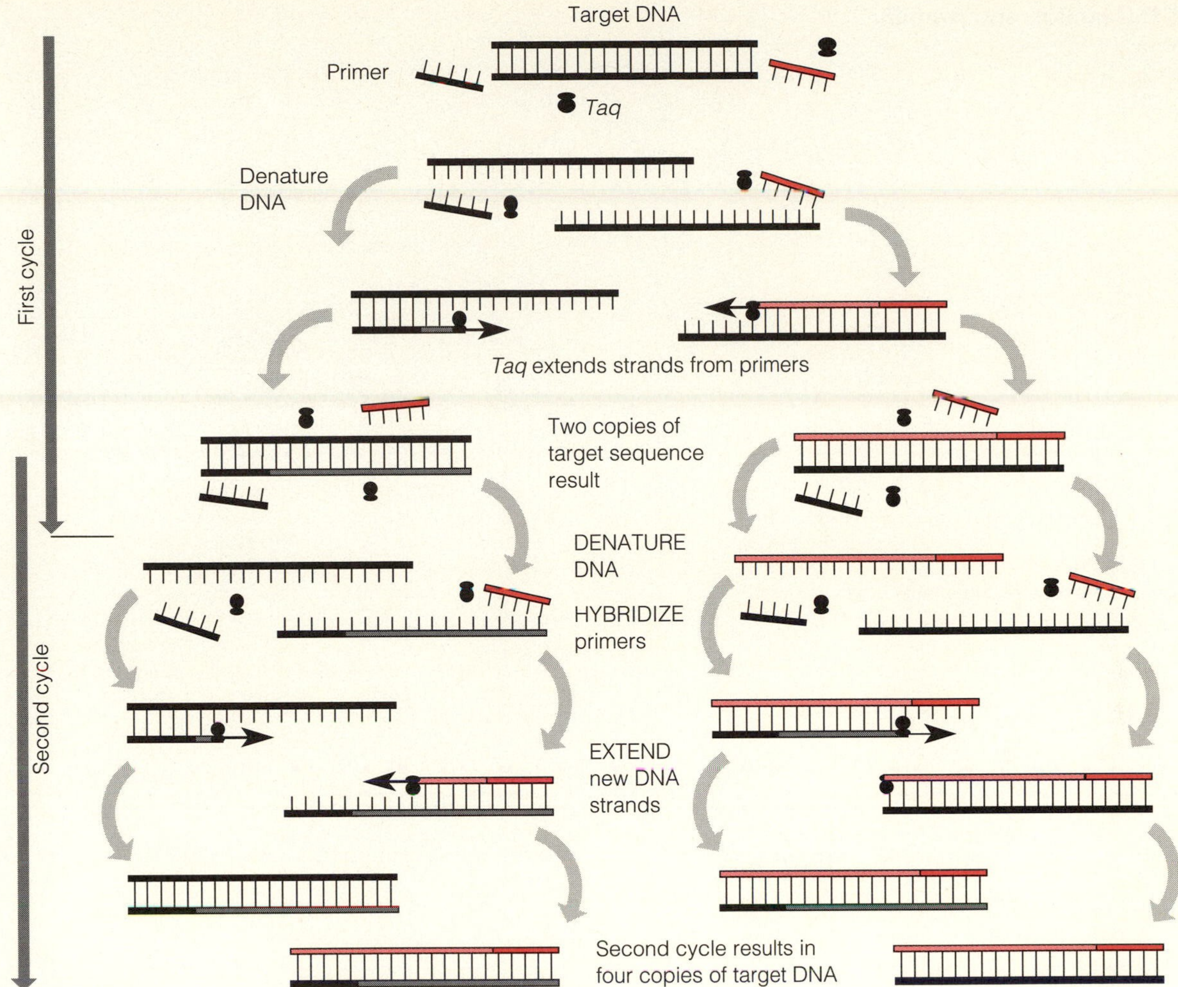

Polymerase Chain Reaction

Set Up PCR Reaction and Amplify

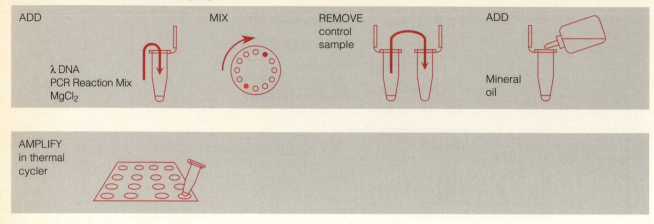

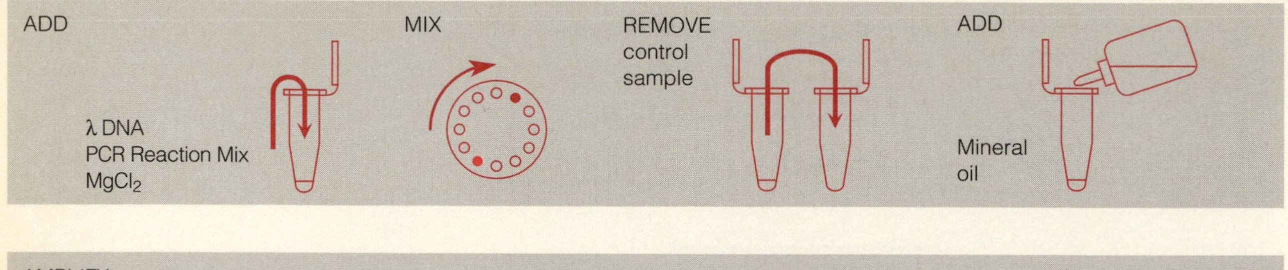

Laboratory 18/Part A
PCR Amplification of a λ DNA Fragment

PRELAB NOTES

PCR Innovations

Initially, PCR was a tedious task that involved transferring reactions between several water baths and adding fresh polymerase for each cycle. (The denaturing step also denatured the polymerase.) Two important innovations were responsible for automating PCR. First, a heat-stable DNA polymerase was isolated from the bacterium *Thermus aquaticus*, which inhabits hot springs. *Taq* polymerase remains active despite repeated heating during many cycles of amplification; thus, it needs to be added only once at the beginning of the PCR reaction. Second, automated DNA thermal cyclers were invented that control the repetitive temperature changes required for PCR.

Manual Cycling

In this laboratory, the target region to be amplified (1,106 bp) is found within the λ genome (48,502 bp). Although amplification using a thermal cycler is the method of choice, manual cycling using water baths is a viable option for this experiment. The small size of the λ genome enables the primers to rapidly locate and bind to their complementary sequences. Partly due to the increased efficiency of the λ system, amplification works well using just two temperatures per cycle. Samples can be rotated between a 55°C water bath (annealing) and a boiling water bath (denaturing). Primer extension occurs as the reaction heats from the *annealing temperature* to the *denaturing temperature*.

During the 20-second incubation in the boiling water bath, samples heat up to a denaturing temperature of approximately 96°C. *Take care not to overincubate the samples in boiling water. A prolonged incubation can bring the sample temperature to boiling, which reduces and quickly destroys the activity of the* Taq *polymerase.*

A simple boiling water bath consists of a beaker of water on a hot plate. Obtain a floating test tube rack or fashion one by poking holes in a piece of Styrofoam; then attach a handle, so that the rack can be easily transferred between water baths.

Primer Hybridization Temperature

Primers are hybridized to template DNA in the range of 50–65°C. Three variables determine the temperature and time required for primer annealing: (1) concentration of primers, (2) length of the primer sequences, and (3) the base composition of the primers.

Generally, a suitable annealing temperature is about 5°C below the T_m of the primers, where T_m (Temperature melting) is defined as that temperature at which 50% of the duplex DNA molecules are denatured. The T_m can be estimated from the primer sequence by assigning 4°C for each G or C base and 2°C for each A or T base. Ultimately, the optimum annealing temperature should be determined experimentally.

Reactant Concentrations

Remember that PCR is essentially DNA replication in a test tube and is therefore subject to the limitations of any enzymatic reaction. Although the PCR reaction includes substrates (primers and deoxynucleotides) in vast excess, a "plateau effect" is observed, during the late PCR cycles, where the rate of product accumulation falls below the theoretical exponential rate. The plateau can be influenced by several factors. These include: (1) depletion of substrates (dNTPs and primers), (2) enzyme or substrate instability, (3) end-product inhibition, (4) competition by nonspecific products ("primer-dimer"), and (5) incomplete denaturation at a high product concentration.

PCR is especially sensitive to magnesium ions, which can affect primer annealing, DNA denaturing temperature, and the activity/fidelity of strand extension by the *Taq* polymerase. Therefore, it is wise to determine the optimum $MgCl_2$ concentration for each DNA sequence to be amplified. Typically, the best results are obtained with a free magnesium concentration in the range of 0.5–2.5 mM. Since deoxynucleotide triphosphates (dNTPs) quantitatively bind Mg^{++}, the amount of free Mg^{++} equals the concentration of $MgCl_2$ minus the total concentration of dNTPs.

PCR and Contamination

PCR is an extremely sensitive technique, capable of detecting a single molecule of DNA. It is critical to observe proper laboratory technique to avoid contamination of the PCR reactions with sample-to-sample contamination or with amplification products from prior reactions.

Ideally, PCR reactions are set up in one location and the post-PCR analysis is performed in another location. Experimenters should always wear disposable gloves and, if possible, use micropipet tips that can contain an aerosol blocking filter.

PCR Mix and Nonspecific Priming

PCR reaction mix—containing PCR buffer, dNTPs, primers, and *Taq* polymerase—will be stable for at least 1 year *in the absence of Mg++*. Magnesium ions can activate contaminating nucleases that can degrade primers. Target DNA and $MgCl_2$ are each added separately to the reaction mix to initiate the reaction. Nonspecific priming can begin almost immediately in the presence of magnesium ions, so thermal cycling should begin immediately upon the addition of $MgCl_2$.

For Further Information

The protocols presented here are based on the following published methods:

Innis, M. A., Gelfand, D. H., Sninsky, J. J., and White, T. J., eds. 1990. *PCR Protocols—A Guide to Methods and Applications.* Academic Press.

Saiki, R. K., S. J. Scharf, F. Faloona, K. B. Mullis, G. T. Horn, H. A. Erlich, and N. Arnheim. 1985. Enzymatic amplification of β-globin sequences and restriction site analysis for diagnosis of sickle cell anemia. *Science* 239: 1350.

PCR AMPLIFICATION OF A λ DNA FRAGMENT

Reagents	Supplies and Equipment
PCR reaction mix	10–100-µl micropipettor + tips
0.05 ng/µl λ DNA	0.5–10-µl micropipettor + tips
12.5 mM $MgCl_2$	0.5-ml PCR tubes
mineral oil	beaker for waste/used tips
	disposable gloves
	DNA thermal cycler
	or
	55°C water bath and boiling water bath

Set Up the PCR Reaction and Amplify

(45 minutes for manual cycling; 150 minutes for automated cycling)

PCR is very sensitive! Always wear disposable gloves to help prevent contamination.

1. Use a permanent marker to label the lid of the PCR tube with your name.

A label placed on the side of the tube may come off during thermal cycling.

2. Use the matrix below as a checklist while adding reagents to the 0.5-ml PCR reaction tube. *Use a fresh pipet tip for each reagent.*

Tube	λ DNA	PCR Reaction Mix	$MgCl_2$
Name	40 µl	40 µl	20 µl

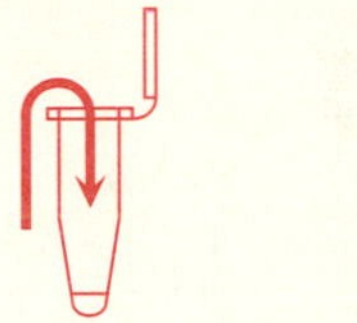

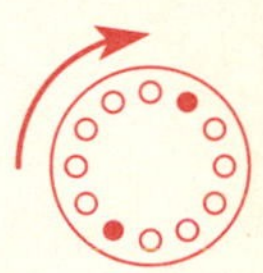

To microfuge, first place the PCR tube in an empty 1.5-ml tube.

3. After adding all of the reagents, close the tube tops and mix the solution.
4. *Use a fresh tip* to transfer 10 µl from the PCR reaction tube into a fresh 1.5-ml tube labeled "PCR Control" with your name. This sample will remain unamplified; store it on ice or in the freezer until you are ready for gel electrophoresis (Part C, Section II).
5. Carefully add one drop of mineral oil from the dropper bottle to the remaining reaction in the PCR tube. *Do not touch the dropper to the PCR tube; subsequent reactions may be contaminated with your sample.*
6. Close the cap of the tube tightly, and amplify the sample using either automated or manual cycling.

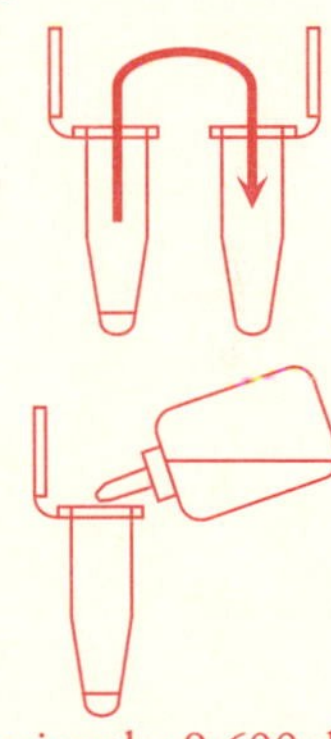

If you are using the 9,600 thermal cycler from the Perkin-Elmer Corporation, you do not need to add mineral oil to the samples.

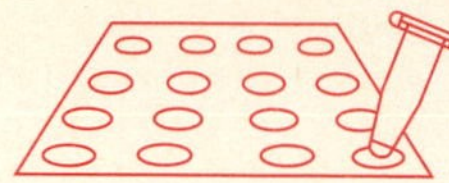

7 a. *For automated cycling:* Recheck the program, then start.

20 cycles of step file (linkage to soak file optional):

94°C for 1 minute

60°C for 45 seconds

72°C for 1 minute

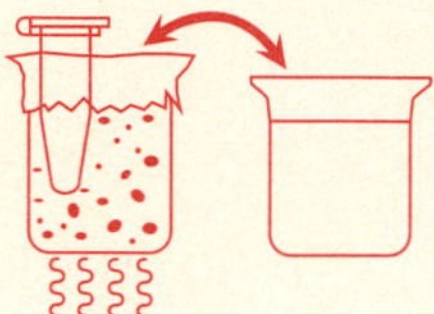

b. *For manual cycling:* Complete 20 cycles with the tubes in a floating rack:

Immerse the tubes in a boiling water bath for 20 seconds.

Immediately immerse the tubes in a 55°C water bath for 1 minute.

Take care not to overincubate the samples in the boiling water bath; prolonged exposure to 100°C gradually destroys the activity of the Taq *polymerase.*

Use a checklist to mark off each cycle as it is completed; otherwise, it is easy to lose track of the number of cycles.

Extract PCR Product with Phenol/Chloroform

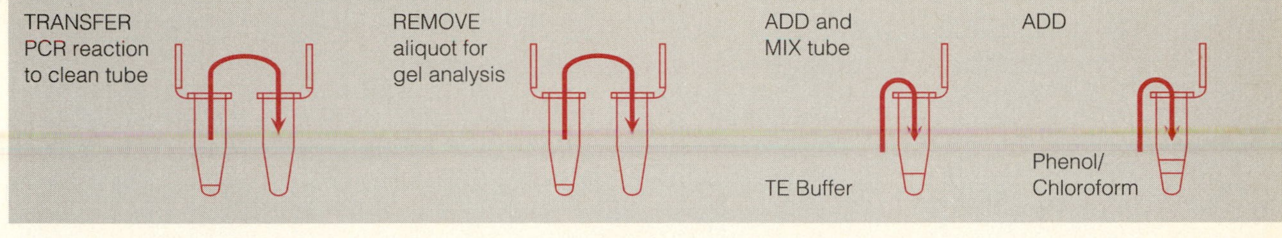

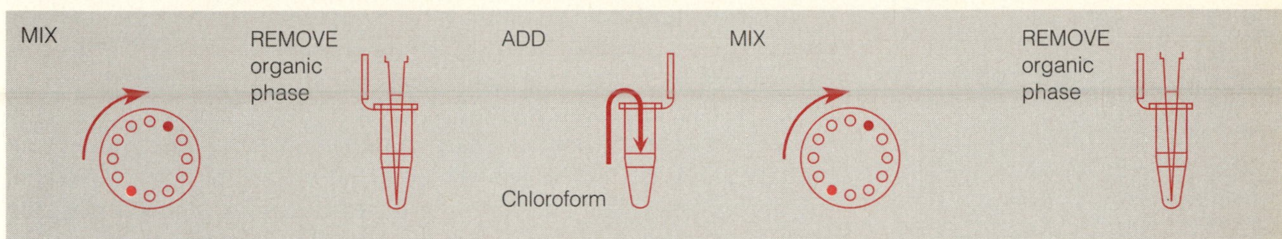

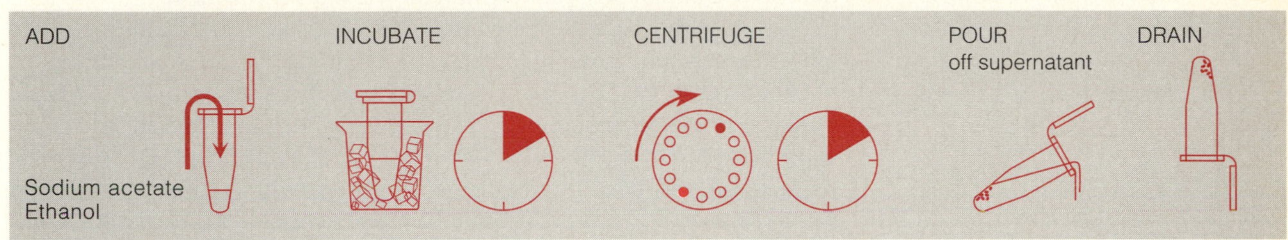

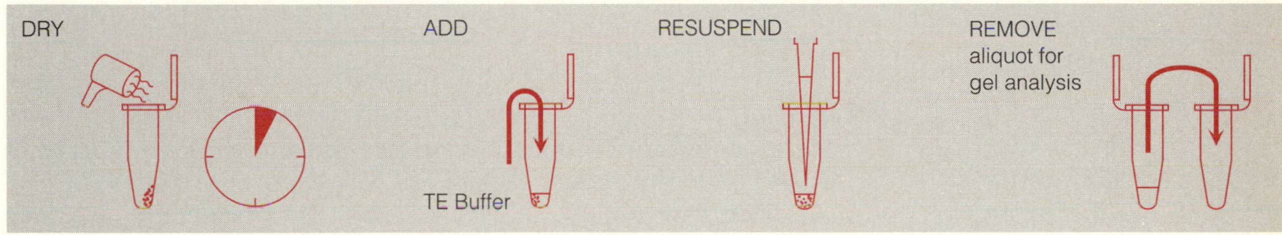

Laboratory 18/Part B
Purification of a λ PCR Product

PRELAB NOTES

Phenol/Chloroform Extraction

Taq polymerase and dNTPs must be removed from the amplified λ DNA before restriction digestion. Otherwise, the polymerase carried over in the λ sample adds deoxynucleotides (also remaining in the sample) to fill in the single-stranded "sticky" ends generated by *Bam*HI and *Hin*dIII. These blunt-ended fragments do not ligate into pBLU.

Phenol/chloroform extraction is the traditional method used for separating proteins from nucleic acids; however, phenol is extremely corrosive, and chloroform is a mutagen and suspected carcinogen. Phenol is primarily responsible for dissolving protein in the organic phase; the chloroform helps to separate the organic phase from the aqueous phase, which contains the DNA. Residual phenol dissolved in the aqueous phase can inhibit subsequent enzymatic reactions and is removed by a second extraction with chloroform alone. Small amounts of remaining chloroform are lost during the ethanol precipitation and drying steps that follow.

Typical reagent-grade phenol must be redistilled and equilibrated with buffer before being used in extractions. Considering the small amount of phenol used in this laboratory, it is well worth the additional expense of purchasing molecular biology–grade phenol that has been redistilled and equilibrated. When chloroform is mentioned, in the context of organic extractions, this usually refers to a mixture of chloroform and isoamyl alcohol (24:1 v/v). The addition of isoamyl alcohol helps to reduce foaming during extractions.

Responsible Handling of Phenol/Chloroform

Phenol is corrosive and can cause severe burns. Work under a chemical hood, and wear gloves, safety glasses, and protective clothing when you are mixing phenol/chloroform and preparing aliquots. If phenol comes in contact with your skin, rinse with plenty of water, then wash with soap and water. *Do not wash with ethanol; it can speed the absorption of phenol through the skin.*

Collect the organic phase from extraction into a clearly marked container. Pool the experiment waste into a capped bottle for disposal with other organic solvents.

Column Chromatography

Alternatively, the λ PCR product can be separated from the dNTPs by spin-column chromatography. This type of separation uses a centrifuge to quickly drive the solution containing the PCR product through the column matrix while the dNTPs are retained on the column. Column resins suitable for this application are Sephadex G-50 and Bio-Gel P-60. Ready-made spin columns can be purchased, or you can make them yourself.

To prepare a spin column:

1. Plug the bottom of a 1-ml disposable syringe with a small quantity of glass wool.
2. Place the syringe into a 15-ml tube.

3. Fill the syringe with Sephadex G-50 or Bio-Gel P-60 that has been equilibrated in a TE buffer. Use a 9-inch Pasteur pipet to add a slurry of gel resin to the bottom of the syringe. As the TE drips from the tip, the resin will pack in the syringe column. Continue adding gel slurry until the column is full.
4. Centrifuge a 15 ml tube containing the syringe at 1,000 × *g* for 1 minute at room temperature. Following centrifugation, the resin will pack and partially dehydrate. If necessary, add more resin and recentrifuge until the packed column is about 0.9 ml.
5. Assemble the spin column for use by placing the packed syringe within a clean 15-ml tube containing a clean 1.5-ml tube (to catch the filtrate).

Spin columns can be stored at 4°C for at least 1 month. Fill the syringe with TE buffer, wrap it with Parafilm, and store it upright.

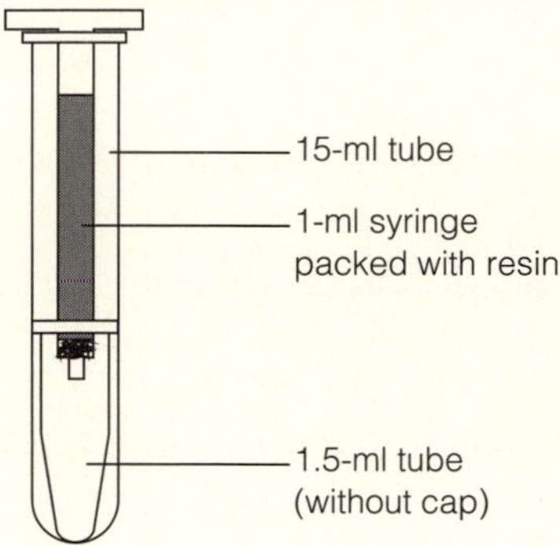

For Further Information

The protocols presented here are based on the following published method:

Sambrook, J., Fritsch, E. F., and T. Maniatis. 1989. *Molecular Cloning—A Laboratory Manual*, 2nd ed. Cold Spring Harbor: Cold Spring Harbor Press.

PURIFICATION OF A λ PCR PRODUCT

Reagents	Equipment and Supplies
chloroform	10–100-µl micropipettor + tips
phenol/chloroform	0.5–10-µl micropipettor + tips
sodium acetate	1.5-ml tubes
95% ethanol, ice-cold	beaker for waste/used tips
TE buffer	disposable gloves
	Kimwipes or paper towels
Alternatively:	microfuge
1-ml disposable syringe	test tube rack
glass wool	
Sephadex G-50 or Bio-Gel P-60	*Alternatively:*
TE buffer	clinical centrifuge (500–1,000 × *g*)

Extract PCR Product with Phenol/Chloroform

(40 minutes)

1. Retrieve your PCR reaction from the thermal cycler or water bath.
2. Use a permanent marker to label two 1.5-ml reaction tubes "PCR extraction" and "PCR product."
3. Transfer as much sample as possible from the PCR reaction tube to the 1.5-ml tube labeled "PCR extraction" (80–90 µl). *Do not transfer any mineral oil from the PCR tube.*

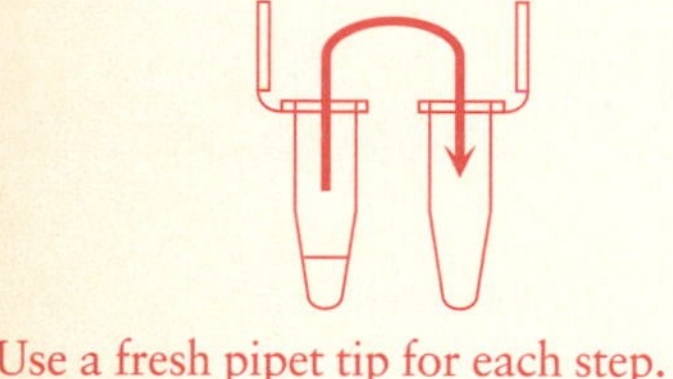

Use a fresh pipet tip for each step.

   a. Dip the pipet tip through the mineral oil on top of the PCR reaction tube, and carefully withdraw sample.
   b. Before transferring, use a Kimwipe or paper towel to remove any mineral oil clinging to the outside of the pipet tip.
   c. Discard the original PCR reaction tube with the remaining mineral oil.

The additional volume provided by the TE buffer makes the extraction easier to perform.

4. Transfer 10 ml of the sample from the PCR extraction tube to the tube labeled "PCR product." Store the "PCR product" tube in the freezer or on ice until you are ready for gel electrophoresis (Part C).
5. Add 100 µl of TE buffer to the PCR extraction tube. Tap the tube with your finger several times to mix.
6. Add an equal volume (about 180 µl) of phenol/chloroform to the PCR extraction tube.
7. Close the cap of the tube, and mix by rapidly inverting the tube several times to form a homogeneous emulsion.

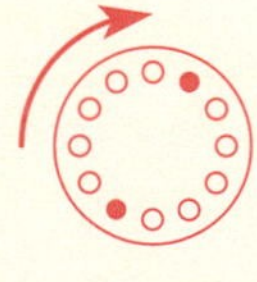

8. Place the extraction tube and a balance tube in a microfuge rotor and spin for 30 seconds to separate the phases.

CAUTION

Phenol is corrosive and can cause severe burns. Wear gloves, safety glasses, and protective clothing when you are working with phenol/chloroform. If phenol comes into contact with your skin, rinse with plenty of water, then wash with soap and water. *Do not use ethanol.*

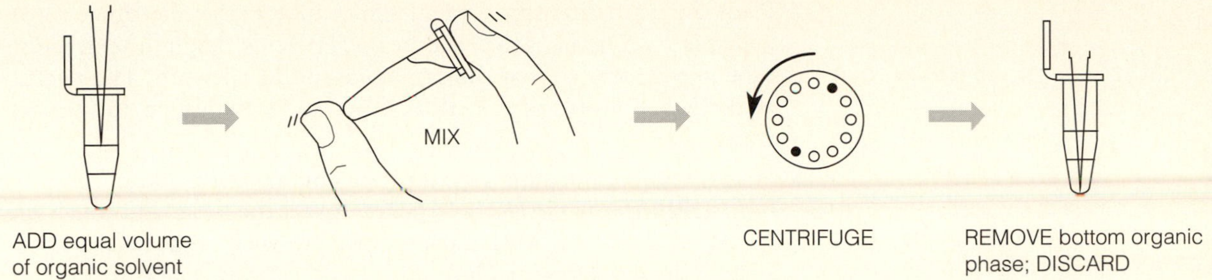

9. Remove the organic (bottom) phase:
   a. Set the micropipettor to 190 µl, insert a pipet tip through the aqueous (top) phase, and position it at the very bottom of the tube.
   b. Withdraw the entire organic phase until a small amount of the aqueous phase enters the pipet tip.
   c. Expel the organic phase into a labeled waste container for disposal by your instructor.
10. Add 100 µl of chloroform to the remaining aqueous phase.
11. Close the tube cap, and mix by rapidly inverting the tube several times to form a homogeneous emulsion.

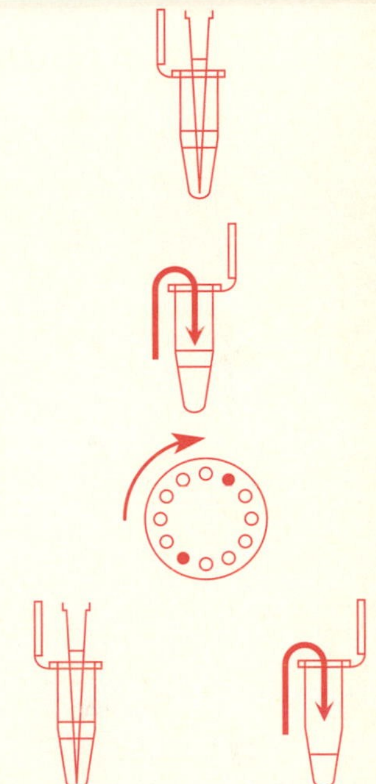

12. Place extraction tube and a balance tube in the microfuge rotor, and spin for 30 seconds to separate the phases.
13. Remove the organic (bottom) phase into a waste container, as described in step 9.
14. Add 10 µl of sodium acetate to the aqueous phase. Close the cap of the tube, and tap the tube with your finger to mix.
15. Add 400 µl (about two volumes) of ice-cold ethanol to the aqueous phase. Close the tube cap and mix by rapidly inverting several times.
16. Incubate the tube on ice (or in a freezer) for 10 minutes.

Sodium acetate adjusts the monovalent cation concentration to a level suitable for ethanol precipitation of nucleic acids in the following step.

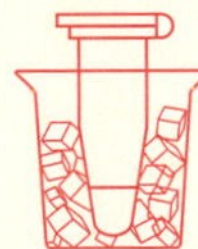

Stop Point

Store the DNA in ethanol at –20°C until you are ready to continue.

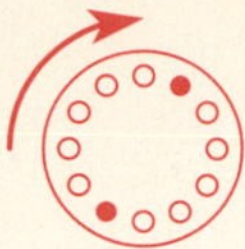

17. Place the extraction tube and a balance tube in the microfuge rotor, and spin for 5 minutes to pellet the nucleic acids. Align the extraction tube in the rotor so that the cap hinges point outward. The nucleic acid residue, visible or not, will collect under the hinge during centrifugation.

The pellet will be very small and difficult to see. It will most likely appear either as a tiny, teardrop-shaped smear or small particles on the bottom of the tube.

18. Pour off the supernatant from the tube into a waste beaker. *Take care not to disturb the nucleic acid pellet.* Invert the tube, and touch the mouth of the tube to a clean paper towel, to wick off as much as possible of the remaining alcohol.

19. Dry the nucleic acid pellet using one of following methods:
    a. Direct a stream of warm air from a hair dryer over the open end of the tube for about 3 minutes. Take care not to blow the pellet out of the tube.

    or

    b. Close the caps, and pulse the tubes in the microfuge to pool the remaining ethanol. *Carefully* use a micropipettor to draw off the ethanol. Let the pellets air dry at room temperature for 10 minutes.

Air bubbles cast in the tube wall can be mistaken for ethanol droplets.

20. Be sure that the nucleic acid pellet is dry and that all ethanol has evaporated before proceeding on to step 21. Hold the tube up to the light, and confirm that no ethanol droplets remain and that the nucleic acid pellet, if visible, appears white and flaky. If ethanol is still evaporating, you can detect an alcohol odor by sniffing the mouth of the tube.
21. Add 50 µl of TE buffer to the extraction tube. Resuspend the pellets by smashing them with the pipet tip and pipetting in and out vigorously. Rinse down the side of tube several times, concentrating on the area where the pellet should have formed during centrifugation (beneath the cap hinge). Check to ensure that all of the DNA has dissolved and that no particles remain in the tip or on the side of the tube.

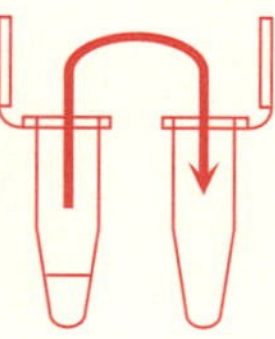

22. Transfer 10 µl of solution from the PCR extraction to a fresh tube labeled "purified PCR product."

Stop Point

Freeze the purified PCR product tube, along with the PCR product tube from step 4 and the PCR control tube from Part A, at –20°C until you are ready to continue. Thaw the solutions at room temperature before using them.

Alternative Method: Purify the PCR Product by Column Chromatography

(15 minutes)

1. Retrieve your PCR reaction from the thermal cycler or water bath.
2. Use a permanent marker to label two 1.5-ml reaction tubes "PCR chromatography" and "PCR product."

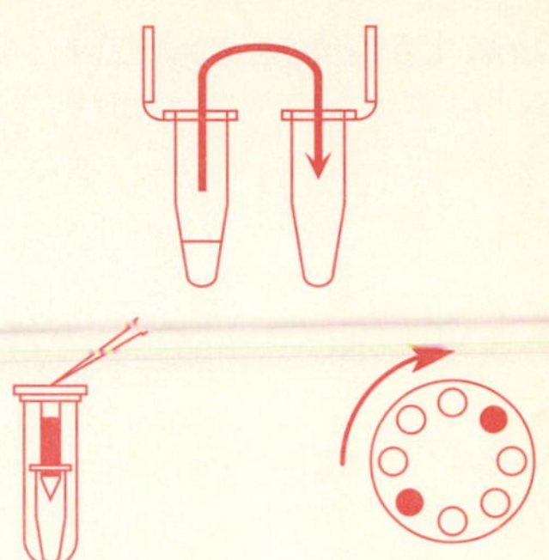

3. Transfer as much sample as possible from the PCR reaction tube to the 1.5-ml tube labeled "PCR chromatography" (80–90 μl). Do not transfer any mineral oil from the PCR tube; use a Kimwipe or a paper towel to remove any mineral oil clinging to the outside of the pipet tip. Discard the original PCR reaction tube with the remaining mineral oil.
4. Transfer 10 ml of the sample from the "PCR chromatography" tube to the tube labeled "PCR product." Store the "PCR product" tube in the freezer or on ice until you are ready for gel electrophoresis (Part C).
5. Load the contents of the "PCR chromatography" tube into the assembled spin column.
6. Centrifuge at 1,000 × *g* for 2 minutes at room temperature, collecting effluent from the column into the decapped 1.5-ml tube.

Stop Point

Freeze the purified PCR reaction tube, along with the PCR product tube from step 4 and the PCR control tube from Part A, at –20°C until you are ready to continue. Thaw the solutions at room temperature before using them.

I. Cast 1.0% Agarose Gel

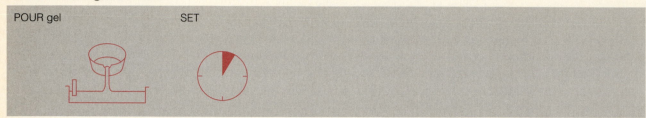

II. Load Gel and Electrophorese

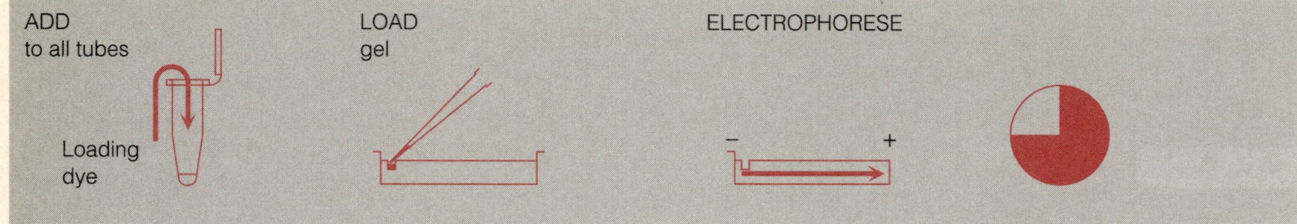

III. Stain Gel, View, and Photograph

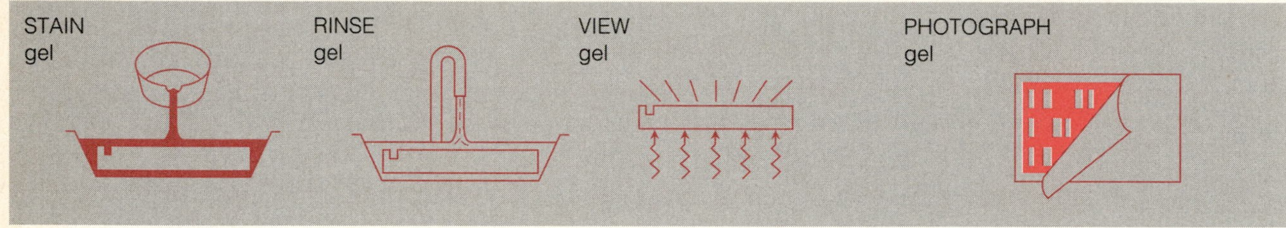

Laboratory 18/Part C
Electrophoresis of a λ PCR Product

PRELAB NOTES

DNA Size Markers

It is advisable to include DNA size markers on each gel. Markers help to verify that the PCR product is the correct size, and they also serve as a control to aid in trouble-shooting problems with amplification, electrophoresis, and staining. For example, a very faint PCR product, compared to the DNA markers, suggests that additional amplification cycles are needed.

For Further Information

The protocol presented here is based on the following published methods:

Helling, R. B., H. M. Goodman, and H. W. Boyer. 1974. Analysis of *Eco*RI fragments of DNA from lambdoid bacteriophages and other viruses by agarose-gel electrophoresis. *Journal of Virology* 14: 1235.

Sharp, P. A., B. Sugden, and J. Sambrook. 1973. Detection of two restriction endonuclease activities in *Haemophilus parainfluenzae* using analytical agarose–ethidium bromide electrophoresis. *Biochemistry* 12: 3055.

ELECTROPHORESIS OF A λ PCR PRODUCT

Reagents

"PCR control," "PCR product," and "purified PCR product"
predigested λ/*Bam*HI + *Hin*dIII
distilled water
loading dye
1% agarose
1× Tris/Borate/EDTA (TBE) buffer
1 μg/ml ethidium bromide

for decontamination:
0.05 M $KMnO_4$
0.25 N HCl
0.25 N NaOH

Equipment and Supplies

0.5–10-μl micropipettor + tips
1.5-ml tubes
aluminum foil
beaker for agarose
beaker for waste/used tips
camera and film (optional)
disposable gloves
electrophoresis box
masking tape
microfuge
Parafilm or waxed paper (optional)
permanent marker
power supply
test tube rack
transilluminator
60°C water bath (for agarose)

I. Cast 1.0% Agarose Gel

Refer to Laboratory 3, "DNA Restriction and Electrophoresis," for detailed instructions on casting and loading gel.

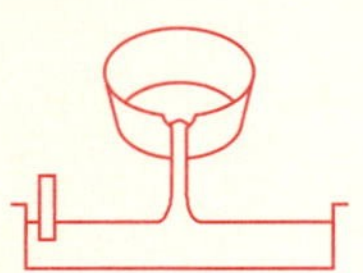

(15 minutes)

1. Carefully pour agarose solution into the gel-casting tray to fill it to a depth of about 5 mm. The gel should cover about one-third of the height of the comb teeth.
2. After the agarose solidifies, place the casting tray into the electrophoresis box and set up for electrophoresis.

Stop Point

Cover the electrophoresis box, and save the gel until you are ready to continue. The gel will remain in good condition for several days if it is completely submerged in buffer.

II. Load Gel and Electrophorese

(10 minutes; then 30–60 minutes of electrophoresis)

1. Collect the tubes labeled "PCR control," "PCR product," and "purified PCR product" from the freezer, and allow them to thaw at room temperature.
2. Obtain an aliquot of λ DNA cut with *Bam*HI + *Hin*dIII, to serve as size markers.
3. Add 1 μl of loading dye to each tube. Close the tube tops, and mix the solution.

4. Load 10 μl of each sample into a separate well in the gel, as shown in the diagram below. *Use a fresh pipet tip* for each sample.

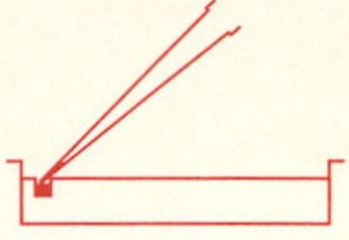

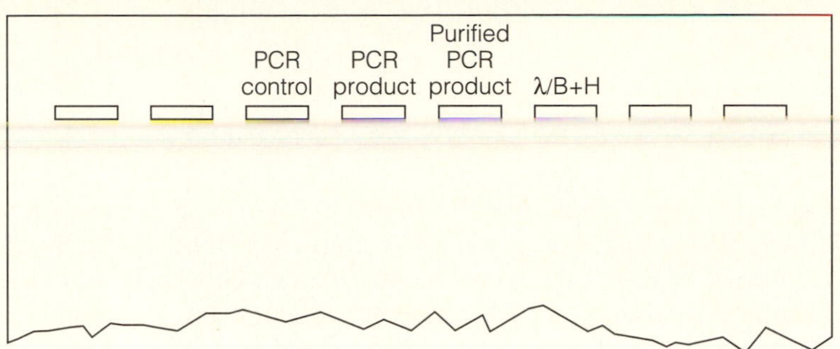

5. Electrophorese at 130 volts until the bromophenol blue bands have moved 50–60 mm from the wells.

6. Turn off the power supply, remove the gel-casting tray from the electrophoresis box, and transfer the gel to a disposable weigh boat (or other shallow tray) for staining.

III. Stain Gel with Ethidium Bromide, View, and Photograph

(10–15 minutes)

Staining can be performed by an instructor in a controlled area, when students are not present.

CAUTION

Review the section on "Ethidium Bromide Staining and Responsible Handling" in Laboratory 3. Wear disposable gloves when staining, viewing, and photographing the gel and during cleanup. Confine all staining to a restricted sink area.

1. Flood the gel with ethidium bromide solution (1 μg/ml), and allow it to stain for 5–10 minutes.
2. View the gel under an ultraviolet transilluminator or other UV light source.

CAUTION

Ultraviolet light can damage your eyes. Never look directly at an unshielded UV light source without eye protection. View only through a filter or safety glasses that absorb the harmful wavelengths.

3. Make an exposure of 1–2 seconds with your camera set at f/8. Develop the print for the recommended time (approximately 45 seconds at room temperature).

4. Take the time for proper cleanup:
   a. Wipe down the camera, the transilluminator, and the staining area.

b. Decontaminate the gel and any staining solution that will not be reused.

c. Wash your hands before leaving the laboratory.

RESULTS AND DISCUSSION

Observe the photograph of your stained gel. The PCR control (unamplified) lane should have no evidence of amplified DNA; the PCR product and purified PCR product lanes should each exhibit a single fragment corresponding to about 1,100 bp. Depending upon how far your samples have electrophoresed, each product lane may also display a diffuse band of low-molecular-weight product, called "primer-dimer," that migrates well ahead of the bromophenol blue marker. Primer-dimer results when single-stranded primers partially hybridize to form a short double-stranded sequence that primes extension by *Taq* polymerase. The primer-dimer may be reduced in intensity in the purified PCR product lane.

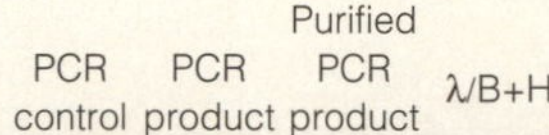

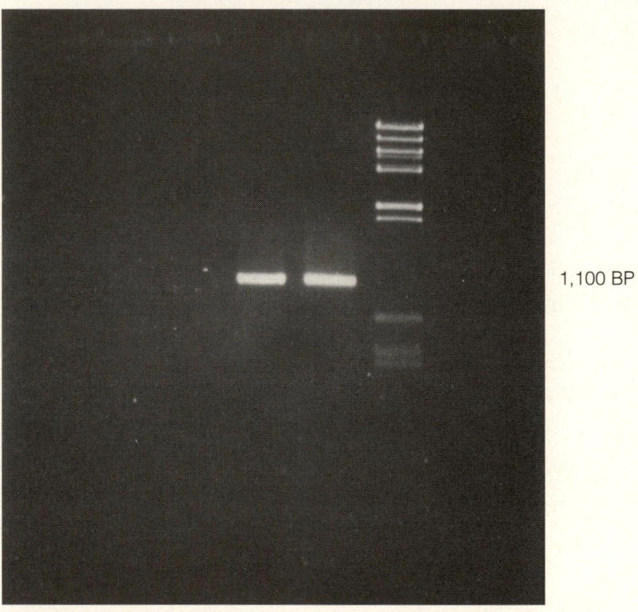

Ideal Gel

1. What is the purpose of the phenol/chloroform extraction? What is the purpose of the ethanol precipitation?
2. What is the purpose of the spin-column chromatography step?
3. Draw a schematic diagram to illustrate the first three cycles of your PCR reaction. It may be easier to use different colored pens or pencils.
4. What would be the result of replacing the *Taq* polymerase with *E. coli* DNA polymerase?

5. What would be the ramifications of setting up a PCR reaction that contains the correct amount of the upstream primer but only 1/50 of the correct amount of the downstream primer?
6. What result could be expected if you substituted human DNA for λ DNA in the PCR reaction?
7. List some advantages and disadvantages of PCR over molecular cloning using *E. coli*.
8. What is the molar ratio of primer to template DNA (λ) in the PCR reaction?
9. Use the following information to calculate the number of target DNA molecules that were present in the 2 ng of λ DNA added to the PCR reaction.

   λ DNA has 48,502 base pairs.

   One base pair has an average molecular weight of 650 grams/mole.

   There are 6.02×10^{23} molecules per mole.
10. Calculate the number of copies of the PCR product produced (in μg) following 20 cycles of amplification.

FOR FURTHER RESEARCH

1. Design a series of experiments to determine the optimal conditions for the amplification of a new DNA sequence, testing variables such as cycling temperatures and times and reactant concentrations.
2. Compare the PCR product yields of manual and automated thermal cycling.

LABORATORY 19

Recombination of a PCR Product and Plasmid pBLU

This laboratory continues the unit to clone a PCR-amplified λ DNA fragment. In Part A, "Restriction Digest of a PCR Product and pBLU," the amplified 1,106-bp amplification product from Laboratory 18 and plasmid pBLU are digested in separate restriction reactions with the enzymes *Bam*HI and *Hin*dIII. Following incubation at 37°C, samples of the digests are electrophoresed in an agarose gel to confirm proper cutting. The double digest removes 209 bp from one end of the PCR product and 113 bp from the other end, releasing a single 784-bp fragment with *Bam*HI and *Hin*dIII ends. Plasmid pBLU has one recognition site for each enzyme, located in the polylinker region. Double digest of pBLU removes a 30-bp sequence between the *Bam*HI and *Hin*dIII sites, leaving a 5,400-bp plasmid "backbone."

In Part B, "Ligation of PCR and pBLU Restriction Fragments," the restriction digests of the PCR product and pBLU are heated to destroy *Bam*HI and *Hin*dIII activities. A sample from each reaction is mixed with DNA ligase plus ATP and incubated at room temperature. Complementary *Bam*HI and *Hin*dIII sticky ends hydrogen-bond to align the restriction fragments. Ligase catalyzes the formation of phosphodiester bonds that covalently link the DNA fragments to form stable recombinant-DNA molecules.

I. Set Up Restriction Digests

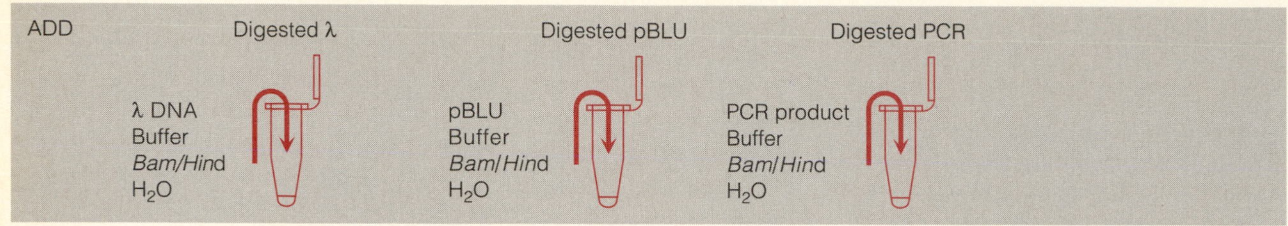

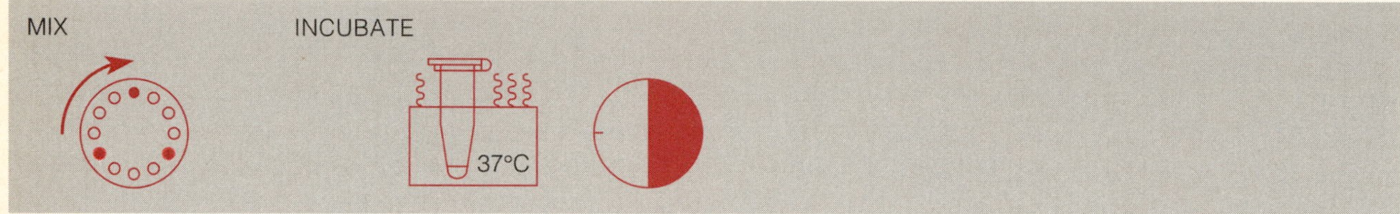

II. Cast 1.0% Agarose Gel

III. Load Gel and Electrophorese

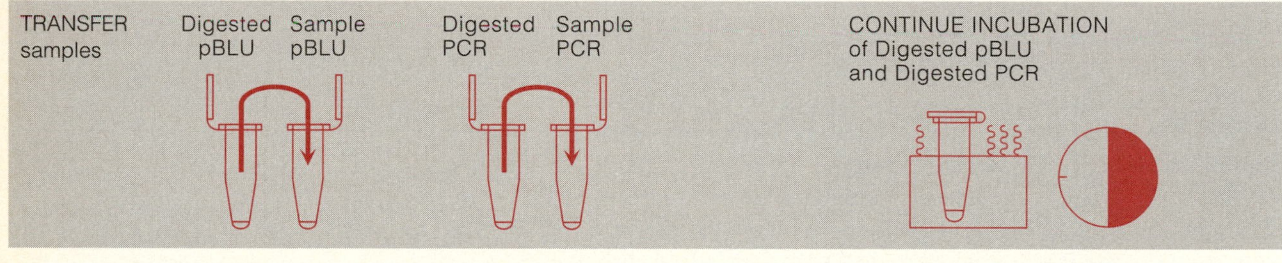

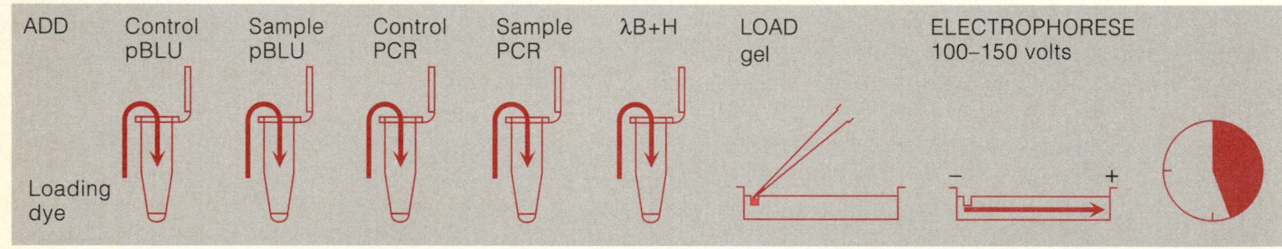

IV. Stain, View, and Photograph

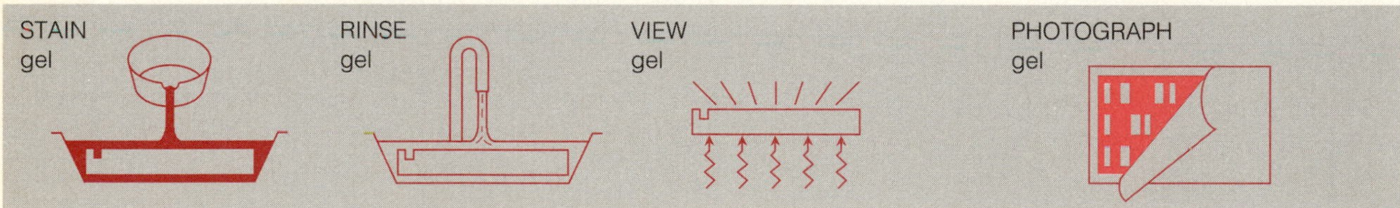

Laboratory 19/Part A
Restriction Digest of a PCR Product and pBLU

PRELAB NOTES

Review the prelab notes in Laboratory 3, "DNA Restriction and Electrophoresis."

Plasmid Substitution

Analysis of the recombinant molecules in this experimental stream is straightforward and relies only on detection of the single 784-bp amplification fragment. Therefore, other plasmid vectors with appropriate *Bam*HI and *Hin*dIII restriction sites can be substituted for pBLU. See the prelab notes in Laboratory 14A, "Restriction Digest of λ and Plasmid pBLU," for a discussion of the advantages of plasmid pBLU.

The Prudent Control

In Section III, samples of the restriction digests are electrophoresed prior to ligation, to confirm complete cutting by the endonucleases. This prudent control is standard experimental procedure. If you are pressed for time, omit electrophoresis, and ligate DNA directly following the restriction digest. However, be sure to pretest the activity of the *Bam*HI and *Hin*dIII, and incubate long enough for complete digestion.

For Further Information

The protocol presented here is based on the following published methods:

Sharp, P. A., B. Sugden, and J. Sambrook. 1973. Detection of two restriction endonuclease activities in *Haemophilus parainfluenzae* using analytical agarose–ethidium bromide electrophoresis. *Biochemistry* 12: 3055.

Helling, R. B., H. M. Goodman, and H. W. Boyer. 1974. Analysis of *Eco*RI fragments of DNA from lambdoid bacteriophages and other viruses by agarose-gel electrophoresis. *Journal of Virology* 14: 1235.

Messing, J. and J. Vieira. 1982. The pUC Plasmids, an M13mp7-derived system for insertion mutagenesis and sequencing with synthetic universal primers. *Gene* 19: 259.

RECOMBINATION OF A PCR PRODUCT AND PLASMID pBLU

Reagents

For Digest:
purified PCR product, from Laboratory 18B
0.1 μg/μl λ DNA
0.15 μg/μl pBLU
BamHI/HindIII
10× restriction buffer
distilled water

For electrophoresis:
purified PCR product
0.1 μg/μl pBLU
loading dye
1.0% agarose
1× Tris/Borate/EDTA (TBE) buffer
1 μg/ml ethidium bromide

For decontamination:
0.05 M $KMnO_4$
0.25 N HCl
0.25 N NaOH

Equipment and Supplies

10–100-μl micropipettor + tips
0.5–10-μl micropipettor + tips
1.5-ml tubes
aluminum foil
beaker for agarose
beaker for waste/used tips
camera and film (optional)
disposable gloves
electrophoresis box
masking tape
microfuge (optional)
permanent marker
power supply
test tube rack
transilluminator
37°C water bath
60°C water bath (for agarose)

I. Set Up Restriction Digests

(10 minutes; then 30+ minutes of incubation)

Refer to Laboratory 3, "DNA Restriction and Electrophoresis," for detailed instructions on setting up reactions.

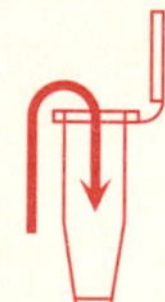

1. Obtain the labeled reaction tubes containing λ and pBLU. Use a permanent marker to add the label "digested" to each tube. *Add additional reagents directly to these tubes.* Label another tube "digested PCR."
2. Use the matrix below as a checklist while adding reagents to each reaction. Read down the columns, adding the same reagent to all appropriate tubes. *Use a fresh pipet tip for each reagent.*

Tube	λ DNA	pBLU	Purified PCR Product	10x Buffer	*Bam*HI/*Hin*dIII	H_2O
digested λ	6 μl (preadded)	—	—	1 μl	2 μl	1 μl
digested pBLU	—	20 μl (preadded)	—	4 μl	4 μl	12 μl
digested PCR	—	—	20 μl	4 μl	4 μl	12 μl

3. After adding all of the reagents, close the tube tops. Pool and mix the reagents by either pulsing them in a microfuge or by sharply tapping the bottom of the tube on the lab bench.
4. Place the reaction tubes in a 37°C water bath, and incubate them for 30 minutes or longer.

Stop Point

After a full 30-minute incubation (or longer), freeze the reactions at –20°C until you are ready to continue. Thaw the reactions before proceeding on to Section III, step 1.

II. Cast 1.0% Agarose Gel

(15 minutes)

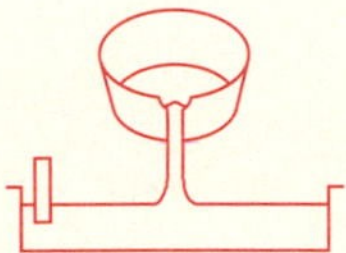

1. Carefully pour agarose solution into the gel-casting tray to fill it to a depth of about 5 mm. The gel should cover only about one-third of the height of the comb teeth.
2. After the agarose solidifies, place the casting tray into the electrophoresis box, and set up for electrophoresis.

Stop Point

Cover the electrophoresis box, and save the gel until you are ready to continue. The gel will remain in good condition for several days, if it is completely submerged in buffer.

III. Load Gel and Electrophorese

(10 minutes; then 30+ minutes of electrophoresis)

Refer to Laboratory 3, "DNA Restriction and Electrophoresis," for detailed instructions on casting and loading gel.

Only a fraction of the *Bam*HI/*Hin*dIII digests of pBLU and the PCR product are electrophoresed to check that the DNAs are completely cut. These restricted samples are electrophoresed along with uncut pBLU, uncut PCR product, and λ/*Bam*HI +*Hin*dIII digest, as controls.

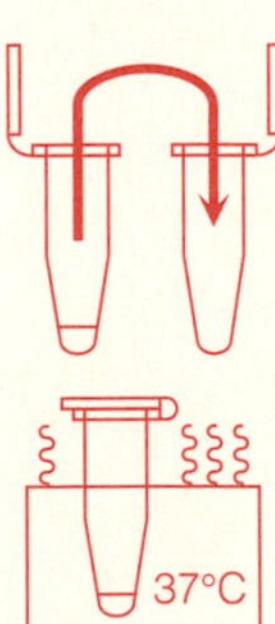

1. Use a permanent marker to label two clean 1.5-ml tubes "sample pBLU" and "sample PCR." Remove the digested pBLU and digested PCR tubes from the 37°C water bath. Transfer 10 µl of plasmid from the digested pBLU tube into the sample pBLU tube. Transfer 10 µl of DNA from the digested PCR tube into the sample PCR tube.
2. *Immediately return the digested pBLU and digested pPCR tubes to the water bath, and continue incubating them at 37°C during electrophoresis.*
3. Add 5 µl of purified PCR product (saved from Laboratory 18) to clean a 1.5-ml tube labeled "control PCR."
4. Obtain a 1.5-ml tube containing 0.1 µg/µl of pBLU. Use a permanent marker to add the label "control" to the tube.

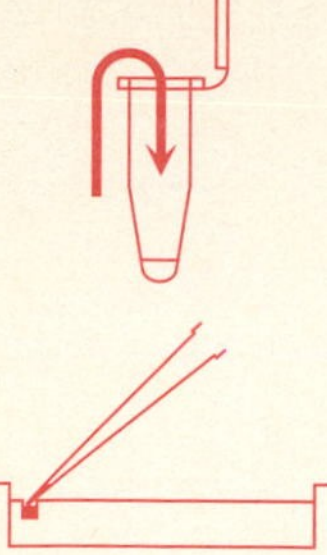

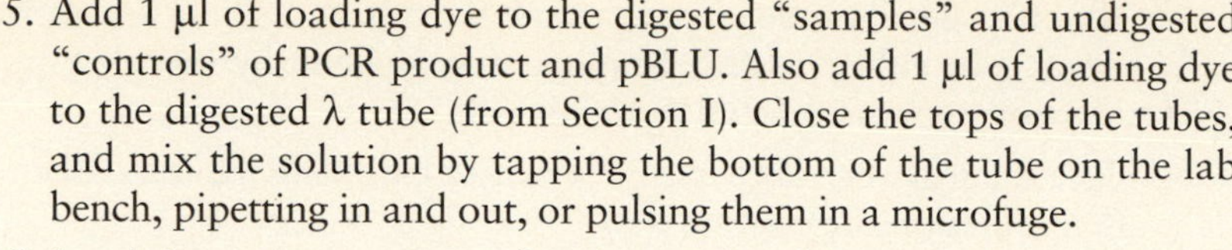

5. Add 1 µl of loading dye to the digested "samples" and undigested "controls" of PCR product and pBLU. Also add 1 µl of loading dye to the digested λ tube (from Section I). Close the tops of the tubes, and mix the solution by tapping the bottom of the tube on the lab bench, pipetting in and out, or pulsing them in a microfuge.

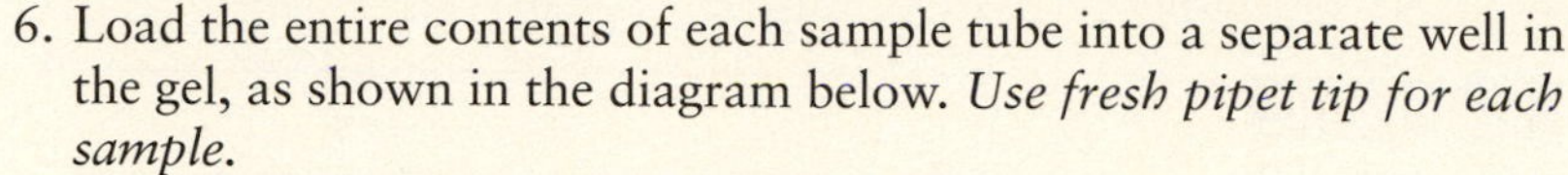

6. Load the entire contents of each sample tube into a separate well in the gel, as shown in the diagram below. *Use fresh pipet tip for each sample.*

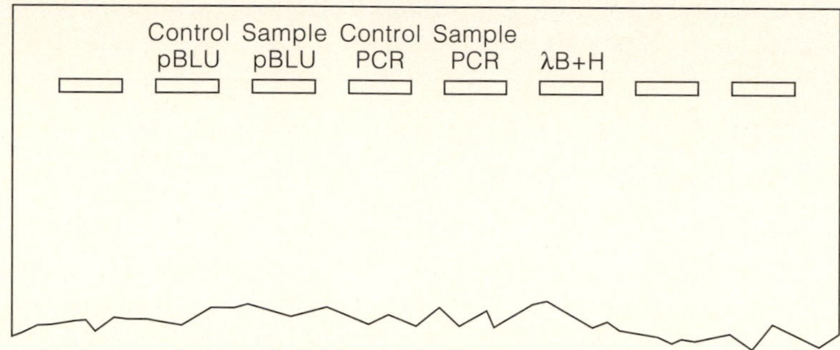

7. Electrophorese at 50–125 volts, until the bromophenol blue bands have moved 50–60 mm from the wells.
8. Turn off the power supply, remove the gel-casting tray from the electrophoresis box, and transfer the gel to a disposable weigh boat (or other shallow tray) for staining.

IV. Stain Gel with Ethidium Bromide, View, and Photograph

(10–15 minutes)

CAUTION

Review the section on "Ethidium Bromide Staining and Responsible Handling" in Laboratory 3. Wear disposable gloves when staining, viewing, and photographing the gel, and during clean up. Confine all staining to a restricted sink area.

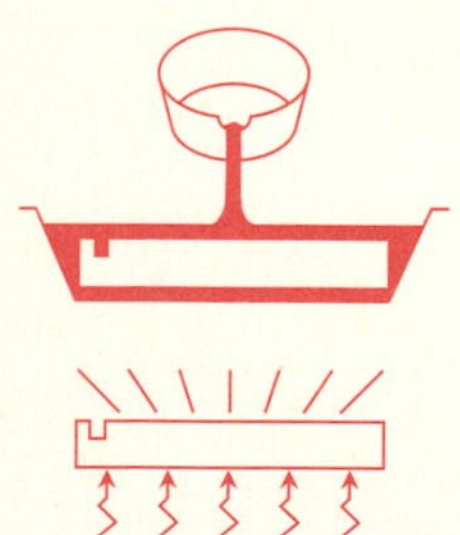

1. Flood the gel with ethidium bromide solution (1 µg/ml), and allow it to stain for 5–10 minutes. Destain, if desired.
2. View the gel under an ultraviolet transilluminator or other UV light source.

CAUTION

Ultraviolet light can damage your eyes. Never look directly at an unshielded UV light source without eye protection. View only through a filter or safety glasses that absorb the harmful wavelengths.

3. Make an exposure of 1–2 seconds with your camera set at f/8. Develop the print for the recommended time (approximately 45 seconds at room temperature).
4. Take the time for proper cleanup:
   a. Wipe down the camera, the transilluminator, and the staining area.
   b. Decontaminate the gel and any staining solution that will not be reused.
   c. Wash your hands before leaving the laboratory.

RESULTS AND DISCUSSION

Compare your stained gel with the ideal gel below, and determine whether the pBLU and PCR DNAs have been completely digested by *Bam*HI and *Hin*dIII.

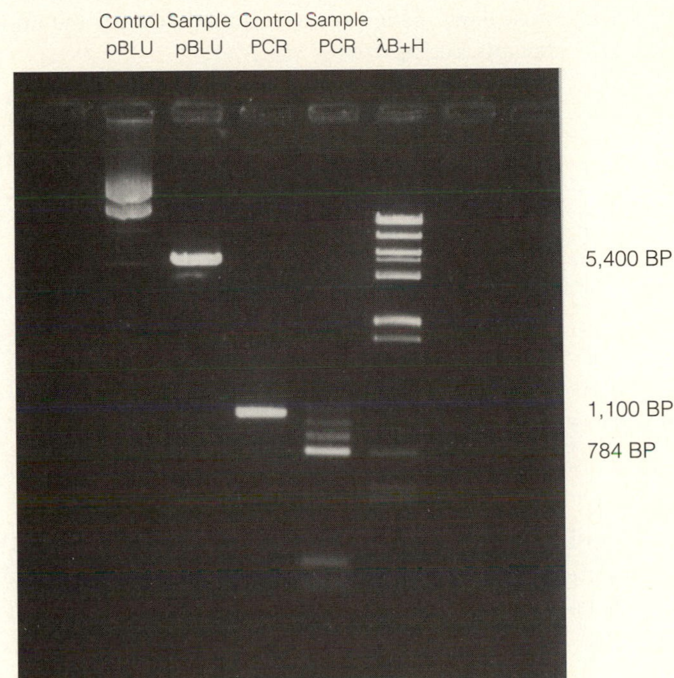

Ideal Gel

1. The sample pBLU lane should show a *single band* of 5,400 bp, usually in an intermediate position between the supercoiled and relaxed circle forms in the control pBLU lane. The presence of additional bands in the sample pBLU lane indicates some degree of incomplete digestion. (The digest actually produces an additional 30-bp fragment, which quickly runs off of the end of the gel during electrophoresis.)
2. The sample PCR lane should exhibit one band of 784 bp and two smaller fragments of approximately 209 bp and 113 bp. Additional bands of higher molecular weight are commonly present, indicating

that a portion of the PCR product is undigested or only partially digested. Partial digestion by *Bam*HI yields a 993-bp fragment; partial digestion by *Hin*dIII yields an 897-bp fragment. Undigested DNA comigrates with the 1,106-bp amplification product in the uncut PCR lane. Our experience tells us that these partial or undigested PCR products cannot be restricted by adding additional enzyme or by incubating for a longer time. These undigestable products are formed during the later PCR cycles, when free nucleotides become the limiting reagents. Under these conditions, the *Taq* polymerase can exhibit a higher rate of nucleotide misincorporation, resulting in PCR products with incorrect sequences that are not recognized by the restriction enzymes.

3. The sample λ lane should show at least nine distinct fragments, with the 784-bp fragment comigrating with the predominant fragment in the cut PCR lane.
4. If the pBLU digest looks complete or nearly complete—and the PCR product appears at least 50% digested—continue on to Part B, "Ligation of PCR and pBLU Restriction Fragments." The reactions will most likely have progressed farther with the additional incubation during electrophoresis.
5. If either digest looks very incomplete, however, add another 1 µl of *Bam*HI/*Hin*dIII solution, and incubate for an additional 20 minutes. Then, continue on to Part B. Remember that some level of incomplete digestion of the PCR product is to be expected and cannot be remedied by additional incubation time or by additional enzyme.
6. Offer an explanation of why the PCR product often is only partially digested by *Bam*HI and *Hin*dIII.
7. How do incomplete digests of pBLU or the PCR product affect the results of the ligation experiment?

FOR FURTHER RESEARCH

Taq DNA polymerase adds an extra adenosine to the 3′ end of each DNA strand that it synthesizes, producing a single nucleotide overhang at each end of the duplex DNA molecule. Use this feature to devise a cloning strategy for PCR products that does not require restriction digestion.

Ligate PCR Product and pBLU

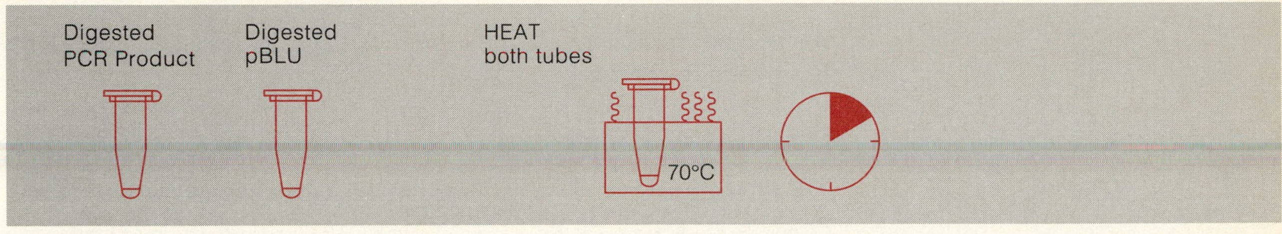

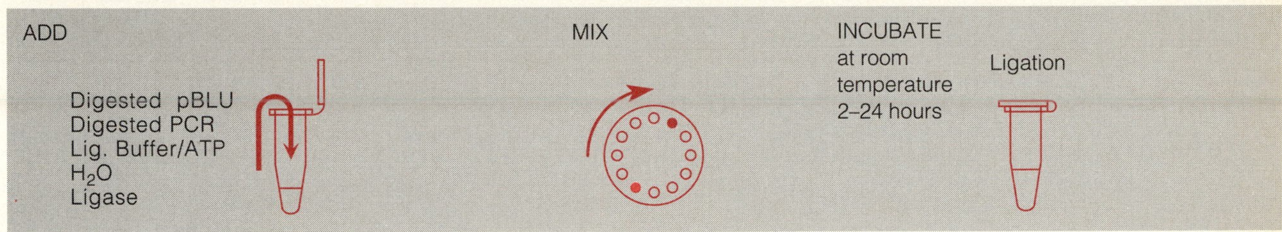

Laboratory 19/Part B
Ligation of PCR and pBLU Restriction Fragments

PRELAB NOTES

Review the prelab notes in Laboratory 7B, "Ligation of pAMP and pKAN Restriction Fragments."

For Further Information

The protocol presented here is based on the following published method:

Cohen, S. N., A. C. Y. Chang, and H. W. Boyer. 1973. Construction of biologically functional bacterial plasmids in vitro. *Proceedings of the National Academy of Sciences USA* 70: 3240.

LIGATION OF PCR AND pBLU RESTRICTION FRAGMENTS

Reagents	Supplies and Equipment
digested PCR (from Part A)	10–100-µl micropipettor + tips
digested pBLU (from Part A)	0.5–10-µl micropipettor + tips
10× ligation buffer/ATP	1.5-ml tubes
T4 DNA ligase	beaker for waste/used tips
distilled water	disposable gloves
	microfuge (optional)
	test tube rack

Ligate PCR Product and pBLU

(30 minutes; then 2+ hours of ligation)

1. Incubate the digested pBLU and digested PCR tubes in a 70°C water bath for 10 minutes.

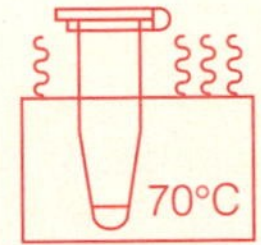

Step 1 is critical: Heat denaturation inactivates the restriction enzymes.

2. Label a clean 1.5-ml tube "ligation."
3. Use the matrix below as a checklist while adding reagents to the reaction. *Use a fresh pipet tip for each reagent.*

Tube	Digested pBLU	Digested PCR	Buffer/ATP	H_2O	DNA Ligase
Ligation	4 µl	10 µl	2 µl	2 µl	2 µl

4. After adding all of the above reagents, close the top of the tube. Pool and mix the reagents by either pulsing them in a microfuge or by sharply tapping the bottom of the tube on the lab bench.
5. Incubate the reaction at room temperature for 2–24 hours.
6. If time permits, ligation can be confirmed by electrophoresing 5 µl of the ligation reaction along with *Bam*HI/*Hin*dIII digests of pBLU and the purified PCR product. The lane containing ligated DNA should show multiple bands of high-molecular-weight DNA near the top of the gel lane; the restriction products from Laboratory 19A should not be visible.

For brief ligations of 2–4 hours, be sure to use high-concentration ligase with at least 5 Weiss units/µl or 300–500 cohesive-end units/µl.

Stop Point

Freeze the ligation at –20°C until you are ready to continue. Thaw the reactions before proceeding on to Laboratory 20.

RESULTS AND DISCUSSION

Ligation of the digested PCR product and pBLU will theoretically produce identical plasmids, all containing the same 784-bp *Bam*HI/*Hin*dIII fragment amplified from the λ genome. This precision demonstrates the utility of PCR cloning. However, other molecular species will be generated also, and some of these may produce ampicillin-resistant transformants.

1. Make a scale drawing of the simple recombinant plasmid that contains the 784-bp fragment. Include the total base-pair size, the fragment bp sizes, and the locations of the *Bam*HI site and the *Hin*dIII site, the origin, the ampicillin-resistance gene, and the *lacZ* gene.
2. Would you expect to ligate any of the partially digested fragments into pBLU? Explain your answer.
3. The difficulty with PCR cloning is that enough DNA sequence must be known to synthesize the primers to bracket a target gene. Name two potential sources of DNA sequence information for a target gene.
4. What limitation of PCR makes it incapable of amplifying most eukaryotic genes?
5. How would you clone your entire PCR fragment (i.e., without cutting off the ends with a restriction enzyme)?

FOR FURTHER RESEARCH

Use complementation PCR cloning to isolate a gene from *E. coli.* Remember that PCR cannot easily amplify sequences over 3,000 base pairs, so choose a small gene.

1. Select an *E. coli* gene that has been sequenced and for which an auxotroph is known. (Auxotrophs are strains that lack an enzyme needed for the synthesis of a critical nutrient, such as an amino acid or vitamin.)
2. Based upon the known sequence, select a set of primers that will amplify the gene of interest. Have the primers synthesized commercially or at a university facility.
3. Use a crude extract of total DNA made from wild-type *E. coli* as the template for a PCR reaction. Then, clone the PCR product into a plasmid and transform the ligation mixture into the auxotrophic strain.
4. First select for transformants by antibiotic resistance. Then test for clones containing an intact target gene, by replica plating onto a defined media lacking the single nutrient missing in the auxotroph. Determine the percentage of recombinant clones that can complement the mutation.
5. Do plasmid minipreps of several positive clones, to compare the inserts that complement versus those that do not.

LABORATORY 20

Transformation of *E. coli* with PCR Product

In Part A, "Classic Procedure for Preparing Competent Cells," *E. coli* cells are made competent to uptake plasmid DNA by successive washes and incubation with cold calcium chloride. In Part B, "Transformation of *E. coli* with Recombinant PCR/pBLU Plasmids," competent cells are transformed with the ligation products from Laboratory 19. Samples of transformed cells are spread on LB plates containing ampicillin and the lactose analog X-gal and incubated overnight at 37°C. Ampicillin in the medium selects for cells that have taken up recombinant plasmids and express the ampicillin-resistance gene of pBLU. Cells that take up a religated pBLU plasmid also express the *lacZ* gene product β-galactosidase, which hydrolyzes X-gal to form a dye that turns the colony blue.

The *Bam*HI/*Hin*dIII digest of pBLU in Laboratory 19 removes a 30-base-pair fragment from the *lacZ* gene but leaves the origin of replication and the ampicillin-resistance gene intact. Thus, all pBLU recombinants replicate and express ampicillin resistance. However, only recombinant plasmids containing the 30-base-pair fragment regenerate an intact *lacZ* gene; transformants of this molecule metabolize X-gal and give rise to blue colonies. Insertion of the restricted PCR product into the pBLU backbone disrupts the *lacZ* gene; transformants of hybrid pBLU/PCR product constructs do not metabolize X-gal and thus give rise to white (uncolored) colonies.

PCR cloning has the advantage of producing a collection of recombinant plasmids that all theoretically contain the desired insert. A high percentage of the recombinant plasmids do, in fact, contain the 784-bp *Bam*HI/*Hin*dIII fragment, amplified by PCR.

Prepare Competent Cells

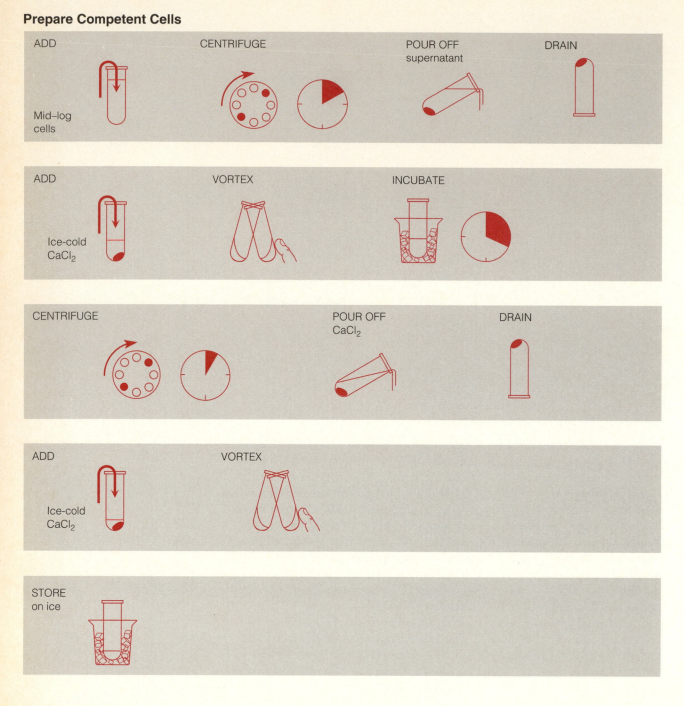

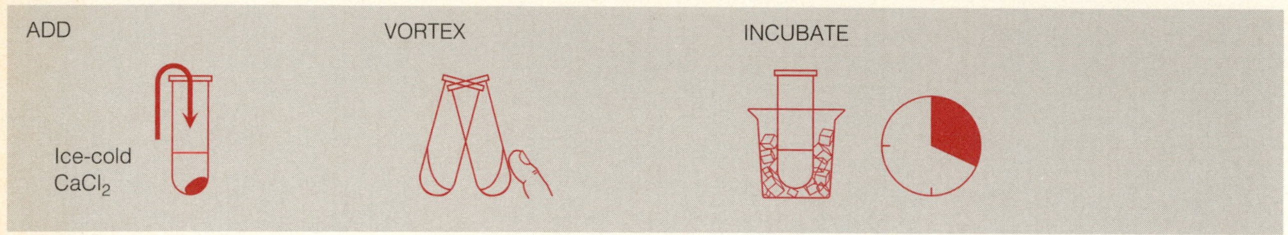

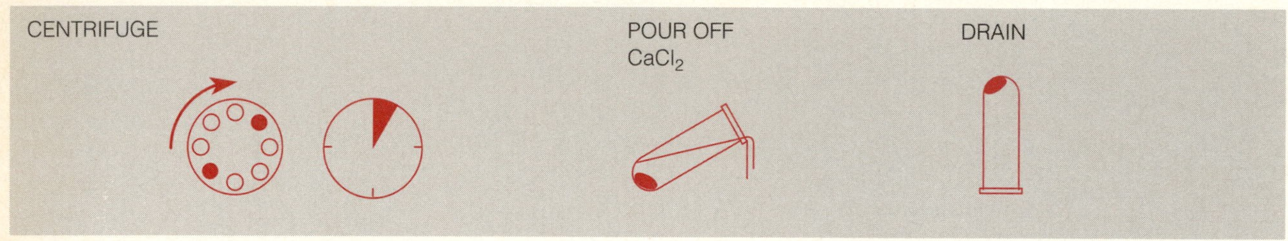

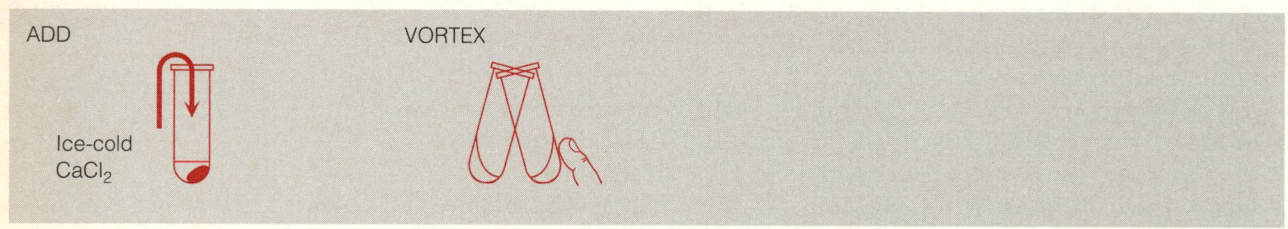

Laboratory 20/Part A
Classic Procedure for Preparing Competent Cells

PRELAB NOTES

Review the prelab notes in Laboratories 2A–C, 5, and 8A regarding *E. coli* culture and transformation.

E. coli Strains

It is essential to use a *lacZ* minus *E. coli* strain that is unable to metabolize X-gal. The protocols in this unit have been tested and optimized with the *lacZ* minus strain JM101. Other *lacZ* minus strains commonly used for molecular biological studies should give comparable results. However, the growth properties of other *E. coli* strains in suspension culture may differ significantly—for example, the time needed to reach mid-log phase and the cell number represented by specific optical densities.

Competent Cell Yield

If competent cells are being prepared in quantity for group use, remember that 100 ml of mid-log culture yields 10 ml of competent cells, enough for 50 transformations (200 µl each).

For Further Information

The protocol presented here is based on the following published methods:

Hanahan, D. 1983. Studies on transformation of *Escherichia coli* with plasmids. *Journal of Molecular Biology* 166: 557.

Hanahan, D. Techniques for transformation of *E. coli.* 1987. In D. M. Glover (ed.), *DNA Cloning: A Practical Approach*. vol. 1. Oxford: IRL Press.

Mandel, M. and A. Higa. 1970. Calcium-dependent bacteriophage DNA infection. *Journal of Molecular Biology* 53: 159.

CLASSIC PROCEDURE FOR PREPARING COMPETENT CELLS

Culture and Reagent	Supplies and Equipment
10 ml mid-log JM101 culture	5- or 10-ml sterile pipets
50 mM calcium chloride ($CaCl_2$)	pipet aid or bulb
	100–1,000-µl micropipettor + tips (or 1-ml pipet)
	beaker of crushed or cracked ice
	beaker for waste/used tips
	10%-bleach solution or disinfectant
	burner flame
	clean paper towels
	clinical centrifuge (500–1,000 × *g*)
	test tube rack

Prepare Competent Cells

(40–50 minutes)

This entire experiment must be performed under sterile conditions, using sterile tubes and pipet tips.

1. Place a sterile tube of 50 mM $CaCl_2$ solution on ice.
2. Obtain a 15-ml tube with 10 ml of mid-log cells, and label it with your name.
3. *Securely close the cap,* and place the tube of cells in a *balanced* configuration with other classmates' tubes in the rotor of a clinical centrifuge. Centrifuge at 500–1,000 × *g* for 10 minutes to pellet the cells on the bottom of the culture tube.

"Blank" tubes with 10 ml of water can be used for balance, if needed. 500–1,000 × *g* corresponds to 2,000–3,000 rpm for most tabletop clinical centrifuges. A tight cell pellet should be visible at the bottom of the tube. If the pellet appears very loose or unconsolidated, centrifuge for another 5 minutes.

4. *Take care not to disturb the cell pellet.* Sterilely pour off the supernatant into a waste beaker for later disinfection. Invert the culture tube, and touch the mouth of the tube to a clean paper towel, to wick off as much as possible of the remaining supernatant. Flame the mouth of the culture tube after removing the top and again before replacing the top.
5. Sharply tap the bottom of the tube on the lab bench several times to loosen the cell pellet.

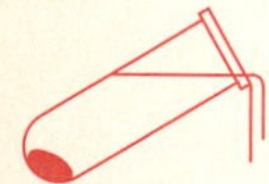

Organize materials on the lab bench and plan out your sterile manipulations. If you are working with a partner, one person can handle the pipet, and the other person can remove the cap and flame the mouth of the tube.

6. Use a sterile 5- or 10-ml pipet to add 5 ml of ice-cold calcium chloride to the culture tube. Briefly flame the pipet cylinder before use. Flame the mouth of the calcium chloride supply tube and the culture tube after removing the top and again before replacing the top.

7. *Close the cap of the culture tube tightly.* Vortex immediately to *completely* resuspend the pelleted cells. Periodically, during vortexing, hold the tube up to the light to check on its progress. *The finished cell suspension should be homogeneous, with no visible clumps of cells.* If a mechanical vortexer is unavailable, "finger vortex" as follows:

   Hold the upper part of the tube securely between your thumb and index finger. Vigorously tap the bottom of the tube with the index finger of the opposite hand, to create a vortex that lifts the cell pellet off of the bottom of the tube.

The cell pellet will become increasingly difficult to resuspend the longer it sits in the calcium chloride. Complete cell suspension is probably the most important variable for obtaining competent cells.

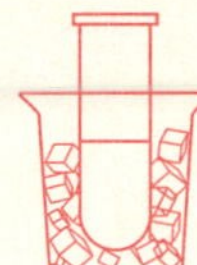

8. Return the tube to ice, and incubate it for 20 minutes.
9. Following incubation, respin the cells in a clinical centrifuge for 5 minutes at 500–1,000 × *g*.
10. *Take special care not to disturb the diffuse cell pellet.* Sterilely pour off the supernatant into a waste beaker for later disinfection. Invert the culture tube, and touch the mouth of the tube to a clean paper towel, to wick off as much as possible of the remaining supernatant. Flame the mouth of the culture tube after removing the top and again before replacing the top.

The calcium chloride treatment alters the adhering properties of the *E. coli* cell membranes. At this point, the cell pellet typically appears diffuse or ring-shaped and should disperse more easily.

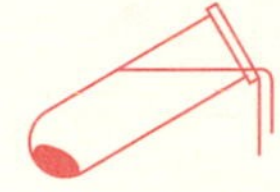

11. Use a 100–1,000-µl micropipettor (or 1-ml pipet) to sterilely add 1,000 µl (1 ml) of fresh, ice-cold $CaCl_2$ to the tube. Flame the mouth of the culture tube after removing the top and again before replacing the top.
12. *Close the cap of the culture tube tightly.* Vortex to *completely* resuspend the pelleted cells. *The finished cell suspension should be homogeneous, with no visible clumps of cells.*

The diffuse cell pellet should resuspend very easily; do not overvortex.

Stop Point

Store the cells in a beaker of ice in the refrigerator (at approximately 0°C) until ready for use. "Seasoning" at 0°C for as long as 24 hours increases the competency of the cells five- to tenfold.

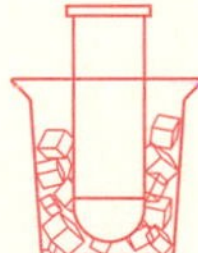

13. Take the time for proper cleanup:
    a. Segregate for proper disposal culture plates and tubes, pipets, and micropipet tips that have come into contact with *E. coli.*
    b. Disinfect the mid-log culture, the tips, and the supernatant from steps 4 and 10 with a 10%-bleach solution or disinfectant (such as Lysol).
    c. Wipe down the lab bench with soapy water, a 10%-bleach solution, or disinfectant.
    d. Wash your hands before leaving the laboratory.

Transform *E. coli* with Recombinant PCR/pBLU Plasmids

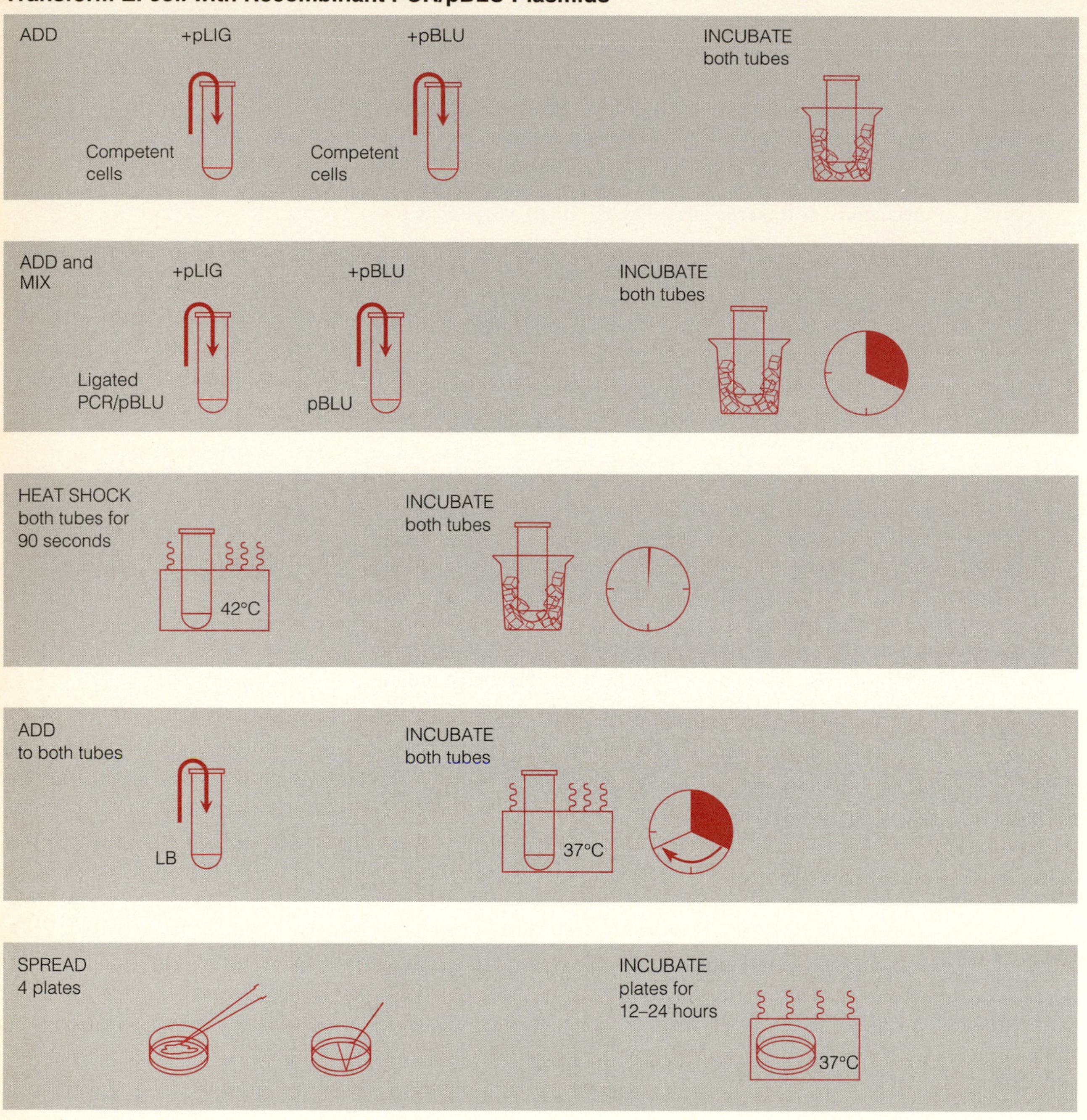

Laboratory 20/Part B
Transformation of *E. coli* with Recombinant PCR/pBLU Plasmids

PRELAB NOTES

Review the prelab notes in Laboratory 5, "Rapid Colony Transformation of *E. coli* with Plasmid DNA."

X-Gal

The chromogenic substrate X-gal (5-bromo-4-chloro-3-indoyl-—D-galactoside) is chemically related to X-phosphate, one of the substrates for color development in Laboratory 13, "Southern Hybridization of λ DNA." Like antibiotics, X-gal is inactivated by heat. When preparing agar plates containing X-gal, be sure to cool the agar solution until the container can be held comfortably in the hand (approximately 60°C) before adding the reagent. X-gal is expensive; one economical alternative to adding X-gal to an entire batch of culture medium is to sterilely spread 50 µl of 20 mg/ml stock solution over the surface of a prepared culture plate. Allow the X-gal solution to diffuse into the medium overnight or incubate it at 37°C for several hours before use.

Cell Plating

In this laboratory, cells are spread directly onto LB agar plates. One advantage of PCR cloning is that replica plating onto a nylon membrane, as in Laboratory 15B, is unnecessary.

For Further Information

The protocol presented here is based on the following published methods:

Cohen, S. N., A. C. Y. Chang, L. Hsu. 1972. Nonchromosomal antibiotic resistance in bacteria: Genetic transformation of *Escherichia coli* by r-factor DNA. *Proceedings of the National Academy of Sciences* 69: 2110.

Messing, J. and J. Vieira. 1982. The pUC plasmids, an M13mp7-derived system for insertion mutagenesis and sequencing with synthetic universal primers. *Gene* 19: 259.

TRANSFORMATION OF *E. COLI* WITH RECOMBINANT PCR/pBLU PLASMIDS

Culture, Media, and Reagents

competent *E. coli* cells (from Part A)
ligated DNA PCR/pBLU (from Lab 19B)
0.005 μg/μl pBLU
LB broth
four LB/amp + X-gal plates

Supplies and Equipment

100–1,000-μl micropipettor + tips
10–100-μl micropipettor + tips
0.5–10-μl micropipettor + tips
2 15-ml culture tubes
beaker of crushed or cracked ice
beaker for waste/used tips
"biobag" or heavy-duty trash bag
10%-bleach solution or disinfectant
burner flame
cell spreader
permanent marker
test tube rack
37°C incubator
37°C shaking water bath (optional)
42°C water bath

Transform *E. coli* with Recombinant PCR/pBLU Plasmids

(70–90 minutes)

This entire experiment must be performed under sterile conditions, using sterile tubes and pipet tips.

1. Use a permanent marker to label two sterile 15-ml culture tubes, as follows:

   +pLIG = ligated PCR/pBLU

   +pBLU = pBLU control

2. Add 200 μl of competent cells to each tube. Return the remaining competent cells to ice, and store them at 0°C.
3. Place both tubes on ice.

Use fresh tip to pipet each DNA sample. Optimally, flame the mouth of the culture tubes after removing the cap and again before replacing the cap.

Store the remainder of the ligated DNA at 4°C in preparation for electrophoretic analysis in Laboratory 21.

4. Add 10 μl of ligated PCR/pBLU *directly into the cell suspension* in the +pLIG tube.
5. Add 10 μl of 0.005 μg/μl pBLU *directly into the cell suspension* in the +pBLU tube.
6. Tap each tube lightly with your finger to mix the solution. Avoid making bubbles in the suspension or splashing the suspension up the sides of the tubes.
7. Return both tubes to ice for 20 minutes.

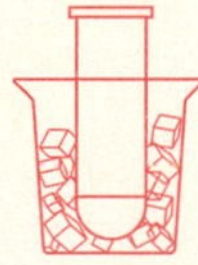

8. While the cells are incubating on ice, use a permanent marker to label four LB/amp + X-gal plates with your name and the date. Divide the plates into pairs, and mark them as follows:

   +pLig   Mark one plate L20 and the other plate "L100."

   +pBLU   Mark one plate "B20" and the other plate "B100."

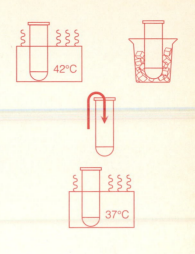

9. Following the 20-minute incubation on ice, heat shock the cells in both tubes. *The cells must receive a sharp and distinct shock.*
   a. Carry the ice beaker to a water bath. Remove the tubes from the ice, and immediately immerse them in a 42°C water bath for 90 seconds.
   b. Immediately return both tubes to ice, and let them stand for at least 1 more minute.
10. Add 800 µl of LB broth to each tube. Lightly tap the tubes with your finger to mix the solution.
11. Allow the cells to recover by incubating both tubes at 37°C in a shaking water bath (with moderate agitation) for 20–40 minutes.

Stop Point

Cells can be allowed to recover for several hours. A longer recovery period will help to compensate for poor ligation or cells of low competence.

The recovery step increases the number of transformants expressing the resistance protein by 2 to 4 fold, but it can be omitted to save time.

12. Use the matrix below as a checklist as the +pLIG and +pBLU cells are spread on plates in the following steps.

Plate	LB	+pLIG	+pBLU
L20	80 µl	20 µl	—
L100	—	100 µl	—
B20	80 µl	—	20 µl
B100	—	—	100 µl

13. Add 80 µl of sterile LB broth to the centers of the L20 and B20 plates. Then *use fresh pipet tips* to:
   a. Add 20 µl of cells from the +pLig tube to the LB broth on the L20 plate.
   b. Add 20 µl of cells from the +pBLU tube to the LB broth on the B20 plate.
14. *Use a fresh tip* to add 100 µl of cells from the +pLig tube to the L100 plate.
15. *Use a fresh tip* to add 100 µl of cells from the +pBLU tube to the B100 plate.
16. Sterilize a cell spreader, and spread cells over the surface of each of the four plates in succession.
   a. Dip the spreader into the ethanol beaker and briefly pass it through a burner flame to ignite the alcohol. Allow the alcohol to burn off *away from the burner flame.*

The added LB broth increases the total volume of liquid on each plate to help you evenly spread the small volume (20 µl) of transformed cells on the L20 and B20 plates.

Do not allow the cells to sit too long on the plates before spreading them. The object is to evenly distribute and separate the transformed cells on the plate surface, so that each gives rise to a distinct colony of clones.

CAUTION

Be extremely careful not to ignite the ethanol in the beaker.

The spreader, submerged in alcohol, is already sterile. The only purpose of the flame is to burn off the alcohol before spreading the cells. The spreader will become too hot if it is held directly in the burner flame for several seconds and can kill *E. coli* cells on the plate.

b. Lift the plate lid, like a clam shell, only enough to allow spreading.

c. Cool the spreader by touching it to the agar surface away from the cells or to condensed water on the plate lid.

d. Touch the spreader to the cell suspension, and gently drag it back and forth several times across the agar surface. Rotate the plate one-quarter turn, and repeat the spreading motion. Be careful not to gouge into the agar.

e. Replace the plate lid. Return the cell spreader to the ethanol without flaming.

17. Reflame the spreader once more before setting it down.

18. Let the gel set for several minutes to allow the cell suspensions to be absorbed.

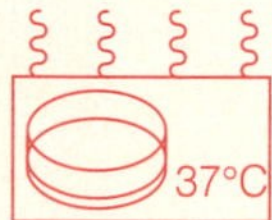

19. Wrap the plates together with tape, place the plates upside down in a 37°C incubator, and incubate for 12–24 hours. Save either the L100 or the L20 plate to begin Laboratory 21.

20. After the initial incubation, store the plates at 4°C to arrest *E. coli* growth and to slow the growth of any contaminating microbes.

21. Take the time for proper cleanup:

a. Segregate for proper disposal culture plates and tubes, pipets, and micropipettor tips that have come into contact with *E. coli.*

b. Disinfect the overnight cell suspension, the tubes, and the tips with a 10%-bleach solution or disinfectant (such as Lysol).

c. Wipe down the lab bench with soapy water, a 10%-bleach solution, or disinfectant.

d. Wash your hands before leaving the laboratory.

RESULTS AND DISCUSSION

Obtain separate counts of the numbers of blue and white colonies growing on each plate, using a permanent marker to mark the bottom of the culture plate as each colony is counted. A total of 50–200 colonies should be observed on the L20 experimental plate, with about equal numbers of blue and white colonies. Approximately fivefold more colonies should be observed on the B20 control plate—all, or nearly all, of these will be blue. The corresponding L100 and B100 plates should contain about fivefold more colonies; however, an extended recovery period (step 11) would most likely inflate these expected results.

Blue colonies can be pale in color and difficult to distinguish from white colonies, but refrigerating the plates for 1 hour or longer can help to intensify the color. If the plates are overincubated at 37°C or left at room temperature for several days, small white "satellite" colonies may be observed growing in a circle around large, resistant colonies. Satellite colonies are clones of ampicillin-sensitive cells that begin to grow as the antibiotic is gradually reduced to below the selection threshold.

1. Record your observations of each plate in the matrix below. Were the results as you expected? Explain possible reasons for variations from the expected results.

	White Colonies	**Blue Colonies**
L20		
L100		
B20		
B100		

2. Calculate the percentage of recombinant clones that contain the 784-bp fragment.
3. Is it possible for a white colony *not* to contain the expected 784-bp insert? Is it possible for a blue colony *to* contain the 784-bp insert?
4. Calculate the transformation efficiency of the control pBLU transformation (the B20 or B100 plate): the number of antibiotic-resistant colonies per microgram of plasmid DNA. The object is to determine the mass of plasmid DNA that was spread on the plate and is, therefore, responsible for the number of transformants observed.
   a. Determine *the total mass (in μg) of pBLU* used in the transformation = the concentration of pBLU × the volume of pBLU.
   b. Determine *the fraction of cell suspension spread* onto the B20 or the B100 plate = the volume of cell suspension spread/total volume of cell suspension.
   c. Determine *the mass of pBLU in the cell suspension spread* onto the B20 or the B100 plate = the total mass of pBLU (a) × the fraction of cell suspension spread (b).

   d. Express *the transformation efficiency* in scientific notation as the number of colonies per μg of pBLU = the number of colonies on the B20 or the B100 plate/mass of pBLU in the cell suspension spread (c).

5. Calculate the transformation efficiency of the experimental transformation with ligated DNA (L20 or L100 plate).
   a. Determine *the total mass (in μg) of pBLU* used in the initial restriction digest (Laboratory 19A) = the concentration of pBLU × the volume of pBLU.
   b. Determine *the fraction of pBLU digest* used in the ligation reaction (Laboratory 19B) = the volume of pBLU digest in the ligation/the total volume of pBLU digest.
   c. Determine *the fraction of ligation reaction* used for transformation = the volume of ligation reaction in the transformation/the total volume of ligation reaction.
   d. Determine *the fraction of cell suspension spread* onto the L20 or the L100 plate = the volume of cell suspension spread/the total volume of cell suspension.
   e. Determine *the mass of pBLU in the cell suspension spread* onto the L20 or the L100 plate = the total mass of pBLU (a) × fraction b × fraction c × fraction d.
   f. Express *the transformation efficiency* in scientific notation as the number of colonies per μg of pBLU: = the total number of colonies on the L20 or the L100 plate/the mass of pBLU in the cell suspension spread (e).

FOR FURTHER RESEARCH

Use PCR to screen a library of λ restriction fragments (prepared according to Laboratories 14 and 15) for the 784-bp *Bam*HI/*Hin*dIII insert. Prepare a crude plasmid extract, using a modified miniprep procedure (Laboratory 6):

1. Select several white colonies, and resuspend each in 50 μl of GTE in a 1.5-ml tube.
2. Lyse the cells by adding 100 μl of 1% SDS/0.2N NaOH to each tube. Invert the tubes several times to mix the solution. Stand the tubes on ice for 5 minutes.
3. Precipitate the chromosomal DNA and cell debris by adding 75 μl of KOAc/acetic acid solution. Invert the tubes several times to mix. Stand the tubes on ice for 5 minutes.
4. Microcentrifuge the tubes for 5 minutes to pellet the cell debris.
5. Use 2 μl of each supernatant for PCR amplification.
6. Electrophorese PCR products in separate lanes of a 0.8% agarose gel, along with a *Bam*HI/*Hin*dIII digest of λ.

LABORATORY 21

The Purification and Identification of PCR/pBLU Recombinants

The growth of *E. coli* cells on the "L" LB/amp + X-gal plates in Laboratory 20 confirms that they have been transformed to an ampicillin-resistant phenotype. In Laboratory 19B, the restricted PCR product was the only fragment available for ligation into the pBLU plasmid. Restriction analysis of plasmid DNA from transformants is sufficient to confirm that the cloned insert is, indeed, the 784-bp λ *Bam*HI/*Hin*dIII fragment. There is a high probability that at least one of a pair of randomly selected white transformants will contain the target sequence.

In Part A, "Plasmid Minipreparation of PCR/pBLU Recombinants," plasmid DNA is isolated from overnight cultures from two different white colonies taken from an L20 or L100 LB/amp + X-gal Plate (Laboratory 20). In Part B, "Restriction Analysis of Purified PCR/pBLU Recombinants," samples of the plasmids isolated in Part A and control samples of pBLU and λ DNA are incubated with *Bam*HI and *Hin*dIII. The digested samples and the uncut controls are coelectrophoresed in an agarose gel. Comigration of 784-bp *Bam*HI/*Hin*dIII fragments in the lanes of miniprep DNA and the pBLU/λ controls and an evaluation of the relative sizes of uncut, supercoiled DNAs, provides evidence of the structure, size, and number of plasmids present in each of the transformed colonies.

Isolate Plasmid DNA

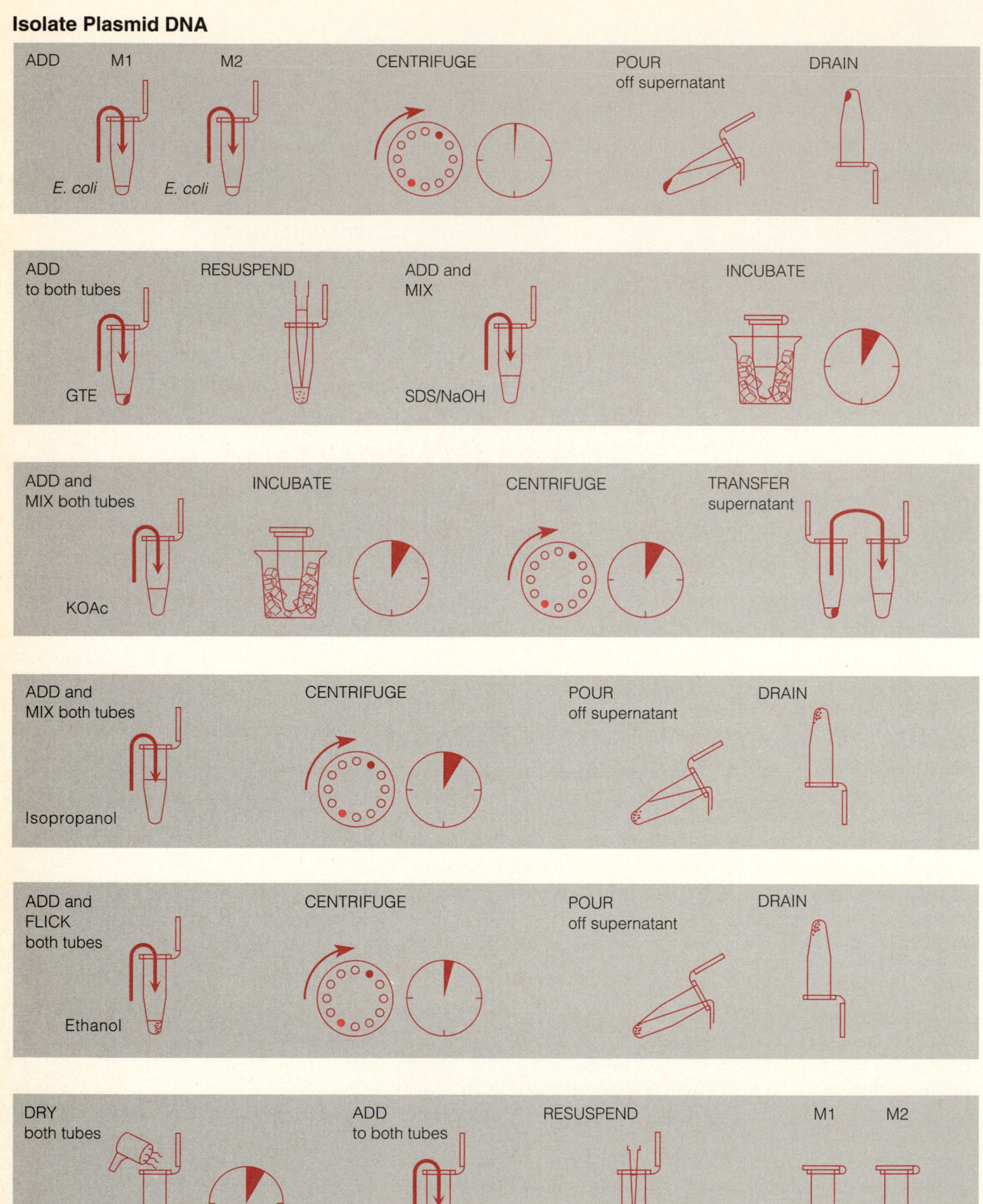

PRELAB NOTES

Review the prelab notes in Laboratory 6A, "Plasmid Minipreparation of pAMP."

For Further Information

The protocol presented here is based on the following published methods:

Birnboim, H. C. and J. Doly. 1979. A rapid alkaline extraction method for screening recombinant plasmid DNA. *Nucleic Acids Research* 7: 1513.

Ish-Horowicz, D. and J. F. Burke. 1981. Rapid and efficient cosmid cloning. *Nucleic Acids Research* 9: 2989.

PLASMID MINIPREPARATION OF PCR/pBLU RECOMBINANTS

Reagents	Supplies and Equipment
two *E. coli* /PCR/pBLU overnight cultures	100–1,000-µl micropipettor + tips
glucose/Tris/EDTA (GTE)	10–100-µl micropipettor + tips
SDS/sodium hydroxide (SDS/NaOH)	0.5–10-µl micropipettor + tips
potassium acetate/acetic acid (KOAc)	1.5-ml tubes
isopropanol	beaker of crushed ice
95% ethanol	beaker for waste/used tips
Tris/EDTA (TE)	10%-bleach solution or disinfectant (such as Lysol)
	hair dryer
	microfuge
	paper towels
	permanent marker
	test tube rack

Isolate Plasmid DNA

(50 minutes)

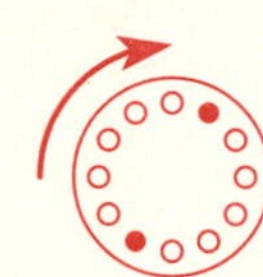

Double all centrifuge times, if you are using a small microfuge that generates approximately 2,000 × g.

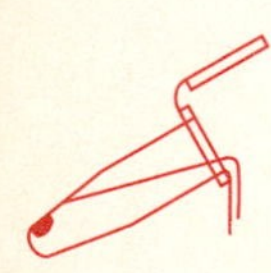
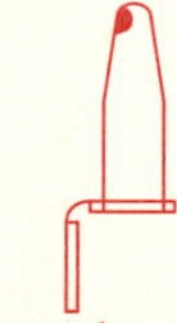

Accurate pipetting is essential to good plasmid yield: The volumes of reagents are precisely calibrated so that the sodium hydroxide added in step 6 is neutralized by the acetic acid in step 8.

Do not overmix. Excessive agitation shears the single-stranded DNA and decreases plasmid yield.

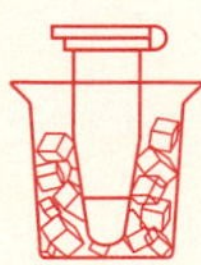

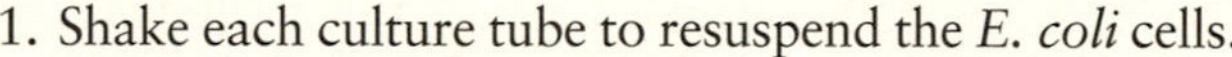

1. Shake each culture tube to resuspend the *E. coli* cells.

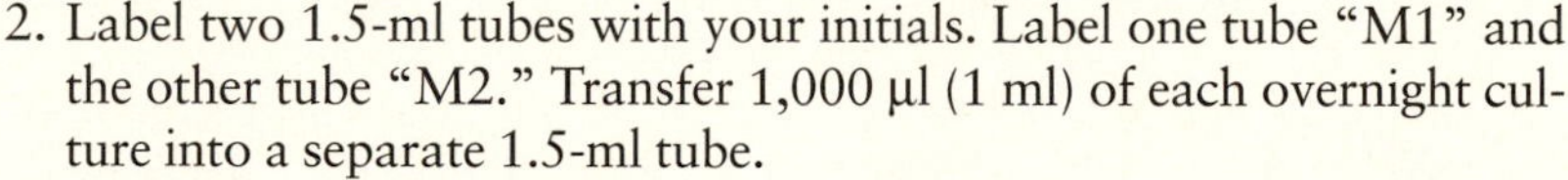

2. Label two 1.5-ml tubes with your initials. Label one tube "M1" and the other tube "M2." Transfer 1,000 µl (1 ml) of each overnight culture into a separate 1.5-ml tube.
3. Close the caps, and place the tubes in a *balanced* configuration in a microfuge rotor. Spin for 1 minute to pellet the cells.
4. Pour off the supernatant from both tubes into a waste beaker for later disinfection. *Take care not to disturb the cell pellets.* Invert the tubes, and touch the mouths of the tubes to a clean paper towel, to wick off as much as possible of the remaining supernatant.
5. Add 100 µl of *ice-cold* GTE solution to each tube. Resuspend the pellets by pipetting the solution in and out several times. Hold the tubes up to the light to verify that the suspension is homogeneous and that no visible clumps of cells remain.
6. Add 200 µl of SDS/NaOH solution to each tube. Close the caps, and mix the solutions by rapidly inverting the tubes about five times.
7. Stand the tubes on ice for 5 minutes. The suspension will become relatively clear.
8. Add 150 µl of *ice-cold* KOAc solution to each tube. Close the caps, and mix the solutions by rapidly inverting the tubes about five times. A white precipitate will immediately appear.
9. Stand the tubes on ice for 5 minutes.

10. Place the tubes in a balanced configuration in a microfuge rotor, and spin for 5 minutes to pellet the precipitate along the side of the tube.

Precipitate typically collects along the side of the tube, rather than as a tight pellet.

11. Transfer 400 µl of supernatant from "M1" into a clean 1.5-ml tube labeled "M1." Transfer 400 µl of supernatant from "M2" into a clean 1.5-ml tube labeled "M2." *Avoid pipetting the precipitate;* wipe off any precipitate that is clinging to the outside of the tip before expelling the supernatant. Discard old tubes containing precipitate.

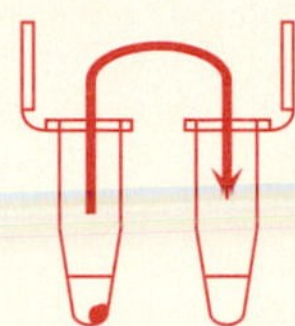

12. Add 400 µl of isopropanol to each tube of supernatant. Close the caps, and mix the solution by rapidly inverting the tubes about five times. *Stand the tubes at room temperature for only 2 minutes.*

Do step 12 quickly, and make sure that the microfuge will be immediately available for step 13. Isopropanol preferentially precipitates nucleic acids rapidly; however, proteins and other cellular components remaining in solution will also begin to precipitate with time.

13. Place tubes in a *balanced* configuration in the microfuge rotor, and spin for 5 minutes to pellet the nucleic acids. Align the tubes in the rotor so that the cap hinges point outward. The nucleic acid residue, visible or not, will collect under the hinge during centrifugation.

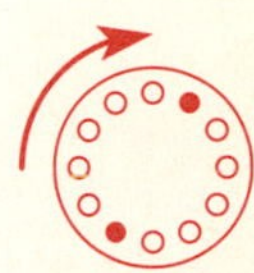

14. Pour off the supernatant from both tubes. *Take care not to disturb the nucleic acid pellets.* Wick off as much as possible of the remaining alcohol on a paper towel.

The pellet will most likely appear either as a tiny, teardrop-shaped smear or small particles on the bottom of each tube. However, pellet size is not a valid predictor of plasmid yield or quality. A large pellet is composed primarily of RNA, salt, and cellular debris, carried over from the original precipitate; a small pellet, or one impossible to see, often means a cleaner preparation.

15. Add 200 µl of 95% ethanol to each tube, and close the caps. Flick the tubes several times to wash the nucleic acid pellets.

Nucleic acid pellets are not soluble in ethanol and will not resuspend during washing.

Stop Point

Store the DNA in ethanol at –20°C until you are ready to continue.

16. Place the tubes in a *balanced* configuration in the microfuge rotor, and spin for 2–3 minutes to recollect the nucleic acid pellets.
17. Pour off the supernatant from both tubes. *Take care not to disturb the nucleic acid pellets.* Wick off as much as possible of the remaining alcohol on a paper towel.
18. Dry the nucleic acid pellets, using one of following methods:
    a. Direct a stream of warm air from a hair dryer over the open ends of the tubes for about 3 minutes. *Take care not to blow the pellets out of the tubes.*

    or

    b. Close the caps, and pulse the tubes in the microfuge to pool the remaining ethanol. *Carefully* use a micropipettor to draw off the ethanol. Let the pellets air dry at room temperature for 10 minutes.

Air bubbles cast in the tube wall can be mistaken for ethanol droplets.

19. Be sure that the nucleic acid pellet is dry and that all ethanol has evaporated before proceeding on to step 20. Hold each tube up to the light to confirm that no ethanol droplets remain and that the nucleic acid pellet, if visible, appears white and flaky. If ethanol is still evaporating, you can detect an alcohol odor by sniffing the mouth of the tube.

If you are using a 0.5–10-µl micropipettor, set it to 7.5 µl, and pipet twice.

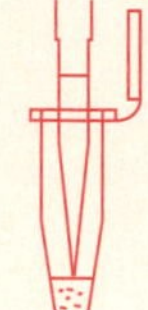

20. Add 15 µl of TE to each tube. Resuspend the pellets by smashing them with the pipet tip and pipetting in and out vigorously. Rinse down the side of the tube several times, concentrating on the area where the pellet should have formed during centrifugation (beneath the cap hinge). Make sure that all DNA has dissolved and that no particles remain in the tip or on the side of the tube.
21. Keep the two DNA/TE solutions *separate. Do not* pool them into one tube!

Stop Point

Freeze the DNA/TE solutions at –20°C until you are ready to continue. Thaw before using.

22. Take the time for proper cleanup:
    a. Segregate for proper disposal the culture tubes and micropipettor tips that have come into contact with *E. coli.*
    b. Disinfect the overnight cultures, the tips, and the supernatant from step 4 with a 10%-bleach solution or disinfectant (such as Lysol).
    c. Wipe down the lab bench with soapy water, a 10%-bleach solution, or disinfectant.
    d. Wash your hands before leaving the laboratory.

RESULTS AND DISCUSSION

See the "Results and Discussion" section of Laboratory 6A, "Plasmid Minipreparation of pAMP," for a detailed discussion of the biochemistry of the alkaline lysis method for plasmid purification.

I. Set Up Restriction Digests

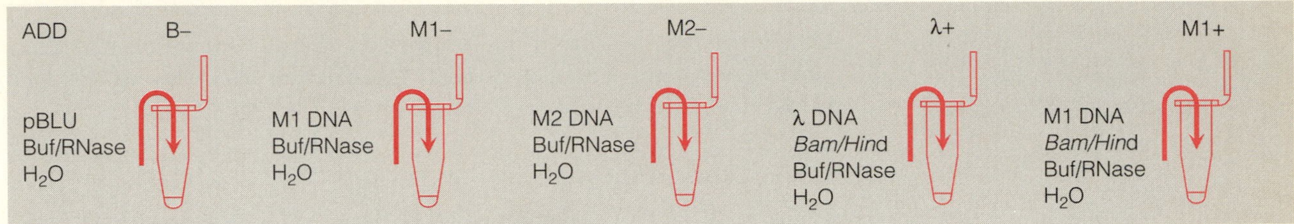

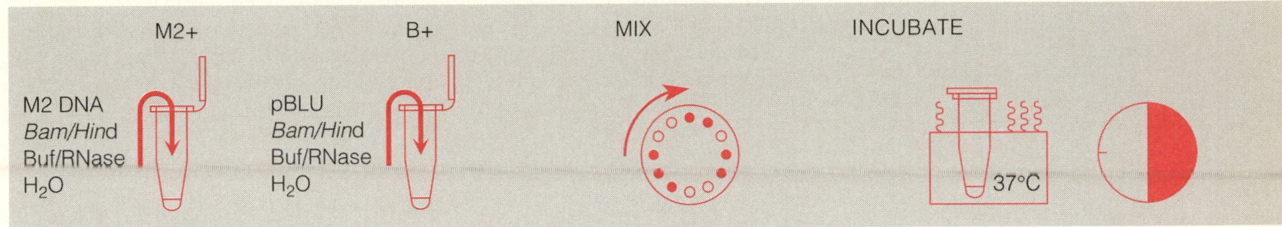

II. Cast 0.8% Agarose Gel

III. Load Gel and Electrophorese

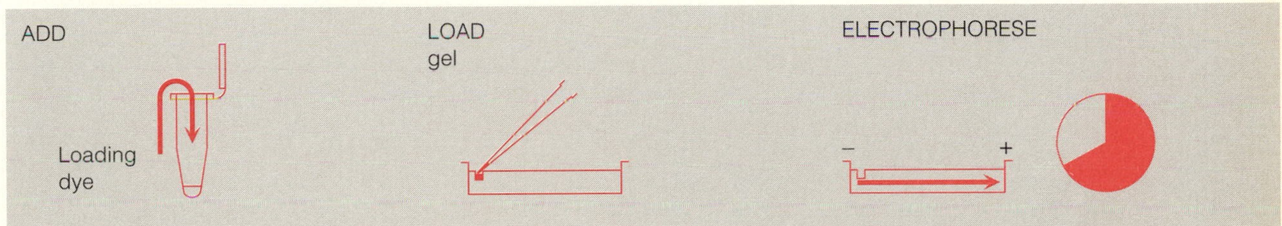

IV. Stain, View, and Photograph Gel

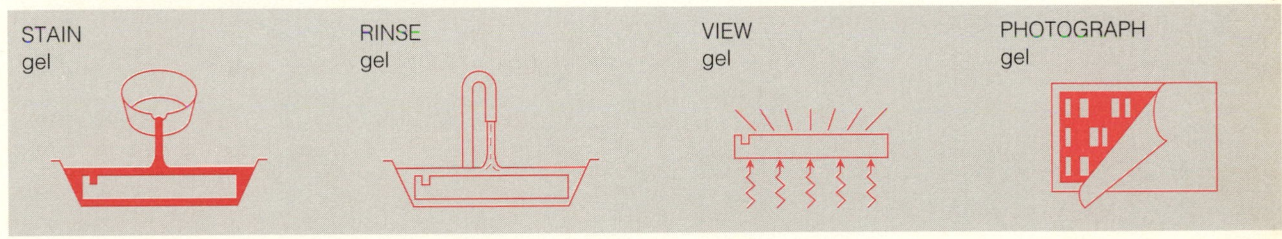

Laboratory 21/Part B
Restriction Analysis of Purified PCR/pBLU Recombinants

PRELAB NOTES

Review the prelab notes in Laboratory 3, "DNA Restriction and Electrophoresis," and Laboratory 6B, "Restriction Analysis of Purified pAMP."

For Further Information

Helling, R. B., H. M. Goodman, and H. W. Boyer. 1974. Analysis of *EcoRI* fragments of DNA from lambdoid bacteriophages and other viruses by agarose-gel electrophoresis. *Journal of Virology* 14: 235.

Sharp, P. A., B. Sugden, and J. Sambrook. 1973. Detection of two restriction endonuclease activities in *Haemophilus parainfluenzae* using analytical agarose–ethidium bromide electrophoresis. *Biochemistry* 12: 3055.

RESTRICTION ANALYSIS OF PCR/pBLU RECOMBINANTS

Reagents	Supplies and Equipment
M1 and M2 miniprep DNA/TE	0.5–10-µl micropipettor + tips
0.1 µg/µl pBLU	1.5-ml tubes
0.1 µg/µl λ DNA	aluminum foil
*Bam*HI/*Hin*dIII	beaker for agarose
5× restriction buffer/RNase	beaker for waste/used tips
distilled water	camera and film (optional)
loading dye	disposable gloves
0.8% agarose	electrophoresis box
1× Tris/Borate/EDTA (TBE) buffer	masking tape
1 µg/ml ethidium bromide	microfuge (optional)
	Parafilm or waxed paper (optional)
for decontamination:	permanent marker
0.05 M $KMnO_4$	plastic wrap (optional)
0.25 N HCl	power supply
0.25 N NaOH	test tube rack
	transilluminator
	37°C water bath
	60°C water bath (for agarose)

I. Set Up Restriction Digests

(10 minutes; then 30–40 minutes of incubation)

Refer to Laboratory 3, "DNA Restriction and Electrophoresis," for detailed instructions on setting up digests.

1. Use a permanent marker to label seven 1.5-ml tubes, as shown in the matrix below. Restriction reactions will be performed in these tubes.
2. Use the matrix below as a checklist while adding reagents to each reaction. Read down each column, adding the same reagent to all appropriate tubes. *Use a fresh pipet tip for each reagent.*

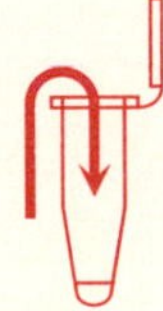

Tube	pBLU	M1	M2	λ	Buffer/ RNase	*Bam*HI/ *Hin*dIII	H_2O
B–	5 µl	—	—	—	2 µl	—	3 µl
M1–	—	5 µl	—	—	2 µl	—	3 µl
M2–	—	—	5 µl	—	2 µl	—	3 µl
λ+	—	—	—	5 µl	2 µl	2 µl	1 µl
M1+	—	5 µl	—	—	2 µl	2 µl	1 µl
M2+	—	—	5 µl	—	2 µl	2 µl	1 µl
B+	5 µl	—	—	—	2 µl	2 µl	1 µl

3. After adding all of the reagents, close the tube tops and mix.
4. Place the reactions in a 37°C water bath, and incubate for 30–40 minutes only.

Do not overincubate. During a longer incubation, the DNases in the miniprep

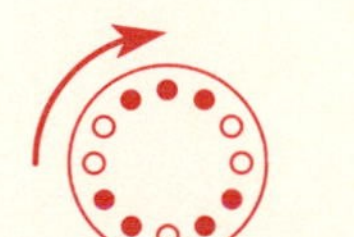

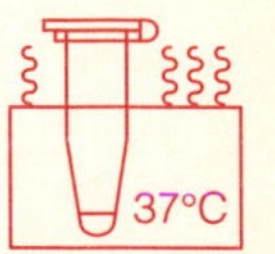

Stop Point

Following incubation, freeze reactions at –20°C until you are ready to continue. Thaw the reactions before continuing on to Section III, step 1.

II. Cast 0.8% Agarose Gel

(15 minutes)

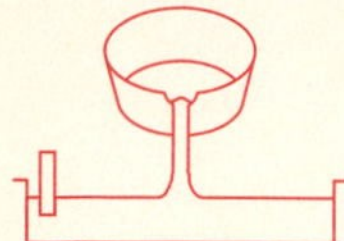

1. Carefully pour enough agarose solution into the gel-casting tray to fill it to a depth of about 5 mm. The gel should cover only about one-third of the height of the comb teeth.
2. After the agarose solidifies, place the casting tray into the electrophoresis box and set up for electrophoresis.

Stop Point

Cover the electrophoresis box and save the gel until you are ready to continue. The gel will remain in good condition for several days, if it is completely submerged in buffer.

III. Load Gel and Electrophorese

Refer to Laboratory 3, "DNA Restriction and Electrophoresis," for detailed instructions on casting and loading gel.

(10 minutes; then 30–45 minutes of electrophoresis)

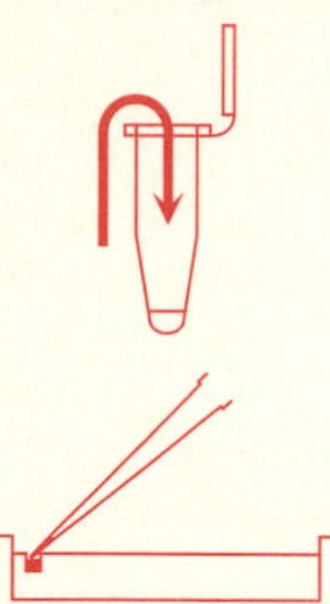

1. Remove the reaction digests from the 37°C water bath.
2. Add 1 µl of loading dye to each reaction tube. Close the tops of the tubes and mix the solution.
3. Load 10 µl from each reaction tube into a separate well in the gel, as shown in the diagram below. Use a *fresh pipet tip* for each reaction.

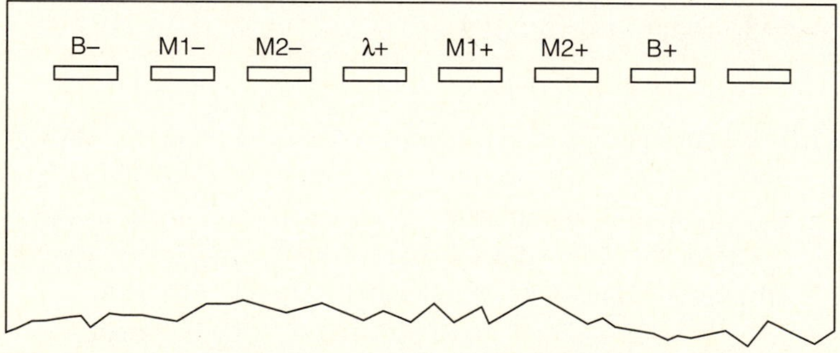

4. Electrophorese at 50–125 volts, until the bromophenol blue bands have moved 50–60 mm from the wells.

The 784-bp λ fragment migrates just behind the bromophenol blue marker; take care not to run it off of the end of the gel. However, if time allows, electrophorese until the bromophenol blue bands near the end of the gel. This allows maximum separation of uncut plasmids, which is important when differentiating a large "superplasmid" from a double transformation of two smaller plasmids.

5. Turn off the power supply, remove the casting tray from the electrophoresis box, and transfer the gel to a disposable weigh boat (or other shallow tray) for staining.

Stop Point

Cover the electrophoresis box, and save the gel until you are ready to continue. The gel can be stored in a zip-lock plastic bag and refrigerated overnight for viewing or photographing the next day. However, over a longer period, the DNA will diffuse through the gel, and the bands will either become indistinct or disappear entirely.

IV. Stain Gel with Ethidium Bromide, View, and Photograph

(10–15 minutes)

Staining can be performed by an instructor in a controlled area, when students are not present.

CAUTION

Review the section on "Ethidium Bromide Staining and Responsible Handling" in Laboratory 3. Wear disposable gloves when staining, viewing, and photographing gel and during cleanup. Confine all staining to a restricted sink area.

Band intensity and contrast increase dramatically if the gel is destained for 15–30 minutes in tap water. More simply, rinse and drain the gels, stack the staining trays, cover the top of the gel, and let it set overnight at room temperature.

1. Flood the gel with ethidium bromide solution (1 μg/ml), and allow it to stain for 5–10 minutes.
2. View the gel under an ultraviolet transilluminator or other UV light source.

CAUTION

Ultraviolet light can damage your eyes. Never look directly at an unshielded UV light source without eye protection. View only through a filter or safety glasses that absorb the harmful wavelengths.

3. Make an exposure of 1–2 seconds with your camera set at f/8. Develop the print for the recommended time (approximately 45 seconds at room temperature).
4. Take the time for proper cleanup:
   a. Wipe down the camera, the transilluminator, and the staining area.
   b. Decontaminate the gel and any staining solution that will not be reused.
   c. Wash your hands before leaving the laboratory.

RESULTS AND DISCUSSION

Refer to the "Results and Discussion" section of Laboratory 6B, "Restriction Analysis of Purified pAMP," for more details about interpreting miniprep gels and plasmid conformation.

Observe the photograph of your stained gel, and determine which lanes contain control DNAs (pBLU and λ) and which contain minipreps M1 and M2. Even if you have not followed the prescribed loading order, the miniprep lanes can be distinguished by the following characteristics:

- A background "smear" of degraded and partially digested chromosomal DNA, plasmid DNA, and RNA
- Undissolved material and high-molecular-weight DNA "trapped" at the front edge of the sample well
- A "cloud" of low-molecular-weight RNA at a position corresponding to 100–200 bp.
- The presence of high-molecular-weight bands of uncut plasmid in lanes of digested miniprep DNA

  Remember these facts when considering possible structures of the recombined plasmids in M1 and M2:
- Every replicating plasmid must have an origin of replication. Recombinant plasmids with more than one origin replicate normally.
- Adjacent restriction fragments in a recombinant plasmid must be ligated at like restriction sites: *Bam*HI to *Bam*HI and *Hin*dIII to *Hin*dIII. Recall that, in Laboratory 17, "Purification and Identification of λ Clones," numerous kinds of recombinant plasmids could potentially be identified. The situation is much simpler in this laboratory, however. Because the 784-bp fragment is the only insert available to be cloned into the restricted pBLU, all recombinant plasmids should be composed of even numbers of these two DNA sequences.
- Repeated copies of the same restriction fragment cannot exist adjacent to one another; that is, they must alternate with other fragments. Adjacent duplicate fragments form "inverted repeats"; in which the sequences, one on either side of the restriction site, are complementary along the entire length of the duplicated fragment. Molecules with such inverted repeats cannot replicate properly. As the plasmid opens up to allow access to DNA polymerase, the single-stranded regions on either side of the restriction site base pair to one another to form a large "hairpin loop," which fouls replication.

Use questions 1–6 to interpret each pair of miniprep results (M1+/– and M2+/–).

1. Examine the photograph of your stained gel. Compare your gel to the ideal gel on the next page, and label the size of the fragments in each lane.

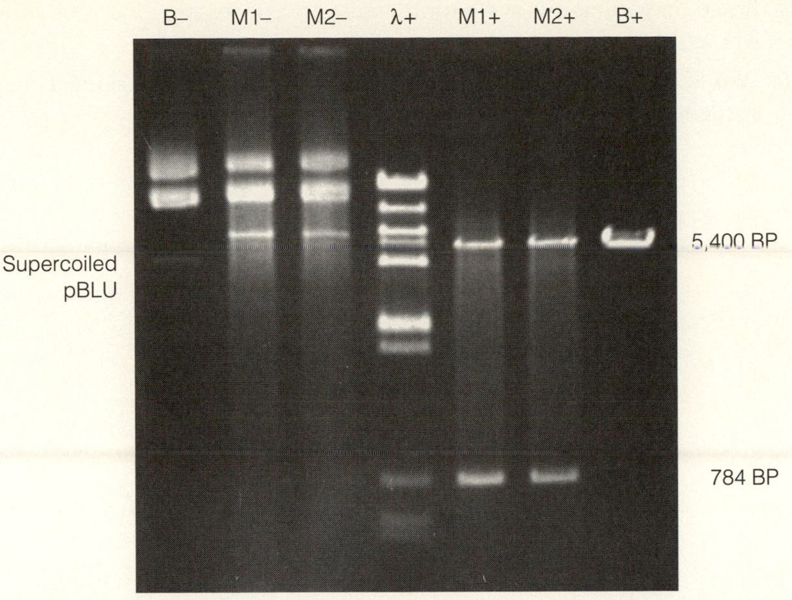

Ideal Gel

2. Label the band in the cut pBLU lane "5,400 bp." Then label the bands that appear in the cut λ lane. It is easiest to start with the smallest fragment at the bottom of the gel and then progress toward the sample well. Remember that you must account for 13 *Bam*HI/*Hin*dIII restriction fragments. Note that several thick bands are actually composed of multiple fragments, and the smallest fragment (125 bp) runs either off of the end of the gel or is too faint to be seen. A 14th fragment of 9,866 bp arises from digesting circular λ molecules joined at the COS sites.
3. *The cut miniprep lane (M+) provides information about the number and types of restriction fragments in the construct.* Every digested miniprep must contain the 5,400-bp pBLU backbone with the amp^r gene and the 784-bp λ fragment generated by restriction digestion of the PCR product. Locate these bands by comparing the cut miniprep lanes with the cut pBLU and λ lanes.
4. *The uncut miniprep lane (M–) provides information about the overall size of the construct.* Compare the M– lane (uncut miniprep) with the B– lane (uncut pBLU). Remember that the uncut plasmid can assume several conformations, but the fastest moving form is supercoiled.
   a. Locate the band that has migrated furthest in the B– lane; this is the supercoiled form of pBLU.
   b. Now, examine the band furthest down the M– lane. If this band is only slightly higher in the gel, then the molecule is most likely a simple recombinant containing the 784-bp fragment.
   c. Supercoiled, uncut miniprep DNA higher on the gel suggests a larger construct—most likely a four-piece superplasmid consisting of alternating units of the pBLU backbone and the 784-bp fragment.

5. Based on your above evaluation, make scale restriction maps of your M1 and M2 plasmids.
6. Would you expect the 784-bp fragment to have the same DNA sequence in each recombinant plasmid?

FOR FURTHER RESEARCH

Investigate the use of PCR and sequenced tagged sites (STSs) in the Human Genome Project.

Human DNA Fingerprinting

LABORATORY 22

Detection of an *Alu* Insertion Polymorphism by Polymerase Chain Reaction

In this experiment, polymerase chain reaction (PCR) is used to amplify a short region from chromosome 8, to look for an insertion of a short DNA sequence called *Alu* within the tissue plasminogen activator (TPA) gene. Although DNA from various individuals is more alike than different, many regions of human chromosomes exhibit a great deal of diversity. Such variable sequences are termed "polymorphic" (meaning *many forms*) and provide the basis for the diagnosis of genetic disease, forensic identification, and paternity testing.

The *Alu* family of short interspersed, repeated DNA elements are distributed throughout primate genomes. *Alu* elements are approximately 300 bp in length and derive their name from a single recognition site for the endonuclease *Alu*I located near the middle of the *Alu* element. Over the last 65 million years, the *Alu* element has amplified via an RNA-mediated transposition process to a copy number of about 500,000—comprising an estimated 5% of the human genome. *Alu* elements are thought to be derived from the 7SL RNA gene that encodes the RNA component of the signal recognition particle that functions in protein synthesis.

An estimated 500–2,000 *Alu* elements are restricted mostly to the human genome. A few of these have inserted recently, within the last 1 million years, and are not fixed in the human species. One such *Alu* element, called TPA-25, is found within an intron of the tissue plasminogen activator gene. This insertion is dimorphic: It is present in some individuals but not in others. Because the *Alu* sequence is found within an intron, it does not affect expression of the TPA gene, and it is phenotypically neutral. PCR can be used to screen individuals for the presence (or absence) of the TPA-25 insertion.

In this experiment, oligonucleotide primers, flanking the insertion site, are used to amplify a 400-bp fragment when TPA-25 is present and a 100-bp fragment when it is absent. Each of the three possible genotypes—homozygotes for the presence of TPA-25 (400-bp fragment only), homozygous for the absence of TPA-25 (100-bp fragment only), and heterozygotes (400-bp and 100-bp fragments)—are distinguished following electrophoresis in an agarose gel.

The source of template DNA is a sample of several thousand cells obtained by saline mouthwash (*a bloodless and noninvasive procedure).* The cells are collected by centrifugation and resuspended in a solution containing the resin "Chelex," which binds metal ions that inhibit the PCR reaction. The cells are then lyzed by boiling and centrifuged to remove cell debris. A sample of the supernatant containing genomic DNA is mixed with *Taq* polymerase, olignucleotide primers, the four deoxynucleotides, and the cofactor Mg^{++}. Temperature cycling is used to denature the target DNA, anneal the primers, and extend a complementary DNA strand. The "upstream primer," '5-GTAAGAGTTCCGTAACAGGACAGCT-3', brackets one side of the TPA locus; the "downstream primer," '5-CCC-CACCCTAGGAGAACTTCTCTTT-3' brackets the other side. The size of the amplification product(s) depends on the presence or absence of the *Alu* insertion at the TPA-25 locus on each copy of chromosome 8.

To compare the genotypes from several individuals, aliquots of the amplified sample and those of other experimenters are loaded into the wells of an agarose gel—along with DNA size markers and an unamplified control. Following electrophoresis and staining, amplification products appear as distinct bands in the gel; the distance moved from the well indicates the presence or absence of the TPA-25 insertion. A fuzzy band may be visible at the bottom of the gel lane. This band is called primer-dimer and is not an amplified human allele; it results from an association between the primers, giving rise to a primer-primer amplification product. The unamplified control should not display bands. It provides assurance that bands seen in the amplified gel lanes are true PCR products. A λ/*Bam*HI+*Hin*dIII digest is a convenient size marker, because it produces 493-bp and 125-bp bands that can be compared to the amplification products to check that they are in the expected size range. One or two bands are visible in each amplified sample, indicating that an individual is either homozygous or heterozygous for the *Alu* insertion.

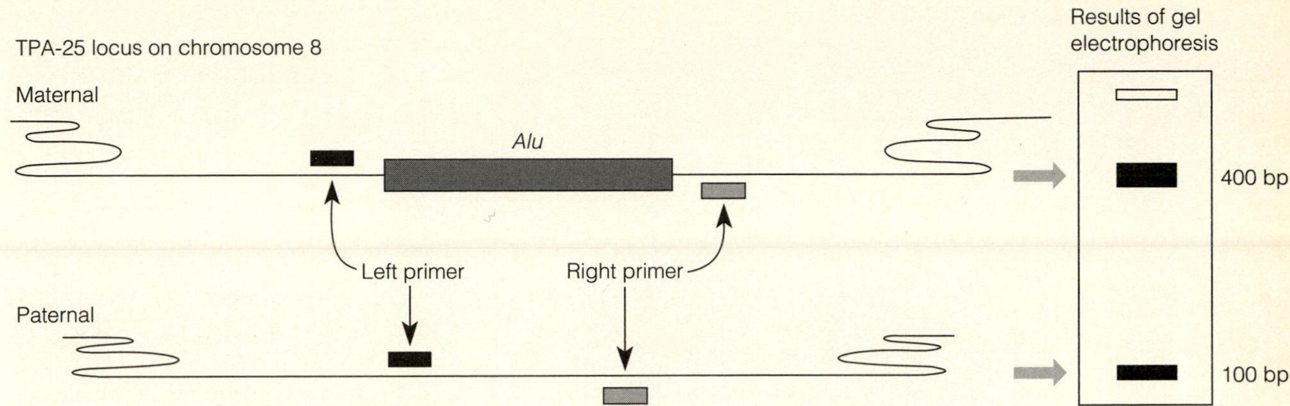

The TPA 25 locus contains a dimorphism reflecting the presence or absence of a 300-bp *Alu* sequence. When the *Alu* sequence is absent, amplification by PCR yields a 100-bp product. If the *Alu* sequence is present, the size of the PCR product is increased to 400 bp.

I. Isolate Cheek Cell DNA

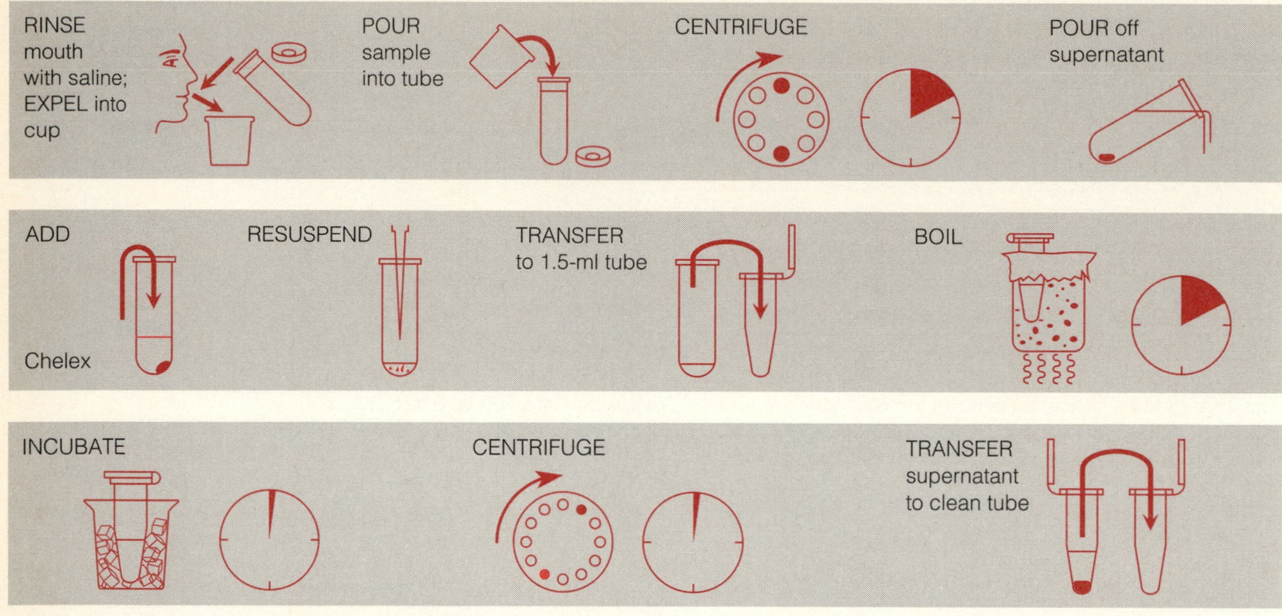

II. Set Up PCR Reaction and Amplify

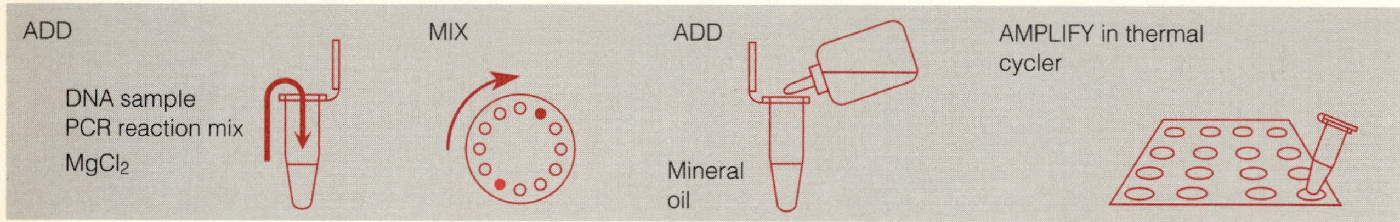

III. Cast 2.0% Agarose Gel

IV. Load Gel and Electrophorese

V. Stain Gel, View, and Photograph

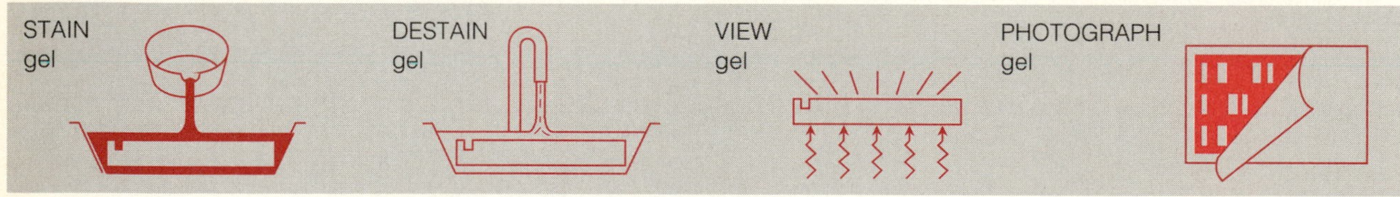

Laboratory 22
Detection of an *Alu* Insertion Polymorphism by Polymerase Chain Reaction

PRELAB NOTES

Review the prelab notes in Laboratory 18A, "PCR Amplification of a λ DNA Fragment."

The Use and Preparation of Chelex

The cell lysate obtained by boiling cheek cells is extremely crude and includes various cellular components that can interfere with PCR amplification. Iron derived from red blood cells and other heavy-metal ions are known to inhibit the *Taq* polymerase. Furthermore, magnesium ions are cofactors for cellular nucleases that can degrade genomic DNA. Therefore, the cell extract is treated with a negatively charged resin, Chelex, which binds the positive metal ions.

Chelex binds metal ions most effectively under alkaline conditions (pH 10.5). To ensure the proper pH, prepare a 10% (w/v) Chelex suspension using a 50-mM Tris base. Adjust the pH to 10.5 with concentrated sodium hydroxide.

The Effects of Food Particles

Although it is not advisable to eat immediately before conducting the experiment, food particles rinsed out with the mouthwash appear to have little effect on the amplification. However, fruit and vegetable particles—notably, from apples—can clog the pipet tips and make cell resuspension extremely difficult.

Centrifuge Requirements

A clinical centrifuge that develops 500–1,000 × *g* is sufficient for pelleting cheek cells.

Storing Cheek Cell Lysates

To limit DNA degradation by nucleases, cheek cell lysates should be stored on ice or frozen at –20°C until you are ready to set up the PCR reactions. Samples can be stored for weeks at –20°C, either as controls or as comparison alleles for future experiments.

PCR Mix and Nonspecific Priming

PCR reaction mix—composed of PCR buffer, dNTPs, primers, and *Taq* polymerase—will remain stable for at least 1 year *in the absence of Mg++ ions.* Target DNA and magnesium chloride are each added separately to the reaction mix, to initiate the reaction. Nonspecific priming begins almost immediately upon the addition of Mg^{++} ions, so be sure that the thermal cycler is programmed and that all experimenters have set up the PCR reactions coordinately (Section II). Work quickly, and initiate thermal cycling as soon as possible after adding the reagents. To ensure maximum specificity, some experimenters employ a "hot start" technique: PCR reactions missing a vital component, such as template DNA, are initiated by addition of the missing component to the reaction tube at an elevated temperature.

Thermal Cycling

Amplification of TPA-25 alleles from crude cell extracts is biochemically demanding and requires the precision of automated thermal cycling. Manual thermal cycling can work but is not recommended.

DNA Size Markers

It is advisable to include DNA size markers on each gel. Markers help to verify that the PCR product is the correct size, and they also serve as a control to aid in trouble-shooting problems with amplification, electrophoresis, and staining. For example, a very faint PCR product, compared to the DNA markers, suggests that additional amplification cycles are needed. A *Bam*HI/*Hin*dIII digest of λ DNA provides a convenient size marker for this experiment.

Student Allele Database

An archival database of student TPA-25 alleles, workspace for statistical analysis, and a shared bulletin board are maintained at the DNA Learning Center of the Cold Spring Harbor Laboratory. Student allele data can be evaluated there and compared to data from around the world. The student allele database can be reached at the World Wide Web address: (URL) http: //darwin.cshl.org.

Safety of the Saline Mouthwash

Cell collection using a sterile saline mouthwash is painless, bloodless, and noninvasive. The risk of spreading an infectious agent via the mouthwash procedure is probably far less than from natural atomizing processes, such as coughing or sneezing. Several elements of the experiment further minimize any risk of spreading an infectious agent that might be present in mouthwash samples:

- Each experimenter works only with his or her own sample.
- The sample is effectively sterilized during a 10-minute incubation in boiling water.
- There is no culturing of the mouthwash sample that might enrich it for pathogens.

Disclosure and Confidentiality

The TPA-25 dimorphism was specifically selected for use in this laboratory because it is phenotypically neutral. TPA-25 insertions reside within an intron and do not affect expression of the tissue plasminogen activator protein; they have no known relationship to disease states, sex determination, or any other human phenotype. Although there is no chance of disclosing phenotypic information about the experimenters, the confidentiality of student TPA-25 genotypes can be maintained by identifying student samples only by numbers on gels. Student submissions to the student allele database have no personal identifiers.

However, TPA-25 insertions are inherited in a Mendelian fashion and can give indications about family relationships. Technically, a single-locus dimorphism, such as TPA-25, could never definitively prove or disprove relatedness; however, to avoid the possibility of discovering inconsistent

TPA-25 inheritance, it is best not to generate genotypes from siblings or other family members.

For Further Information

The protocol presented here is based on the following published methods:

Batzer, M. A. and Deininger, P. L. 1991. A human-specific subfamily of *Alu* sequences. *Genomics* 9: 481.

Lench, N., P. Stainer, and R. Williamson. 1988. Simple non-invasive method to obtain DNA for gene analysis. *The Lancet* June 18, 1988: 1356.

Perna, N. T., Batzer, M. A., Deininger, P. L., and Stoneking, M. 1992. *Alu* insertion polymorphism: A new type of marker for human population studies. *Human Biology* 64: 641.

Saiki, R. K., S. J. Scharf, F. Faloona, K. B. Mullis, G. T. Horn, H. A. Erlich, and N. Arnheim. 1985. Enzymatic amplification of β-globin sequences and restriction site analysis for diagnosis of sickle cell anemia. *Science* 239: 1350.

Sharp, P. A., B. Sugden, and J. Sambrook. 1973. Detection of two restriction endonuclease activities in *Haemophilus parainfluenzae* using analytical agarose–ethidium bromide electrophoresis. *Biochemistry* 12: 3055.

Singer-Sam, J., R. L. Tanguay, and A. D. Riggs. 1989. Use of Chelex to improve the PCR signal from a small number of cells. *Amplifications* 1,3: 11.

DETECTION OF AN *Alu* INSERTION POLYMORPHISM BY POLYMERASE CHAIN REACTION

Reagents

10% Chelex
0.9% sodium chloride
PCR reaction mix
25 mM $MgCl_2$
mineral oil
λ/*Bam*HI + *Hin*dIII
distilled water
loading dye
2% agarose
1× Tris/Borate/EDTA (TBE) buffer
1 μg/ml ethidium bromide

for decontamination:
0.05 M $KMnO_4$
0.25 N HCl
0.25 N NaOH

Equipment and Supplies

100–1,000-μl micropipettor + tips
10–100-μl micropipettor + tips
0.5–10-μl micropipettor + tips
15-ml culture tube
1.5-ml tubes
0.5-ml PCR tubes
beaker for waste/used tips
boiling water bath
disposable gloves
DNA thermal cycler
forceps
paper cup
60°C water bath (for agarose)

I. Isolate Cheek Cell DNA

(30 minutes)

1. Use a permanent marker to label your name on a 15-ml culture tube containing saline solution.
2. Pour all of the saline solution into your mouth, and vigorously swish for 10 seconds. *Save the empty 15-ml tube for reuse in the next step.*
3. Expel the saline mouthwash into a paper cup. Then, carefully pour the saline mouthwash from the paper cup back into the 15-ml tube from step 1.
4. *Securely close the cap of the tube,* and place the mouthwash tube in a *balanced* configuration with other tubes in the rotor of a clinical centrifuge. Centrifuge at 500–1,000 × *g* for 10 minutes to pellet the cells on the bottom of the culture tube.

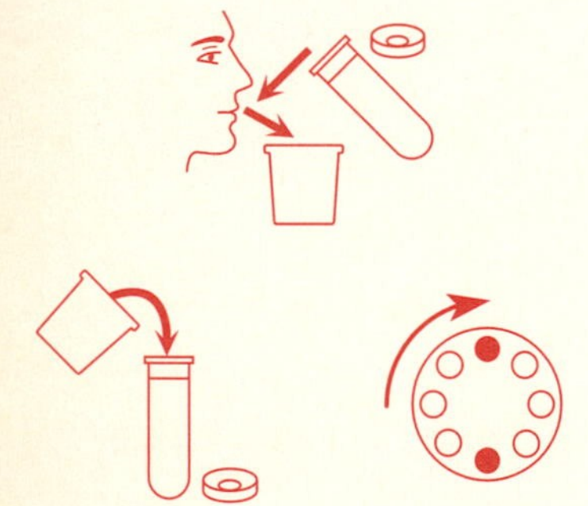

"Blank" tubes with 10 ml of water can be used for balance, if needed. 500–1,000 × *g* corresponds to 2,000–3,000 rpm for most tabletop clinical centrifuges. A small cell pellet should be visible at the bottom of the tube.

5. *Being careful not to disturb the cell pellet,* pour off as much supernatant as possible into the sink or paper cup. Place the tube with the mouthwash cell pellet on ice.

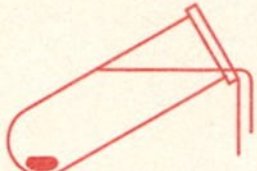

If the cell pellet does not adhere to the centrifuge tube, pour off as much supernatant as possible and then use a 100–1,000-µl micropipettor to carefully remove additional supernatant.

6. Use a micropipettor to add 500 µl of 10% Chelex to the cell pellet:
   a. Pipet Chelex solution in and out of the pipet tip several times to suspend the Chelex beads.
   b. Before the Chelex has had a chance to settle, transfer 500 µl to the culture tube.

7. Resuspend the cells in the Chelex by pipetting up and down several times. Hold the tube up to the light to confirm that no visible clumps of cells remain.

Make sure that the resin beads are well suspended before adding Chelex to the cheek cells.

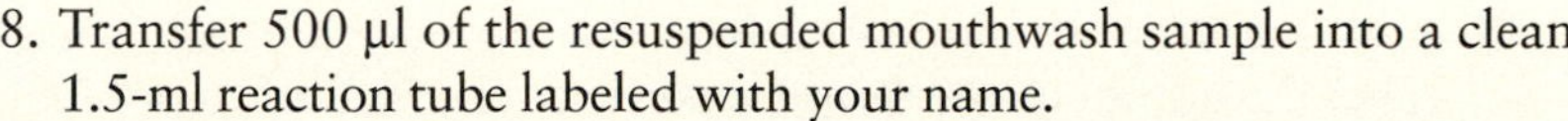

8. Transfer 500 µl of the resuspended mouthwash sample into a clean 1.5-ml reaction tube labeled with your name.

If food particles clog the pipet tip, use scissors to snip off the end of the tip, and resuspend the sample.

9. Incubate the 1.5-ml sample tube in a boiling water bath for 10 minutes.

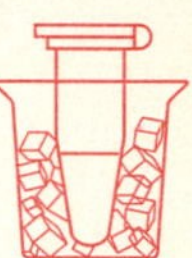

10. Following incubation, use forceps to remove the sample tube from the boiling water bath, and cool the tube on ice for approximately 1 minute.

11. Place the sample tube in a *balanced* configuration in a microfuge rotor, and spin for 30 seconds to pellet the Chelex beads at the bottom of the tube.

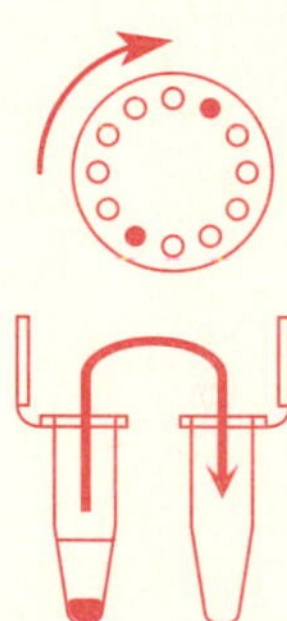

12. Transfer 200 µl of the supernatant to a fresh 1.5-ml tube labeled with your name, and place the tube on ice. *Avoid transferring any of the Chelex pellet.*

Stop Point

Store the cheek cell DNA samples on ice or freeze them at –20°C until you are ready to continue. Samples can be stored for weeks at –20°C, either as controls or as comparison alleles for future experiments.

PCR is very sensitive! Always wear disposable gloves to help prevent contamination.

A label placed on the side of the tube may come off during thermal cycling.

Nonspecific priming begins almost immediately upon the addition of magnesium, so work quickly, and initiate thermal cycling as soon as possible after adding the reagents. It is not necessary to mix the reagents, as long as they are consolidated at the bottom of the tube; heat from the first denaturing cycling will adequately mix the reagents.

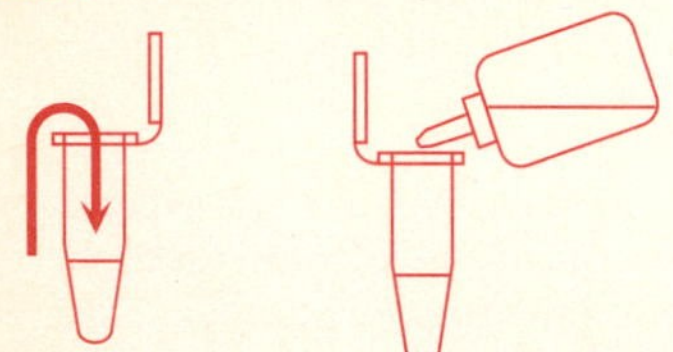

Some thermal cyclers do not require that mineral oil be added to the samples.

II. Set Up PCR Reaction and Amplify

(15 minutes)

1. Use a permanent marker to label the cap of a 0.5-ml PCR tube with your initials.
2. Use the matrix below as a checklist while adding reagents to the 0.5-ml PCR tube. *Use a fresh pipet tip for each reagent.*

Tube	Cheek Cell DNA	PCR Reaction Mix	$MgCl_2$
Name	5 µl	40 µl	5 µl

3. Add one drop of mineral oil from the dropper bottle to the PCR tube. Do not touch the dropper bottle to the PCR tube; subsequent reactions can become contaminated with your preparation.
4. Close the tube cap tightly, and amplify the sample using automated thermal cycling. Recheck the program, then start.

   30 cycles of step file (linkage to 4°C soak file optional):

   94°C for 1 minute

   58°C for 2 minutes

   72°C for 2 minutes

Refer to Laboratory 3, "DNA Restriction and Electrophoresis," for detailed instructions on casting and loading gel.

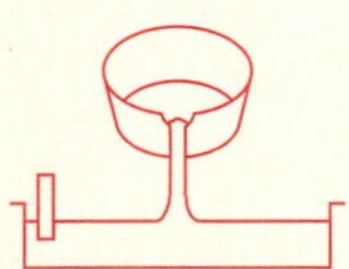

III. Cast 2.0% Agarose Gel

(15 minutes)

1. Carefully pour agarose solution into the gel-casting tray to fill it to a depth of about 5 mm. The gel should cover about one-third of the height of the comb teeth.
2. After the agarose solidifies, place the casting tray into the electrophoresis box, and set up for electrophoresis.

Stop Point

Cover the electrophoresis box, and save the gel until you are ready to continue. The gel will remain in good condition for several days, if it is completely submerged in buffer.

IV. Load Gel and Electrophorese

(20 minutes; then 30 minutes of electrophoresis)

1. Use a permanent marker to label a clean 1.5-ml tube with your name.
2. Transfer 10 µl of amplification product from the PCR reaction tube into the clean 1.5-ml tube. *Take care not to transfer any mineral oil into the clean tube.*

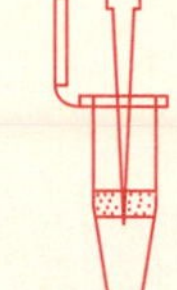

   a. Dip the pipet tip through the mineral oil on top of the PCR reaction tube, and carefully withdraw 10 µl of amplification product.
   b. Before transferring, use a Kimwipe or a paper towel to remove any mineral oil clinging to the outside of the pipet tip.

Remaining amplified samples can be stored for weeks at –20°C, either as controls or as comparison alleles for future experiments.

3. Designate one person to load the unamplified control and DNA size markers *for each gel,* as described below:
   a. Obtain one unamplified control and one aliquot of λ/*Bam*HI+*Hin*dIII size markers from your instructor.
   b. Add 1 µl of loading dye to the unamplified control sample and to the λ/*Bam*HI+*Hin*dIII size markers. Close the tube tops and mix the solution.

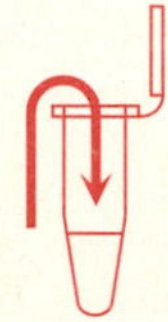

   c. Load 10 µl of λ/*Bam*HI+*Hin*dIII size markers into an outside lane of gel and 10 µl of unamplified control into the adjacent gel lane.
4. Add 1 µl of loading dye to the 10 µl of your PCR sample. Close the top of the tube and mix the solution.
5. Load 10 µl of your PCR reaction into a sample well in the gel. Note your lane position, counting from left to right on the gel.

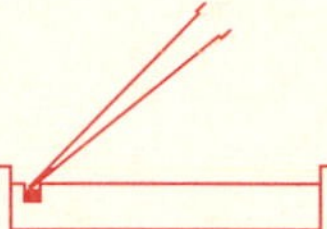

Other experimenters' samples are loaded in lanes of the same gel. Prepare a key to identify all samples.

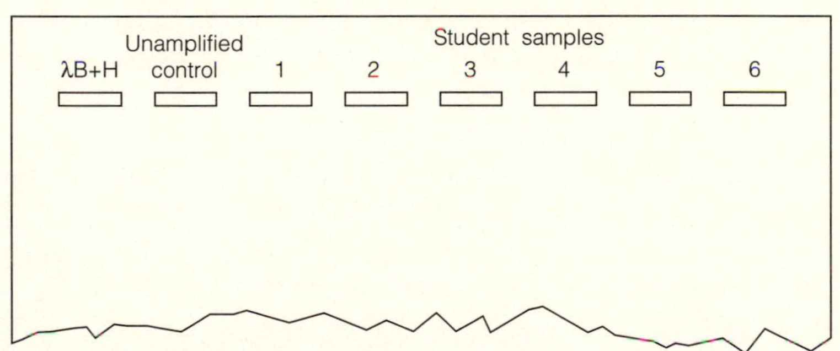

6. Electrophorese at 100 volts for about 30 minutes. Adequate separation has occurred when the bromophenol blue bands have moved 30 mm from the wells.

7. Turn off the power supply, remove the casting tray from the electrophoresis box, and transfer the gel to a disposable weigh boat (or other shallow tray) for staining.

V. Stain Gel with Ethidium Bromide, View, and Photograph

Staining can be performed by an instructor in a controlled area, when students are not present.

(10–15 minutes)

CAUTION

Review the section on "Ethidium Bromide Staining and Responsible Handling" in Laboratory 3. Wear disposable gloves when staining, viewing, and photographing gel and during cleanup. Confine all staining to a restricted sink area.

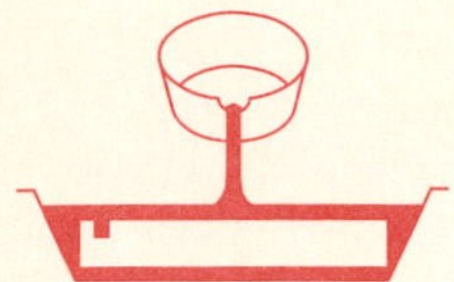

1. Flood the gel with ethidium bromide solution (1 µg/ml), and allow it to stain for 20–30 minutes.
2. Rinse and destain the gel in tap water for 20–30 minutes.
3. View the gel under an ultraviolet transilluminator or other UV light source.

The relatively high agarose concentration used in this experiment requires slightly extended staining and destaining times. Destaining leaches unbound ethidium bromide from the gel, thus increasing the contrast between the stained alleles and the "background" ethidium bromide in the gel.

CAUTION

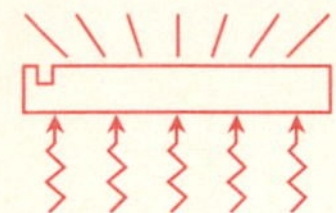

Ultraviolet light can damage your eyes. Never look directly at an unshielded UV light source without eye protection. View only through a filter or safety glasses that absorb the harmful wavelengths.

4. Make an exposure of 1–2 seconds with your camera set at f/8. Develop the print for the recommended time (approximately 45 seconds at room temperature).
5. Take the time for proper cleanup:
   a. Wipe down the camera, the transilluminator, and the staining area.
   b. Decontaminate the gel and any staining solution that will not be reused.
   c. Wash your hands before leaving the laboratory.

RESULTS AND DISCUSSION

1. Observe the photograph of the stained gel containing your sample and those from other experimenters. Orient the photograph with the sample wells at the top. First, ascertain whether you can see a diffuse (fuzzy) band of primer-dimer appearing at the same position in each lane toward the bottom of the gel. Primer-dimer is not amplified human DNA but is an artifact of the PCR reaction that results from the primers amplifying themselves. Excluding primer-dimer, interpret the allele bands in each lane of the gel:

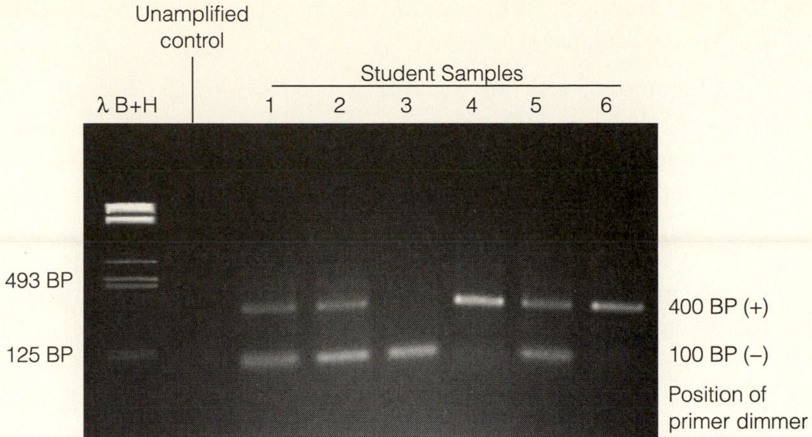

Ideal Gel

a. *No bands are visible:* This usually results from an error during sample preparation, such as losing the cell pellet or using a Chelex solution that is too acidic.

b. *One band is visible:* Compare its migration to that of the 493-bp and 125-bp bands in the λ/*Bam*HI+*Hin*dIII lane. If the PCR product migrates slightly ahead of the 493-bp band, that individual is then homozygous for the TPA-25 *Alu* insertion (+/+). If the PCR product migrates well ahead of the 493-bp band and just ahead of the 125-bp band, then that individual is homozygous for the *absence* of the TPA-25 *Alu* insertion (–/–).

c. *Two bands are visible:* Compare the migration of each PCR product to that of the 493-bp and 125-bp bands in the λ/*Bam*HI+*Hin*dIII lane. Confirm that one PCR product corresponds to a size of about 400 bp and that the other PCR product corresponds to a size of about 100 bp. This individual is *heterozygous* for the TPA-25 *Alu* insertion (+/-).

d. *Three or more bands are visible:* The one or two bright bands are most likely the true alleles. Additional bands can occur when the primers bind nonspecifically to chromosomal loci other than TPA-25 and give rise to nonspecific amplification products.

2. Determine the genotype distribution for the class—that is, count how many students are (+/+), (+/–), and (–/–).

3. An allelic frequency is a ratio comparing the number of copies of a particular allele to the total number of alleles present. This means that the homozygous state for the allele is counted twice, and the heterozygous state is counted once. Imagine a class of 100 students, listing their genoytpe distribution as follows:

+/+	20
+/–	50
–/–	30

Because humans are diploid, the total number of alleles present is 2 × 100 = 200. The allelic frequency for the presence of the TPA-25 insertion (+) is calculated as

2 × 20 (homozygotes) + 50 (hetrozygotes) / 200 = 90 / 200 = 0.45

Likewise, the allelic frequency for the absence of the TPA-25 insertion (–) is calculated as

2 × 30 (homozygotes) + 50 (heterozygotes) / 200 = 110 / 200 = 0.55

Using the genotype distribution from your class, calculate the allelic frequencies for the presence and absence of the TPA-25 insertion.

4. If a population is genetically stable, the allelic frequencies will remain constant from one generation to the next. Such a population is said to be in Hardy-Weinberg equilibrium. Once the allelic frequencies have been determined, the distribution of genotypes is described by the equation

$$p^2 + 2pq + q^2 = 1$$

where p and q represent the allelic frequencies, p^2 and q^2 represent the homozygous genotype frequencies, and 2pq represents the heterozygous genotype frequency. For example, in step 3 the allelic frequency for the TPA-25 insertion (+) was calculated to be 0.45, while the allelic frequency for the absence of the TPA-25 insertion (–) was 0.55. If we let p = presence of TPA-25 insertion and q = absence of TPA-25 insertion, then the expected genotype is as follows:

expected genotypic frequency for TPA-25 insertion
= $p^2 = (0.45)^2 = 0.2025$ or 20.25 out of 100 students

expected genotypic frequency for the absence of the TPA-25 insertion
= $q^2 = (0.55)^2 = 0.3025$ or 30.25 out of 100 students

Use the allelic frequencies calculated for your class in step 3 to determine the expected genotype frequencies. How do they compare with the actual genotype frequencies? How can you account for differences?

5. Differences between observed genotype frequencies and expected genotype frequencies may be due to chance or may just indicate that the sample population is not in Hardy-Weinberg equilibrium. To help you decide which of these explanations is more likely, we employ a statistic called the chi square (χ^2). This analysis takes into account that smaller samples will deviate more from the ideal expected result than larger samples. The χ^2 is defined as the sum of the (d^2/e) factors, where d is the deviation of the observed from the expected and e is the expected value. Referring to the example presented in step 3, χ^2 is calculated as follows:

Observed – Expected = Deviation	d^2	d^2/e
20 – 20.25 = –0.25	0.0625	0.003125
50 – 49.50 = 0.50	0.2500	0.005000
30 – 30.25 = –0.25	0.0625	0.002083

$$\chi^2 = 0.003125 + 0.005000 + 0.002083 = 0.010208$$

To interpret the meaning of this χ^2 value, we locate it on the χ^2 table on the next page. When using a chi-square (χ^2) table, the variable *n* refers to *degrees of freedom,* defined as the number of phenotypic classes minus one. In our example, there are two phenotypic classes, the presence and the absence of the TPA-25 insertion. Therefore, we use one degree of freedom.

Chart of values for chi-square (χ^2)

n	p = 0.99	0.98	0.95	0.90	0.80	0.70	0.50	0.30	0.20	0.10	0.05	0.02	0.01
1	0.000157	0.00628	0.00393	0.0158	0.0642	0.148	0.455	1.074	1.642	2.706	3.841	5.412	6.635

As seen in the table, our χ^2 value falls between probability values of 0.90 and 0.95. This means that, 90–95% of the time, deviations this great or greater would be expected by chance alone. Therefore, the hypothesis that our example population is in Hardy-Weinberg equilibrium still stands.

Again, using the data from your class, calculate a χ^2 value, and test the hypothesis that the class as a population is in Hardy-Weinberg equilibrium to within a 5% level of significance.

6. Under what conditions does a population come into Hardy-Weinberg equilibrium? Does your class population meet these criteria? Where might you find populations that meet the conditions necessary for Hardy-Weinberg equilibrium?
7. What conditions for Hardy-Weinberg equilibrium does the TPA-25 polymorphism satisfy?
8. How would you determine the error rate of *Taq* polymerase, without sequencing individual amplification products?

LABORATORY 23

Detection of a VNTR Polymorphism by Polymerase Chain Reaction

In this experiment, polymerase chain reaction (PCR) is used to amplify a short nucleotide sequence from chromosome 1, to create a personal DNA fingerprint. Although the DNA from various individuals is more alike than different, many regions of human chromosomes exhibit a great deal of diversity. Such variable sequences are termed "polymorphic" (meaning *many forms*) and provide the basis for the diagnosis of genetic disease, forensic identification, and paternity testing. Many DNA polymorphisms are found within the estimated 90% of the human genome that does not code for protein. A special type of polymorphism, called a variable number of tandem repeats (VNTR), is composed of repeated copies of a DNA sequence that lie adjacent to one another on the chromosome. The VNTR amplified in this experiment, D1S80, is located in a noncoding region of chromosome 1 and has a repeat unit of 16 base pairs. At the D1S80 locus, most individuals have alleles containing between 14 and 40 repeats, which are inherited in Mendelian fashion on the maternal and paternal copies of chromosome 1.

The source of template DNA for amplification is a sample of several thousand cheek cells obtained by saline mouthwash (*a bloodless and non-invasive procedure*). The cells are collected by centrifugation and resuspended in a solution containing the resin "Chelex," which binds metal ions that inhibit the PCR reaction. The cells are then lyzed by boiling and centrifuged to remove cell debris. A sample of the supernatant containing chromosomal DNA is mixed with *Taq* polymerase, oligonucleotide primers, the four deoxynucleotides, and the cofactor Mg^{++}. As in Labs 18A and 22, temperature cycling is used to denature the target DNA, anneal the primers, and extend a complementary DNA strand. The "upstream" primer, '5-GAAACTGGCCTCCAAACACTGCCCGCCG-3', brackets one side of the D1S80 locus; the "downstream" primer, '5-GTCTTGTTG-GAGATGCACGTGCCCCTTGC-3', brackets the other side. The size of the amplification product(s) depends on the number of copies of the VNTR sequence at D1S80 on each copy of chromosome 1.

To compare a variety of D1S80 alleles, aliquots of the amplified sample and those from other experimenters are loaded into wells of an agarose or polyacrylamide gel along with DNA size markers and an unamplified control. Following electrophoresis and staining, different alleles appear as distinct bands in the gel; the distance moved from the well is inversely proportional to the number of repeat units at D1S80. One or two bands are visible in each lane, indicating that an individual is either homozygous or heterozygous for the D1S80 locus.

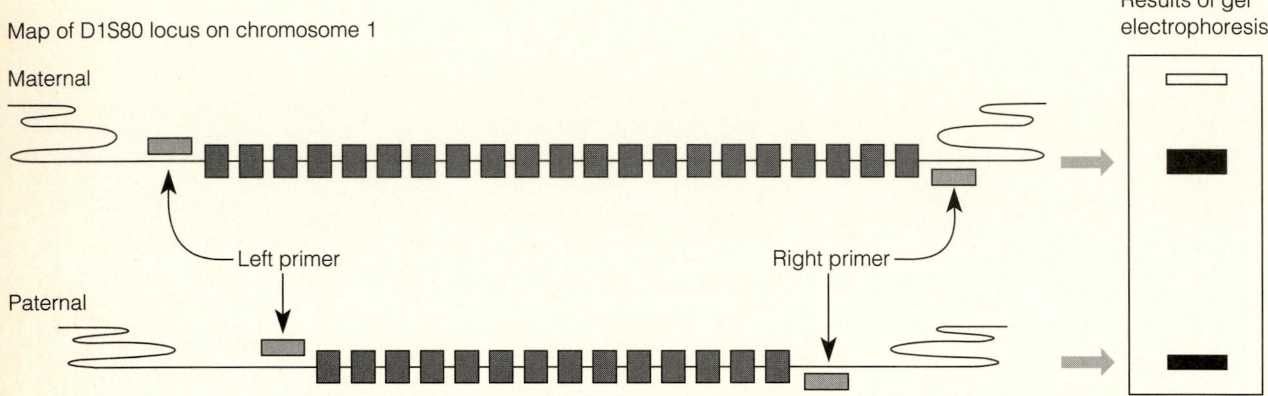

I. Isolate Cheek Cell DNA

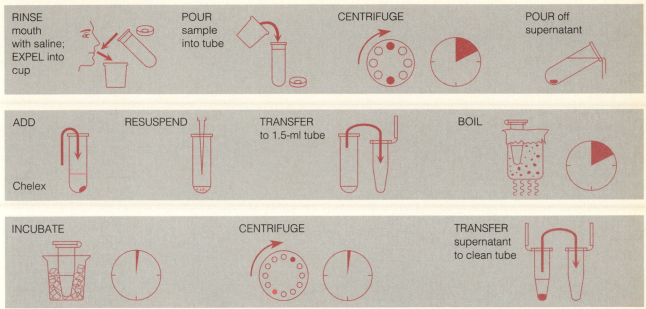

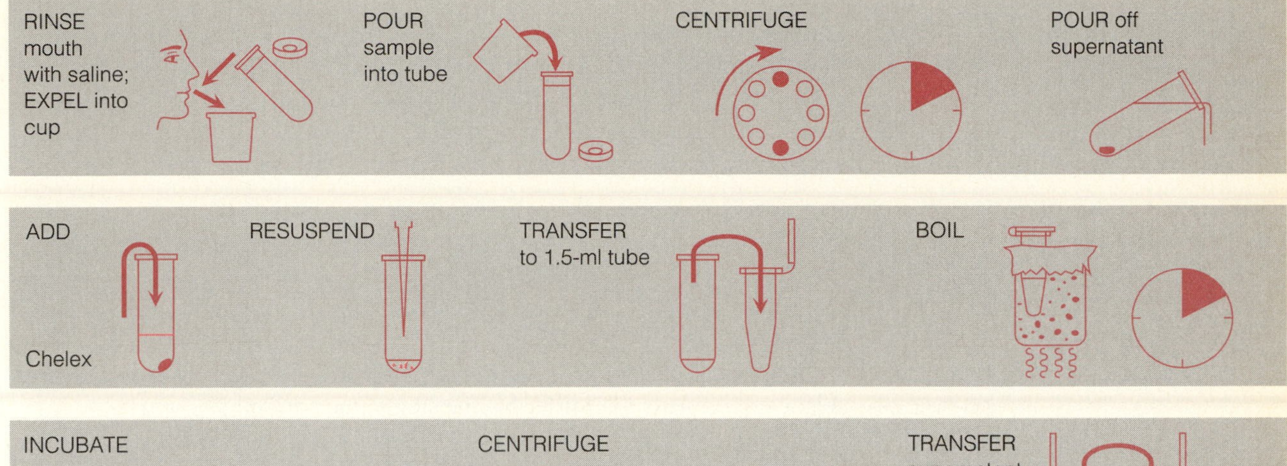

II. Set Up PCR Reaction and Amplify

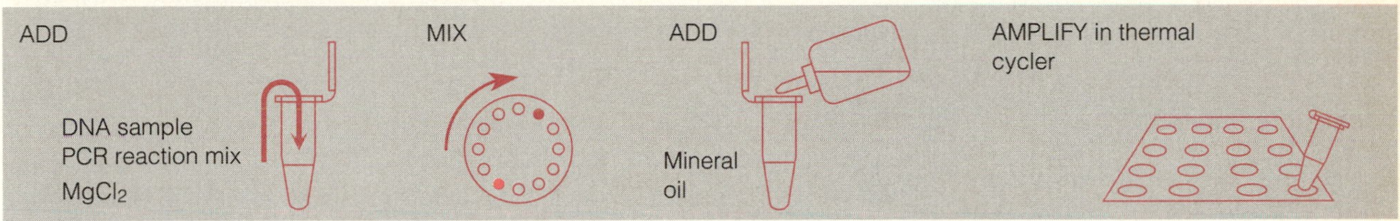

III. Cast Agarose or Polyacrylamide Gel

IV. Load Gel and Electrophorese

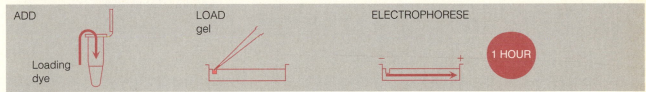

V. Stain Gel, View, and Photograph

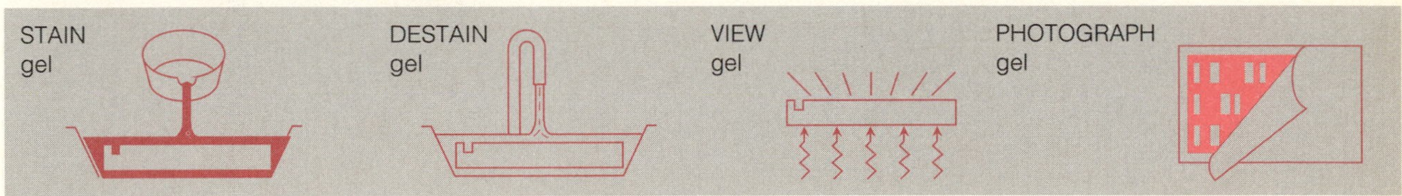

PRELAB NOTES

Review the prelab notes in Laboratory 18A, "PCR Amplification of a λ DNA Fragment" and Laboratory 22, "Detection of an *Alu* Insertion Polymorphism by Polymerase Chain Reaction."

Electrophoresis Options and Limitations

D1S80 is one of several single-locus polymorphisms used in forensic DNA typing. Although the 29 known D1S80 alleles range in size from 369 to 801 base pairs, individual alleles can differ by as little as one repeat unit (16 base pairs). Only polyacrylamide gel electrophoresis is capable of reproducibly resolving alleles with such a small size difference. (Recall that, under the conditions for DNA sequencing, polyacrylamide can even resolve fragments that differ by one single nucleotide.)

Agarose gel electrophoresis is a simple alternative. Although it cannot resolve alleles that differ by one single repeat unit, it does provide adequate separation to illustrate genetic diversity. The poor resolving ability of agarose does result in an apparently higher rate of homozygosity: Heterozygotes with alleles of similar size show one single band on an agarose gel.

The high-resolution agarose NuSieve 3:1, by FMC BioProducts, offers an excellent compromise, providing the simplicity of agarose with allele separation approaching that of polyacrylamide.

DNA Ladders

A DNA "ladder" composed of regularly spaced fragments is run, along with student samples, to aid in the identification of D1S80 alleles amplified by PCR. A ladder of 1–34 repeats of a 123-bp sequence of the rat prolactin gene is suitable for use in agarose systems. A ladder of actual D1S80 alleles, included with a forensic kit from the Perkin-Elmer Corporation, should be used in polyacrylamide systems where the object is to determine accurate D1S80 genotypes.

Thermal Cycling

Amplification of D1S80 alleles from crude cell extracts is biochemically demanding and requires the precision of automated thermal cycling. Manual thermal cycling is not recommended.

The Safety of the Saline Mouthwash

Cell collection using sterile saline mouthwash is painless, bloodless, and noninvasive. The risk of spreading an infectious agent by the mouthwash procedure is far less likely than from natural atomizing processes, such as coughing or sneezing. Several elements further minimize the risk of spreading an infectious agent that might be present in mouthwash samples:

- Each experimenter works only with his or her own sample.
- The sample is effectively sterilized during a 10-minute incubation in boiling water.
- There is no culturing of the mouthwash sample that might enrich for pathogens.

Disclosure and Confidentiality

A VNTR polymorphism was specifically selected for use in this laboratory because it is phenotypically neutral. D1S80 alleles result from chance recombination and do not encode protein: They have no known relationship to disease states, sex determination, or any other human phenotype. Although there is no chance of disclosing phenotypic information about the experimenters, the confidentiality of student D1S80 genotypes can be maintained by identifying student samples only by the numbers on the gels. Student submissions to the student allele database have no personal identifiers.

However, D1S80 polymorphisms are inherited in a Mendelian fashion and can give indications about family relationships. Technically, a single-locus polymorphism, such as D1S80, could never definitively prove or disprove relatedness; however, to avoid the possibility of discovering inconsistent D1S80 inheritance, it is best not to generate genotypes from siblings or other family members.

For Further Information

The protocol presented here is based on the following published methods:

Allen, R., B. Budowle, R. Chakraborty, A. Giusti, and A. Eisenberg. 1991. Analysis of the VNTR locus *AmpliFLP* D1S80 by the PCR followed by high resolution PAGE. *American Journal of Human Genetics* 48: 137.

Gill, P. A., A. J. Jeffreys, and D. J. Werrett. 1985. Forensic application of "DNA fingerprints." *Nature* 318: 577.

Lench, N., P. Stanier, and R. Williamson. 1988. Simple non-invasive method to obtain DNA for gene analysis. *The Lancet* 1: 1356.

Nakamura, Y., M. Carlson, K. Krapco, and R. White. 1988. Isolation and mapping of a polymorphic DNA sequence (pMCT118) on chromosome 1p[D1S80]. *Nucleic Acids Research* 16: 9364.

Saiki, R. K., S. J. Scharf, F. Faloona, K. B. Mullis, G. T. Horn, H. A. Erlich, and N. Arnheim. 1985. Enzymatic amplification of beta-globin sequences and restriction site analysis for diagnosis of sickle cell anemia. *Science* 239: 1350.

Sharp, P. A., B. Sugden, and J. Sambrook. 1973. Detection of two restriction endonuclease activities in *Haemophilus parainfluenzae* using analytical agarose–ethidium bromide electrophoresis. *Biochemistry* 12: 3055.

Singer-Sam, J., R. L. Tanguay, and A. D. Riggs. 1989. Use of Chelex to improve the PCR signal from a small number of cells. *Amplifications* 1,3: 11.

DETECTION OF A VNTR POLYMORPHISM BY POLYMERASE CHAIN REACTION

Reagents

10% Chelex
0.9% sodium chloride
PCR Mix
25 mM of magnesium chloride
mineral oil
DNA size markers
distilled water
loading dye
1.5% agarose *or* 2.5% NuSieve 3:1 *or* 6% polyacrylamide
1× Tris/Borate/EDTA (TBE) buffer
1 µg/ml ethidium bromide

for decontamination:
0.05 M $KMnO_4$
0.25 N HCl
0.25 N NaOH

Equipment and Supplies

100–1,000-µl micropipettor + tips
10–100-µl micropipettor + tips
0.5–10-µl micropipettor + tips
15-ml culture tube
1.5-ml tubes
0.5-ml PCR tubes
beaker for waste/used tips
boiling water bath
DNA thermal cycler
disposable gloves
forceps
paper cup
60°C water bath (for agarose)

I. Isolate Cheek Cell DNA

(30 minutes)

1. Use a permanent marker to label a 15-ml culture tube containing saline solution with your name.
2. Pour all of the saline solution into your mouth, and vigorously swish for 10 seconds. *Save the empty 15-ml tube for reuse in the next step.*
3. Expel the saline mouthwash into a paper cup. Then carefully pour the saline mouthwash from the paper cup back into the 15-ml tube from step 1.
4. *Securely close the cap,* and place the mouthwash tube in a *balanced* configuration with other tubes in the rotor of a clinical centrifuge. Centrifuge at 500–1,000 × *g* for 10 minutes to pellet the cells on the bottom of the culture tube.

"Blank" tubes containing 10 ml of water can be used for balance, if needed. 500–1,000 × *g* corresponds to 2,000–3,000 rpm for most tabletop clinical centrifuges. A small cell pellet should be visible at the bottom of the tube.

5. *Taking care not to disturb the cell pellet,* pour off as much supernatant as possible into a sink or paper cup. Place the tube with the mouthwash cell pellet on ice.

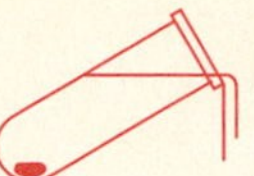

If the cell pellet does not adhere to the centrifuge tube, pour off as much supernatant as possible, then use a 100–1,000-µl micropipettor to carefully remove the remaining supernatant.

6. Use a micropipettor to add 500 µl of 10% Chelex to the cell pellet:
   a. Pipet the Chelex solution in and out of the pipet tip several times to suspend the Chelex beads.
   b. Before the Chelex has had a chance to settle, transfer the 500 µl to the culture tube.
7. Resuspend the cells in the Chelex by pipetting up and down several times. Hold the tube up to the light to confirm that no visible clumps of cells remain.

Make sure that the resin beads are well suspended before adding Chelex to the cheek cells.

8. Transfer 500 µl of the resuspended mouthwash sample into a clean 1.5-ml reaction tube labeled with your name.

If food particles clog the pipet tip, use scissors to snip off the end of the tip, and resuspend the sample.

9. Incubate the 1.5-ml sample tube in a boiling water bath for 10 minutes.

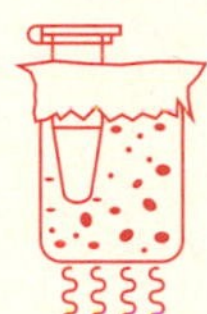

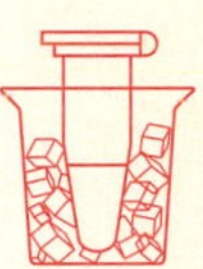

10. Following incubation, use forceps to remove the sample tube from the boiling water bath, and cool the tube on ice for approximately 1 minute.
11. Place the sample tube in a *balanced* configuration in a microfuge rotor, and spin for 30 seconds to pellet the Chelex beads at the bottom of the tube.

12. Transfer 200 µl of the supernatant to a fresh 1.5-ml tube labeled with your name, and place the tube on ice. *Avoid transferring any of the Chelex pellet.*

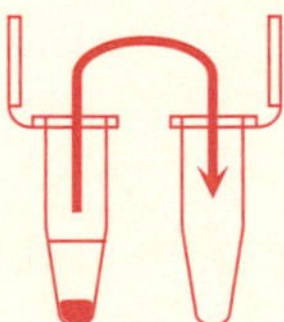

Stop Point

Store the cheek cell DNA samples on ice or freeze them at –20°C until you are ready to continue. The samples can be stored for months at –20°C, to be used either as controls or as comparison alleles for future experiments.

PCR is very sensitive! Always wear disposable gloves to help prevent contamination.

A label placed on the side of the tube will come off during thermal cycling.

Nonspecific priming begins almost immediately upon the addition of magnesium, so work quickly, and initiate thermal cycling as soon as possible after adding the reagents. You do not need to mix the reagents, as long as they are consolidated at the bottom of the tube. Heat from the first denaturing cycling will adequately mix the reagents.

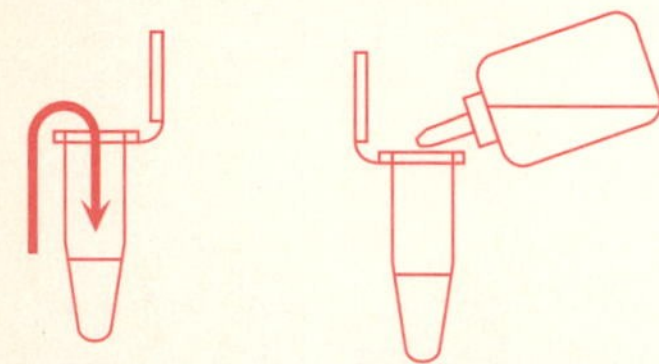

II. Set Up PCR Reaction and Amplify

(15 minutes)

1. Use a permanent marker to label the cap of a 0.5-ml PCR tube with your initials.
2. Use the matrix below as a checklist while adding reagents to the 0.5-ml PCR tube. *Use a fresh pipet tip for each reagent.*

Tube	Cheek Cell DNA	PCR Reaction Mix	$MgCl_2$
Name	5 µl	40 µl	5 µl

Some thermal cyclers do not require that mineral oil be added to the samples.

3. Add one drop of mineral oil from the dropper bottle to the PCR tube. Do not touch the dropper bottle to the PCR tube; subsequent reactions may become contaminated with your preparation.
4. Close the cap of the tube tightly, and amplify the sample using automated thermal cycling. Recheck your program, and then start.

   30 cycles of step file; link to the time-delay file for 10 minutes at 72°C

   94°C for 1 minute

   65°C for 1 minute

   72°C for 1 minute

The time-delay file (10 minutes at 72°C) is a "polishing" step that gives the *Taq* polymerase a last chance to fully extend any incomplete amplification products.

Refer to Laboratory 3, "DNA Restriction and Electrophoresis," for detailed instructions on casting and loading gel. Gels of relatively high percentage—1.5% agarose or 2.5% NuSieve—effectively separate the small D1S80 alleles, which have less than 1,000 bp.

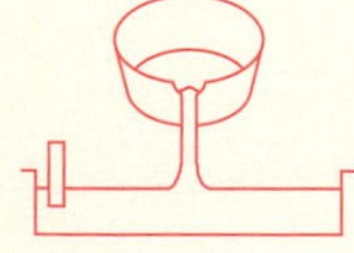

III. Cast Agarose or Polyacrylamide Gel

(15 minutes)

Cast 1.5% Agarose or 2.5% NuSieve Gel

1. Carefully pour agarose solution into the gel-casting tray to fill it to a depth of about 8 mm. The gel should cover about *two-thirds* the height of the comb teeth. Allow the agarose to solidify.
2. Place the casting tray into an electrophoresis box, and set up for electrophoresis.

Stop Point

Cover the electrophoresis box and save the gel until you are ready to continue. The gel will remain in good condition for several days, if it is completely submerged in buffer.

Alternative: Cast 6% Polyacrylamide Gel

1. Follow the manufacturer's instructions for preparing a 6% polyacrylamide gel.

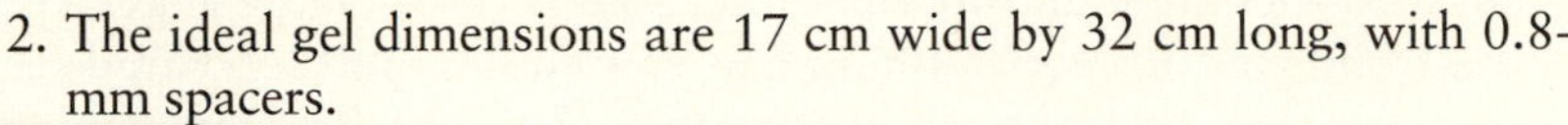

CAUTION

The monomeric form of acrylamide is a neurotoxin and must be weighed out while wearing a mask and gloves.

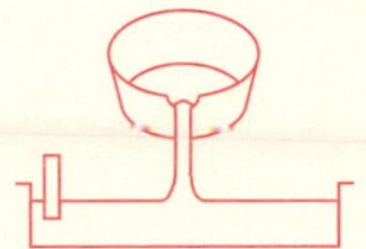

2. The ideal gel dimensions are 17 cm wide by 32 cm long, with 0.8-mm spacers.

Stop Point

Polymerized gel can be stored at room temperature for as long as 24 hours. Pipet TBE buffer around the gel comb, and wrap the polymerized gel/glass plate sandwich in plastic wrap to prevent it from drying.

IV. Load Gel and Electrophorese

(20 minutes; then 60–120 minutes of electrophoresis)

1. Use a permanent marker to label a clean 1.5-ml tube with your name.
2. Transfer 20 μl of amplification product from the PCR reaction tube into a clean 1.5-ml tube. *Take care not to transfer any mineral oil into the clean tube.*
   a. Dip the pipet tip through the mineral oil layer on top of the PCR reaction tube, and carefully withdraw 20 μl of amplification product.
   b. Before transferring the solution, use a Kimwipe or paper towel to remove any mineral oil clinging to the outside of the pipet tip.

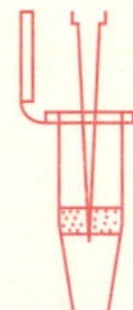

Remaining amplified samples can be stored for months at –20°C, to be used either as controls or as comparison alleles for future experiments.

3. Add 2 μl of loading dye to the 20 μl of your PCR sample. Close the top of the tube, and mix the solution.

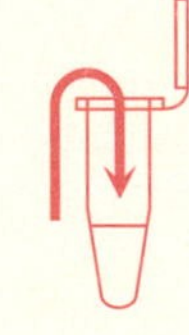

4. Load 20 μl of DNA size markers in an outside gel lane and 20 μl of unamplified control into an adjacent gel lane.

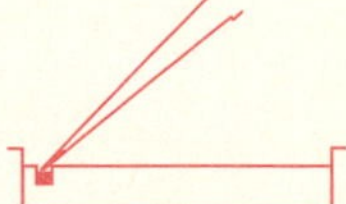

5. Load 20 μl of your PCR reaction into a sample well in the gel. Note your lane position, counting from left to right on the gel.

Samples from other experimenters are loaded in lanes of the same gel. Prepare a key to identify all samples.

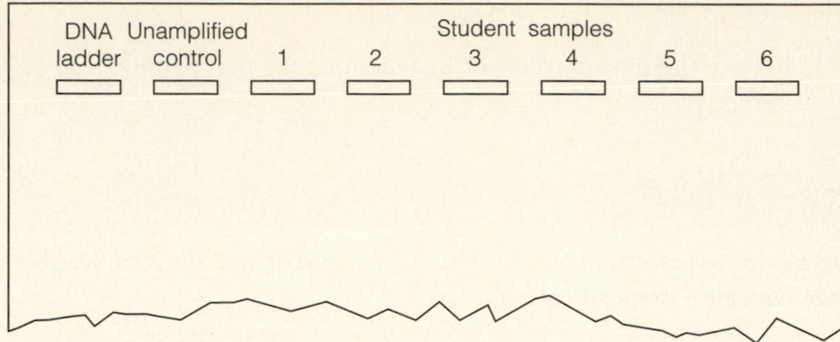

For optimum allele separation, electrophorese the agarose gels for approximately 2 hours and 15 minutes.

6. Electrophorese at 100 volts for about 1 hour (agarose gels) or at 1,000 volts for about 1 hour and 45 minutes (polyacrylamide gels). Adequate separation has occurred when the bromophenol blue bands have moved 50–60 mm from the wells.
7. Turn off the power supply, remove the gel-casting tray from the electrophoresis box (or the polyacrylamide gel from the glass plates), and transfer the gel to a disposable weigh boat (or other shallow tray) for staining.

V. Stain Gel with Ethidium Bromide, View, and Photograph

Staining can be performed by an instructor in a controlled area, when students are not present.

(10–15 minutes)

CAUTION

Review the section on "Ethidium Bromide Staining and Responsible Handling" in Laboratory 3. Wear disposable gloves when staining, viewing, and photographing gel and during cleanup. Confine all staining to a restricted sink area.

The relatively thick gel and high agarose concentration used in this experiment require extended staining and destaining times. Destaining leaches unbound ethidium bromide from the gel, thus increasing the contrast between the stained alleles and the "background" ethidium bromide in the gel.

1. Flood the gel with ethidium bromide solution (1 µg/ml), and allow it to stain for 20–30 minutes.
2. Rinse and destain the gel in tap water for 20–30 minutes.
3. View the gel under an ultraviolet transilluminator or other UV light source.

CAUTION

Ultraviolet light can damage your eyes. Never look directly at an unshielded UV light source without eye protection. View only through a filter or safety glasses that absorb the harmful wavelengths.

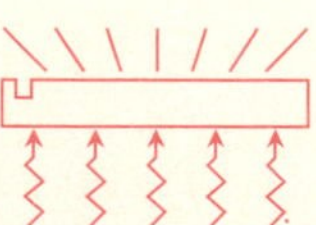

4. Make an exposure of 1–2 seconds with your camera set at f/8. Develop the print for the recommended time (approximately 45 seconds at room temperature).
5. Take the time for proper cleanup:
   a. Wipe down the camera, the transilluminator, and the staining area.

b. Decontaminate the gel and any staining solution that will not be reused.

c. Wash your hands before leaving the laboratory.

RESULTS AND DISCUSSION

1. Observe the photograph of the stained gel that contains your sample and those from other individuals. Orient the photograph with the sample wells at the top. First, ascertain whether you can see a diffuse (fuzzy) band of primer-dimer that appears at the same position in each lane, toward the bottom of the gel. Primer-dimer is not amplified human DNA but is an artifact of the PCR reaction that results from the primers amplifying themselves. Excluding primer-dimer, interpret the allele bands in each lane of the gel:

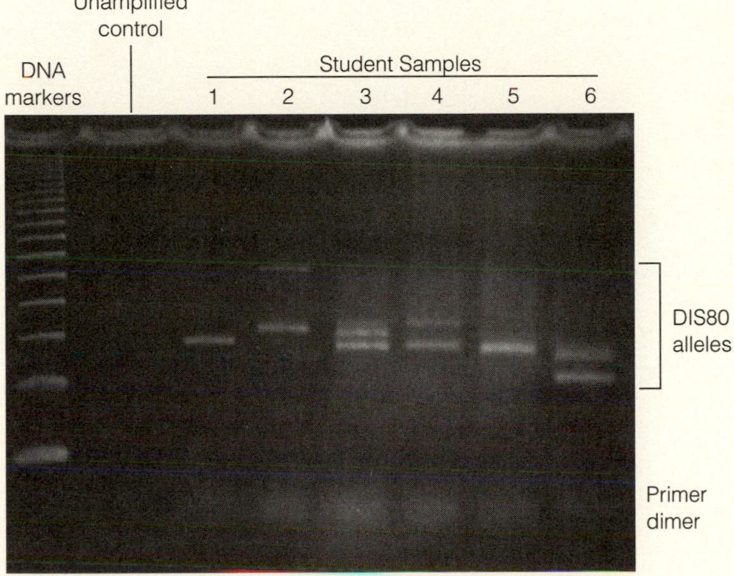

Ideal gel

a. *No bands are visible:* This usually results from an error during the sample preparation, such as losing the cell pellet or using a Chelex solution that is too acidic.

b. *One band is visible:* The simplest explanation is that the individual is homozygous at the D1S80 locus, having inherited the same allele on both maternal and paternal chromosome 1. There is always a strong possibility that two different, but similarly sized, alleles of a hyterozygote cannot be resolved by the agarose gel system. Another possibility is that a larger allele (with many repeats) has failed to amplify efficiently.

c. *Two bands are visible:* The individual is heterozygous at the D1S80 locus. Often, the larger allele amplifies less efficiently and appears to be less intense than the smaller one.

d. *Three or more bands are visible:* The two brightest bands are most likely the true alleles. Additional bands can occur when the primers bind nonspecifically to chromosomal loci other than D1S80 and give rise to additional amplification products.

2. Population studies have identified 29 different alleles at the D1S80 locus that determine 435 possible genotypes. The chart below gives D1S80 allele frequencies in American populations. Determine the number of different alleles represented among your classmates and the percentage of heterozygous individuals. How does your class data compare with that of the general population? How can you account for the differences?

D1S80 Allele Frequencies from Three U.S. Population Groups*, **

Allele	U.S. Caucasians (n = 400)	African Americans (n = 400)	Hispanic Americans (n = 400)
14	—	—	0.003
15	—	—	—
16	—	—	0.003
17	—	0.048	0.013
18	0.238	0.098	0.263
19	0.010	0.003	0.005
20	0.040	0.033	0.020
21	0.018	0.115	0.025
22	0.030	0.088	0.028
23	0.008	0.023	0.003
24	0.348	0.193	0.318
25	0.040	0.023	0.055
26	0.015	0.008	0.010
27	0.013	0.013	0.008
28	0.063	0.153	0.050
29	0.053	0.055	0.055
30	0.008	0.008	0.055
31	0.080	0.048	0.058
32	0.013	0.005	0.003
33	0.003	0.005	—
34	0.003	0.073	0.008
35	0.003	—	—
36	0.005	0.003	—
37	0.008	—	0.003
38	—	—	0.005
39	—	—	—
40	0.003	0.003	0.010
41	—	—	—
>41	0.005	0.010	0.005

* Roche Molecular Systems
** Allele Frequency (n = number of alleles typed)

3. Based on your results, do you think this protocol could be used to link a suspect to a crime or to establish a paternity relationship? Why? How could you modify the experiment to improve your ability to positively identify individuals?
4. The sizes of alleles run on polyacrylamide gels can be determined directly by comparison with D1S80 size markers (allelic ladder). Allele sizes can be estimated in agarose gels by simply comparing their positions to the ladder of size markers included in one lane of each gel. However, a more accurate size determination can be obtained by graphing the function that determines the migration of linear DNA fragments through an agarose gel

$$D = \frac{1}{\log^{10} MW}$$

where D equals the distance migrated and MW equals the molecular weight of the fragment. For simplicity, biologists often substitute base-pair length for molecular weight in this calculation.

   a. The fragments in the marker ladder are multiples of a 123-base-pair repeat. Orient your gel photo with the wells at the top and, working from bottom to top, assign base-pair sizes to the first 8 bands that appear in the ladder on your gel: 123, 246, 369, 492, 615, 738, 861, and 984 bp. Carefully measure the distance (in mm) that each marker fragment migrated from the sample well. Measure from the front edge of the well to the leading edge of each band.
   b. Set up semilog graph paper with the distance migrated as the x (arithmetic) axis and base-pair length as the y (logarithmic) axis. Then, plot the distance migrated *versus* base-pair length for each marker fragment. Connect the data points.
   c. Measure and record the distances migrated by various alleles. To determine the base-pair size of an allele, first locate the distance that it migrated on the x axis. Then, use a ruler to draw a vertical line from this point to its intersection with the marker data line. Now, extend a horizontal line from this point to the y axis. The number on the y axis is the calculated base-pair size of the allele.
   d. Compare the largest and smallest allele observed in your class with the known range of most D1S80 alleles: 369–801 base pairs.
   e. Estimate the size of the primer-dimer.

FOR FURTHER RESEARCH

1. Research the application of the Hardy-Weinberg equilibrium to the analysis of human DNA fingerprinting. Why is it difficult to apply to D1S80?
2. Investigate other DNA polymorphisms in human DNA that can be used to
   a. establish identity.
   b. investigate evolutionary relationships.
   c. detect disease-causing mutations.

Appendices

APPENDIX 1

Equipment, Supplies, and Reagents

EQUIPMENT

Equipment Needed Per Lab	CBS Cat. #	Lab Number																						
Material needed		1	2	3	4	5	6	7	8	9	10	11	12	13	14	15	16	17	18	19	20	21	22	23
Cell spreader	21-5820					*			*							*					*			
Clinical centrifuge	21-4075								*							*					*			
Electro. gel chamber	21-3668			*	*		*	*			*	*	*	*	*			*	*	*			*	*
Electric power supply	21-3673			*	*		*	*			*	*	*	*	*			*	*	*			*	*
Forceps	21-5616													*		*	*			*				
Hybridization containers	21-5592																							
	21-5594													*			*							
Incubator	21-5868		*			*	*		*	*	*						*				*			
Inoculating loop	21-5826		*			*	*		*	*	*													
Microfuge	21-4048	*		*	*		*	*				*	*	*	*			*	*	*	*	*	*	*
Micropipet 1–10	21-4640	*		*	*	*	*	*	*		*	*	*	*	*			*	*	*	*	*	*	*
Micropipet 10–100	21-4642			*	*	*	*	*	*		*	*	*	*		*		*	*	*	*	*	*	*
Micropipet 100–1,000	21-4644	*				*	*		*		*				*	*		*			*	*	*	*
Pipet aid	21-4690	*	*			*	*		*		*					*								
Oven	70-1531													*			*							
Shaking water bath	21-6258		*				*		*		*			*		*	*				*			
Spectrophotometer	65-3300								*							*								
Test tube rack	21-5572	*	*	*	*	*	*	*	*		*	*	*	*	*	*		*	*	*	*	*	*	*
Thermocycler	21-6270																		*				*	*
UV camera system	21-3678			*	*		*	*				*	*	*	*				*	*			*	*
White light system	21-3680			*	*		*	*				*	*	*					*					
Water bath	21-6254			*	*	*	*	*	*		*	*	*	*	*	*	*	*	*	*	*		*	*

CONSUMABLE MATERIALS CHECKLIST BY LAB

Lab 1: Measurements, Micropipetting, and Sterile Techniques

Description	Cat. #
10-ml pipets	21-4642
50-ml conical tube	21-5100
15-ml culture tube	21-5080
1.5-ml microcentrifuge tubes	21-5222

Lab 2: Bacterial Culture Techniques

Description	Cat. #
MM294 culture	21-1530
MM294/pAMP culture	21-1540
LB agar plates(prepoured)	21-6610
or (melt-n-pour)	21-6620
LB/amp agar plates(prepoured)	21-6611
(melt-n-pour)	21-6621
(dehydrated)	21-6700
50-ml conical tube	21-5100
10-ml pipets	21-4642
Luria broth (prepared, sterile)	21-6660
(dehydrated)	21-6710

Lab 3: DNA Restriction and Electrophoresis

Description	Cat. #
Lambda DNA	21-1414
*Eco*RI	21-1670
*Bam*HI	21-1660
*Hin*dIII	21-1690
Loading dye	21-8200
Microcentrifuge tubes	21-5222
Agarose	21-7080
TBE buffer	21-9027
Ethidium bromide	21-7420
or Methylene blue	21-8290
Staining trays	21-5590
Film	21-3679
(Also available in kit form/ 6 stations/kit)	
Restriction Analysis kit with Ethidium Bromide	21-1150
Restriction Analysis kit with Carolina Blu™	21-1151

Lab 4: Effects of DNA Methylation on Restriction

Description	Cat. #
Lambda DNA	21-1414
*Eco*RI	21-1670
*Hin*dIII	21-1690
*Eco*RI methylase	21-1671
Loading dye	21-8200
Microcentrifuge tubes	21-5222
Agarose	21-7080
TBE buffer	21-9025
Ethidium bromide	21-7420
or Methylene blue	21-8290
Staining trays	21-5590
Film (20 exposures)	21-3679

Lab 5: Rapid Colony Transformation of *E. coli* with Plasmid DNA

Description	Cat. #
MM294 culture	21-1530
50 mM calcium chloride	21-1320
0.005-µg/µl pAMP	21-1438
LB broth (prepared, sterile)	21-6660
(dehydrated powder)	21-6710
LB agar plates(prepoured)	21-6610
or (melt-n-pour)	21-6620
(dehydrated)	21-6700
LB/amp agar plates(pre-poured)	21-6611
(melt-n-pour)	21-6621
Transformation tubes, 15-ml	21-5080
Also available in kit form/ 6 stations/kit	
Colony Transformation Kit	21-1142

Lab 6: Purification and Identification of Plasmid DNA

Description	Cat. #
MM294/pAMP	21-1540
GTE, SDS/NaOH, potasium acetate/acetic acid isopropanol, ethanol(Plasmid Miniprep Reagent Kit)	21-1310
TE buffer	21-9026
Microcentrifuge tubes, 1.5-ml	21-5222
*Bam*HI	21-1660
*Hin*dIII	21-1690
pAMP plasmid	21-1430
RNase	21-1745
Loading dye	21-8200
Agarose	21-7080
TBE buffer	21-9025

Ethidium bromide	21-7420
or Methylene blue	21-8290
Staining trays	21-5590
Film (20 exposures)	21-3679

Lab 7: Recombination of Antibiotic-Resistance Genes

Description	Cat. #
pAMP plasmid	21-1433
pKAN plasmid	21-1443
*Bam*HI	21-1660
*Hin*dIII	21-1690
T-4 ligase	21-1740
Ligation buffer/ATP provided with enzyme	
Loading dye	21-8200
Agarose	21-7080
TBE buffer	21-9027
Ethidium bromide	21-7420
or Methylene blue	21-8290
Staining trays	21-5590
Film (20 exposures)	21-3679

Lab 8: Transformation of *E. coli* with Recombinant DNA

Description	Cat. #
MM294 culture	21-1530
Luria broth	21-6660
50-mM calcium chloride	21-1320
Transformation tubes	21-5080
pAMP plasmid	21-1438
pKAN plasmid	21-1445
Luria broth (prepared, sterile)	21-6660
(dehydrated)	21-6710
LB/amp plates(prepoured)	21-6611
(melt-n-pour)	21-6621
LB/kan plates(prepoured)	21-6612
(melt-n-pour)	21-6622
LB/amp+kan plates(prepoured)	21-6613
(melt-n-pour)	21-6623

Lab 9: Replica Plating to Identify Mixed *E. coli* Populations

Description	Cat. #
LB/amp plate (prepoured)	21-6611
(melt-n-pour)	21-6621
LB/kan plate (prepoured)	21-6612
(melt-n-pour)	21-6622

Lab 10: Purification and Identification of Recombinant Plasmid DNA

Description	Cat. #
GTE, SDS/NaOH, potasium acetate/acetic acid isopropanol, ethanol(Plasmid Miniprep Reagent Kit)	21-1310
TE buffer	21-9026
pAMP	21-1429
pKAN	21-1439
*Hind*III	21-1690
*Bam*HI	21-1660
RNase	21-1745
Loading dye	21-8200
Agarose	21-7080
TBE buffer	21-9025
Ethidium bromide	21-7420
or Methylene blue	21-8290
Staining trays	21-5590
Film (20 exposures)	21-3679

Lab 11: Restriction Mapping of the λ Chromosome

Description	Cat. #
Lambda DNA	21-1414
ApaI	21-1650
*Eco*0109	21-1668
*Sna*BI	21-1710
Microcentrifuge tubes	21-5222
Loading dye	21-8200
Agarose	21-7080
TBE buffer	21-9027
Ethidium bromide	21-7420
or Methylene blue	21-8290
Film (20 exposures)	21-3679
Staining trays	21-5590
or kit	21-1172 Ethidium bromide stain
	21-1173 Methylene blue

Lab 12: Restriction Mapping of the Plasmid pBR322

Description	Cat. #
pBR322 plasmid	21-1450
Lambda DNA	21-1414
*Eco*RI	21-1670
*Hinc*II	21-1680
*Hind*III	21-1690
*Pvu*II	21-1700
Microcentrifuge tubes	21-5222

Loading dye	21-8200
Agarose	21-7080
TBE buffer	21-9027
Ethidium bromide	21-7420
or Methylene blue	21-8290
Staining trays	21-5590
Film (20 exposures)	21-3679
Staining trays	21-5590

Lab 13: Southern Hybridization of λ DNA

Description	Cat. #
Lambda DNA	21-1414
*Bam*HI	21-1660
*Eco*RI	21-1670
*Hin*dIII	21-1690
Microcentrifuge tubes	21-5222
Loading dye	21-8200
Agarose	21-7080
TBE buffer	21-9025
Ethidium bromide	21-7420
Staining trays	21-5590
Film (20 exposures)	21-3679
Denaturation buffer, 2×, 500-ml	21-7350
Neutralization/wash buffer, 2×, 1,000-ml	21-8500
SSC buffer, 20×, 500-ml	21-8830
Nylon membrane, 10× 15-cm, 20 sheets	21-5612
Parafilm	21-5600
Filter paper, 3 mm × 46 cm × 57 cm	21-5610
Staining trays	21-5590
Prehybridization buffer	21-8650
Labeled probe, 3-μg	21-1495
Genius™ Nonradioactive Detection Kit	21-1770
TE buffer	21-9026

Lab 14: Construction of a Genomic Library of λ DNA

Description	Cat. #
Lambda DNA	21-1414
pBLU plasmid	21-1421
*Bam*HI	21-1660
*Hin*dIII	21-1690
Microcentrifuge tubes	21-5222
Loading dye	21-8200
Agarose	21-7080
TBE buffer	21-9027
Ethidium bromide	21-7420
Staining trays	21-5590

Film (20 exposures)	21-3679
T4 ligase	21-1740

Lab 15: Transformation of *E. coli* with λ Library

Description	Cat. #
JM101 culture	21-1561
pBLU plasmid	21-1427
Calcium chloride, 50-mM	21-1310
10-ml pipets	21-4626
Culture tubes, 15-ml	21-5080
Luria broth (prepared, sterile)	21-6660
(dehydrated)	21-6710
LB/amp/X-gal plates	21-6624
Nylon membranes, circular, 82-mm	21-5615

Lab 16: Colony Hybridization of the λ Library

Description	Cat. #
Nylon membranes, circular, 82-mm	21-5615
Filter paper (Whatman)	21-5610
LB/amp plates (prepoured)	21-6611
(melt-n-pour)	21-6621
Denaturation buffer, 2×, 500-ml	21-7350
Neutralization/wash buffer, 2×, 1,000-ml	21-8500
Prehybridization buffer	21-8650
Labeled probe, 3-μg	21-1495
Genius™ Nonradioactive Detection Kit	21-1770
TE buffer	21-9026

Lab 17: Purification and Identification of λ Clones

Description	Cat. #
Luria broth (prepared, sterile)	21-6660
(dehydrated)	21-6710
GTE, SDS/NaOH, potasium acetate/acetic acid isopropanol, ethanol (Plasmid Miniprep Reagent Kit)	21-1310
TE buffer	21-9026
*Bam*HI	21-1660
*Hin*dIII	21-1690
RNase	21-1745
Lambda DNA	21-1414
pBLU plasmid	21-1408
Microcentrifuge tubes	21-5222
Loading dye	21-8200
Agarose	21-7080
TBE buffer	21-9027

Ethidium bromide	21-7420
Staining trays	21-5590
Film (20 exposures)	21-3679

Lab 18: Amplification and Purification of a λ DNA Fragment

Description	Cat. #
Lambda PCR Amplification Kit	21-1220
Phenol/chloroform/isoamyl alcohol solution	21-8596
Chloroform solution	21-7320
3 M Sodium acetate	21-8800
TE buffer	21-9026
Ethanol	21-7410
Lambda DNA	21-1414
*Bam*HI	21-1660
*Hin*d III	21-1690
Microcentrifuge tubes	21-5222
Loading dye	21-8200
Agarose	21-7080
TBE buffer	21-9027
Ethidium bromide	21-7420
Staining trays	21-5590
Film (20 exposures)	21-3679

Lab 19: Recombination of a PCR Product and Plasmid pBLU

Description	Cat. #
Lambda DNA	21-1414
pBLU plasmid	21-1421
*Bam*HI	21-1660
*Hin*dIII	21-1690
Microcentrifuge tubes	21-5222
Loading dye	21-8200
Agarose	21-7080
TBE buffer	21-9027
Ethidium bromide	21-7420
Staining trays	21-5590
Film (20 exposures)	21-3679
T4 ligase	21-1740

Lab 20: Transformation of *E. coli* with PCR Product

Description	Cat. #
JM101 culture	21-1561
Calcium chloride, 50-mM	21-1310

LB amp/X-gal plates	21-6624
Luria broth (prepared, sterile)	21-6660
(dehydrated)	21-6710

Lab 21: The Purification and Identification of PCR/pBLU Recombinants

Description	Cat. #
GTE, SDS/NaOH, potasium acetate/acetic acid isopropanol, ethanol(Plasmid Miniprep Reagent Kit)	21-1310
TE buffer	21-9026
*Bam*HI	21-1660
*Hind*III	21-1690
RNase	21-1745
Lambda DNA	21-1414
pBLU plasmid	21-1408
Microcentrifuge tubes	21-5222
Loading dye	21-8200
Agarose	21-7080
TBE buffer	21-9027
Ethidium bromide	21-7420
Staining trays	21-5590
Film (20 exposures)	21-3679

Lab 22: Detection of an *Alu* Insertion Polymorphism by Polymerase Chain Reaction

Description	Cat. #	Qty./Group
Taq polymerase	21-1750	
Deoxynucleotide mix	21-1760	
PCR primers for *Alu* insertion	21-1500	
PCR tubes	21-5240	
Chelex	21-7310	
Lambda DNA	21-1414	
*Bam*HI	21-1660	
*Hind*III	21-1690	
Loading dye	21-8200	
Agarose	21-7080	
TBE buffer	21-9027	
Ethidium bromide	21-7420	
Staining trays	21-5590	
Film (20 exposures)	21-3679	
15-ml tubes	21-5080	

Lab 23: Detection of a VNTR Polymorphism by Polymerase Chain Reaction

Human DNA fingerprinting by PCR (D1S80)	21-1226
Loading dye	21-8200
Agarose	21-7080
TBE buffer	21-9027
Ethidium bromide	21-7420
Staining trays	21-5590
Film (20 exposures)	21-3679
15-ml tubes	21-5080

APPENDIX 2

Recipes for Media, Reagents, and Stock Solutions

The success of the laboratories in this book depends on the use of uncontaminated reagents. Follow the recipes with care, and pay scrupulous attention to cleanliness. Use a clean spatula for each ingredient, or carefully pour each ingredient from the bottle.

The recipes are organized in eight sections. Stock solutions that are used in more than one laboratory are listed once, according to their first use.

I. Bacterial Culture

4 N sodium hydroxide (NaOH)
10 mg/ml ampicillin
10 mg/ml kanamycin
20 mg/ml X-gal (5-bromo-4-chloro-3-indoyl-β-D-galactoside)
Luria-Bertani (LB) broth
LB broth + antibiotic
LB agar plates
LB agar + antibiotic
LB agar + antibiotic + X-gal
Stab cultures

II. DNA Restriction

1 M Tris (pH 7.6, 8.0, and 8.3)
5 M sodium chloride (NaCl)

1 M magnesium chloride ($MgCl_2$)
1 M dithiothreitol (DTT)
10× compromise restriction buffer
New England Biolabs buffer #1
New England Biolabs buffer #4
30 mM *S*-adenosyl methionine (SAM)
2× restriction buffer
0.05% glacial acetic acid
5 mg/ml RNase A (pancreatic RNase)
5× restriction buffer/RNase

III. Gel Electrophoresis

10× Tris/Borate/EDTA (TBE) electrophoresis buffer
1× Tris/Borate/EDTA (TBE) electrophoresis buffer
0.8%, 1.0%, 1.5%, and 2% agarose
2.5% NuSieve agarose
Loading dye
5 mg/ml ethidium bromide stock solution
1 μg/ml ethidium bromide staining solution
0.2% Methylene blue stock solution
0.025% Methylene blue staining solution

IV. Southern/Colony Hybridization

Denaturation buffer
Neutralization buffer
10× sodium chloride/sodium citrate (SSC) buffer
2× SSC buffer
2× SSC buffer + 0.1% SDS
10% *N*-lauroylsarcosine, sodium salt
Prehybridization buffer
Hybridization buffer
Wash buffer 1
Wash buffer 2
Wash buffer 3
Antibody/enzyme conjugate solution
Color development solution
Tris/EDTA (TE) buffer

V. DNA Ligation

0.1 M adenosine triphosphate (ATP)
10× ligation buffer + ATP

VI. Bacterial Transformation

1 M calcium chloride ($CaCl_2$)

50 mM calicium chloride ($CaCl_2$)

VII. Plasmid Minipreparation

0.5 M ethylene diamine tetraacetic acid, disodium salt (EDTA)

Glucose/Tris/EDTA (GTE)

5 M potassium acetate (KOAc)

Potassium acetate/acetic acid

10% sodium dodecyl sulfate (SDS)

1% SDS/0.2 N NaOH

VIII. Polymerase Chain Reaction

0.9% sodium chloride (NaCl)

50 mM Tris

10% Chelex

Chloroform/isoamyl alcohol

Phenol (equilibrated)

Phenol/chloroform/isoamyl alcohol

3 M sodium acetate pH 5.2

1 M potassium chloride (KCl)

10× PCR buffer II

Deoxynucleotide mixture

12.5 mM and 25 mM $MgCl_2$

ABOUT BUFFERS

1. Solid reagents are typically dissolved in a volume of deionized or distilled water equivalent to 70–80% of the finished volume of buffer. This leaves room for the addition of acids or base to adjust the pH. Finally, water is added to bring the solution up to the final volume.
2. When appropriate, the final concentration of each liquid reagent is given in the right-hand column of the reagent list.
3. Buffers are used as 2×, 5×, or 10× solutions. Buffers are diluted when mixed with other reagents, to produce a working concentration of 1×.
4. The commercial enzymes used for these laboratories are all supplied with appropriate buffers; these should be used unless otherwise noted.
5. Storage temperatures of 4°C and –20°C refer to normal refrigerator and freezer temperatures, respectively.

I. Bacterial Culture

4 N Sodium Hydroxide (NaOH)

Makes 100 ml. Store at room temperature (indefinitely).

1. Slowly add 16 g of NaOH pellets (m.w. 40.00) to 80 ml of deionized or distilled water, stirring. The solution will become very warm.
2. When the NaOH pellets are completely dissolved, add water to make a final volume of 100 ml.

10 mg/ml Ampicillin

Makes 100 ml. Store at –20°C for 1 year or at 4°C for 3 months.

1. Add 1 g of ampicillin (sodium salt, m.w. = 371.40) to 100 ml of deionized or distilled water in a clean 250-ml flask. (The sodium salt dissolves readily; however, the free-acid form is difficult to dissolve.)
2. Stir to dissolve.
3. Prewash a 0.45- or 0.22-micron sterile filter (Nalgene or Corning) by drawing through 50–100 ml of deionized or distilled water. Pass the ampicillin solution through the filter.
4. Dispense 10-ml aliquots in sterile 15-ml tubes (Falcon 2059 or the equivalent), and freeze at –20°C.

10 mg/ml Kanamycin

Makes 100 ml. Store at –20°C for 1 year or at 4°C for 3 months.

1. Add 1.0 g of kanamycin sulfate (m.w. = 582.60) to 100 ml of deionized or distilled water in a clean 250-ml flask.
2. Stir to dissolve.
3. Prewash a 0.45- or 0.22-micron sterile filter (Nalgene or Corning) by drawing through 50–100 ml of deionized or distilled water. Pass the kanamycin solution through the washed filter.
4. Dispense 10-ml aliquots in sterile 15-ml tubes (Falcon 2059 or the equivalent), and freeze at –20°C.

20 mg/ml X-Gal (5-Bromo-4-chloro-3-indolyl-β-D-galactoside)

Makes 100 ml. Store in the dark at –20°C for 1 year.

1. Add 0.5 g of X-gal (m.w. 408.63) to 25 ml of dimethylformamide (m.w. 73.09) in a clean 100-ml flask.
2. Stir to dissolve. It is not necessary to sterilize the solution.
3. Dispense 1-ml aliquots in 1.5-ml tubes, and freeze at –20°C.

LB broth can be considered sterile as long as the solution remains clear; cloudiness is a sign of contamination by microbes. Always swirl the solution to check for bacterial or fungal cells that may have settled at the bottom of the flask or bottle.

Luria-Bertani (LB) Broth

Makes 1 liter. Store at room temperature (indefinitely).

1. Weigh out:

   10 g of tryptone

   5 g of yeast extract

   10 g of NaCl (m.w. 58.44)

   (Alternatively, use 25 g of premix, containing these ingredients.)

2. Add all the ingredients to a clean 2-l flask that has been rinsed with deionized or distilled water.
3. Add 1 l of deionized or distilled water to the flask.
4. Add 0.5 ml of 4-M NaOH.
5. Stir to dissolve the dry ingredients, preferably using a magnetic stir bar.
6. If you are preparing for mid-log the cultures, split the solution into four 250-ml aliquots in 1-l flasks. Plug the top with cotton or foam, and cover it with aluminum foil. (Alternatively, cover with *only* aluminum foil.) Autoclave for 20 minutes at 121°C.
7. If you are preparing the solution for general use in transformations, make 100-ml aliquots in sterile 150–250-ml bottles, using one of the following methods:
   a. Put on the caps loosely. Autoclave for 15–20 minutes at 121°C. (To help guard against breakage, autoclave the bottles in a shallow pan with a small amount of water.)
   b. Prewash a 0.45- or 0.22-micron sterile filter (Nalgene or Corning) by drawing through 50–100 ml of deionized or distilled water. Pass the LB solution through the filter, and aliquot into sterile bottles.

LB Broth + Antibiotic

Makes 100 ml. Store at 4°C for 3 months.

1. Sterilely add 1 ml of 10 mg/ml antibiotic to 100 ml of *cooled* LB broth.
2. Swirl to mix.

LB Agar Plates

Makes 35–40 plates. Store at 4°C or at room temperature for 3 months.

1. Weigh out:
   10 g of tryptone
   5 g of yeast extract
   10 g of NaCl (m.w. 58.44)
   15 g of agar
   (Alternatively, use 40 g of premix, containing these ingredients.)
2. Add all ingredients to a clean 2-l flask that has been rinsed with deionized or distilled water.
3. Add 1 l of deionized or distilled water.
4. Add 0.5 ml of 4 N NaOH.
5. Stir to dissolve the dry ingredients, preferably using a magnetic stir bar. Any undissolved material will dissolve during autoclaving.
6. Cover the mouth of the flask with aluminum foil, and autoclave the solution for 15 minutes at 121°C.
7. During autoclaving, the agar may settle to the bottom of the flask. Swirl to mix the agar evenly.

8. Let the solution cool just until the flask can be held in your bare hands (55–60°C). (If the solution cools too long and the agar begins to solidify, remelt it by briefly autoclaving for 5 minutes or less.)
9. While the agar is cooling, mark the bottoms of the culture plates with the date and a description of the media (e.g., LB). If you are using presterilized polystyrene plates, carefully cut the end of the plastic sleeve, and save it for storing the poured plates. Spread the plates out on the lab bench.
10. When the agar flask is cool enough to hold, lift the lid of the culture plate just wide enough to pour the solution. Do not set the lid down on the lab bench. Quickly pour in agar to just cover the plate bottom (approximately 3 mm). Tilt the plate to spread the agar, and immediately replace the lid.
11. Continue pouring plates. Occasionally, flame the mouth of the flask to maintain sterility.
12. To remove bubbles from the surface of the poured agar, briefly touch the surface with a burner flame while the agar is still liquid.
13. Allow the agar to solidify undisturbed.
14. If possible, incubate the plates lidside down for several hours or overnight at 37°C. This dries the agar, limiting condensation when the plates are stored under refrigeration; it also allows any contaminated plates to be readily detected.
15. Stack the plates in their original plastic sleeves for storage.

Antibiotic-containing plates can be made quickly by evenly spreading 200 ml of 10-mg/ml antibiotic on the surface of an LB agar plate. Allow the antibiotic to absorb into the agar for 10–20 minutes before using. Outdated antibiotic plates can be refurbished in this manner.

LB Agar + Antibiotic

Makes 30–45 plates. Store at 4°C for 2 months.

1. Follow the above recipe for LB agar plates, through step 9.
2. When the agar flask is cool enough to hold, sterilely add 10 ml of 10-mg/ml antibiotic. Ampicillin and kanamycin are destroyed by heat; therefore, it is essential to cool the agar before adding the antibiotic. (For LB/amp+kan plates, add 10 ml each of 10-mg/ml ampicillin and kanamycin.)
3. Swirl the flask to mix the antibiotic.
4. Resume with step 10 above.

LB Agar + Ampicillin + X-Gal

X-gal is expensive. Therefore, we recommend spreading a stock solution of X-gal on the surface of premade agar plates, rather than incorporating the chemical throughout the entire plate. Store at 4°C for 2 months.

1. Prepare LB + ampicillin plates, as described above.
2. Add 40 µl of the 20-mg/ml X-gal stock solution to a sterile LB plate + ampicillin.
3. Use a sterile spreading rod to evenly distribute the X-gal over the entire surface of the plate.
4. Incubate the plate(s) at 37°C until all the liquid has evaporated (about 3 or 4 hours).

Stab Cultures

Makes 30–40 stabs. Store in the dark at room temperature for 1 year.

This recipe is the same as for the LB agar above, but it has a lower percentage of agar, which makes the stab easier to use. Standard LB agar or 4 g of premix can also be used.

1. Weigh out:
   1.0 g of tryptone
   0.5 g of yeast extract
   1.0 g of NaCl (m.w. 58.44)
   1.0 g of agar
2. Add all ingredients to a clean 250-ml flask that has been rinsed with deionized or distilled water.
3. Add 100 ml of deionized or distilled water.
4. Add 50 µl of 4N NaOH.
5. Stir while heating to dissolve the dry ingredients, preferably using a magnetic hot plate and stir bar.
6. Pour the dissolved solution into 4-ml vials (15 × 45 mm) to fill the vial two-thirds full, and loosely replace the caps.
7. Autoclave the vials for 15 minutes at 121°C. (Alternatively, sterilized agar can be poured into presterilized vials.)
8. Allow the agar to solidify undisturbed.
9. Before storing the solution, tighten the caps securely.
10. Seal the caps with Parafilm to help reduce the danger of contamination.

To inoculate:

1. Sterilely scrape up a cell mass from a single colony of the desired genotype.
2. Stab an inoculating loop several times into the agar.
3. Loosely replace the cap, and incubate the stab overnight at 37°C.
4. Following incubation, tighten the cap, and store in the dark at room temperature. Wrap the cap with Parafilm for long-term storage.

II. DNA Restriction

1 M Tris (pH 7.6, pH 8.0, and pH 8.3)

Makes 100 ml. Store at room temperature (indefinitely).

1. Dissolve 12.1 g of Tris base (m.w. 121.10) in 70 ml of deionized or distilled water.
2. Adjust the pH by slowly adding concentrated hydrochloric acid (HCl); monitor with a pH meter.
3. Add deionized or distilled water to make a total of 100 ml of solution.

CAUTION:

Avoid inhaling Tris powder; wear a mask over your nose and mouth.

Notes:

a. A yellow solution indicates poor-quality Tris. Discard the solution, and obtain Tris from another source.

b. Many types of electrodes do not accurately measure the pH of Tris solutions; check with the manufacturer to determine whether your electrode is Tris-compatible.

c. The pH of a Tris solution is temperature-dependent; make pH measurements at room temperature.

5 M Sodium Chloride (NaCl)

Makes 100 ml. Store at room temperature (indefinitely).

1. Dissolve 29.2 g of NaCl (m.w. 58.44) in 70 ml of deionized or distilled water.
2. Add deionized or distilled water to make 100 ml of total solution.

1 M Magnesium Chloride ($MgCl_2$)

Makes 100 ml. Store at room temperature (indefinitely).

1. Dissolve 20.3 g of $MgCl_2$ (6-hydrate, m.w. 203.30) in 80 ml of deionized or distilled water.
2. Add deionized or distilled water to make 100 ml of total solution.

1 M Dithiothreitol (DTT)

Makes 10 ml. Store at –20°C (indefinitely).

1. Dissolve 1.5 g of DTT (m.w. 154.25) in 8 ml of deionized or distilled water.
2. Add deionized or distilled water to make 10 ml of total solution.
3. Dispense into 1-ml aliquots in 1.5-ml tubes.

Do not autoclave DTT or solutions containing it.

10× Compromise Restriction Buffer

Makes 1 ml. Store at –20°C (indefinitely).

100 μl of 1 M Tris pH 8.0	(100 mM)
100 μl of 1 M $MgCl_2$	(100 mM)
200 μl of 5 M NaCl	(1 M)
7 μl of 14.3 M βME	(100 mM)
593 μl of deionized water	

or

100 μl of 1 M Tris pH 8.0	(100 mM)
100 μl of 1 M $MgCl_2$	(100 mM)
200 μl of 5 M NaCl	(1 M)
10 μl of 1 M DTT	(10 mM)
590 μl of deionized water	

Mix the ingredients in a 1.5-ml tube.

Notes:

a. β-mercaptoethanol (βME), also called 2-mercaptoethanol, is a 14.3-M liquid at room temperature.

b. Compromise buffer provides salt conditions that allow relatively high activity of a number of restriction enzymes, including most of those used in these laboratories. Consider using the buffer provided by the manufacturer, when you are using more expensive enzymes in critical experiments.

c. DTT used in the second recipe can precipitate from the solution with repeated freezing and thawing. Vortex vigorously to redissolve the DTT.

10× NEBuffer 1 (yellow)

Makes 1 ml. Store at –20°C (indefinitely).

100 µl 1 M Tris propane pH 7.0	(100 mM)
100 µl 1 M $MgCl_2$	(100 mM)
10 µl 1 M DTT	(10 mM)
790 µl distilled water	

10× NEBuffer 4 (green)

Makes 1 ml. Store at –20°C (indefinitely).

200 µl 1 M Tris-acetate pH 7.9	(200 mM)
100 µl 1 M magnesium acetate	(100 mM)
500 µl 1 M potassium acetate	(500 mM)
10 µl 1 M DTT	(10 mM)
190 µl distilled water	

30 mM *S*-Adenosyl Methionine (SAM)

Makes 1 ml. Store at –20°C for 1 year.

1. Obtain 1 M sulfuric acid (H_2SO_4), or prepare by carefully adding 1 part concentrated acid (18 M) to 17 parts of deionized or distilled water.
2. Prepare a 5-mM solution by adding 5 µl of 1 M H_2SO_4 to 995 µl of deionized or distilled water.
3. Add 900 µl of a 5-mM H_2SO_4 solution to 100 µl of 100% ethanol.
4. Dissolve 15.8 mg of SAM (iodide salt, grade I, m.w. = 526.3) in 1 ml of a 5-mM H_2SO_4/10% ethanol solution.
5. Dispense 100-µl aliquots in 1.5-ml tubes. Use each solution once, and then discard it.

Notes:

a. SAM is very unstable, and its activity diminishes rapidly with freezing and thawing. Store SAM solution on ice at all times; these solutions have a half-life of about 30 minutes at room temperature.

b. 30 mM of SAM is often supplied with M.*Eco*RI methylase.

2× Restriction Buffer

Makes 1 ml. Store at –20°C (indefinitely).

Mix a 200-µl 10× restriction buffer with 800 µl of deionized or distilled water.

0.05% Glacial Acetic Acid

Makes 50 ml. Store at room temperature (indefinitely).

1. Add 5 µl of glacial acetic acid to 50 ml of deionized or distilled water.
2. The solution should be approximately pH 4.0.

5 mg/ml RNase A (Pancreatic RNase)

Makes 20 ml. Store at –20°C (indefinitely).

1. Dissolve 100 mg of RNase A in 20 ml of 0.05% glacial acetic acid, and transfer the solution to a 50-ml conical tube.
2. Place the tube in a boiling water bath for 15 minutes.
3. Cool the solution, and neutralize it by adding 120 µl of 1M Tris (pH 8.0).
4. Dispense 1-ml aliquots in 1.5-ml tubes.

Notes:

a. Use only RNase A from bovine pancreas.

b. Dissolving RNase in acetic acid prevents subsequent precipitation of the RNase. The solution can be prepared by simply dissolving RNase in deionized or distilled water; however, the RNase will occasionally precipitate from solution, and activity will be lost.

c. The boiling step destroys the activity of DNases that can contaminate the preparation. The RNase activity will not be affected by boiling.

5× Restriction Buffer/RNase

Makes 1 ml. Store at –20°C for several months.

500 µl of 10× restriction buffer

100 µl of 5-mg/ml RNase

400 µl of water

Mix in a 1.5-ml tube.

III. Gel Electrophoresis

10× Tris/Borate/EDTA (TBE) Electrophoresis Buffer

Makes 1 liter. Store at room temperature (indefinitely).

1 g of NaOH (m.w. 40.00)

108 g of Tris base (m.w. 121.10)

55 g of boric acid (m.w. 61.83)

7.4 g of ethylene diamine tetraacetic acid (EDTA, disodium salt, m.w. 372.24)

1. Add all dry ingredients to 700 ml of deionized or distilled water in a 2-l flask.
2. Stir to dissolve, preferably using a magnetic stir bar.
3. Add deionized water to bring the total solution to 1 liter.

1× TBE Electrophoresis Buffer

Makes 10 liters. Store at room temperature (indefinitely).

1. Into a spigoted carboy, add 9 liters of deionized or distilled water to 1 liter of 10× TBE electrophoresis buffer.
2. Stir to mix.

0.8%, 1.0%, 1.5%, and 2.0% Agarose

Makes 200 ml. Use fresh, or store jelled at room temperature for several weeks.

1. Add 1.6 g (0.8%), 2.0 g (1.0%), 3.0 g (1.5%), or 4.0 g (2.0%) of agarose (electrophoresis grade) to 200 ml of 1× TBE electrophoresis buffer in a 600-ml beaker or Erlenmeyer flask.
2. Stir to suspend the agarose.
3. Cover the container with aluminum foil, and heat in a boiling water bath (use a double boiler) or on a hot plate, until all the agarose is dissolved (approximately 10 minutes).

   or

   Heat *uncovered* in a microwave oven at the high setting, until all the agarose is dissolved (3–5 minutes per container).
4. Swirl the solution, and check the bottom of the container to ensure that all the agarose has dissolved. (Just before complete dissolution, particles of agarose will appear as translucent grains.) Reheat for several minutes, if necessary.
5. Cover the solution with aluminum foil, and place it in a hot water bath (at about 60°C) until you are ready to use it.

Notes:

a. Samples of agarose powder can be preweighed and stored in capped test tubes until you are ready to use them.

b. If you are microwaving agarose in a beaker, a "skin" of solidifed agarose may form on the surface. Remove this skin before pouring a gel.

c. Solidified agarose can be stored at room temperature and then remelted in a boiling water bath (15–20 minutes) or in a microwave oven (3–5 minutes per container) before use. Always loosen the cap when remelting agarose in a bottle.

d. When remelting agarose, evaporation will cause the concentration to increase. If necessary, compensate by adding a small volume of water.

2.5% NuSieve Agarose

Makes 200 ml. Use fresh, or store jelled at room temperature for several days.

1. Add 5 g of NuSieve agarose to a 200-ml 1× TBE electrophoresis buffer.
2. Stir to suspend the agarose.
3. Cover the container with aluminum foil, and heat in a boiling water bath (use a double boiler) or on a hot plate, until all the agarose is dissolved (approximately 20 minutes).

   or

   Heat uncovered in a microwave oven at the high setting, until all the agarose is dissolved (3–5 minutes per container).
4. Swirl the solution, and check the bottom of the container to ensure that all the agarose has dissolved. (Just before complete dissolution, particles of agarose will appear as translucent grains.) Reheat for several minutes, if necessary.
5. Cover the solution with aluminum foil, and place it in a hot water bath (at about 60°C) until you are ready to use it. Remove any "skin" of solidified agarose from the surface before pouring.

Loading Dye

Makes 100 ml. Store at room temperature (indefinitely).

0.25 g of bromophenol blue (m.w. 669.96)
0.25 g of xylene cyanol (m.w. 538.60)
50.00 g of sucrose (m.w. 342.30) or 50 ml of glycerol
1.00 ml of 1 M Tris (pH 8.0)

If you are using sucrose:

1. Dissolve bromophenol blue, xylene cyanol, sucrose, and Tris in 60 ml of deionized or distilled water.
2. Add deionized or distilled water to make 100 ml of total solution.

If you are using glycerol:

1. Dissolve xylene cyanol, bromophenol blue, and Tris in 49 ml of deionized or distilled water.
2. Stir in 50 ml of glycerol to make 100 ml of total solution.

5 mg/ml Ethidium Bromide Stock Solution

Makes 50 ml. Store in the dark at room temperature or at 4°C (indefinitely).

CAUTION

Ethidium bromide is a mutagen by the Ames microsome assay and a suspected carcinogen. Because mixing the 5 mg/ml solution from the concentrated powder poses the greatest hazard, we recommend obtaining ready-mixed 5 mg/ml solution from a supplier. If you choose to mix ethidium bromide solution, handle the powder carefully to avoid creating dust. Wear rubber gloves and a mask that covers your nose and mouth. Observe the proper disposal methods explained on page 38.

1. Dissolve 250 mg of ethidium bromide in 50 ml of deionized or distilled water in a nonbreakable, screw-top container (preferably opaque).
2. Label the bottle "CAUTION: Ethidium Bromide. Mutagen and cancer suspect agent. Wear rubber gloves when handling."

Ethidium bromide is light-sensitive; store it in a dark container, or else wrap the container in aluminum foil.

1 μl/ml Ethidium Bromide Staining Solution

Makes 500 ml. Store in the dark at room temperature (indefinitely).

CAUTION

Ethidium bromide is a mutagen by the Ames microsome assay and a suspected carcinogen. Wear rubber gloves when preparing and using ethidium bromide solution. Observe the proper disposal methods explained on page 38.

1. Add 100 μl of a 5 mg/ml ethidium bromide solution to 500 ml of deionized or distilled water.
2. Store in unbreakable bottles (preferably opaque). Label the bottles "CAUTION: Ethidium Bromide. Mutagen and cancer suspect agent. Wear rubber gloves when handling."

Ethidium bromide is light-sensitive; store it in a dark container, or else wrap the container in aluminum foil.

0.2% Methylene Blue Stock Solution

Makes 100 ml. Store at room temperature (indefinitely).

1. Weigh out 0.2 g of methylene blue-trihydrate (m.w. 373.9), and add it to 100 ml of distilled water.
2. Stir until it is completely dissolved.

0.025% Methelyene Blue Staining Solution

Makes 500 ml. Store at room temperature (indefinitely).

1. Add 62.5 ml of 0.2% methelyene blue stock solution to 437.5 ml of distilled water.
2. Stir to mix the solution.

IV. Southern/Colony Hybridization

Denaturation Buffer

Makes 1 liter. Store at room temperature (indefinitely).

Review the comments regarding Tris solutions on pages 393–394.

1. Add the following dry ingredients to 700 ml of deionized or distilled water in a 2-l flask:

   20 g of NaOH (m.w. 40.00) (0.5 M)

   87.7 g of NaCl (m.w. 58.44) (1.5 M)
2. Stir to dissolve, preferably using a magnetic stir bar.
3. Add water to bring the total solution to 1 liter (1,000 ml).

Neutralization Buffer

Makes 1 liter. Store at room temperature (indefinitely).

1. Add the following dry ingredients to 700 ml of deionized or distilled water in a 2-l flask:

   60.6 g of Tris (m.w. 121.10) (0.5 M)

   87.7 g of NaCl (m.w. 58.44) (1.5 M)
2. Stir to dissolve, preferably using a magnetic stir bar.
3. Adjust the pH to 7.5 by slowly adding concentrated HCl; monitor with a pH meter.
4. Add water to bring the total solution to 1 liter (1,000 ml).

10× SSC Buffer

Makes 1 liter. Store at room temperature (indefinitely).

1. Add the following dry ingredients to 700 ml of deionized or distilled water:

   87.7 g NaCl (m.w. 58.44) (1.5 M)

   44.1 g sodium citrate (m.w. 294.1) (0.15 M)
2. Stir to dissolve, preferably using a magnetic stir bar.
3. Adjust pH to approximately 7.0 by adding NaOH solution.
4. Add water to bring the total solution to 1 liter (1,000 ml).

2× SSC Buffer

Makes 1 liter. Store at room temperature (indefinitely).

1. Add 200 ml of 10× SSC buffer to 800 ml of deionized or distilled water.
2. Stir to mix.

2× SSC Buffer + 0.1% SDS

Makes 1 liter. Store at room temperature (indefinitely).

A precipitate may form at colder temperatures. Warm the solution, and shake it gently to dissolve the precipitate.

1. Mix 10 ml of 10% SDS with 990 ml of 2× SSC buffer.
2. Stir to mix.

10% *N*-lauroylsarcosine, Sodium Salt

Makes 100 ml. Store at room temperature.

1. Dissolve 10 g of *N*-lauroylsarcosine, sodium salt (m.w. 293.4) in 80 ml of deionized or distilled water.
2. Add water to make 100 ml of total solution.

Prehybridization Buffer

Makes 100 ml. Store at –20°C for 1 year.

1. Mix the following in a 200-ml flask:

50 ml of 10× SSC buffer	(5× SSC)
1 ml of 10% N-lauroylsarcosine	(0.1% w/v)
200 µl of 10% SDS	(0.02% w/v)
30 ml of distilled water	

2. Add 1 g of blocking reagent (supplied with the Genius Detection Kit).
3. Dissolve the blocking reagent by stirring at 50–70°C, preferably using a magnetic stir bar. The solution will become translucent.
4. Add water to make 100 ml of total solution.

Prepare the solution in advance, and cool it to below 40°C before use.

Hybridization Buffer

Makes 100 ml. Store at –20°C for 6 months.

Add a labeled oligonucleotide probe to 100 ml of prehybridization buffer, so that the probe concentration is about 10 ng/ml.

Wash Buffer 1

Makes 1 liter. Store at room temperature (indefinitely).

1. Mix the following in a 2-l flask:

100 ml of 1 M Tris pH 7.6	(100 mM)
30 ml of 5 M NaCl	(150 mM)
870 ml of deionized or distilled water	

2. Stir to mix.

Wash Buffer 2

Makes 1 liter. Store at –20°C for 1 year.

1. Add 10 g of blocking reagent (supplied with the Genius Detection Kit) to 1,000 ml of wash buffer 1 in a 2-l flask.
2. Dissolve the blocking reagent by stirring it at 50–70°C, preferably using a magnetic stir bar. The solution will become translucent.

Prepare the solution in advance, and cool it to below 40°C before use.

Wash Buffer 3

Makes 1 liter. Store at room temperature (indefinitely).

1. Add 12.1 g of Tris (m.w. 121.1) to a 2-l flask containing 700 ml of deionized or distilled water.
2. Stir to dissolve, preferably using a magnetic stir bar.

3. Add 20 ml of 5 M NaCl.
4. Add 50 ml of 1 M $MgCl_2$.
5. Adjust the pH by slowly adding concentrated HCl; monitor with a pH meter.
6. Add water to make 1000 ml of total solution.

Antibody/Enzyme Conjugate

Makes 100 ml. Store at 4°C for only 12 hours.

Add 10 µl of the antibody/enzyme conjugate (supplied with the Genius Detection Kit) to a flask containing 100 ml of wash buffer 2, and mix.

Color Development Solution

Makes 10 ml. Prepare fresh solution.

Once the color development solution has been mixed, it should be stored in the dark.

To a flask containing 10 ml of wash buffer 3, add 45 µl of nitroblue tetrazolium salt and 35 µl of X-phosphate solution; both of these are supplied with the Genius Detection Kit.

Tris/EDTA (TE) Buffer

Makes 100 ml. Store at room temperature (indefinitely.)

1 ml of 1 M Tris (pH 8.0)	(10 mM)
200 µl of 0.5 M EDTA	(1 mM)
99 ml of deionized water	

Mix the ingredients.

V. DNA Ligation

0.1 M Adenosine Triphosphate (ATP)

Makes 5 ml. Store at –20°C for 1 year.

1. Dissolve 0.3 g of ATP (disodium salt, m.w. 605.19) in 5 ml of deionized or distilled water.
2. Dispense 500-µl aliquots into 1.5-ml tubes.

10× Ligation Buffer + ATP

Makes 1,000 µl. Store at –20°C for 1 month.

600 µl of 1 M Tris, pH 7.6	(600 mM)
100 µl of 1 M $MgCl_2$	(100 mM)
70 µl of 1 M DTT	(70 mM)
100 µl of 0.1 M ATP	(100 mM)
130 µl of deionized water	

Mix the ingredients in a 1.5-ml tube.

Notes:

a. DTT can precipitate from the solution with repeated freezing and thawing. Vortex vigorously to redissolve.

b. ATP loses activity in a dilute solution; be sure to use fresh solution.

VI. Bacterial Transformation

1M Calcium Chloride (CaCl2)

Makes 100 ml. Store at room temperature (indefinitely).

1. Dissolve 11.1 g of anhydrous $CaCl_2$ (m.w. 110.99) or 14.7 g of the dihydrate (m.w. 146.99) in 80 ml deionized or distilled water.
2. Add deionized or distilled water to make 100 ml of total solution.

50 mM Calcium Chloride (CaCl2)

Makes 500 ml. Store at 4°C or at room temperature (indefinitely).

1. Mix 50 ml of 1M $CaCl_2$ with 950 ml of deionized water.
2. Prerinse a 0.45- or 0.22-micron sterile filter by drawing through it 50–100 ml of deionized or distilled water.
3. Pass the $CaCl_2$ solution through a prerinsed filter.
4. Dispense aliquots into presterilized 50-ml conical tubes or autoclaved 150–250-ml bottles.

Alternatively, dispense 100-ml aliquots into 150–250-ml bottles, and autoclave for 15 minutes at 121°C. With storage at 4°C, the solution will be precooled and ready for making competent cells.

VII. Plasmid Minipreparation

0.5 M Ethylene Diamine Tetraacetic Acid (EDTA) (pH 8.0)

Makes 100 ml. Store at room temperature (indefinitely).

1. Add 18.6 g of EDTA (disodium salt, m.w. 372.24) to 80 ml of deionized or distilled water.
2. Adjust the solution to pH 8.0 by slowly adding sodium hydroxide. Monitor with a pH meter.
3. Mix vigorously using a magnetic stirrer or by hand. EDTA will dissolve only after the pH has reached 8.0 or higher.

Use only the disodium salt of EDTA.

Glucose/Tris/EDTA (GTE)

Makes 100 ml. Store at 4°C or at room temperature (indefinitely).

0.9 g of glucose (m.w. 180.16)	(50 mM)
2.5 ml of 1 M Tris (pH 8.0)	(25 mM)
2 ml of 0.5 M EDTA	(10 mM)
94.5 ml of deionized water	

When stored at 4°C, the solution will be precooled and ready for minipreparations.

5 M Potassium Acetate (KOAc)

Makes 200 ml. Store at room temperature (indefinitely).

1. Add 98.1 g of potassium acetate (m.w. 98.14) to 160 ml of deionized water.
2. Add deionized or distilled water to make 200 ml of total solution.

Potassium Acetate/Acetic Acid

Makes 100 ml. Store at 4°C or at room temperature (indefinitely).

Add 60 ml of 5 M potassium acetate and 11.5 ml of glacial acetic acid to 28.5 ml of deionized or distilled water.

The sharp odor of acetic acid distinguishes the finished KOAc/acetic acid solution from the KOAc stock; the two can be easily confused. When stored at 4°C, the solution will be precooled and ready for minipreparations.

10% Sodium Dodecyl Sulfate (SDS)

Makes 100 ml. Store at room temperature (indefinitely).

1. Dissolve 10 g of electrophoresis-grade SDS (m.w. 288.37) in 80 ml of deionized water.
2. Add deionized or distilled water to make 100 ml of total solution.

CAUTION

SDS is the same as sodium lauryl sulfate. Avoid inhaling SDS powder; wear a mask over your nose and mouth.

1% SDS/0.2 N NaOH

Always use fresh SDS/NaOH solution. A precipitate can form at colder temperatures. Warm the solution and shake it gently, to dissolve the precipitate.

Makes 10 ml. Store at room temperature for several days.

Mix 1 ml of 10% SDS and 0.5 ml of 4 N NaOH into 8.5 ml of distilled water.

VIII. Polymerase Chain Reaction

0.9% NaCl

Makes 100 ml. Store at room temperature (indefinitely, if sterile).

If the solution is to be used immediately, it does not need to be sterile. However, for long-term storage, we recommend sterilizing by autoclaving it for 15 minutes at 121°C or passing it through a 0.45- or 0.22-micron sterile filter.

1. Weigh out 0.9 g of NaCl (m.w. 58.44).
2. Add it to a clean 200-ml flask containing 90 ml of deionized or distilled water.
3. Stir to dissolve the NaCl, preferably using a magnetic stir bar.
4. Add water to bring the total solution to 100 ml.

50 mM Tris

Makes 100 ml. Store at room temperature (indefinitely).

Dissolve 0.6 g of Tris base (m.w. 121.10) in 100 ml of deionized or distilled water. The solution will have a pH of about 10.

10% Chelex

Makes 10 ml. Store at room temperature for 3 months.

1. Weigh out 1 g of Chelex 100 (100–200 mesh, sodium form, from BioRad).
2. Add 50 mM Tris to the dry Chelex, to make 10 ml of solution.
3. Adjust the pH to 11, using concentrated NaOH.

Chloroform/Isoamyl Alcohol

Makes 100 ml. Store at room temperature (indefinitely).

Isoamyl alcohol is also called isopentyl alcohol.

Add 4 ml of isoamyl alcohol (m.w. 88.15) to a clean bottle containing 96 ml of chloroform (m.w. 119.38), and mix.

Phenol (Equilibrated)

Makes 100 ml. Store at 4°C for 1 month.

1. Remove the liquified phenol (m.w. 94.11) from the freezer, warm it to room temperature, and then melt it at 68°C.
2. Add 50 ml of the melted phenol to a light-tight bottle.
3. Add 50 ml of 0.5 M Tris pH 8.0, and stir for 15 minutes, preferably using a magnetic stir bar.
4. Allow the two phases to separate, and aspirate as much of the upper aqueous phase as possible.
5. Add 50 ml of 0.1 M Tris pH 8.0 to the phenol, and stir for 15 minutes.
6. Remove the upper aqueous phase, as described in step 4.
7. Repeat extractions with 50 ml 0.1 M Tris pH 8.0 until the pH of the phenolic phase is greater than 7.8 (as measured with pH paper).
8. Add a final 50 ml of 0.1 M Tris pH 8.0 to the phenol.

CAUTION

Phenol is corrosive and can cause severe burns. Be sure to handle phenol under a chemical hood; wear protective clothing, gloves, and safety glasses. Rinse with a large volume of water any area of skin that comes into contact with phenol. Wash with soap and water. Do not use ethanol.

We do not recommend using crystalline phenol, because it must be redistilled to remove oxidation products that can damage DNA.

Phenol/Chloroform/Isoamyl Alcohol

Makes 100 ml. Store at 4°C for 1 month.

Add 50 ml of equilibrated phenol to a clean, light-tight bottle containing 50 ml of chloroform/isoamyl alcohol, and mix.

3 M Sodium Acetate

Makes 100 ml. Store at room temperature (indefinitely).

1. Dissolve 40.8 g of sodium acetate (m.w. 136.08) in 70 ml of deionized or distilled water.
2. Adjust the pH to 5.2 by adding glacial acetic acid; monitor with a pH meter.
3. Add water to bring the total solution to 100 ml.

1 M Potassium Chloride (KCl)

Makes 100 ml. Store at room temperature (indefinitely).

1. Dissolve 7.46 g of potassium chloride in 70 ml of deionized or distilled water.
2. Add water to bring the total solution to 100 ml.

10× PCR Buffer II

Makes 1 ml. Store at –20°C (indefinitely).

100 µl of 1 M Tris pH 8.3	(100 mM)
500 µl of 1 M KCl	(500 mM)
400 µl of deionized water	

Mix the ingredients in a 1.5-ml tube.

Deoxynucleotide Mixture

Makes 1 ml. Store at –20°C for 1 year.

10 mM dATP	125 µl	(1.25 mM)
10 mM dCTP	125 µl	(1.25 mM)
10 mM dGTP	125 µl	(1.25 mM)
10 mM dTTP	125 µl	(1.25 mM)
deionized H_2O	500 µl	

Mix in a 1.5-ml tube.

12.5 mM and 25 mM MgCl2

Makes 1 ml. Store at –20°C for 1 year.

Mix 12.5 µl of 1M $MgCl_2$ with 988 µl of deionized water. (12.5 mM)

or

Mix 25 µl of 1M $MgCl_2$ with 975 µl of deionized water. (25 mM)

APPENDIX 3

Restriction Map Data for pAMP, pKAN, and pBLU

A. RESTRICTION MAP DATA FOR pAMP, pKAN, AND pBLU

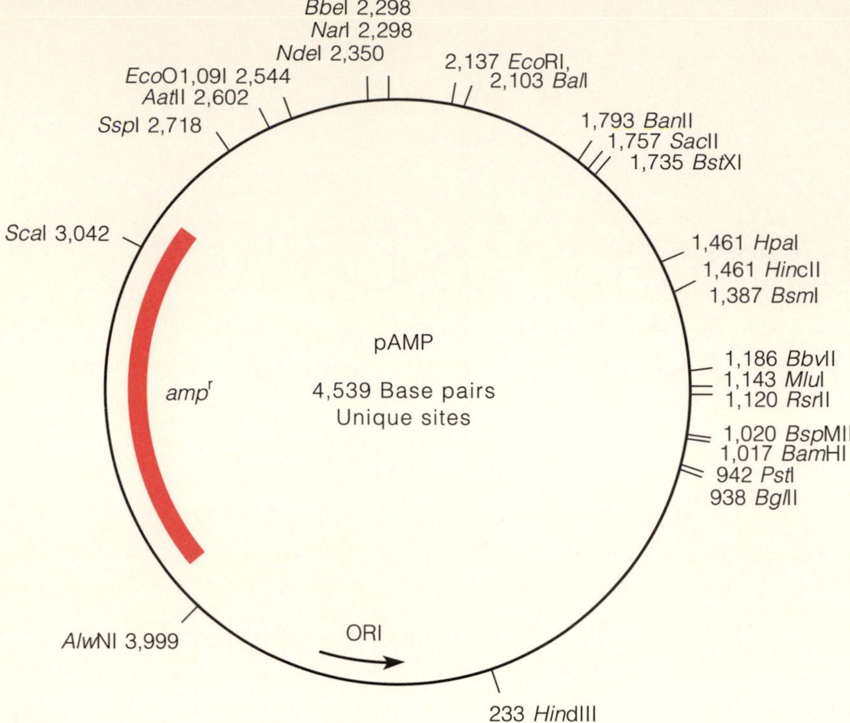

Restriction Map for pAMP

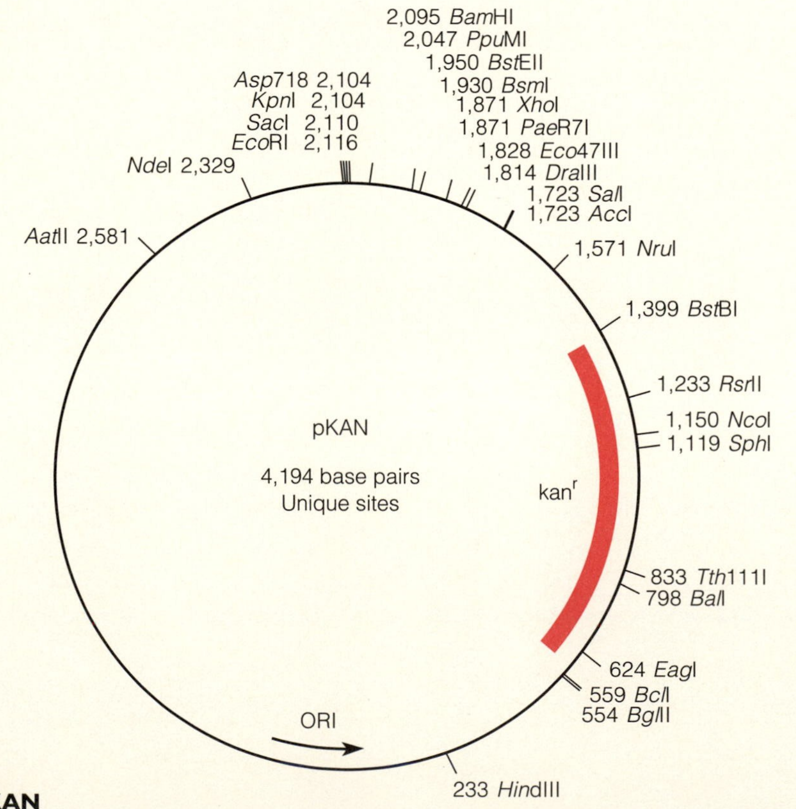

Restriction Map for pKAN

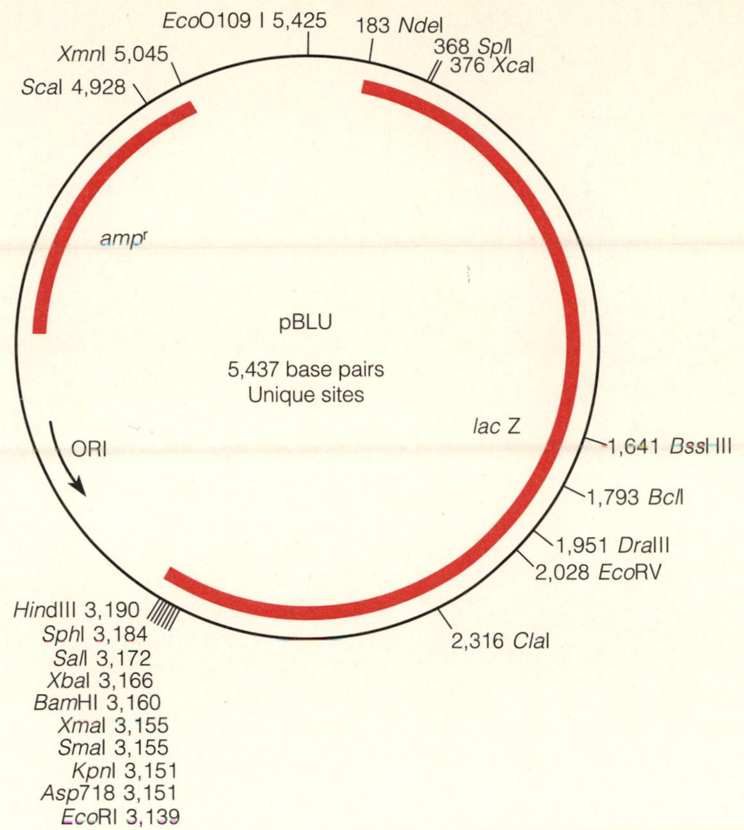

Restriction Map for pBLU

B. CONSTRUCTION OF pAMP, pKAN, AND pBLU

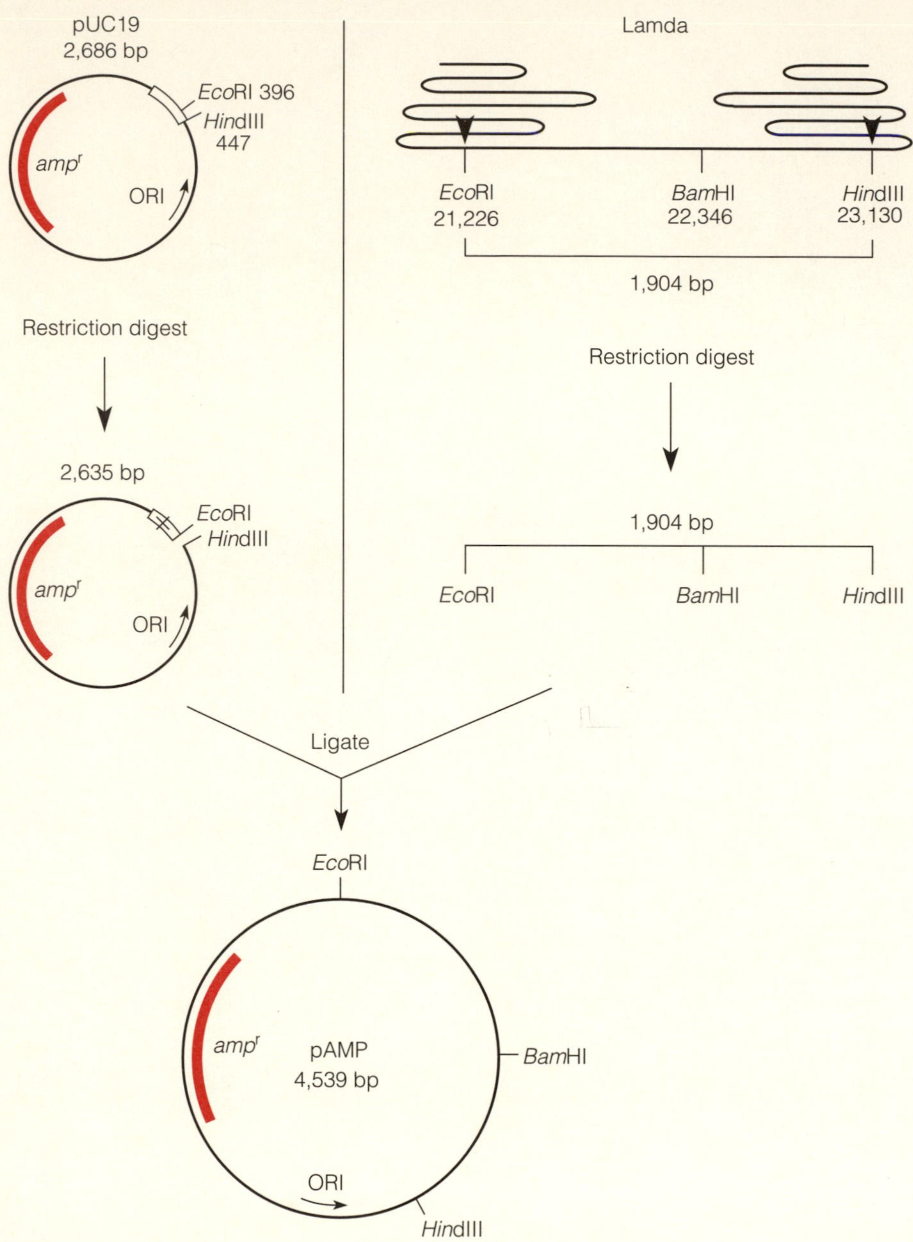

Construction of pAMP

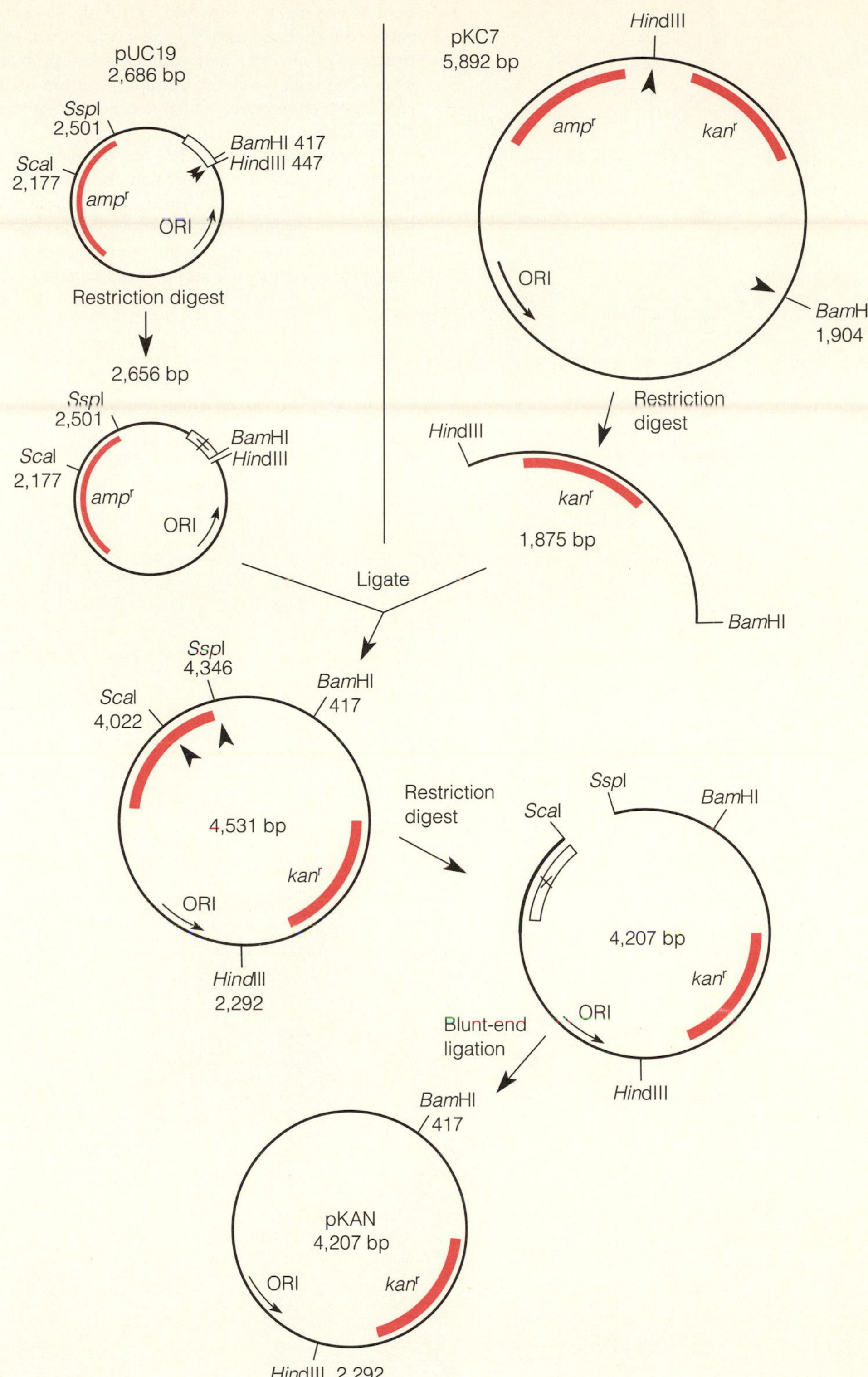

Construction of pKAN

pUC19 was partially digested with *Pvu*II. The plasmid was religated to include a single *Pst*I linker at the *Pvu*II (628) site. The presence of the *Pst*I linker within the *lac* i gene blocks its expression. The plasmid was then double-digested with *Nde*I and *Hin*dIII, which removes a fragment containing the α fragment of the *lac* Z gene.

In parallel, plasmid YEp353 was digested with *Nde*I and *Hin*dIII, producing a fragment that contains all but the extreme 5′ portion of the *lac* Z gene. This fragment is then ligated into the *Nde*I/*Hin*dIII-digested pUC19 plasmid to produce pBLU. This places the complete *lac* Z gene next to its own promoter and allows for the expression of β-galactosidase from the plasmid.

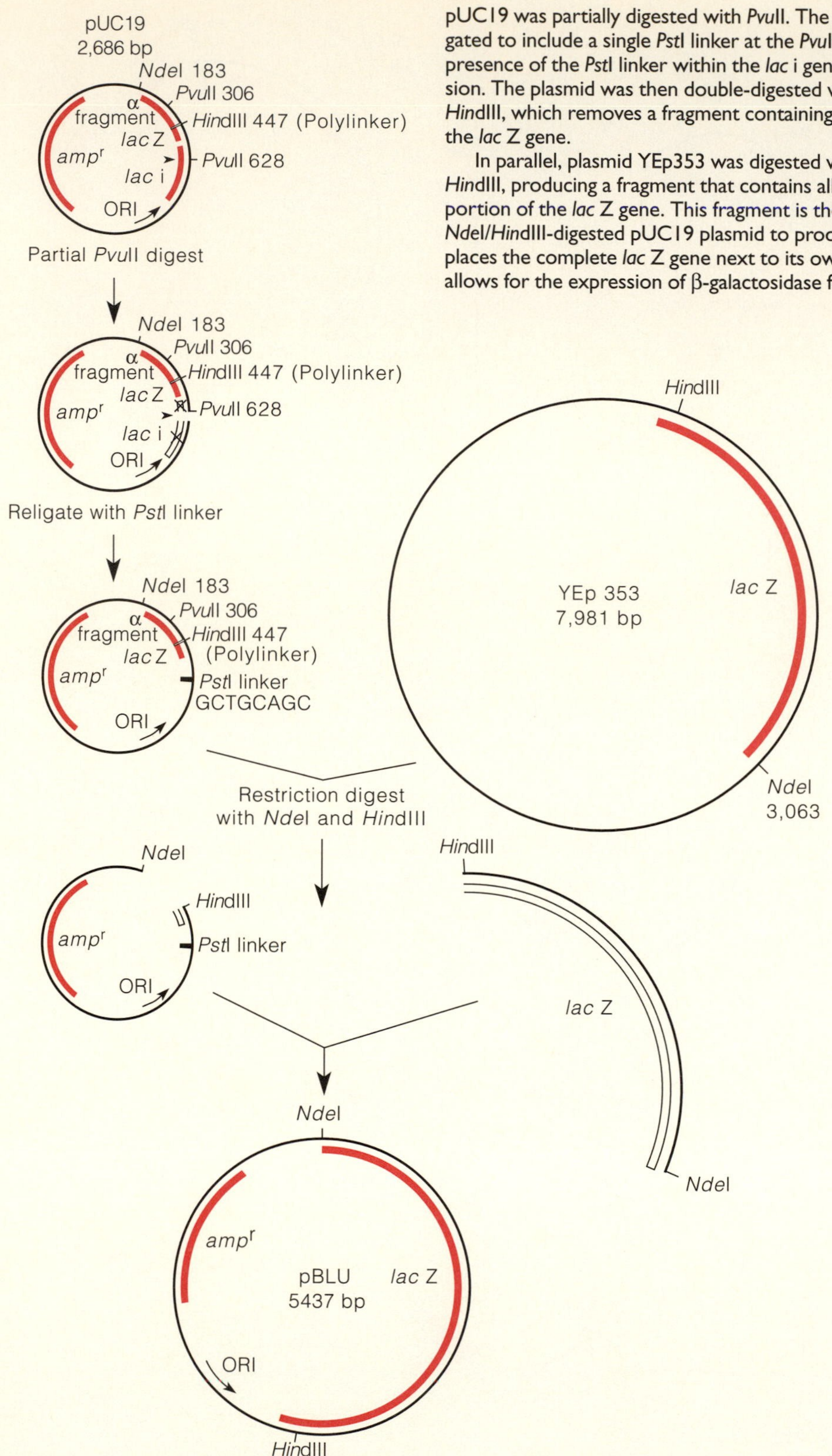

Construction of pBLU

APPENDIX 4

Sequence Data for pAMP, pKAN, pBLU, and Bacteriophage λ

A. NUCLEOTIDE SEQUENCE OF pAMP (4,539 bp COMPLETE SEQUENCE)

1	GCGCCCAATA	CGCAAACCGC	CTCTCCCCGC	GCGTTGGCCG	ATTCATTAAT	GCAGCTGGCA
61	CGACAGGTTT	CCCGACTGGA	AAGCGGGCAG	TGAGCGCAAC	GCAATTAATG	TGAGTTAGCT
121	CACTCATTAG	GCACCCCAGG	CTTTACACTT	TATGCTTCCG	GCTCGTATGT	TGTGTGGAAT
181	TGTGAGCGGA	TAACAATTTC	ACACAGGAAA	CAGCTATGAC	CATGATTACG	CCAAGCTTGC
241	AGAAACGACT	TTTTTAAAGG	ACGGTTATCA	CATTCAAACA	TTAATTTTTT	ATGATAAACA
301	ATTCCATCCA	ATTGATTTAA	TCAATACAAC	ATTTGAAGAT	CAAGCAGATA	AATATATTTT
361	TTGGCGTTAT	GCAGCTGACA	GAGCCAAAAT	AACAAATGCC	TATGGCTTCA	TTTGGATATC
421	AGAGCTATGG	CTCAGAAAAG	CAAGCATCTA	CTCCAATAAA	CCAATACATA	CAATGCCAAT
481	TATAGATGAA	AGACTTCAGG	TAATTGGAAT	TGATTCAAAT	AATAATCAAA	AATGTATTTC
541	ATGGAAAATA	GTTAGAGAAA	ACGAAGAAAA	AAAACCGACT	TTAGAAATAT	CAACAGCAGA
601	CTCAAAACAT	GACGAAAAAC	CATATTTCAT	GCGTTCAGTC	TTAAAAGCAA	TTGGCGGTGA
661	TGTAAACACT	ATGAACAATT	GAGTCATAGA	ACTTCCATTA	TTCTCCTGAA	GATAATAATC
721	GCCAAATAAA	CCAATACTCA	GCTTTACAAT	ATACTAACTA	ACCGCAGAAC	GTTATTTCAT
781	ACAACGTTTC	TGCGGCATAT	CACAAAACGA	TTACTCCATA	ACAGGGACAG	CAGGCCACTC
841	AATATCAGGT	GCAGTTGATG	TATCAACACG	GTTCAGCAAC	ACCCGATACT	TCTTCCAGGC
901	TTCCAGCAAC	GAGGTTTCTT	CCTTCGTTGC	AATTTCCAGA	TCTGCAGCAT	CCTGAAGCGG
961	CGCAATATGC	TCACTGGCTA	CCTGCATCAG	GCTTTTTTTT	GTTTCTTCCG	CCTCCCGGAT
1021	CCGGAACAGT	TTTTCTGCTT	CCGTATCCTT	CACCCAGGCT	GTGCCGTTCC	ACTTCTGATA
1081	TTCCCCTCCC	GGCGATAACC	AGGTAAAATT	TTCCGGTAAC	GGACCGAGTT	CAGAAATAAA
1141	TAACGCGTCG	CCGGAAGCCA	CGTCATAGAC	GGTTTTACCC	CGATGGTCTT	CAACGAGATG
1201	CCACGATGCC	TCATCACTGT	TGAAAACAGC	CACAAAGCCA	GCCGGAATAT	CTGGCGGTGC
1261	AATATCGGTA	CTGTTTGCAG	GCAGACCGGT	ATGAGGCGGA	ATATATGCGT	CACCTTCACC
1321	AATAAATTCA	TTAGTTCCGG	CCAGCAGATT	ATAAATTTTT	ATGGTCCGTG	GTTGTTCACT
1381	CATTCTGAAT	GCCATTATGC	AAGCCTCACA	ATATAGTTAA	ATGCAATGTT	TTTGACGGTG
1441	TTTTCCGCGT	TACCCGCAGC	GTTAACGGTG	ATGGTGTGTC	CGTGTGAACC	AATACTGAAA
1501	GAATGGGCAT	GAGCACCGAT	AACAACCGGA	TGCTGGTGCG	CACCAATACC	AACTGTATGC
1561	GCATGTGCAC	CGGCACTCAC	GGCTGTACCG	GACAATGAGT	GACTGTGGCT	GCCCTGACTG
1621	TCCGTTTTCG	ATAAATAAGC	AATACCCTGT	GTGCTGGTTC	CTTTAACTGT	GGATAAACTT
1681	CCTGTAATGG	TTGCTGTTCC	ATACTGACTC	CAGCCAGAAC	TGTTCATCCT	TAAACCACTT
1741	GTGTGGGCAT	GAGCACCCGC	GGCCCCTGTT	GAACCGCTCA	GACTGTGAGC	ATGAGCCCCC
1801	GTGTTATTCG	TCGATTTGGT	GCCGTAATCG	AAACTGCCTG	TTGTTTTCGT	CCCGTAATCA
1861	AACGACGATG	TGGTTTTCGT	CCCCAAATCC	GTACCGGATG	CACTGGCACT	GTGGGTGTGC
1921	GACTTAATTC	CATCCTGTTC	CTGAGACAAT	ACAGCACGAC	CGCTGGCGGG	TTTCCCCTTG
1981	ATTGTCCAGC	CTCGCATATC	AGGAAGCACA	CCCGATGGAT	ACGCGACAGC	AAGTTTTGGG
2041	TAGGCTGATT	TGTCAAACGC	CTGCCCCTGC	ATCAGGACGT	AGCCAGACGG	AACGATATCT
2101	GATGGCCACG	GGATCGGCGC	ACCTGCCGGA	AAGGCCGAAT	TCACTGGCCG	TCGTTTTACA
2161	ACGTCGTGAC	TGGGAAAACC	CTGGCGTTAC	CCAACTTAAT	CGCCTTGCAG	CACATCCCCC
2221	TTTCGCCAGC	TGGCGTAATA	GCGAAGAGGC	CCGCACCGAT	CGCCCTTCCC	AACAGTTGCG
2281	CAGCCTGAAT	GGCGAATGGC	GCCTGATGCG	GTATTTTCTC	CTTACGCATC	TGTGCGGTAT
2341	TTCACACCGC	ATATGGTGCA	CTCTCAGTAC	AATCTGCTCT	GATGCCGCAT	AGTTAAGCCA
2401	GCCCCGACAC	CCGCCAACAC	CCGCTGACGC	GCCCTGACGG	GCTTGTCTGC	TCCCGGCATC
2461	CGCTTACAGA	CAAGCTGTGA	CCGTCTCCGG	GAGCTGCATG	TGTCAGAGGT	TTTCACCGTC
2521	ATCACCGAAA	CGCGCGAGAC	GAAAGGGCCT	CGTGATACGC	CTATTTTTAT	AGGTTAATGT
2581	CATGATAATA	ATGGTTTCTT	AGACGTCAGG	TGGCACTTTT	CGGGGAAATG	TGCGCGGAAC
2641	CCCTATTTGT	TTATTTTTCT	AAATACATTC	AAATATGTAT	CCGCTCATGA	GACAATAACC
2701	CTGATAAATG	CTTCAATAAT	ATTGAAAAAG	GAAGAGTATG	AGTATTCAAC	ATTTCCGTGT
2761	CGCCCTTATT	CCCTTTTTTG	CGGCATTTTG	CCTTCCTGTT	TTTGCTCACC	CAGAAACGCT
2821	GGTGAAAGTA	AAAGATGCTG	AAGATCAGTT	GGGTGCACGA	GTGGGTTACA	TCGAACTGGA
2881	TCTCAACAGC	GGTAAGATCC	TTGAGAGTTT	TCGCCCCGAA	GAACGTTTTC	CAATGATGAG

```
2941  CACTTTTAAA  GTTCTGCTAT  GTGGCGCGGT  ATTATCCCGT  ATTGACGCCG  GGCAAGAGCA
3001  ACTCGGTCGC  CGCATACACT  ATTCTCAGAA  TGACTTGGTT  GAGTACTCAC  CAGTCACAGA
3061  AAAGCATCTT  ACGGATGGCA  TGACAGTAAG  AGAATTATGC  AGTGCTGCCA  TAACCATGAG
3121  TGATAACACT  GCGGCCAACT  TACTTCTGAC  AACGATCGGA  GGACCGAAGG  AGCTAACCGC
3181  TTTTTTGCAC  AACATGGGGG  ATCATGTAAC  TCGCCTTGAT  CGTTGGGAAC  CGGAGCTGAA
3241  TGAAGCCATA  CCAAACGACG  AGCGTGACAC  CACGATGCCT  GTAGCAATGG  CAACAACGTT
3301  GCGCAAACTA  TTAACTGGCG  AACTACTTAC  TCTAGCTTCC  CGGCAACAAT  TAATAGACTG
3361  GATGGAGGCG  GATAAAGTTG  CAGGACCACT  TCTGCGCTCG  GCCCTTCCGG  CTGGCTGGTT
3421  TATTGCTGAT  AAATCTGGAG  CCGGTGAGCG  TGGGTCTCGC  GGTATCATTG  CAGCACTGGG
3481  GCCAGATGGT  AAGCCCTCCC  GTATCGTAGT  TATCTACACG  ACGGGGAGTC  AGGCAACTAT
3541  GGATGAACGA  AATAGACAGA  TCGCTGAGAT  AGGTGCCTCA  CTGATTAAGC  ATTGGTAACT
3601  GTCAGACCAA  GTTTACTCAT  ATATACTTTA  GATTGATTTA  AAACTTCATT  TTTAATTTAA
3661  AAGGATCTAG  GTGAAGATCC  TTTTTGATAA  TCTCATGACC  AAAATCCCTT  AACGTGAGTT
3721  TTCGTTCCAC  TGAGCGTCAG  ACCCCGTAGA  AAAGATCAAA  GGATCTTCTT  GAGATCCTTT
3781  TTTTCTGCGC  GTAATCTGCT  GCTTGCAAAC  AAAAAAACCA  CCGCTACCAG  CGGTGGTTTG
3841  TTTGCCGGAT  CAAGAGCTAC  CAACTCTTTT  TCCGAAGGTA  ACTGGCTTCA  GCAGAGCGCA
3901  GATACCAAAT  ACTGTCCTTC  TAGTGTAGCC  GTAGTTAGGC  CACCACTTCA  AGAACTCTGT
3961  AGCACCGCCT  ACATACCTCG  CTCTGCTAAT  CCTGTTACCA  GTGGCTGCTG  CCAGTGGCGA
4021  TAAGTCGTGT  CTTACCGGGT  TGGACTCAAG  ACGATAGTTA  CCGGATAAGG  CGCAGCGGTC
4081  GGGCTGAACG  GGGGGTTCGT  GCACACAGCC  CAGCTTGGAG  CGAACGACCT  ACACCGAACT
4141  GAGATACCTA  CAGCGTGAGC  TATGAGAAAG  CGCCACGCTT  CCCGAAGGGA  GAAAGGCGGA
4201  CAGGTATCCG  GTAAGCGGCA  GGGTCGGAAC  AGGAGAGCGC  ACGAGGGAGC  TTCCAGGGGG
4261  AAACGCCTGG  TATCTTTATA  GTCCTGTCGG  GTTTCGCCAC  CTCTGACTTG  AGCGTCGATT
4321  TTTGTGATGC  TCGTCAGGGG  GGCGGAGCCT  ATGGAAAAAC  GCCAGCAACG  CGGCCTTTTT
4381  ACGGTTCCTG  GCCTTTTGCT  GGCCTTTTGC  TCACATGTTC  TTTCCTGCGT  TATCCCCTGA
4441  TTCTGTGGAT  AACCGTATTA  CCGCCTTTGA  GTGAGCTGAT  ACCGCTCGCC  GCAGCCGAAC
4501  GACCGAGCGC  AGCGAGTCAG  TGAGCGAGGA  AGCGGAAGA
```

B. NUCLEOTIDE SEQUENCE OF pKAN (4,194 bp COMPLETE SEQUENCE)

```
   1  AGCGCCCAAT  ACGCAAACCG  CCTCTCCCCG  CGCGTTGGCC  GATTCATTAA  TGCAGCTGGC
  61  ACGACAGGTT  TCCCGACTGG  AAAGCGGGCA  GTGAGCGCAA  CGCAATTAAT  GTGAGTTAGC
 121  TCACTCATTA  GGCACCCCAG  GCTTTACACT  TTATGCTTCC  GGCTCGTATG  TTGTGTGGAA
 181  TTGTGAGCGG  ATAACAATTT  CACACAGGAA  ACAGCTATGA  CCATGATTAC  GCCAAGCTTC
 241  ACGCTGCCGC  AAGCACTCAG  GGCGCAAGGG  CTGCTAAAGG  AAGCGGAACA  CGTAGAAAGC
 301  CAGTCCGCAG  AAACGGTGCT  GACCCCGGAT  GAATGTCAGC  TACTGGGCTA  TCTGGACAAG
 361  GGAAAACGCA  AGCGCAAAGA  GAAAGCAGGT  AGCTTGCAGT  GGGCTTACAT  GGCGATAGCT
 421  AGACTGGGCG  GTTTTATGGA  CAGCAAGCGA  ACCGGAATTG  CCAGCTGGGG  CGCCCTCTGG
 481  TAAGGTTGGG  AAGCCCTGCA  AAGTAAACTG  GATGGCTTTC  TTGCCGCCAA  GGATCTGATG
 541  GCGCAGGGGA  TCAAGATCTG  ATCAAGAGAC  AGGATGAGGA  TCGTTTCGCA  TGATTGAACA
 601  AGATGGATTG  CACGCAGGTT  CTCCGGCCGC  TTGGGTGGAG  AGGCTATTCG  GCTATGACTG
 661  GGCACAACAG  ACAATCGGCT  GCTCTGATGC  CGCCGTGTTC  CGGCTGTCAG  CGCAGGGGCG
 721  CCCGGTTCTT  TTTGTCAAGA  CCGACCTGTC  CGGTGCCCTG  AATGAACTGC  AGGACGAGGC
 781  AGCGCGGCTA  TCGTGGCTGG  CCACGACGGG  CGTTCCTTGC  GCAGCTGTGC  TCGACGTTGT
 841  CACTGAAGCG  GGAAGGGACT  GGCTGCTATT  GGGCGAAGTG  CCGGGGCAGG  ATCTCCTGTC
 901  ATCTCACCTT  GCTCCTGCCG  AGAAAGTATC  CATCATGGCT  GATGCAATGC  GGCGGCTGCA
 961  TACGCTTGAT  CCGGCTACCT  GCCCATTCGA  CCACCAAGCG  AAACATCGCA  TCGAGCGAGC
1021  ACGTACTCGG  ATGGAAGCCG  GTCTTGTCGA  TCAGGATGAT  CTGGACGAAG  AGCATCAGGG
1081  GCTCGCGCCA  GCCGAACTGT  TCGCCAGGCT  CAAGGCGCGC  ATGCCCGACG  GCGAGGATCT
1141  CGTCGTGACC  CATGGCGATG  CCTGCTTGCC  GAATATCATG  GTGGAAAATG  GCCGCTTTTC
```

```
1201  TGGATTCATC  GACTGTGGCC  GGCTGGGTGT  GGCGGACCGC  TATCAGGACA  TAGCGTTGGC
1261  TACCCGTGAT  ATTGCTGAAG  AGCTTGGCGG  CGAATGGGCT  GACCGCTTCC  TCGTGCTTTA
1321  CGGTATCGCC  GCTCCCGATT  CGCAGCGCAT  CGCCTTCTAT  CGCCTTCTTG  ACGAGTTCTT
1381  CTGAGCGGGA  CTCTGGGGTT  CGAAATGACC  GACCAAGCGA  CGCCCAACCT  GCCATCACGA
1441  GATTTCGATT  CCACCGCCGC  CTTCTATGAA  AGGTTGGGCT  TCGGAATCGT  TTTCCGGGAC
1501  GCCGGCTGGA  TGATCCTCCA  GCGCGGGGAT  CTCATGCTGG  AGTTCTTCGC  CCACCCCGGG
1561  CTCGATCCCC  TCGCGAGTTG  GTTCAGCTGC  TGCCTGAGGC  TGGACGACCT  CGCGGAGTTC
1621  TACCGGCAGT  GCAAATCCGT  CGGCATCCAG  GAAACCAGCA  GCGGCTATCC  GCGCATCCAT
1681  GCCCCCGAAC  TGCAGGAGTG  GGGAGGCACG  ATGGCCGCTT  TGGTCGACCC  GGACGGGACG
1741  CTCCTGCGCC  TGATACAGAA  CGAATTGCTT  GCAGGCATCT  CATGAGTGTG  TCTTCCCGTT
1801  TTCCGCCTGA  GGTCACTGCG  TGGATGGAGC  GCTGGCGCCT  GCTGCGCGAC  GGCGAGCTGC
1861  TCACCACCCA  CTCGAGCTGG  ATACTTCCCG  TCCGCCAGGG  GGACATGCCG  GCGATGCTGA
1921  AGGTCGCGCG  CATTCCCGAT  GAAGAGGCCG  GTTACCGCCT  GTTGACCTGG  TGGGACGGGC
1981  AGGGCGCCGC  CCGAGTCTTC  GCCTCGGCGG  CGGGCGCTCT  GCTCATGGAG  CGCGCGTCCG
2041  GGGCCGGGGA  CCTTGCACAG  ATAGCGTGGT  CCGGCCAGGA  CGACGAGGCT  TGCAGGATCC
2101  CCGGGTACCG  AGCTCGAATT  CACTGGCCGT  CGTTTTACAA  CGTCGTGACT  GGGAAAACCC
2161  TGGCGTTACC  CAACTTAATC  GCCTTGCAGC  ACATCCCCCT  TTCGCCAGCT  GGCGTAATAG
2221  CGAAGAGGCC  CGCACCGATC  GCCCTTCCCA  ACAGTTGCGC  AGCCTGAATG  GCGAATGGCG
2281  CCTGATGCGG  TATTTTCTCC  TTACGCATCT  GTGCGGTATT  TCACACCGCA  TATGGTGCAC
2341  TCTCAGTACA  ATCTGCTCTG  ATGCCGCATA  GTTAAGCCAG  CCCCGACACC  CGCCAACACC
2401  CGCTGACGCG  CCCTGACGGG  CTTGTCTGCT  CCCGGCATCC  GCTTACAGAC  AAGCTGTGAC
2461  CGTCTCCGGG  AGCTGCATGT  GTCAGAGGTT  TTCACCGTCA  TCACCGAAAC  GCGCGAGACG
2521  AAAGGGCCTC  GTGATACGCC  TATTTTTATA  GGTTAATGTC  ATGATAATAA  TGGTTTCTTA
2581  GACGTCAGGT  GGCACTTTTC  GGGGAAATGT  GCGCGGAACC  CCTATTTGTT  TATTTTTCTA
2641  AATACATTCA  AATATGTATC  CGCTCATGAG  ACAATAACCC  TGATAAATGC  TTCAATAATA
2701  CTCACCAGTC  ACAGAAAAGC  ATCTTACGGA  TGGCATGACA  GTAAGAGAAT  TATGCAGTGC
2761  TGCCATAACC  ATGAGTGATA  ACACTGCGGC  CAACTTACTT  CTGACAACGA  TCGGAGGACC
2821  GAAGGAGCTA  ACCGCTTTTT  TGCACAACAT  GGGGGATCAT  GTAACTCGCC  TTGATCGTTG
2881  GGAACCGGAG  CTGAATGAAG  CCATACCAAA  CGACGAGCGT  GACACCACGA  TGCCTGTAGC
2941  AATGGCAACA  ACGTTGCGCA  AACTATTAAC  TGGCGAACTA  CTTACTCTAG  CTTCCCGGCA
3001  ACAATTAATA  GACTGGATGG  AGGCGGATAA  AGTTGCAGGA  CCACTTCTGC  GCTCGGCCCT
3061  TCCGGCTGGC  TGGTTTATTG  CTGATAAATC  TGGAGCCGGT  GAGCGTGGGT  CTCGCGGTAT
3121  CATTGCAGCA  CTGGGGCCAG  ATGGTAAGCC  CTCCCGTATC  GTAGTTATCT  ACACGACGGG
3181  GAGTCAGGCA  ACTATGGATG  AACGAAATAG  ACAGATCGCT  GAGATAGGTG  CCTCACTGAT
3241  TAAGCATTGG  TAACTGTCAG  ACCAAGTTTA  CTCATATATA  CTTTAGATTG  ATTTAAAACT
3301  TCATTTTTAA  TTTAAAAGGA  TCTAGGTGAA  GATCCTTTTT  GATAATCTCA  TGACCAAAAT
3361  CCCTTAACGT  GAGTTTTCGT  TCCACTGAGC  GTCAGACCCC  GTAGAAAAGA  TCAAAGGATC
3421  TTCTTGAGAT  CCTTTTTTTC  TGCGCGTAAT  CTGCTGCTTG  CAAACAAAAA  AACCACCGCT
3481  ACCAGCGGTG  GTTTGTTTGC  CGGATCAAGA  GCTACCAACT  CTTTTTCCGA  AGGTAACTGG
3541  CTTCAGCAGA  GCGCAGATAC  CAAATACTGT  CCTTCTAGTG  TAGCCGTAGT  TAGGCCACCA
3601  CTTCAAGAAC  TCTGTAGCAC  CGCCTACATA  CCTCGCTCTG  CTAATCCTGT  TACCAGTGGC
3661  TGCTGCCAGT  GGCGATAAGT  CGTGTCTTAC  CGGGTTGGAC  TCAAGACGAT  AGTTACCGGA
3721  TAAGGCGCAG  CGGTCGGGCT  GAACGGGGGG  TTCGTGCACA  CAGCCCAGCT  TGGAGCGAAC
3781  GACCTACACC  GAACTGAGAT  ACCTACAGCG  TGAGCTATGA  GAAAGCGCCA  CGCTTCCCGA
3841  AGGGAGAAAG  GCGGACAGGT  ATCCGGTAAG  CGGCAGGGTC  GGAACAGGAG  AGCGCACGAG
3901  GGAGCTTCCA  GGGGGAAACG  CCTGGTATCT  TTATAGTCCT  GTCGGGTTTC  GCCACCTCTG
3961  ACTTGAGCGT  CGATTTTTGT  GATGCTCGTC  AGGGGGGCGG  AGCCTATGGA  AAAACGCCAG
4021  CAACGCGGCC  TTTTTACGGT  TCCTGGCCTT  TTGCTGGCCT  TTTGCTCACA  TGTTCTTTCC
4081  TGCGTTATCC  CCTGATTCTG  TGGATAACCG  TATTACCGCC  TTTGAGTGAG  CTGATACCGC
4141  TCGCCGCAGC  CGAACGACCG  AGCGCAGCGA  GTCAGTGAGC  GAGGAAGCGG  AAGA
```

C. NUCLEOTIDE SEQUENCE OF pBLU (5,437 bp COMPLETE SEQUENCE)

1	TCGCGCGTTT	CGGTGATGAC	GGTGAAAACC	TCTGACACAT	GCAGCTCCCG	GAGACGGTCA
61	CAGCTTGTCT	GTAAGCGGAT	GCCGGGAGCA	GACAAGCCCG	TCAGGGCGCG	TCAGCGGGTG
121	TTGGCGGGTG	TCGGGGCTGG	CTTAACTATG	CGGCATCAGA	GCAGATTGTA	CTGAGAGTGC
181	ACCATATGGA	AACCGTCGAT	ATTCAGCCAT	GTGCCTTCTT	CCGCGTGCAG	CAGATGGCGA
241	TGGCTGGTTT	CCATCAGTTG	CTGTTGACTG	TAGCGGCTGA	TGTTGAACTG	GAAGTCGCCG
301	CGCCACTGGT	GTGGGCCATA	ATTCAATTCG	CGCGTCCCGC	AGCGCAGACC	GTTTTCGCTC
361	GGGAAGACGT	ACGGGGTATA	CATGTCTGAC	AATGGCAGAT	CCCAGCGGTC	AAAACAGGCG
421	GCAGTAAGGC	GGTCGGGATA	GTTTTCTTGC	GGCCCTAATC	CGAGCCAGTT	TACCCGCTCT
481	GCTACCTGCG	CCAGCTGGCA	GTTCAGGCCA	ATCCGCGCCG	GATGCGGTGT	ATCGCTCGCC
541	ACTTCAACAT	CAACGGTAAT	CGCCATTTGA	CCACTACCAT	CAATCCGGTA	GGTTTTCCGG
601	CTGATAAATA	AGGTTTTCCC	CTGATGCTGC	CACGCGTGAG	CGGTCGTAAT	CAGCACCGCA
661	TCAGCAAGTG	TATCTGCCGT	GCACTGCAAC	AACGCTGCTT	CGGCCTGGTA	ATGGCCCGCC
721	GCCTTCCAGC	GTTCGACCCA	GGCGTTAGGG	TCAATGCGGG	TCGCTTCACT	TACGCCAATG
781	TCGTTATCCA	GCGGTGCACG	GGTGAACTGA	TCGCGCAGCG	GCGTCAGCAG	TTGTTTTTTA
841	TCGCCAATCC	ACATCTGTGA	AAGAAAGCCT	GACTGGCGGT	TAAATTGCCA	ACGCTTATTA
901	CCCAGCTCGA	TGCAAAAATC	CATTTCGCTG	GTGGTCAGAT	GCGGCATGGC	GTGGGACGCG
961	GCGGGGAGCG	TCACACTGAG	GTTTTCCGCC	AGACGCCACT	GCTGCCAGGC	GCTGATGTGC
1021	CCGGCTTCTG	ACCATGCGGT	CGCGTTCGGT	TGCACTACGC	GTACTGTGAG	CCAGAGTTGC
1081	CCGGCGCTCT	CCGGCTGCGG	TAGTTCAGGC	AGTTCAATCA	ACTGTTTACC	TTGTGGAGCG
1141	ACATCCTGAG	GCACTTCACC	GCTTGCCAGC	GGCTTACCAT	CCAGCGCCAC	CATCCAGTGC
1201	AGGAGCTCGT	TATCGCTATG	ACGGAACAGG	TATTCGCTGG	TCACTTCGAT	GGTTTGCCCG
1261	GATAAACGGA	ACTGGAAAAA	CTGCTGCTGG	TGTTTTGCTT	CCGTCAGCGC	TGGATGCGGC
1321	GTGCGGTCGG	CAAAGACCAG	ACCGTTCATA	CAGAACTGGC	GATCGTTCGG	CGTATCGCCA
1381	AAATCACCGC	CGTAAGCCGA	CCACGGGTTG	CCGTTTTCAT	CATATTTAAT	CAGCGACTGA
1441	TCCACCCAGT	CCCAGACGAA	GCCGCCCTGT	AAACGGGGAT	ACTGACGAAA	CGCCTGCCAG
1501	TATTTAGCGA	AACCGCCAAG	ACTGTTACCC	ATCGCGTGGG	CGTATTCGCA	AAGGATCAGC
1561	GGGCGCGTCT	CTCCAGGTAG	CGAAAGCCAT	TTTTTGATGG	ACCATTTCGG	CACAGCCGGG
1621	AAGGGGTGGT	CTTCATCCAC	GCGCGCGTAC	ATCGGGCAAA	TAATATCGGT	GGCCGTGGTG
1681	TCGGCTCCGC	CGCCTTCATA	CTGCACCGGG	CGGGAAGGAT	CGACAGATTT	GATCCAGCGA
1741	TACAGCGCGT	CGTGATTAGC	GCCGTGGCCT	GATTCATTCC	CCAGCGACCA	GATGATCACA
1801	CTCGGGTGAT	TACGATCGCG	CTGCACCATT	CGCGTTACGC	GTTCGCTCAT	CGCCGGTAGC
1861	CAGCGCGGAT	CATCGGTCAG	ACGATTCATT	GGCACCATGC	CGTGGGTTTC	AATATTGGCT
1921	TCATCCACCA	CATACAGGCC	GTAGCGTTCG	CACAGCGTGT	ACCACAGCGG	ATGGTTCGGA
1981	TAATGCGAAC	AGCGCACGGC	TTTAAAGTTG	TTCTGCTTCA	TCAGCAGGAT	ATCCTGCACC
2041	ATCGTCTGCT	CATCCATGAC	CTGACCATGC	AGAGGATGAT	GCTCGTGACG	GTTAACGCCT
2101	CGAATCAGCA	ACGGCTTGCC	GTTCAGCAGC	AGCAGACCAT	TTTCAATCCG	CACCTCGCGG
2161	AAACCGACAT	CGCAGGCTTC	TGCTTCAATC	AGCGTGCCGT	CGGCGGTGTG	CAGTTCAACC
2221	ACCGCACGAT	AGAGATTCGG	GATTTCGGCG	CTCCACAGTT	TCGGGTTTTC	CAGGTTCAGA
2281	CGTAGTGTGA	CGCGATCGGC	ATCACCACCA	CGCTCATCGA	TAATTTCACC	GCCGAAAGGC
2341	GCGGTGCCGC	TGGCGACCTG	CGTTTCACCC	TGCCATAAAG	AAACTGTTAC	CCGTAGGTAG
2401	TCACGCAACT	CGCCGCACAT	CTGAACTTCA	GCCTCCAGTA	CAGCGCGGCT	TAAATCATCA
2461	TTAAAGCGAG	TGGCAACATG	GAAATCGCTG	ATTTGTGTAG	TCGGTTTATG	CAGCAACGAG
2521	ACGTCACGGA	AAATGCCGCT	CATCCGCCAC	ATATCCTGAT	CTTCCAGATA	ACTGCCTTCA
2581	CTCCAACGCA	GCACCATCAC	CGCGAGGCGG	TTTTCTCCGG	CGCGTAAAAA	AGCGCTCAGG
2641	TCAAATTCAG	ACGGCAAACG	ACTGTCCTGG	CCGTAACCGA	CCCAGCGCCC	GTTGCACCAC
2701	AGATGAAACG	CCGAGTTAAC	GCCATCAAAA	ATAATTCGCG	TCTGGCCTTC	CTGTAGCCAG
2761	CTTTCATCAA	CATTAAATGT	GAGCGAGTAA	CAACCCGTCG	GATTCTCCGT	GGGAACAAAC
2821	GGCGGATTGA	CCGTAATGGG	ATAGGTTACG	TTGGTGTAGA	TGGGCGCATC	GTAACCGTGC
2881	ATCTGCCAGT	TTGAGGGGAC	GACGACAGTA	TCGGCCTCAG	GAAGATCGCA	CTCCAGCCAG

2941 CTTTCCGGCA CCGCTTCTGG TGCCGGAAAC CAGGCAAAGC GCCATTCGCC ATTCAGGCTG
3001 CGCAACTGTT GGGAAGGGCG ATCGGTGCGG GCCTCTTCGC TATTACGCCA GCTGGCGAAA
3061 GGGGGATGTG CTGCAAGGCG ATTAAGTTGG GTAACGCCAG GGTTTTCCCA GTCACGACGT
3121 TGTAAAACGA CCGCCAGTGA ATTCGAGCTC GGTACCCGGG GATCCTCTAG AGTCGACCTG
3181 CAGGCATGCA AGCTTGGCGT AATCATGGTC ATAGCTGTTT CCTGTGTGAA ATTGTTATCC
3241 GCTCACAATT CCACACAACA TACGAGCCGG AAGCATAAAG TGTAAAGCCT GGGGTGCCTA
3301 ATGAGTGAGC TAACTCACAT TAATTGCGTT GCGCTCACTG CCCGCTTTCC AGTCGGGAAA
3361 CCTGTCGTGC CAGGCTGCAG CCTGCATTAA TGAATCGGCC AACGCGCGGG GAGAGGCGGT
3421 TTGCGTATTG GGCGCTCTTC CGCTTCCTCG CTCACTGACT CGCTGCGCTC GGTCGTTCGG
3481 CTGCGGCGAG CGGTATCAGC TCACTCAAAG GCGGTAATAC GGTTATCCAC AGAATCAGGG
3541 GATAACGCAG GAAAGAACAT GTGAGCAAAA GGCCAGCAAA AGGCCAGGAA CCGTAAAAAG
3601 GCCGCGTTGC TGGCGTTTTT CCATAGGCTC CGCCCCCCTG ACGAGCATCA CAAAAATCGA
3661 CGCTCAAGTC AGAGGTGGCG AAACCCGACA GGACTATAAA GATACCAGGC GTTTCCCCCT
3721 GGAAGCTCCC TCGTGCGCTC TCCTGTTCCG ACCCTGCCGC TTACCGGATA CCTGTCCGCC
3781 TTTCTCCCTT CGGGAAGCGT GGCGCTTTCT CATAGCTCAC GCTGTAGGTA TCTCAGTTCG
3841 GTGTAGGTCG TTCGCTCCAA GCTGGGCTGT GTGCACGAAC CCCCCGTTCA GCCCGACCGC
3901 TGCGCCTTAT CCGGTAACTA TCGTCTTGAG TCCAACCCGG TAAGACACGA CTTATCGCCA
3961 CTGGCAGCAG CCACTGGTAA CAGGATTAGC AGAGCGAGGT ATGTAGGCGG TGCTACAGAG
4021 TTCTTGAAGT GGTGGCCTAA CTACGGCTAC ACTAGAAGGA CAGTATTTGG TATCTGCGCT
4081 CTGCTGAAGC CAGTTACCTT CGGAAAAAGA GTTGGTAGCT CTTGATCCGG CAAACAAACC
4141 ACCGCTGGTA GCGGTGGTTT TTTTGTTTGC AAGCAGCAGA TTACGCGCAG AAAAAAAGGA
4201 TCTCAAGAAG ATCCTTTGAT CTTTTCTACG GGGTCTGACG CTCAGTGGAA CGAAAACTCA
4261 CGTTAAGGGA TTTTGGTCAT GAGATTATCA AAAAGGATCT TCACCTAGAT CCTTTTAAAT
4321 TAAAAATGAA GTTTTAAATC AATCTAAAGT ATATATGAGT AAACTTGGTC TGACAGTTAC
4381 CAATGCTTAA TCAGTGAGGC ACCTATCTCA GCGATCTGTC TATTTCGTTC ATCCATAGTT
4441 GCCTGACTCC CCGTCGTGTA GATAACTACG ATACGGGAGG GCTTACCATC TGGCCCCAGT
4501 GCTGCAATGA TACCGCGAGA CCCACGCTCA CCGGCTCCAG ATTTATCAGC AATAAACCAG
4561 CCAGCCGGAA GGGCCGAGCG CAGAAGTGGT CCTGCAACTT TATCCGCCTC CATCCAGTCT
4621 ATTAATTGTT GCCGGGAAGC TAGAGTAAGT AGTTCGCCAG TTAATAGTTT GCGCAACGTT
4681 GTTGCCATTG CTGCAGGCAT CGTGGTGTCA CGCTCGTCGT TTGGTATGGC TTCATTCAGC
4741 TCCGGTTCCC AACGATCAAG GCGAGTTACA TGATCCCCCA TGTTGTGCAA AAAAGCGGTT
4801 AGCTCCTTCG GTCCTCCGAT CGTTGTCAGA AGTAAGTTGG CCGCAGTGTT ATCACTCATG
4861 GTTATGGCAG CACTGCATAA TTCTCTTACT GTCATGCCAT CCGTAAGATG CTTTTCTGTG
4921 ACTGGTGAGT ACTCAACCAA GTCATTCTGA GAATAGTGTA TGCGGCGACC GAGTTGCTCT
4981 TGCCCGGCGT CAACACGGGA TAATACCGCG CCACATAGCA GAACTTTAAA AGTGCTCATC
5041 ATTGGAAAAC GTTCTTCGGG GCGAAAACTC TCAAGGATCT TACCGCTGTT GAGATCCAGT
5101 TCGATGTAAC CCACTCGTGC ACCCAACTGA TCTTCAGCAT CTTTTACTTT CACCAGCGTT
5161 TCTGGGTGAG CAAAAACAGG AAGGCAAAAT GCCGCAAAAA AGGGAATAAG GGCGACACGG
5221 AAATGTTGAA TACTCATACT CTTCCTTTTT CAATATTATT GAAGCATTTA TCAGGGTTAT
5281 TGTCTCATGA GCGGATACAT ATTTGAATGT ATTTAGAAAA ATAAACAAAT AGGGGTTCCG
5341 CGCACATTTC CCCGAAAAGT GCCACCTGAC GTCTAAGAAA CCATTATTAT CATGACATTA
5401 ACCTATAAAA ATAGGCGTAT CACGAGGCCC TTTCGTC

D. NUCLEOTIDE SEQUENCE OF BACTERIOPHAGE λ

1 GGGCGGCGAC CTCGCGGGTT TTCGCTATTT ATGAAAATTT TCCGGTTTAA GGCGTTTCCG
61 TTCTTCTTCG TCATAACTTA ATGTTTTTAT TTAAAATACC CTCTGAAAAG AAAGGAAACG
121 ACAGGTGCTG AAAGCGAGGC TTTTTGGCCT CTGTCGTTTC CTTTCTCTGT TTTTGTCCGT
181 GGAATGAACA ATGGAAGTCA ACAAAAAGCA GCTGGCTGAC ATTTTCGGTG CGAGTATCCG
241 TACCATTCAG AACTGGCAGG AACAGGGAAT GCCCGTTCTG CGAGGCGGTG GCAAGGGTAA
301 TGAGGTGCTT TATGACTCTG CCGCCGTCAT AAAATGGTAT GCCGAAAGGG ATGCTGAAAT
361 TGAGAACGAA AAGCTGCGCC GGGAGGTTGA AGAACTGCGG CAGGCCAGCG AGGCAGATCT

421	CCAGCCAGGA	ACTATTGAGT	ACGAACGCCA	TCGACTTACG	CGTGCGCAGG	CCGACGCACA
481	GGAACTGAAG	AATGCCAGAG	ACTCCGCTGA	AGTGGTGGAA	ACCGCATTCT	GTACTTTCGT
541	GCTGTCGCGG	ATCGCAGGTG	AAATTGCCAG	TATTCTCGAC	GGGCTCCCCC	TGTCGGTGCA
601	GCGGCGTTTT	CCGGAACTGG	AAAACCGACA	TGTTGATTTC	CTGAAACGGG	ATATCATCAA
661	AGCCATGAAC	AAAGCAGCCG	CGCTGGATGA	ACTGATACCG	GGGTTGCTGA	GTGAATATAT
721	CGAACAGTCA	GGTTAACAGG	CTGCGGCATT	TTGTCCGCGC	CGGGCTTCGC	TCACTGTTCA
781	GGCCGGAGCC	ACAGACCGCC	GTTGAATGGG	CGGATGCTAA	TTACTATCTC	CCGAAAGAAT
841	CCGCATACCA	GGAAGGGCGC	TGGGAAACAC	TGCCCTTTCA	GCGGGCCATC	ATGAATGCGA
901	TGGGCAGCGA	CTACATCCGT	GAGGTGAATG	TGGTGAAGTC	TGCCCGTGTC	GGTTATTCCA
961	AAATGCTGCT	GGGTGTTTAT	GCCTACTTTA	TAGAGCATAA	GCAGCGCAAC	ACCCTTATCT
1021	GGTTGCCGAC	GGATGGTGAT	GCCGAGAACT	TTATGAAAAC	CCACGTTGAG	CCGACTATTC
1081	GTGATATTCC	GTCGCTGCTG	GCGCTGGCCC	CGTGGTATGG	CAAAAAGCAC	CGGGATAACA
1141	CGCTCACCAT	GAAGCGTTTC	ACTAATGGGC	GTGGCTTCTG	GTGCCTGGGC	GGTAAAGCGG
1201	CAAAAAACTA	CCGTGAAAAG	TCGGTGGATG	TGGCGGGTTA	TGATGAACTT	GCTGCTTTTG
1261	ATGATGATAT	TGAACAGGAA	GGCTCTCCGA	CGTTCCTGGG	TGACAAGCGT	ATTGAAGGCT
1321	CGGTCTGGCC	AAAGTCCATC	CGTGGCTCCA	CGCCAAAAGT	GAGAGGCACC	TGTCAGATTG
1381	AGCGTGCAGC	CAGTGAATCC	CCGCATTTTA	TGCGTTTTCA	TGTTGCCTGC	CCGCATTGCG
1441	GGGAGGAGCA	GTATCTTAAA	TTTGGCGACA	AAGAGACGCC	GTTTGGCCTC	AAATGGACGC
1501	CGGATGACCC	CTCCAGCGTG	TTTTATCTCT	GCGAGCATAA	TGCCTGCGTC	ATCCGCCAGC
1561	AGGAGCTGGA	CTTTACTGAT	GCCCGTTATA	TCTGCGAAAA	GACCGGGATC	TGGACCCGTG
1621	ATGGCATTCT	CTGGTTTTCG	TCATCCGGTG	AAGAGATTGA	GCCACCTGAC	AGTGTGACCT
1681	TTCACATCTG	GACAGCGTAC	AGCCCGTTCA	CCACCTGGGT	GCAGATTGTC	AAAGACTGGA
1741	TGAAAACGAA	AGGGGATACG	GGAAAACGTA	AAACCTTCGT	AAACACCACG	CTCGGTGAGA
1801	CGTGGGAGGC	GAAAATTGGC	GAACGTCCGG	ATGCTGAAGT	GATGGCAGAG	CGGAAAGAGC
1861	ATTATTCAGC	GCCCGTTCCT	GACCGTGTGG	CTTACCTGAC	CGCCGGTATC	GACTCCCAGC
1921	TGGACCGCTA	CGAAATGCGC	GTATGGGGAT	GGGGGCCGGG	TGAGGAAAGC	TGGCTGATTG
1981	ACCGGCAGAT	TATTATGGGC	CGCCACGACG	ATGAACAGAC	GCTGCTGCGT	GTGGATGAGG
2041	CCATCAATAA	AACCTATACC	CGCCGGAATG	GTGCAGAAAT	GTCGATATCC	CGTATCTGCT
2101	GGGATACTGG	CGGGATTGAC	CCGACCATTG	TGTATGAACG	CTCGAAAAAA	CATGGGCTGT
2161	TCCGGGTGAT	CCCCATTAAA	GGGGCATCCG	TCTACGGAAA	GCCGGTGGCC	AGCATGCCAC
2221	GTAAGCGAAA	CAAAAACGGG	GTTTACCTTA	CCGAAATCGG	TACGGATACC	GCGAAAGAGC
2281	AGATTTATAA	CCGCTTCACA	CTGACGCCGG	AAGGGGATGA	ACCGCTTCCC	GGTGCCGTTC
2341	ACTTCCCGAA	TAACCCGGAT	ATTTTTGATC	TGACCGAAGC	GCAGCAGCTG	ACTGCTGAAG
2401	AGCAGGTCGA	AAAATGGGTG	GATGGCAGGA	AAAAAATACT	GTGGGACAGC	AAAAAGCGAC
2461	GCAATGAGGC	ACTCGACTGC	TTCGTTTATG	CGCTGGCGGC	GCTGCGCATC	AGTATTTCCC
2521	GCTGGCAGCT	GGATCTCAGT	GCGCTGCTGG	CGAGCCTGCA	GGAAGAGGAT	GGTGCAGCAA
2581	CCAACAAGAA	AACACTGGCA	GATTACGCCC	GTGCCTTATC	CGGAGAGGAT	GAATGACGCG
2641	ACAGGAAGAA	CTTGCCGCTG	CCCGTGCGGC	ACTGCATGAC	CTGATGACAG	GTAAACGGGT
2701	GGCAACAGTA	CAGAAAGACG	GACGAAGGGT	GGAGTTTACG	GCCACTTCCG	TGTCTGACCT
2761	GAAAAAATAT	ATTGCAGAGC	TGGAAGTGCA	GACCGGCATG	ACACAGCGAC	GCAGGGGACC
2821	TGCAGGATTT	TATGTATGAA	AACGCCCACC	ATTCCCACCC	TTCTGGGGCC	GGACGGCATG
2881	ACATCGCTGC	GCGAATATGC	CGGTTATCAC	GGCGGTGGCA	GCGGATTTGG	AGGGCAGTTG
2941	CGGTCGTGGA	ACCCACCGAG	TGAAAGTGTG	GATGCAGCCC	TGTTGCCCAA	CTTTACCCGT
3001	GGCAATGCCC	GCGCAGACGA	TCTGGTACGC	AATAACGGCT	ATGCCGCCAA	CGCCATCCAG
3061	CTGCATCAGG	ATCATATCGT	CGGGTCTTTT	TTCCGGCTCA	GTCATCGCCC	AAGCTGGCGC
3121	TATCTGGGCA	TCGGGGAGGA	AGAAGCCCGT	GCCTTTTCCC	GCGAGGTTGA	AGCGGCATGG
3181	AAAGAGTTTG	CCGAGGATGA	CTGCTGCTGC	ATTGACGTTG	AGCGAAAACG	CACGTTTACC
3241	ATGATGATTC	GGGAAGGTGT	GGCCATGCAC	GCCTTTAACG	GTGAACTGTT	CGTTCAGGCC
3301	ACCTGGGATA	CCAGTTCGTC	GCGGCTTTTC	CGGACACAGT	TCCGGATGGT	CAGCCCGAAG
3361	CGCATCAGCA	ACCCGAACAA	TACCGGCGAC	AGCCGGAACT	GCCGTGCCGG	TGTGCAGATT
3421	AATGACAGCG	GTGCGGCGCT	GGGATATTAC	GTCAGCGAGG	ACGGGTATCC	TGGCTGGATG
3481	CCGCAGAAAT	GGACATGGAT	ACCCCGTGAG	TTACCCGGCG	GGCGCGCCTC	GTTCATTCAC
3541	GTTTTTGAAC	CCGTGGAGGA	CGGGCAGACT	CGCGGTGCAA	ATGTGTTTTA	CAGCGTGATG

```
3601  GAGCAGATGA  AGATGCTCGA  CACGCTGCAG  AACACGCAGC  TGCAGAGCGC  CATTGTGAAG
3661  GCGATGTATG  CCGCCACCAT  TGAGAGTGAG  CTGGATACGC  AGTCAGCGAT  GGATTTTATT
3721  CTGGGCGCGA  ACAGTCAGGA  GCAGCGGGAA  AGGCTGACCG  GCTGGATTGG  TGAAATTGCC
3781  GCGTATTACG  CCGCAGCGCC  GGTCCGGCTG  GGAGGCGCAA  AAGTACCGCA  CCTGATGCCG
3841  GGTGACTCAC  TGAACCTGCA  GACGGCTCAG  GATACGGATA  ACGGCTACTC  CGTGTTTGAG
3901  CAGTCACTGC  TGCGGTATAT  CGCTGCCGGG  CTGGGTGTCT  CGTATGAGCA  GCTTTCCCGG
3961  AATTACGCCC  AGATGAGCTA  CTCCACGGCA  CGGGCCAGTG  CGAACGAGTC  GTGGGCGTAC
4021  TTTATGGGGC  GGCGAAAATT  CGTCGCATCC  CGTCAGGCGA  GCCAGATGTT  TCTGTGCTGG
4081  CTGGAAGAGG  CCATCGTTCG  CCGCGTGGTG  ACGTTACCTT  CAAAAGCGCG  CTTCAGTTTT
4141  CAGGAAGCCC  GCAGTGCCTG  GGGGAACTGC  GACTGGATAG  GCTCCGGTCG  TATGGCCATC
4201  GATGGTCTGA  AAGAAGTTCA  GGAAGCGGTG  ATGCTGATAG  AAGCCGGACT  GAGTACCTAC
4261  GAGAAAGAGT  GCGCAAAACG  CGGTGACGAC  TATCAGGAAA  TTTTTGCCCA  GCAGGTCCGT
4321  GAAACGATGG  AGCGCCGTGC  AGCCGGTCTT  AAACCGCCCG  CCTGGGCGGC  TGCAGCATTT
4381  GAATCCGGGC  TGCGACAATC  AACAGAGGAG  GAGAAGAGTG  ACAGCAGAGC  TGCGTAATCT
4441  CCCGCATATT  GCCAGCATGG  CCTTTAATGA  GCCGCTGATG  CTTGAACCCG  CCTATGCGCG
4501  GGTTTTCTTT  TGTGCGCTTG  CAGGCCAGCT  TGGGATCAGC  AGCCTGACGG  ATGCGGTGTC
4561  CGGCGACAGC  CTGACTGCCC  AGGAGGCACT  CGCGACGCTG  GCATTATCCG  GTGATGATGA
4621  CGGACCACGA  CAGGCCCGCA  GTTATCAGGT  CATGAACGGC  ATCGCCGTGC  TGCCGGTGTC
4681  CGGCACGCTG  GTCAGCCGGA  CGCGGGCGCT  GCAGCCGTAC  TCGGGGATGA  CCGGTTACAA
4741  CGGCATTATC  GCCCGTCTGC  AACAGGCTGC  CAGCGATCCG  ATGGTGGACG  GCATTCTGCT
4801  CGATATGGAC  ACGCCCGGCG  GGATGGTGGC  GGGGGCATTT  GACTGCGCTG  ACATCATCGC
4861  CCGTGTGCGT  GACATAAAAC  CGGTATGGGC  GCTTGCCAAC  GACATGAACT  GCAGTGCAGG
4921  TCAGTTGCTT  GCCAGTGCCG  CCTCCCGGCG  TCTGGTCACG  CAGACCGCCC  GGACAGGCTC
4981  CATCGGCGTC  ATGATGGCTC  ACAGTAATTA  CGGTGCTGCG  CTGGAGAAAC  AGGGTGTGGA
5041  AATCACGCTG  ATTTACAGCG  GCAGCCATAA  GGTGGATGGC  AACCCCTACA  GCCATCTTCC
5101  GGATGACGTC  CGGGAGACAC  TGCAGTCCCG  GATGGACGCA  ACCCGCCAGA  TGTTTGCGCA
5161  GAAGGTGTCG  GCATATACCG  GCCTGTCCGT  GCAGGTTGTG  CTGGATACCG  AGGCTGCAGT
5221  GTACAGCGGT  CAGGAGGCCA  TTGATGCCGG  ACTGGCTGAT  GAACTTGTTA  ACAGCACCGA
5281  TGCGATCACC  GTCATGCGTG  ATGCACTGGA  TGCACGTAAA  TCCCGTCTCT  CAGGAGGGCG
5341  AATGACCAAA  GAGACTCAAT  CAACAACTGT  TTCAGCCACT  GCTTCGCAGG  CTGACGTTAC
5401  TGACGTGGTG  CCAGCGACGG  AGGGCGAGAA  CGCCAGCGCG  GCGCAGCCGG  ACGTGAACGC
5461  GCAGATCACC  GCAGCGGTTG  CGGCAGAAAA  CAGCCGCATT  ATGGGGATCC  TCAACTGTGA
5521  GGAGGCTCAC  GGACGCGAAG  AACAGGCACG  CGTGCTGGCA  GAAACCCCCG  GTATGACCGT
5581  GAAAACGGCC  CGCCGCATTC  TGGCCGCAGC  ACCACAGAGT  GCACAGGCGC  GCAGTGACAC
5641  TGCGCTGGAT  CGTCTGATGC  AGGGGGCACC  GGCACCGCTG  GCTGCAGGTA  ACCCGGCATC
5701  TGATGCCGTT  AACGATTTGC  TGAACACACC  AGTGTAAGGG  ATGTTTATGA  CGAGCAAAGA
5761  AACCTTTACC  CATTACCAGC  CGCAGGGCAA  CAGTGACCCG  GCTCATACCG  CAACCGCGCC
5821  CGGCGGATTG  AGTGCGAAAG  CGCCTGCAAT  GACCCCGCTG  ATGCTGGACA  CCTCCAGCCG
5881  TAAGCTGGTT  GCGTGGGATG  GCACCACCGA  CGGTGCTGCC  GTTGGCATTC  TTGCGGTTGC
5941  TGCTGACCAG  ACCAGCACCA  CGCTGACGTT  CTACAAGTCC  GGCACGTTCC  GTTATGAGGA
6001  TGTGCTCTGG  CCGGAGGCTG  CCAGCGACGA  GACGAAAAAA  CGGACCGCGT  TTGCCGGAAC
6061  GGCAATCAGC  ATCGTTTAAC  TTTACCCTTC  ATCACTAAAG  GCCGCCTGTG  CGGCTTTTTT
6121  TACGGGATTT  TTTTATGTCG  ATGTACACAA  CCGCCCAACT  GCTGGCGGCA  AATGAGCAGA
6181  AATTTAAGTT  TGATCCGCTG  TTTCTGCGTC  TCTTTTTCCG  TGAGAGCTAT  CCCTTCACCA
6241  CGGAGAAAGT  CTATCTCTCA  CAAATTCCGG  GACTGGTAAA  CATGGCGCTG  TACGTTTCGC
6301  CGATTGTTTC  CGGTGAGGTT  ATCCGTTCCC  GTGGCGGCTC  CACCTCTGAA  TTTACGCCGG
6361  GATATGTCAA  GCCGAAGCAT  GAAGTGAATC  CGCAGATGAC  CCTGCGTCGC  CTGCCGGATG
6421  AAGATCCGCA  GAATCTGGCG  GACCCGGCTT  ACCGCCGCCG  TCGCATCATC  ATGCAGAACA
6481  TGCGTGACGA  AGAGCTGGCC  ATTGCTCAGG  TCGAAGAGAT  GCAGGCAGTT  TCTGCCGTGC
6541  TTAAGGGCAA  ATACACCATG  ACCGGTGAAG  CCTTCGATCC  GGTTGAGGTG  GATATGGGCC
6601  GCAGTGAGGA  GAATAACATC  ACGCAGTCCG  GCGGCACGGA  GTGGAGCAAG  CGTGACAAGT
6661  CCACGTATGA  CCCGACCGAC  GATATCGAAG  CCTACGCGCT  GAACGCCAGC  GGTGTGGTGA
6721  ATATCATCGT  GTTCGATCCG  AAAGGCTGGG  CGCTGTTCCG  TTCCTTCAAA  GCCGTCAAGG
```

6781	AGAAGCTGGA	TACCCGTCGT	GGCTCTAATT	CCGAGCTGGA	GACAGCGGTG	AAAGACCTGG
6841	GCAAAGCGGT	GTCCTATAAG	GGGATGTATG	GCGATGTGGC	CATCGTCGTG	TATTCCGGAC
6901	AGTACGTGGA	AAACGGCGTC	AAAAAGAACT	TCCTGCCGGA	CAACACGATG	GTGCTGGGGA
6961	ACACTCAGGC	ACGCGGTCTG	CGCACCTATG	GCTGCATTCA	GGATGCGGAC	GCACAGCGCG
7021	AAGGCATTAA	CGCCTCTGCC	CGTTACCCGA	AAAACTGGGT	GACCACCGGC	GATCCGGCGC
7081	GTGAGTTCAC	CATGATTCAG	TCAGCACCGC	TGATGCTGCT	GGCTGACCCT	GATGAGTTCG
7141	TGTCCGTACA	ACTGGCGTAA	TCATGGCCCT	TCGGGGCCAT	TGTTTCTCTG	TGGAGGAGTC
7201	CATGACGAAA	GATGAACTGA	TTGCCCGTCT	CCGCTCGCTG	GGTGAACAAC	TGAACCGTGA
7261	TGTCAGCCTG	ACGGGGACGA	AAGAAGAACT	GGCGCTCCGT	GTGGCAGAGC	TGAAAGAGGA
7321	GCTTGATGAC	ACGGATGAAA	CTGCCGGTCA	GGACACCCCT	CTCAGCCGGG	AAAATGTGCT
7381	GACCGGACAT	GAAAATGAGG	TGGGATCAGC	GCAGCCGGAT	ACCGTGATTC	TGGATACGTC
7441	TGAACTGGTC	ACGGTCGTGG	CACTGGTGAA	GCTGCATACT	GATGCACTTC	ACGCCACGCG
7501	GGATGAACCT	GTGGCATTTG	TGCTGCCGGG	AACGGCGTTT	CGTGTCTCTG	CCGGTGTGGC
7561	AGCCGAAATG	ACAGAGCGCG	GCCTGGCCAG	AATGCAATAA	CGGGAGGCGC	TGTGGCTGAT
7621	TTCGATAACC	TGTTCGATGC	TGCCATTGCC	CGCGCCGATG	AAACGATACG	CGGGTACATG
7681	GGAACGTCAG	CCACCATTAC	ATCCGGTGAG	CAGTCAGGTG	CGGTGATACG	TGGTGTTTTT
7741	GATGACCCTG	AAAATATCAG	CTATGCCGGA	CAGGGCGTGC	GCGTTGAAGG	CTCCAGCCCG
7801	TCCCTGTTTG	TCCGGACTGA	TGAGGTGCGG	CAGCTGCGGC	GTGGAGACAC	GCTGACCATC
7861	GGTGAGGAAA	ATTTCTGGGT	AGATCGGGTT	TCGCCGGATG	ATGGCGGAAG	TTGTCATCTC
7921	TGGCTTGGAC	GGGGCGTACC	GCCTGCCGTT	AACCGTCGCC	GCTGAAAGGG	GGATGTATGG
7981	CCATAAAAGG	TCTTGAGCAG	GCCGTTGAAA	ACCTCAGCCG	TATCAGCAAA	ACGGCGGTGC
8041	CTGGTGCCGC	CGCAATGGCC	ATTAACCGCG	TTGCTTCATC	CGCGATATCG	CAGTCGGCGT
8101	CACAGGTTGC	CCGTGAGACA	AAGGTACGCC	GGAAACTGGT	AAAGGAAAGG	GCCAGGCTGA
8161	AAAGGGCCAC	GGTCAAAAAT	CCGCAGGCCA	GAATCAAAGT	TAACCGGGGG	GATTTGCCCG
8221	TAATCAAGCT	GGGTAATGCG	CGGGTTGTCC	TTTCGCGCCG	CAGGCGTCGT	AAAAAGGGGC
8281	AGCGTTCATC	CCTGAAAGGT	GGCGGCAGCG	TGCTTGTGGT	GGGTAACCGT	CGTATTCCCG
8341	GCGCGTTTAT	TCAGCAACTG	AAAAATGGCC	GGTGGCATGT	CATGCAGCGT	GTGGCTGGGA
8401	AAAACCGTTA	CCCCATTGAT	GTGGTGAAAA	TCCCGATGGC	GGTGCCGCTG	ACCACGGCGT
8461	TTAAACAAAA	TATTGAGCGG	ATACGGCGTG	AACGTCTTCC	GAAAGAGCTG	GGCTATGCGC
8521	TGCAGCATCA	ACTGAGGATG	GTAATAAAGC	GATGAAACAT	ACTGAACTCC	GTGCAGCCGT
8581	ACTGGATGCA	CTGGAGAAGC	ATGACACCGG	GGCGACGTTT	TTTGATGGTC	GCCCCGCTGT
8641	TTTTGATGAG	GCGGATTTTC	CGGCAGTTGC	CGTTTATCTC	ACCGGCGCTG	AATACACGGG
8701	CGAAGAGCTG	GACAGCGATA	CCTGGCAGGC	GGAGCTGCAT	ATCGAAGTTT	TCCTGCCTGC
8761	TCAGGTGCCG	GATTCAGAGC	TGGATGCGTG	GATGGAGTCC	CGGATTTATC	CGGTGATGAG
8821	CGATATCCCG	GCACTGTCAG	ATTTGATCAC	CAGTATGGTG	GCCAGCGGCT	ATGACTACCG
8881	GCGCGACGAT	GATGCGGGCT	TGTGGAGTTC	AGCCGATCTG	ACTTATGTCA	TTACCTATGA
8941	AATGTGAGGA	CGCTATGCCT	GTACCAAATC	CTACAATGCC	GGTGAAAGGT	GCCGGGACCA
9001	CCCTGTGGGT	TTATAAGGGG	AGCGGTGACC	CTTACGCGAA	TCCGCTTTCA	GACGTTGACT
9061	GGTCGCGTCT	GGCAAAAGTT	AAAGACCTGA	CGCCCGGCGA	ACTGACCGCT	GAGTCCTATG
9121	ACGACAGCTA	TCTCGATGAT	GAAGATGCAG	ACTGGACTGC	GACCGGGCAG	GGGCAGAAAT
9181	CTGCCGGAGA	TACCAGCTTC	ACGCTGGCGT	GGATGCCCGG	AGAGCAGGGG	CAGCAGGCGC
9241	TGCTGGCGTG	GTTTAATGAA	GGCGATACCC	GTGCCTATAA	AATCCGCTTC	CCGAACGGCA
9301	CGGTCGATGT	GTTCCGTGGC	TGGGTCAGCA	GTATCGGTAA	GGCGGTGACG	GCGAAGGAAG
9361	TGATCACCCG	CACGGTGAAA	GTCACCAATG	TGGGACGTCC	GTCGATGGCA	GAAGATCGCA
9421	GCACGGTAAC	AGCGGCAACC	GGCATGACCG	TGACGCCTGC	CAGCACCTCG	GTGGTGAAAG
9481	GGCAGAGCAC	CACGCTGACC	GTGGCCTTCC	AGCCGGAGGG	CGTAACCGAC	AAGAGCTTTC
9541	GTGCGGTGTC	TGCGGATAAA	ACAAAAGCCA	CCGTGTCGGT	CAGTGGTATG	ACCATCACCG
9601	TGAACGGCGT	TGCTGCAGGC	AAGGTCAACA	TTCCGGTTGT	ATCCGGTAAT	GGTGAGTTTG
9661	CTGCGGTTGC	AGAAATTACC	GTCACCGCCA	GTTAATCCGG	AGAGTCAGCG	ATGTTCCTGA
9721	AAACCGAATC	ATTTGAACAT	AACGGTGTGA	CCGTCACGCT	TTCTGAACTG	TCAGCCCTGC
9781	AGCGCATTGA	GCATCTCGCC	CTGATGAAAC	GGCAGGCAGA	ACAGGCGGAG	TCAGACAGCA
9841	ACCGGAAGTT	TACTGTGGAA	GACGCCATCA	GAACCGGCGC	GTTTCTGGTG	GCGATGTCCC
9901	TGTGGCATAA	CCATCCGCAG	AAGACGCAGA	TGCCGTCCAT	GAATGAAGCC	GTTAAACAGA

9961	TTGAGCAGGA	AGTGCTTACC	ACCTGGCCCA	CGGAGGCAAT	TTCTCATGCT	GAAAACGTGG
10021	TGTACCGGCT	GTCTGGTATG	TATGAGTTTG	TGGTGAATAA	TGCCCCTGAA	CAGACAGAGG
10081	ACGCCGGGCC	CGCAGAGCCT	GTTTCTGCGG	GAAAGTGTTC	GACGGTGAGC	TGAGTTTTGC
10141	CCTGAAACTG	GCGCGTGAGA	TGGGGCGACC	CGACTGGCGT	GCCATGCTTG	CCGGGATGTC
10201	ATCCACGGAG	TATGCCGACT	GGCACCGCTT	TTACAGTACC	CATTATTTTC	ATGATGTTCT
10261	GCTGGATATG	CACTTTTCCG	GGCTGACGTA	CACCGTGCTC	AGCCTGTTTT	TCAGCGATCC
10321	GGATATGCAT	CCGCTGGATT	TCAGTCTGCT	GAACCGGCGC	GAGGCTGACG	AAGAGCCTGA
10381	AGATGATGTG	CTGATGCAGA	AAGCGGCAGG	GCTTGCCGGA	GGTGTCCGCT	TTGGCCCGGA
10441	CGGGAATGAA	GTTATCCCCG	CTTCCCCGGA	TGTGGCGGAC	ATGACGGAGG	ATGACGTAAT
10501	GCTGATGACA	GTATCAGAAG	GGATCGCAGG	AGGAGTCCGG	TATGGCTGAA	CCGGTAGGCG
10561	ATCTGGTCGT	TGATTTGAGT	CTGGATGCGG	CCAGATTTGA	CGAGCAGATG	GCCAGAGTCA
10621	GGCGTCATTT	TTCTGGTACG	GAAAGTGATG	CGAAAAAAAC	AGCGGCAGTC	GTTGAACAGT
10681	CGCTGAGCCG	ACAGGCGCTG	GCTGCACAGA	AAGCGGGGAT	TTCCGTCGGG	CAGTATAAAG
10741	CCGCCATGCG	TATGCTGCCT	GCACAGTTCA	CCGACGTGGC	CACGCAGCTT	GCAGGCGGGC
10801	AAAGTCCGTG	GCTGATCCTG	CTGCAACAGG	GGGGGCAGGT	GAAGGACTCC	TTCGGCGGGA
10861	TGATCCCCAT	GTTCAGGGGG	CTTGCCGGTG	CGATCACCCT	GCCGATGGTG	GGGGCCACCT
10921	CGCTGGCGGT	GGCGACCGGT	GCGCTGGCGT	ATGCCTGGTA	TCAGGGCAAC	TCAACCCTGT
10981	CCGATTTCAA	CAAAACGCTG	GTCCTTTCCG	GCAATCAGGC	GGGACTGACG	GCAGATCGTA
11041	TGCTGGTCCT	GTCCAGAGCC	GGGCAGGCGG	CAGGGCTGAC	GTTTAACCAG	ACCAGCGAGT
11101	CACTCAGCGC	ACTGGTTAAG	GCGGGGGTAA	GCGGTGAGGC	TCAGATTGCG	TCCATCAGCC
11161	AGAGTGTGGC	GCGTTTCTCC	TCTGCATCCG	GCGTGGAGGT	GGACAAGGTC	GCTGAAGCCT
11221	TCGGGAAGCT	GACCACAGAC	CCGACGTCGG	GGCTGACGGC	GATGGCTCGC	CAGTTCCATA
11281	ACGTGTCGGC	GGAGCAGATT	GCGTATGTTG	CTCAGTTGCA	GCGTTCCGGC	GATGAAGCCG
11341	GGGCATTGCA	GGCGGCGAAC	GAGGCCGCAA	CGAAAGGGTT	TGATGACCAG	ACCCGCCGCC
11401	TGAAAGAGAA	CATGGGCACG	CTGGAGACCT	GGGCAGACAG	GACTGCGCGG	GCATTCAAAT
11461	CCATGTGGGA	TGCGGTGCTG	GATATTGGTC	GTCCTGATAC	CGCGCAGGAG	ATGCTGATTA
11521	AGGCAGAGGC	TGCGTATAAG	AAAGCAGACG	ACATCTGGAA	TCTGCGCAAG	GATGATTATT
11581	TTGTTAACGA	TGAAGCGCGG	GCGCGTTACT	GGGATGATCG	TGAAAAGGCC	CGTCTTGCGC
11641	TTGAAGCCGC	CCGAAAGAAG	GCTGAGCAGC	AGACTCAACA	GGACAAAAAT	GCGCAGCAGC
11701	AGAGCGATAC	CGAAGCGTCA	CGGCTGAAAT	ATACCGAAGA	GGCGCAGAAG	GCTTACGAAC
11761	GGCTGCAGAC	GCCGCTGGAG	AAATATACCG	CCCGTCAGGA	AGAACTGAAC	AAGGCACTGA
11821	AAGACGGGAA	AATCCTGCAG	GCGGATTACA	ACACGCTGAT	GGCGGCGGCG	AAAAAGGATT
11881	ATGAAGCGAC	GCTGAAAAAG	CCGAAACAGT	CCAGCGTGAA	GGTGTCTGCG	GGCGATCGTC
11941	AGGAAGACAG	TGCTCATGCT	GCCCTGCTGA	CGCTTCAGGC	AGAACTCCGG	ACGCTGGAGA
12001	AGCATGCCGG	AGCAAATGAG	AAAATCAGCC	AGCAGCGCCG	GGATTTGTGG	AAGGCGGAGA
12061	GTCAGTTCGC	GGTACTGGAG	GAGGCGGCGC	AACGTCGCCA	GCTGTCTGCA	CAGGAGAAAT
12121	CCCTGCTGGC	GCATAAAGAT	GAGACGCTGG	AGTACAAACG	CCAGCTGGCT	GCACTTGGCG
12181	ACAAGGTTAC	GTATCAGGAG	CGCCTGAACG	CGCTGGCGCA	GCAGGCGGAT	AAATTCGCAC
12241	AGCAGCAACG	GGCAAAACGG	GCCGCCATTG	ATGCGAAAAG	CCGGGGGCTG	ACTGACCGGC
12301	AGGCAGAACG	GGAAGCCACG	GAACAGCGCC	TGAAGGAACA	GTATGGCGAT	AATCCGCTGG
12361	CGCTGAATAA	CGTCATGTCA	GAGCAGAAAA	AGACCTGGGC	GGCTGAAGAC	CAGCTTCGCG
12421	GGAACTGGAT	GGCAGGCCTG	AAGTCCGGCT	GGAGTGAGTG	GGAAGAGAGC	GCCACGGACA
12481	GTATGTCGCA	GGTAAAAAGT	GCAGCCACGC	AGACCTTTGA	TGGTATTGCA	CAGAATATGG
12541	CGGCGATGCT	GACCGGCAGT	GAGCAGAACT	GGCGCAGCTT	CACCCGTTCC	GTGCTGTCCA
12601	TGATGACAGA	AATTCTGCTT	AAGCAGGCAA	TGGTGGGGAT	TGTCGGGAGT	ATCGGCAGCG
12661	CCATTGGCGG	GGCTGTTGGT	GGCGGCGCAT	CCGCGTCAGG	CGGTACAGCC	ATTCAGGCCG
12721	CTGCGGCGAA	ATTCCATTTT	GCAACCGGAG	GATTTACGGG	AACCGGCGGC	AAATATGAGC
12781	CAGCGGGGAT	TGTTCACCGT	GGTGAGTTTG	TCTTCACGAA	GGAGGCAACC	AGCCGGATTG
12841	GCGTGGGGAA	TCTTTACCGG	CTGATGCGCG	GCTATGCCAC	CGGCGGTTAT	GTCGGTACAC
12901	CGGGCAGCAT	GGCAGACAGC	CGGTCGCAGG	CGTCCGGGAC	GTTTGAGCAG	AATAACCATG
12961	TGGTGATTAA	CAACGACGGC	ACGAACGGGC	AGATAGGTCC	GGCTGCTCTG	AAGGCGGTGT
13021	ATGACATGGC	CCGCAAGGGT	GCCCGTGATG	AAATTCAGAC	ACAGATGCGT	GATGGTGGCC
13081	TGTTCTCCGG	AGGTGGACGA	TGAAGACCTT	CCGCTGGAAA	GTGAAACCCG	GTATGGATGT

13141 GGCTTCGGTC CCTTCTGTAA GAAAGGTGCG CTTTGGTGAT GGCTATTCTC AGCGAGCGCC
13201 TGCCGGGCTG AATGCCAACC TGAAAACGTA CAGCGTGACG CTTTCTGTCC CCCGTGAGGA
13261 GGCCACGGTA CTGGAGTCGT TTCTGGAAGA GCACGGGGGC TGGAAATCCT TTCTGTGGAC
13321 GCCGCCTTAT GAGTGGCGGC AGATAAAGGT GACCTGCGCA AAATGGTCGT CGCGGGTCAG
13381 TATGCTGCGT GTTGAGTTCA GCGCAGAGTT TGAACAGGTG GTGAACTGAT GCAGGATATC
13441 CGGCAGGAAA CACTGAATGA ATGCACCCGT GCGGAGCAGT CGGCCAGCGT GGTGCTCTGG
13501 GAAATCGACC TGACAGAGGT CGGTGGAGAA CGTTATTTTT TCTGTAATGA GCAGAACGAA
13561 AAAGGTGAGC CGGTCACCTG GCAGGGGCGA CAGTATCAGC CGTATCCCAT TCAGGGGAGC
13621 GGTTTTGAAC TGAATGGCAA AGGCACCAGT ACGCGCCCCA CGCTGACGGT TTCTAACCTG
13681 TACGGTATGG TCACCGGGAT GGCGGAAGAT ATGCAGAGTC TGGTCGGCGG AACGGTGGTC
13741 CGGCGTAAGG TTTACGCCCG TTTTCTGGAT GCGGTGAACT TCGTCAACGG AAACAGTTAC
13801 GCCGATCCGG AGCAGGAGGT GATCAGCCGC TGGCGCATTG AGCAGTGCAG CGAACTGAGC
13861 GCGGTGAGTG CCTCCTTTGT ACTGTCCACG CCGACGGAAA CGGATGGCGC TGTTTTTCCG
13921 GGACGTATCA TGCTGGCCAA CACCTGCACC TGGACCTATC GCGGTGACGA GTGCGGTTAT
13981 AGCGGTCCGG CTGTCGCGGA TGAATATGAC CAGCCAACGT CCGATATCAC GAAGGATAAA
14041 TGCAGCAAAT GCCTGAGCGG TTGTAAGTTC CGCAATAACG TCGGCAACTT TGGCGGCTTC
14101 CTTTCCATTA ACAAACTTTC GCAGTAAATC CCATGACACA GACAGAATCA GCGATTCTGG
14161 CGCACGCCCG GCGATGTGCG CCAGCGGAGT CGTGCGGCTT CGTGGTAAGC ACGCCGGAGG
14221 GGGAAAGATA TTTCCCCTGC GTGAATATCT CCGGTGAGCC GGAGGCTATT TCCGTATGTC
14281 GCCGGAAGAC TGGCTGCAGG CAGAAATGCA GGGTGAGATT GTGGCGCTGG TCCACAGCCA
14341 CCCCGGTGGT CTGCCCTGGC TGAGTGAGGC CGACCGGCGG CTGCAGGTGC AGAGTGATTT
14401 GCCGTGGTGG CTGGTCTGCC GGGGGACGAT TCATAAGTTC CGCTGTGTGC CGCATCTCAC
14461 CGGGCGGCGC TTTGAGCACG GTGTGACGGA CTGTTACACA CTGTTCCGGG ATGCTTATCA
14521 TCTGGCGGGG ATTGAGATGC CGGACTTTCA TCGTGAGGAT GACTGGTGGC GTAACGGCCA
14581 GAATCTCTAT CTGGATAATC TGGAGGCGAC GGGGCTGTAT CAGGTGCCGT TGTCAGCGGC
14641 ACAGCCGGGC GATGTGCTGC TGTGCTGTTT TGGTTCATCA GTGCCGAATC ACGCCGCAAT
14701 TTACTGCGGC GACGGCGAGC TGCTGCACCA TATTCCTGAA CAACTGAGCA AACGAGAGAG
14761 GTACACCGAC AAATGGCAGC GACGCACACA CTCCCTCTGG CGTCACCGGG CATGGCGCGC
14821 ATCTGCCTTT ACGGGGATTT ACAACGATTT GGTCGCCGCA TCGACCTTCG TGTGAAAACG
14881 GGGGCTGAAG CCATCCGGGC ACTGGCCACA CAGCTCCCGG CGTTTCGTCA GAAACTGAGC
14941 GACGGCTGGT ATCAGGTACG GATTGCCGGG CGGGACGTCA GCACGTCCGG GTTAACGGCG
15001 CAGTTACATG AGACTCTGCC TGATGGCGCT GTAATTCATA TTGTTCCCAG AGTCGCCGGG
15061 GCCAAGTCAG GTGGCGTATT CCAGATTGTC CTGGGGGCTG CCGCCATTGC CGGATCATTC
15121 TTTACCGCCG GAGCCACCCT TGCAGCATGG GGGGCAGCCA TTGGGGCCGG TGGTATGACC
15181 GGCATCCTGT TTTCTCTCGG TGCCAGTATG GTGCTCGGTG GTGTGGCGCA GATGCTGGCA
15241 CCGAAAGCCA GAACTCCCCG TATACAGACA ACGGATAACG GTAAGCAGAA CACCTATTTC
15301 TCCTCACTGG ATAACATGGT TGCCCAGGGC AATGTTCTGC CTGTTCTGTA CGGGGAAATG
15361 CGCGTGGGGT CACGCGTGGT TTCTCAGGAG ATCAGCACGG CAGACGAAGG GGACGGTGGT
15421 CAGGTTGTGG TGATTGGTCG CTGATGCAAA ATGTTTTATG TGAAACCGCC TGCGGGCGGT
15481 TTTGTCATTT ATGGAGCGTG AGGAATGGGT AAAGGAAGCA GTAAGGGGCA TACCCCGCGC
15541 GAAGCGAAGG ACAACCTGAA GTCCACGCAG TTGCTGAGTG TGATCGATGC CATCAGCGAA
15601 GGGCCGATTG AAGGTCCGGT GGATGGCTTA AAAAGCGTGC TGCTGAACAG TACGCCGGTG
15661 CTGGACACTG AGGGGAATAC CAACATATCC GGTGTCACGG TGGTGTTCCG GGCTGGTGAG
15721 CAGGAGCAGA CTCCGCCGGA GGGATTTGAA TCCTCCGGCT CCGAGACGGT GCTGGGTACG
15781 GAAGTGAAAT ATGACACGCC GATCACCCGC ACCATTACGT CTGCAAACAT CGACCGTCTG
15841 CGCTTTACCT TCGGTGTACA GGCACTGGTG GAAACCACCT CAAAGGGTGA CAGGAATCCG
15901 TCGGAAGTCC GCCTGCTGGT TCAGATACAA CGTAACGGTG GCTGGGTGAC GGAAAAAGAC
15961 ATCACCATTA AGGGCAAAAC CACCTCGCAG TATCTGGCCT CGGTGGTGAT GGGTAACCTG
16021 CCGCCGCGCC CGTTTAATAT CCGGATGCGC AGGATGACGC CGGACAGCAC CACAGACCAG
16081 CTGCAGAACA AAACGCTCTG GTCGTCATAC ACTGAAATCA TCGATGTGAA ACAGTGCTAC
16141 CCGAACACGG CACTGGTCGG CGTGCAGGTG GACTCGGAGC AGTTCGGCAG CCAGCAGGTG
16201 AGCCGTAATT ATCATCTGCG CGGGCGTATT CTGCAGGTGC CGTCGAACTA TAACCCGCAG
16261 ACGCGGCAAT ACAGCGGTAT CTGGGACGGA ACGTTTAAAC CGGCATACAG CAACAACATG

16321 GCCTGGTGTC TGTGGGATAT GCTGACCCAT CCGCGCTACG GCATGGGGAA ACGTCTTGGT
16381 GCGGCGGATG TGGATAAATG GGCGCTGTAT GTCATCGGCC AGTACTGCGA CCAGTCAGTG
16441 CCGGACGGCT TTGGCGGCAC GGAGCCGCGC ATCACCTGTA ATGCGTACCT GACCACACAG
16501 CGTAAGGCGT GGGATGTGCT CAGCGATTTC TGCTCGGCGA TGCGCTGTAT GCCGGTATGG
16561 AACGGGCAGA CGCTGACGTT CGTGCAGGAC CGACCGTCGG ATAAGACGTG GACCTATAAC
16621 CGCAGTAATG TGGTGATGCC GGATGATGGC GCGCCGTTCC GCTACAGCTT CAGCGCCCTG
16681 AAGGACCGCC ATAATGCCGT TGAGGTGAAC TGGATTGACC CGAACAACGG CTGGGAGACG
16741 GCGACAGAGC TTGTTGAAGA TACGCAGGCC ATTGCCCGTT ACGGTCGTAA TGTTACGAAG
16801 ATGGATGCCT TTGGCTGTAC CAGCCGGGGG CAGGCACACC GCGCCGGGCT GTGGCTGATT
16861 AAAACAGAAC TGCTGGAAAC GCAGACCGTG GATTTCAGCG TCGGCGCAGA AGGGCTTCGC
16921 CATGTACCGG GCGATGTTAT TGAAATCTGC GATGATGACT ATGCCGGTAT CAGCACCGGT
16981 GGTCGTGTGC TGGCGGTGAA CAGCCAGACC CGGACGCTGA CGCTCGACCG TGAAATCACG
17041 CTGCCATCCT CCGGTACCGC GCTGATAAGC CTGGTTGACG GAAGTGGCAA TCCGGTCAGC
17101 GTGGAGGTTC AGTCCGTCAC CGACGGCGTG AAGGTAAAAG TGAGCCGTGT TCCTGACGGT
17161 GTTGCTGAAT ACAGCGTATG GGAGCTGAAG CTGCCGACGC TGCGCCAGCG ACTGTTCCGC
17221 TGCGTGAGTA TCCGTGAGAA CGACGACGGC ACGTATGCCA TCACCGCCGT GCAGCATGTG
17281 CCGGAAAAAG AGGCCATCGT GGATAACGGG GCGCACTTTG ACGGCGAACA GAGTGGCACG
17341 GTGAATGGTG TCACGCCGCC AGCGGTGCAG CACCTGACCG CAGAAGTCAC TGCAGACAGC
17401 GGGGAATATC AGGTGCTGGC GCGATGGGAC ACACCGAAGG TGGTGAAGGG CGTGAGTTTC
17461 CTGCTCCGTC TGACCGTAAC AGCGGACGAC GGCAGTGAGC GGCTGGTCAG CACGGCCCGG
17521 ACGACGGAAA CCACATACCG CTTCACGCAA CTGGCGCTGG GGAACTACAG GCTGACAGTC
17581 CGGGCGGTAA ATGCGTGGGG GCAGCAGGGC GATCCGGCGT CGGTATCGTT CCGGATTGCC
17641 GCACCGGCAG CACCGTCGAG GATTGAGCTG ACGCCGGGCT ATTTTCAGAT AACCGCCACG
17701 CCGCATCTTG CCGTTTATGA CCCGACGGTA CAGTTTGAGT TCTGGTTCTC GGAAAAGCAG
17761 ATTGCGGATA TCAGACAGGT TGAAACCAGC ACGCGTTATC TTGGTACGGC GCTGTACTGG
17821 ATAGCCGCCA GTATCAATAT CAAACCGGGC CATGATTATT ACTTTTATAT CCGCAGTGTG
17881 AACACCGTTG GCAAATCGGC ATTCGTGGAG GCCGTCGGTC GGGCGAGCGA TGATGCGGAA
17941 GGTTACCTGG ATTTTTTCAA AGGCAAGATA ACCGAATCCC ATCTCGGCAA GGAGCTGCTG
18001 GAAAAAGTCG AGCTGACGGA GGATAACGCC AGCAGACTGG AGGAGTTTTC GAAAGAGTGG
18061 AAGGATGCCA GTGATAAGTG GAATGCCATG TGGGCTGTCA AAATTGAGCA GACCAAAGAC
18121 GGCAAACATT ATGTCGCGGG TATTGGCCTC AGCATGGAGG ACACGGAGGA AGGCAAACTG
18181 AGCCAGTTTC TGGTTGCCGC CAATCGTATC GCATTTATTG ACCCGGCAAA CGGGAATGAA
18241 ACGCCGATGT TTGTGGCGCA GGGCAACCAG ATATTCATGA ACGACGTGTT CCTGAAGCGC
18301 CTGACGGCCC CCACCATTAC CAGCGGCGGC AATCCTCCGG CCTTTTCCCT GACACCGGAC
18361 GGAAAGCTGA CCGCTAAAAA TGCGGATATC AGTGGCAGTG TGAATGCGAA CTCCGGGACG
18421 CTCAGTAATG TGACGATAGC TGAAAACTGT ACGATAAACG GTACGCTGAG GGCGGAAAAA
18481 ATCGTCGGGG ACATTGTAAA GGCGGCGAGC GCGGCTTTTC CGCGCCAGCG TGAAAGCAGT
18541 GTGGACTGGC CGTCAGGTAC CCGTACTGTC ACCGTGACCG ATGACCATCC TTTTGATCGC
18601 CAGATAGTGG TGCTTCCGCT GACGTTTCGC GGAAGTAAGC GTACTGTCAG CGGCAGGACA
18661 ACGTATTCGA TGTGTTATCT GAAAGTACTG ATGAACGGTG CGGTGATTTA TGATGGCGCG
18721 GCGAACGAGG CGGTACAGGT GTTCTCCCGT ATTGTTGACA TGCCAGCGGG TCGGGGAAAC
18781 GTGATCCTGA CGTTCACGCT TACGTCCACA CGGCATTCGG CAGATATTCC GCCGTATACG
18841 TTTGCCAGCG ATGTGCAGGT TATGGTGATT AAGAAACAGG CGCTGGGCAT CAGCGTGGTC
18901 TGAGTGTGTT ACAGAGGTTC GTCCGGGAAC GGGCGTTTTA TTATAAAACA GTGAGAGGTG
18961 AACGATGCGT AATGTGTGTA TTGCCGTTGC TGTCTTTGCC GCACTTGCGG TGACAGTCAC
19021 TCCGGCCCGT GCGGAAGGTG GACATGGTAC GTTTACGGTG GGCTATTTTC AAGTGAAACC
19081 GGGTACATTG CCGTCGTTGT CGGGCGGGGA TACCGGTGTG AGTCATCTGA AAGGGATTAA
19141 CGTGAAGTAC CGTTATGAGC TGACGGACAG TGTGGGGGTG ATGGCTTCCC TGGGGTTCGC
19201 CGCGTCGAAA AAGAGCAGCA CAGTGATGAC CGGGGAGGAT ACGTTTCACT ATGAGAGCCT
19261 GCGTGGACGT TATGTGAGCG TGATGGCCGG ACCGGTTTTA CAAATCAGTA AGCAGGTCAG
19321 TGCGTACGCC ATGGCCGGAG TGGCTCACAG TCGGTGGTCC GGCAGTACAA TGGATTACCG
19381 TAAGACGGAA ATCACTCCCG GGTATATGAA AGAGACGACC ACTGCCAGGG ACGAAAGTGC
19441 AATGCGGCAT ACCTCAGTGG CGTGGAGTGC AGGTATACAG ATTAATCCGG CAGCGTCCGT

19501 CGTTGTTGAT ATTGCTTATG AAGGCTCCGG CAGTGGCGAC TGGCGTACTG ACGGATTCAT
19561 CGTTGGGGTC GGTTATAAAT TCTGATTAGC CAGGTAACAC AGTGTTATGA CAGCCCGCCG
19621 GAACCGGTGG GCTTTTTTGT GGGGTGAATA TGGCAGTAAA GATTTCAGGA GTCCTGAAAG
19681 ACGGCACAGG AAAACCGGTA CAGAACTGCA CCATTCAGCT GAAAGCCAGA CGTAACAGCA
19741 CCACGGTGGT GGTGAACACG GTGGGCTCAG AGAATCCGGA TGAAGCCGGG CGTTACAGCA
19801 TGGATGTGGA GTACGGTCAG TACAGTGTCA TCCTGCAGGT TGACGGTTTT CCACCATCGC
19861 ACGCCGGGAC CATCACCGTG TATGAAGATT CACAACCGGG GACGCTGAAT GATTTTCTCT
19921 GTGCCATGAC GGAGGATGAT GCCCGGCCGG AGGTGCTGCG TCGTCTTGAA CTGATGGTGG
19981 AAGAGGTGGC GCGTAACGCG TCCGTGGTGG CACAGAGTAC GGCAGACGCG AAGAAATCAG
20041 CCGGCGATGC CAGTGCATCA GCTGCTCAGG TCGCGGCCCT TGTGACTGAT GCAACTGACT
20101 CAGCACGCGC CGCCAGCACG TCCGCCGGAC AGGCTGCATC GTCAGCTCAG GAAGCGTCCT
20161 CCGGCGCAGA AGCGGCATCA GCAAAGGCCA CTGAAGCGGA AAAAAGTGCC GCAGCCGCAG
20221 AGTCCTCAAA AAACGCGGCG GCCACCAGTG CCGGTGCGGC GAAAACGTCA GAAACGAATG
20281 CTGCAGCGTC ACAACAATCA GCCGCCACGT CTGCCTCCAC CGCGGCCACG AAAGCGTCAG
20341 AGGCCGCCAC TTCAGCACGA GATGCGGTGG CCTCAAAAGA GGCAGCAAAA TCATCAGAAA
20401 CGAACGCATC ATCAAGTGCC GGTCGTGCAG CTTCCTCGGC AACGGCGGCA GAAAATTCTG
20461 CCAGGGCGGC AAAAACGTCC GAGACGAATG CCAGGTCATC TGAAACAGCA GCGGAACGGA
20521 GCGCCTCTGC CGCGGCAGAC GCAAAAACAG CGGCGGCGGG GAGTGCGTCA ACGGCATCCA
20581 CGAAGGCGAC AGAGGCTGCG GGAAGTGCGG TATCAGCATC GCAGAGCAAA AGTGCGGCAG
20641 AAGCGGCGGC AATACGTGCA AAAAATTCGG CAAAACGTGC AGAAGATATA GCTTCAGCTG
20701 TCGCGCTTGA GGATGCGGAC ACAACGAGAA AGGGGATAGT GCAGCTCAGC AGTGCAACCA
20761 ACAGCACGTC TGAAACGCTT GCTGCAACGC CAAAGGCGGT TAAGGTGGTA ATGGATGAAA
20821 CGAACAGAAA AGCCCACTGG ACAGTCCGGC ACTGACCGGA ACGCCAACAG CACCAACCGC
20881 GCTCAGGGGA ACAAACAATA CCCAGATTGC GAACACCGCT TTTGTACTGG CCGCGATTGC
20941 AGATGTTATC GACGCGTCAC CTGACGCACT GAATACGCTG AATGAACTGG CCGCAGCGCT
21001 CGGGAATGAT CCAGATTTTG CTACCACCAT GACTAACGCG CTTGCGGGTA AACAACCGAA
21061 GAATGCGACA CTGACGGCGC TGGCAGGGCT TTCCACGGCG AAAAATAAAT TACCGTATTT
21121 TGCGGAAAAT GATGCCGCCA GCCTGACTGA ACTGACTCAG GTTGGCAGGG ATATTCTGGC
21181 AAAAAATTCC GTTGCAGATG TTCTTGAATA CCTTGGGGCC GGTGAGAATT CGGCCTTTCC
21241 GGCAGGTGCG CCGATCCCGT GGCCATCAGA TATCGTTCCG TCTGGCTACG TCCTGATGCA
21301 GGGGCAGGCG TTTGACAAAT CAGCCTACCC AAAACTTGCT GTCGCGTATC CATCGGGTGT
21361 GCTTCCTGAT ATGCGAGGCT GGACAATCAA GGGGAAACCC GCCAGCGGTC GTGCTGTATT
21421 GTCTCAGGAA CAGGATGGAA TTAAGTCGCA CACCCACAGT GCCAGTGCAT CCGGTACGGA
21481 TTTGGGGACG AAAACCACAT CGTCGTTTGA TTACGGGACG AAAACAACAG GCAGTTTCGA
21541 TTACGGCACC AAATCGACGA ATAACACGGG GGCTCATGCT CACAGTCTGA GCGGTTCAAC
21601 AGGGGCCGCG GGTGCTCATG CCCACACAAG TGGTTTAAGG ATGAACAGTT CTGGCTGGAG
21661 TCAGTATGGA ACAGCAACCA TTACAGGAAG TTTATCCACA GTTAAAGGAA CCAGCACACA
21721 GGGTATTGCT TATTTATCGA AAACGGACAG TCAGGGCAGC CACAGTCACT CATTGTCCGG
21781 TACAGCCGTG AGTGCCGGTG CACATGCGCA TACAGTTGGT ATTGGTGCGC ACCAGCATCC
21841 GGTTGTTATC GGTGCTCATG CCCATTCTTT CAGTATTGGT TCACACGGAC ACACCATCAC
21901 CGTTAACGCT GCGGGTAACG CGGAAAACAC CGTCAAAAAC ATTGCATTTA ACTATATTGT
21961 GAGGCTTGCA TAATGGCATT CAGAATGAGT GAACAACCAC GGACCATAAA AATTTATAAT
22021 CTGCTGGCCG GAACTAATGA ATTTATTGGT GAAGGTGACG CATATATTCC GCCTCATACC
22081 GGTCTGCCTG CAAACAGTAC CGATATTGCA CCGCCAGATA TTCCGGCTGG CTTTGTGGCT
22141 GTTTTCAACA GTGATGAGGC ATCGTGGCAT CTCGTTGAAG ACCATCGGGG TAAAACCGTC
22201 TATGACGTGG CTTCCGGCGA CGCGTTATTT ATTTCTGAAC TCGGTCCGTT ACCGGAAAAT
22261 TTTACCTGGT TATCGCCGGG AGGGGAATAT CAGAAGTGGA ACGGCACAGC CTGGGTGAAG
22321 GATACGGAAG CAGAAAAACT GTTCCGGATC CGGGAGGCGG AAGAAACAAA AAAAAGCCTG
22381 ATGCAGGTAG CCAGTGAGCA TATTGCGCCG CTTCAGGATG CTGCAGATCT GGAAATTGCA
22441 ACGAAGGAAG AAACCTCGTT GCTGGAAGCC TGGAAGAAGT ATCGGGTGTT GCTGAACCGT
22501 GTTGATACAT CAACTGCACC TGATATTGAG TGGCCTGCTG TCCCTGTTAT GGAGTAATCG
22561 TTTTGTGATA TGCCGCAGAA ACGTTGTATG AAATAACGTT CTGCGGTTAG TTAGTATATT
22621 GTAAAGCTGA GTATTGGTTT ATTTGGCGAT TATTATCTTC AGGAGAATAA TGGAAGTTCT

22681 ATGACTCAAT TGTTCATAGT GTTTACATCA CCGCCAATTG CTTTTAAGAC TGAACGCATG
22741 AAATATGGTT TTTCGTCATG TTTTGAGTCT GCTGTTGATA TTTCTAAAGT CGGTTTTTTT
22801 TCTTCGTTTT CTCTAACTAT TTTCCATGAA ATACATTTTT GATTATTATT TGAATCAATT
22861 CCAATTACCT GAAGTCTTTC ATCTATAATT GGCATTGTAT GTATTGGTTT ATTGGAGTAG
22921 ATGCTTGCTT TTCTGAGCCA TAGCTCTGAT ATCCAAATGA AGCCATAGGC ATTTGTTATT
22981 TTGGCTCTGT CAGCTGCATA ACGCCAAAAA ATATATTTAT CTGCTTGATC TTCAAATGTT
23041 GTATTGATTA AATCAATTGG ATGGAATTGT TTATCATAAA AAATTAATGT TTGAATGTGA
23101 TAACCGTCCT TTAAAAAAGT CGTTTCTGCA AGCTTGGCTG TATAGTCAAC TAACTCTTCT
23161 GTCGAAGTGA TATTTTTAGG CTTATCTACC AGTTTTAGAC GCTCTTTAAT ATCTTCAGGA
23221 ATTATTTTAT TGTCATATTG TATCATGCTA AATGACAATT TGCTTATGGA GTAATCTTTT
23281 AATTTTAAAT AAGTTATTCT CCTGGCTTCA TCAAATAAAG AGTCGAATGA TGTTGGCGAA
23341 ATCACATCGT CACCCATTGG ATTGTTTATT TGTATGCCAA GAGAGTTACA GCAGTTATAC
23401 ATTCTGCCAT AGATTATAGC TAAGGCATGT AATAATTCGT AATCTTTTAG CGTATTAGCG
23461 ACCCATCGTC TTTCTGATTT AATAATAGAT GATTCAGTTA AATATGAAGG TAATTTCTTT
23521 TGTGCAAGTC TGACTAACTT TTTTATACCA ATGTTTAACA TACTTTCATT TGTAATAAAC
23581 TCAATGTCAT TTTCTTCAAT GTAAGATGAA ATAAGAGTAG CCTTTGCCTC GCTATACATT
23641 TCTAAATCGC CTTGTTTTTC TATCGTATTG CGAGAATTTT TAGCCCAAGC CATTAATGGA
23701 TCATTTTTCC ATTTTTCAAT AACATTATTG TTATACCAAA TGTCATATCC TATAATCTGG
23761 TTTTTGTTTT TTTGAATAAT AAATGTTACT GTTCTTGCGG TTTGGAGGAA TTGATTCAAA
23821 TTCAAGCGAA ATAATTCAGG GTCAAAATAT GTATCAATGC AGCATTTGAG CAAGTGCGAT
23881 AAATCTTTAA GTCTTCTTTC CCATGGTTTT TTAGTCATAA AACTCTCCAT TTTGATAGGT
23941 TGCATGCTAG ATGCTGATAT ATTTTAGAGG TGATAAAATT AACTGCTTAA CTGTCAATGT
24001 AATACAAGTT GTTTGATCTT TGCAATGATT CTTATCAGAA ACCATATAGT AAATTAGTTA
24061 CACAGGAAAT TTTTAATATT ATTATTATCA TTCATTATGT ATTAAAATTA GAGTTGTGGC
24121 TTGGCTCTGC TAACACGTTG CTCATAGGAG ATATGGTAGA GCCGCAGACA CGTCGTATGC
24181 AGGAACGTGC TGCGGCTGGC TGGTGAACTT CCGATAGTGC GGGTGTTGAA TGATTTCCAG
24241 TTGCTACCGA TTTTACATAT TTTTTGCATG AGAGAATTTG TACCACCTCC CACCGACCAT
24301 CTATGACTGT ACGCCACTGT CCCTAGGACT GCTATGTGCC GGAGCGGACA TTACAAACGT
24361 CCTTCTCGGT GCATGCCACT GTTGCCAATG ACCTGCCTAG GAATTGGTTA GCAAGTTACT
24421 ACCGGATTTT GTAAAAACAG CCCTCCTCAT ATAAAAAGTA TTCGTTCACT TCCGATAAGC
24481 GTCGTAATTT TCTATCTTTC ATCATATTCT AGATCCCTCT GAAAAAATCT TCCGAGTTTG
24541 CTAGGCACTG ATACATAACT CTTTTCCAAT AATTGGGGAA GTCATTCAAA TCTATAATAG
24601 GTTTCAGATT TGCTTCAATA AATTCTGACT GTAGCTGCTG AAACGTTGCG GTTGAACTAT
24661 ATTTCCTTAT AACTTTTACG AAAGAGTTTC TTTGAGTAAT CACTTCACTC AAGTGCTTCC
24721 CTGCCTCCAA ACGATACCTG TTAGCAATAT TTAATAGCTT GAAATGATGA AGAGCTCTGT
24781 GTTTGTCTTC CTGCCTCCAG TTCGCCGGGC ATTCAACATA AAAACTGATA GCACCCGGAG
24841 TTCCGGAAAC GAAATTTGCA TATACCCATT GCTCACGAAA AAAAATGTCC TTGTCGATAT
24901 AGGGATGAAT CGCTTGGTGT ACCTCATCTA CTGCGAAAAC TTGACCTTTC TCTCCCATAT
24961 TGCAGTCGCG GCACGATGGA ACTAAATTAA TAGGCATCAC CGAAAATTCA GGATAATGTG
25021 CAATAGGAAG AAAATGATCT ATATTTTTTG TCTGTCCTAT ATCACCACAA AATGGACATT
25081 TTTCACCTGA TGAAACAAGC ATGTCATCGT AATATGTTCT AGCGGGTTTG TTTTTATCTC
25141 GGAGATTATT TTCATAAAGC TTTTCTAATT TAACCTTTGT CAGGTTACCA ACTACTAAGG
25201 TTGTAGGCTC AAGAGGGTGT GTCCTGTCGT AGGTAAATAA CTGACCTGTC GAGCTTAATA
25261 TTCTATATTG TTGTTCTTTC TGCAAAAAAG TGGGGAAGTG AGTAATGAAA TTATTTCTAA
25321 CATTTATCTG CATCATACCT TCCGAGCATT TATTAAGCAT TTCGCTATAA GTTCTCGCTG
25381 GAAGAGGTAG TTTTTTCATT GTACTTTACC TTCATCTCTG TTCATTATCA TCGCTTTTAA
25441 AACGGTTCGA CCTTCTAATC CTATCTGACC ATTATAATTT TTTAGAATGG TTTCATAAGA
25501 AAGCTCTGAA TCAACGGACT GCGATAATAA GTGGTGGTAT CCAGAATTTG TCACTTCAAG
25561 TAAAAACACC TCACGAGTTA AAACACCTAA GTTCTCACCG AATGTCTCAA TATCCGGACG
25621 GATAATATTT ATTGCTTCTC TTGACCGTAG GACTTTCCAC ATGCAGGATT TTGGAACCTC
25681 TTGCAGTACT ACTGGGGAAT GAGTTGCAAT TATTGCTACA CCATTGCGTG CATCGAGTAA
25741 GTCGCTTAAT GTTCGTAAAA AAGCAGAGAG CAAAGGTGGA TGCAGATGAA CCTCTGGTTC
25801 ATCGAATAAA ACTAATGACT TTTCGCCAAC GACATCTACT AATCTTGTGA TAGTAAATAA

25861	AACAATTGCA	TGTCCAGAGC	TCATTCGAAG	CAGATATTTC	TGGATATTGT	CATAAAACAA
25921	TTTAGTGAAT	TTATCATCGT	CCACTTGAAT	CTGTGGTTCA	TTACGTCTTA	ACTCTTCATA
25981	TTTAGAAATG	AGGCTGATGA	GTTCCATATT	TGAAAAGTTT	TCATCACTAC	TTAGTTTTTT
26041	GATAGCTTCA	AGCCAGAGTT	GTCTTTTTCT	ATCTACTCTC	ATACAACCAA	TAAATGCTGA
26101	AATGAATTCT	AAGCGGAGAT	CGCCTAGTGA	TTTTAAACTA	TTGCTGGCAG	CATTCTTGAG
26161	TCCAATATAA	AAGTATTGTG	TACCTTTTGC	TGGGTCAGGT	TGTTCTTTAG	GAGGAGTAAA
26221	AGGATCAAAT	GCACTAAACG	AAACTGAAAC	AAGCGATCGA	AAATATCCCT	TTGGGATTCT
26281	TGACTCGATA	AGTCTATTAT	TTTCAGAGAA	AAAATATTCA	TTGTTTTCTG	GGTTGGTGAT
26341	TGCACCAATC	ATTCCATTCA	AAATTGTTGT	TTTACCACAC	CCATTCCGCC	CGATAAAAGC
26401	ATGAATGTTC	GTGCTGGGCA	TAGAATTAAC	CGTCACCTCA	AAAGGTATAG	TTAAATCACT
26461	GAATCCGGGA	GCACTTTTTC	TATTAAATGA	AAAGTGGAAA	TCTGACAATT	CTGGCAAACC
26521	ATTTAACACA	CGTGCGAACT	GTCCATGAAT	TTCTGAAAGA	GTTACCCCTC	TAAGTAATGA
26581	GGTGTTAAGG	ACGCTTTCAT	TTTCAATGTC	GGCTAATCGA	TTTGGCCATA	CTACTAAATC
26641	CTGAATAGCT	TTAAGAAGGT	TATGTTTAAA	ACCATCGCTT	AATTTGCTGA	GATTAACATA
26701	GTAGTCAATG	CTTTCACCTA	AGGAAAAAAA	CATTTCAGGG	AGTTGACTGA	ATTTTTTATC
26761	TATTAATGAA	TAAGTGCTTA	CTTCTTCTTT	TTGACCTACA	AAACCAATTT	TAACATTTCC
26821	GATATCGCAT	TTTTCACCAT	GCTCATCAAA	GACAGTAAGA	TAAAACATTG	TAACAAAGGA
26881	ATAGTCATTC	CAACCATCTG	CTCGTAGGAA	TGCCTTATTT	TTTTCTACTG	CAGGAATATA
26941	CCCGCCTCTT	TCAATAACAC	TAAACTCCAA	CATATAGTAA	CCCTTAATTT	TATTAAAATA
27001	ACCGCAATTT	ATTTGGCGGC	AACACAGGAT	CTCTCTTTTA	AGTTACTCTC	TATTACATAC
27061	GTTTTCCATC	TAAAAATTAG	TAGTATTGAA	CTTAACGGGG	CATCGTATTG	TAGTTTTCCA
27121	TATTTAGCTT	TCTGCTTCCT	TTTGGATAAC	CCACTGTTAT	TCATGTTGCA	TGGTGCACTG
27181	TTTATACCAA	CGATATAGTC	TATTAATGCA	TATATAGTAT	CGCCGAACGA	TTAGCTCTTC
27241	AGGCTTCTGA	AGAAGCGTTT	CAAGTACTAA	TAAGCCGATA	GATAGCCACG	GACTTCGTAG
27301	CCATTTTTCA	TAAGTGTTAA	CTTCCGCTCC	TCGCTCATAA	CAGACATTCA	CTACAGTTAT
27361	GGCGGAAAGG	TATGCATGCT	GGGTGTGGGG	AAGTCGTGAA	AGAAAAGAAG	TCAGCTGCGT
27421	CGTTTGACAT	CACTGCTATC	TTCTTACTGG	TTATGCAGGT	CGTAGTGGGT	GGCACACAAA
27481	GCTTTGCACT	GGATTGCGAG	GCTTTGTGCT	TCTCTGGAGT	GCGACAGGTT	TGATGACAAA
27541	AAATTAGCGC	AAGAAGACAA	AAATCACCTT	GCGCTAATGC	TCTGTTACAG	GTCACTAATA
27601	CCATCTAAGT	AGTTGATTCA	TAGTGACTGC	ATATGTTGTG	TTTTACAGTA	TTATGTAGTC
27661	TGTTTTTTAT	GCAAAATCTA	ATTTAATATA	TTGATATTTA	TATCATTTTA	CGTTTCTCGT
27721	TCAGCTTTTT	TATACTAAGT	TGGCATTATA	AAAAAGCATT	GCTTATCAAT	TTGTTGCAAC
27781	GAACAGGTCA	CTATCAGTCA	AAATAAAATC	ATTATTTGAT	TTCAATTTTG	TCCCACTCCC
27841	TGCCTCTGTC	ATCACGATAC	TGTGATGCCA	TGGTGTCCGA	CTTATGCCCG	AGAAGATGTT
27901	GAGCAAACTT	ATCGCTTATC	TGCTTCTCAT	AGAGTCTTGC	AGACAAACTG	CGCAACTCGT
27961	GAAAGGTAGG	CGGATCCCCT	TCGAAGGAAA	GACCTGATGC	TTTTCGTGCG	CGCATAAAAT
28021	ACCTTGATAC	TGTGCCGGAT	GAAAGCGGTT	CGCGACGAGT	AGATGCAATT	ATGGTTTCTC
28081	CGCCAAGAAT	CTCTTTGCAT	TTATCAAGTG	TTTCCTTCAT	TGATATTCCG	AGAGCATCAA
28141	TATGCAATGC	TGTTGGGATG	GCAATTTTTA	CGCCTGTTTT	GCTTTGCTCG	ACATAAAGAT
28201	ATCCATCTAC	GATATCAGAC	CACTTCATTT	CGCATAAATC	ACCAACTCGT	TGCCCGGTAA
28261	CAACAGCCAG	TTCCATTGCA	AGTCTGAGCC	AACATGGTGA	TGATTCTGCT	GCTTGATAAA
28321	TTTTCAGGTA	TTCGTCAGCC	GTAAGTCTTG	ATCTCCTTAC	CTCTGATTTT	GCTGCGCGAG
28381	TGGCAGCGAC	ATGGTTTGTT	GTTATATGGC	CTTCAGCTAT	TGCCTCTCGG	AATGCATCGC
28441	TCAGTGTTGA	TCTGATTAAC	TTGGCTGACG	CCGCCTTGCC	CTCGTCTATG	TATCCATTGA
28501	GCATTGCCGC	AATTTCTTTT	GTGGTGATGT	CTTCAAGTGG	AGCATCAGGC	AGACCCCTCC
28561	TTATTGCTTT	AATTTTGCTC	ATGTAATTTA	TGAGTGTCTT	CTGCTTGATT	CCTCTGCTGG
28621	CCAGGATTTT	TTCGTAGCGA	TCAAGCCATG	AATGTAACGT	AACGGAATTA	TCACTGTTGA
28681	TTCTCGCTGT	CAGAGGCTTG	TGTTTGTGTC	CTGAAAATAA	CTCAATGTTG	GCCTGTATAG
28741	CTTCAGTGAT	TGCGATTCGC	CTGTCTCTGC	CTAATCCAAA	CTCTTTACCC	GTCCTTGGGT
28801	CCCTGTAGCA	GTAATATCCA	TTGTTTCTTA	TATAAAGGTT	AGGGGGTAAA	TCCCGGCGCT
28861	CATGACTTCG	CCTTCTTCCC	ATTTCTGATC	CTCTTCAAAA	GGCCACCTGT	TACTGGTCGA
28921	TTTAAGTCAA	CCTTTACCGC	TGATTCGTGG	AACAGATACT	CTCTTCCATC	CTTAACCGGA
28981	GGTGGGAATA	TCCTGCATTC	CCGAACCCAT	CGACGAACTG	TTTCAAGGCT	TCTTGGACGT

29041 CGCTGGCGTG CGTTCCACTC CTGAAGTGTC AAGTACATCG CAAAGTCTCC GCAATTACAC
29101 GCAAGAAAAA ACCGCCATCA GGCGGCTTGG TGTTCTTTCA GTTCTTCAAT TCGAATATTG
29161 GTTACGTCTG CATGTGCTAT CTGCGCCCAT ATCATCCAGT GGTCGTAGCA GTCGTTGATG
29221 TTCTCCGCTT CGATAACTCT GTTGAATGGC TCTCCATTCC ATTCTCCTGT GACTCGGAAG
29281 TGCATTTATC ATCTCCATAA AACAAAACCC GCCGTAGCGA GTTCAGATAA AATAAATCCC
29341 CGCGAGTGCG AGGATTGTTA TGTAATATTG GGTTTAATCA TCTATATGTT TTGTACAGAG
29401 AGGGCAAGTA TCGTTTCCAC CGTACTCGTG ATAATAATTT TGCACGGTAT CAGTCATTTC
29461 TCGCACATTG CAGAATGGGG ATTTGTCTTC ATTAGACTTA TAAACCTTCA TGGAATATTT
29521 GTATGCCGAC TCTATATCTA TACCTTCATC TACATAAACA CCTTCGTGAT GTCTGCATGG
29581 AGACAAGACA CCGGATCTGC ACAACATTGA TAACGCCCAA TCTTTTTGCT CAGACTCTAA
29641 CTCATTGATA CTCATTTATA AACTCCTTGC AATGTATGTC GTTTCAGCTA AACGGTATCA
29701 GCAATGTTTA TGTAAAGAAA CAGTAAGATA ATACTCAACC CGATGTTTGA GTACGGTCAT
29761 CATCTGACAC TACAGACTCT GGCATCGCTG TGAAGACGAC GCGAAATTCA GCATTTTCAC
29821 AAGCGTTATC TTTTACAAAA CCGATCTCAC TCTCCTTTGA TGCGAATGCC AGCGTCAGAC
29881 ATCATATGCA GATACTCACC TGCATCCTGA ACCCATTGAC CTCCAACCCC GTAATAGCGA
29941 TGCGTAATGA TGTCGATAGT TACTAACGGG TCTTGTTCGA TTAACTGCCG CAGAAACTCT
30001 TCCAGGTCAC CAGTGCAGTG CTTGATAACA GGAGTCTTCC CAGGATGGCG AACAACAAGA
30061 AACTGGTTTC CGTCTTCACG GACTTCGTTG CTTTCCAGTT TAGCAATACG CTTACTCCCA
30121 TCCGAGATAA CACCTTCGTA ATACTCACGC TGCTCGTTGA GTTTTGATTT TGCTGTTTCA
30181 AGCTCAACAC GCAGTTTCCC TACTGTTAGC GCAATATCCT CGTTCTCCTG GTCGCGGCGT
30241 TTGATGTATT GCTGGTTTCT TTCCCGTTCA TCCAGCAGTT CCAGCACAAT CGATGGTGTT
30301 ACCAATTCAT GGAAAAGGTC TGCGTCAAAT CCCCAGTCGT CATGCATTGC CTGCTCTGCC
30361 GCTTCACGCA GTGCCTGAGA GTTAATTTCG CTCACTTCGA ACCTCTCTGT TTACTGATAA
30421 GTTCCAGATC CTCCTGGCAA CTTGCACAAG TCCGACAACC CTGAACGACC AGGCGTCTTC
30481 GTTCATCTAT CGGATCGCCA CACTCACAAC AATGAGTGGC AGATATAGCC TGGTGGTTCA
30541 GGCGGCGCAT TTTTATTGCT GTGTTGCGCT GTAATTCTTC TATTTCTGAT GCTGAATCAA
30601 TGATGTCTGC CATCTTTCAT TAATCCCTGA ACTGTTGGTT AATACGCTTG AGGGTGAATG
30661 CGAATAATAA AAAAGGAGCC TGTAGCTCCC TGATGATTTT GCTTTTCATG TTCATCGTTC
30721 CTTAAAGACG CCGTTTAACA TGCCGATTGC CAGGCTTAAA TGAGTCGGTG TGAATCCCAT
30781 CAGCGTTACC GTTTCGCGGT GCTTCTTCAG TACGCTACGG CAAATGTCAT CGACGTTTTT
30841 ATCCGGAAAC TGCTGTCTGG CTTTTTTTGA TTTCAGAATT AGCCTGACGG GCAATGCTGC
30901 GAAGGGCGTT TTCCTGCTGA GGTGTCATTG AACAAGTCCC ATGTCGGCAA GCATAAGCAC
30961 ACAGAATATG AAGCCCGCTG CCAGAAAAAT GCATTCCGTG GTTGTCATAC CTGGTTTCTC
31021 TCATCTGCTT CTGCTTTCGC CACCATCATT TCCAGCTTTT GTGAAAGGGA TGCGGCTAAC
31081 GTATGAAATT CTTCGTCTGT TTCTACTGGT ATTGGCACAA ACCTGATTCC AATTTGAGCA
31141 AGGCTATGTG CCATCTCGAT ACTCGTTCTT AACTCAACAG AAGATGCTTT GTGCATACAG
31201 CCCCTCGTTT ATTATTTATC TCCTCAGCCA GCCGCTGTGC TTTCAGTGGA TTTCGGATAA
31261 CAGAAAGGCC GGGAAATACC CAGCCTCGCT TTGTAACGGA GTAGACGAAA GTGATTGCGC
31321 CTACCCGGAT ATTATCGTGA GGATGCGTCA TCGCCATTGC TCCCCAAATA CAAAACCAAT
31381 TTCAGCCAGT GCCTCGTCCA TTTTTTCGAT GAACTCCGGC ACGATCTCGT CAAAACTCGC
31441 CATGTACTTT TCATCCCGCT CAATCACGAC ATAATGCAGG CCTTCACGCT TCATACGCGG
31501 GTCATAGTTG GCAAAGTACC AGGCATTTTT TCGCGTCACC CACATGCTGT ACTGCACCTG
31561 GGCCATGTAA GCTGACTTTA TGGCCTCGAA ACCACCGAGC CGGAACTTCA TGAAATCCCG
31621 GGAGGTAAAC GGGCATTTCA GTTCAAGGCC GTTGCCGTCA CTGCATAAAC CATCGGGAGA
31681 GCAGGCGGTA CGCATACTTT CGTCGCGATA GATGATCGGG GATTCAGTAA CATTCACGCC
31741 GGAAGTGAAT TCAAACAGGG TTCTGGCGTC GTTCTCGTAC TGTTTTCCCC AGGCCAGTGC
31801 TTTAGCGTTA ACTTCCGGAG CCACACCGGT GCAAACCTCA GCAAGCAGGG TGTGGAAGTA
31861 GGACATTTTC ATGTCAGGCC ACTTCTTTCC GGAGCGGGGT TTTGCTATCA CGTTGTGAAC
31921 TTCTGAAGCG GTGATGACGC CGAGCCGTAA TTTGTGCCAC GCATCATCCC CCTGTTCGAC
31981 AGCTCTCACA TCGATCCCGG TACGCTGCAG GATAATGTCC GGTGTCATGC TGCCACCTTC
32041 TGCTCTGCGG CTTTCTGTTT CAGGAATCCA AGAGCTTTTA CTGCTTCGGC CTGTGTCAGT
32101 TCTGACGATG CACGAATGTC GCGGCGAAAT ATCTGGGAAC AGAGCGGCAA TAAGTCGTCA
32161 TCCCATGTTT TATCCAGGGC GATCAGCAGA GTGTTAATCT CCTGCATGGT TTCATCGTTA

32221 ACCGGAGTGA TGTCGCGTTC CGGCTGACGT TCTGCAGTGT ATGCAGTATT TTCGACAATG
32281 CGCTCGGCTT CATCCTTGTC ATAGATACCA GCAAATCCGA AGGCCAGACG GGCACACTGA
32341 ATCATGGCTT TATGACGTAA CATCCGTTTG GGATGCGACT GCCACGGCCC CGTGATTTCT
32401 CTGCCTTCGC GAGTTTTGAA TGGTTCGCGG CGGCATTCAT CCATCCATTC GGTAACGCAG
32461 ATCGGATGAT TACGGTCCTT GCGGTAAATC CGGCATGTAC AGGATTCATT GTCCTGCTCA
32521 AAGTCCATGC CATCAAACTG CTGGTTTTCA TTGATGATGC GGGACCAGCC ATCAACGCCC
32581 ACCACCGGAA CGATGCCATT CTGCTTATCA GGAAAGGCGT AAATTTCTTT CGTCCACGGA
32641 TTAAGGCCGT ACTGGTTGGC AACGATCAGT AATGCGATGA ACTGCGCATC GCTGGCATCA
32701 CCTTTAAATG CCGTCTGGCG AAGAGTGGTG ATCAGTTCCT GTGGGTCGAC AGAATCCATG
32761 CCGACACGTT CAGCCAGCTT CCCAGCCAGC GTTGCGAGTG CAGTACTCAT TCGTTTTATA
32821 CCTCTGAATC AATATCAACC TGGTGGTGAG CAATGGTTTC AACCATGTAC CGGATGTGTT
32881 CTGCCATGCG CTCCTGAAAC TCAACATCGT CATCAAACGC ACGGGTAATG GATTTTTTGC
32941 TGGCCCCGTG GCGTTGCAAA TGATCGATGC ATAGCGATTC AAACAGGTGC TGGGGCAGGC
33001 CTTTTTCCAT GTCGTCTGCC AGTTCTGCCT CTTTCTCTTC ACGGGCGAGC TGCTGGTAGT
33061 GACGCGCCCA GCTCTGAGCC TCAAGACGAT CCTGAATGTA ATAAGCGTTC ATGGCTGAAC
33121 TCCTGAAATA GCTGTGAAAA TATCGCCCGC GAAATGCCGG GCTGATTAGG AAAACAGGAA
33181 AGGGGGTTAG TGAATGCTTT TGCTTGATCT CAGTTTCAGT ATTAATATCC ATTTTTTATA
33241 AGCGTCGACG GCTTCACGAA ACATCTTTTC ATCGCCAATA AAAGTGGCGA TAGTGAATTT
33301 AGTCTGGATA GCCATAAGTG TTTGATCCAT TCTTTGGGAC TCCTGGCTGA TTAAGTATGT
33361 CGATAAGGCG TTTCCATCCG TCACGTAATT TACGGGTGAT TCGTTCAAGT AAAGATTCGG
33421 AAGGGCAGCC AGCAACAGGC CACCCTGCAA TGGCATATTG CATGGTGTGC TCCTTATTTA
33481 TACATAACGA AAAACGCCTC GAGTGAAGCG TTATTGGTAT GCGGTAAAAC CGCACTCAGG
33541 CGGCCTTGAT AGTCATATCA TCTGAATCAA ATATTCCTGA TGTATCGATA TCGGTAATTC
33601 TTATTCCTTC GCTACCATCC ATTGGAGGCC ATCCTTCCTG ACCATTTCCA TCATTCCAGT
33661 CGAACTCACA CACAACACCA TATGCATTTA AGTCGCTTGA AATTGCTATA AGCAGAGCAT
33721 GTTGCGCCAG CATGATTAAT ACAGCATTTA ATACAGAGCC GTGTTTATTG AGTCGGTATT
33781 CAGAGTCTGA CCAGAAATTA TTAATCTGGT GAAGTTTTTC CTCTGTCATT ACGTCATGGT
33841 CGATTTCAAT TTCTATTGAT GCTTTCCAGT CGTAATCAAT GATGTATTTT TTGATGTTTG
33901 ACATCTGTTC ATATCCTCAC AGATAAAAAA TCGCCCTCAC ACTGGAGGGC AAAGAAGATT
33961 TCCAATAATC AGAACAAGTC GGCTCCTGTT TAGTTACGAG CGACATTGCT CCGTGTATTC
34021 ACTCGTTGGA ATGAATACAC AGTGCAGTGT TTATTCTGTT ATTTATGCCA AAAATAAAGG
34081 CCACTATCAG GCAGCTTTGT TGTTCTGTTT ACCAAGTTCT CTGGCAATCA TTGCCGTCGT
34141 TCGTATTGCC CATTTATCGA CATATTTCCC ATCTTCCATT ACAGGAAACA TTTCTTCAGG
34201 CTTAACCATG CATTCCGATT GCAGCTTGCA TCCATTGCAT CGCTTGAATT GTCCACACCA
34261 TTGATTTTTA TCAATAGTCG TAGTCATACG GATAGTCCTG GTATTGTTCC ATCACATCCT
34321 GAGGATGCTC TTCGAACTCT TCAAATTCTT CTTCCATATA TCACCTTAAA TAGTGGATTG
34381 CGGTAGTAAA GATTGTGCCT GTCTTTTAAC CACATCAGGC TCGGTGGTTC TCGTGTACCC
34441 CTACAGCGAG AAATCGGATA AACTATTACA ACCCCTACAG TTTGATGAGT ATAGAAATGG
34501 ATCCACTCGT TATTCTCGGA CGAGTGTTCA GTAATGAACC TCTGGAGAGA ACCATGTATA
34561 TGATCGTTAT CTGGGTTGGA CTTCTGCTTT TAAGCCCAGA TAACTGGCCT GAATATGTTA
34621 ATGAGAGAAT CGGTATTCCT CATGTGTGGC ATGTTTTCGT CTTTGCTCTT GCATTTTCGC
34681 TAGCAATTAA TGTGCATCGA TTATCAGCTA TTGCCAGCGC CAGATATAAG CGATTTAAGC
34741 TAAGAAAACG CATTAAGATG CAAAACGATA AAGTGCGATC AGTAATTCAA AACCTTACAG
34801 AAGAGCAATC TATGGTTTTG TGCGCAGCCC TTAATGAAGG CAGGAAGTAT GTGGTTACAT
34861 CAAAACAATT CCCATACATT AGTGAGTTGA TTGAGCTTGG TGTGTTGAAC AAAACTTTTT
34921 CCCGATGGAA TGGAAAGCAT ATATTATTCC CTATTGAGGA TATTTACTGG ACTGAATTAG
34981 TTGCCAGCTA TGATCCATAT AATATTGAGA TAAAGCCAAG GCCAATATCT AAGTAACTAG
35041 ATAAGAGGAA TCGATTTTCC CTTAATTTTC TGGCGTCCAC TGCATGTTAT GCCGCGTTCG
35101 CCAGGCTTGC TGTACCATGT GCGCTGATTC TTGCGCTCAA TACGTTGCAG GTTGCTTTCA
35161 ATCTGTTTGT GGTATTCAGC CAGCACTGTA AGGTCTATCG GATTTAGTGC GCTTTCTACT
35221 CGTGATTTCG GTTTGCGATT CAGCGAGAGA ATAGGGCGGT TAACTGGTTT TGCGCTTACC
35281 CCAACCAACA GGGGATTTGC TGCTTTCCAT TGAGCCTGTT TCTCTGCGCG ACGTTCGCGG
35341 CGGCGTGTTT GTGCATCCAT CTGGATTCTC CTGTCAGTTA GCTTTGGTGG TGTGTGGCAG

```
35401   TTGTAGTCCT   GAACGAAAAC   CCCCCGCGAT   TGGCACATTG   GCAGCTAATC   CGGAATCGCA
35461   CTTACGGCCA   ATGCTTCGTT   TCGTATCACA   CACCCCAAAG   CCTTCTGCTT   TGAATGCTGC
35521   CCTTCTTCAG   GGCTTAATTT   TTAAGAGCGT   CACCTTCATG   GTGGTCAGTG   CGTCCTGCTG
35581   ATGTGCTCAG   TATCACCGCC   AGTGGTATTT   ATGTCAACAC   CGCCAGAGAT   AATTTATCAC
35641   CGCAGATGGT   TATCTGTATG   TTTTTTATAT   GAATTTATTT   TTTGCAGGGG   GGCATTGTTT
35701   GGTAGGTGAG   AGATCTGAAT   TGCTATGTTT   AGTGAGTTGT   ATCTATTTAT   TTTTCAATAA
35761   ATACAATTGG   TTATGTGTTT   TGGGGGCGAT   CGTGAGGCAA   AGAAAACCCG   GCGCTGAGGC
35821   CGGGTTATTC   TTGTTCTCTG   GTCAAATTAT   ATAGTTGGAA   AACAAGGATG   CATATATGAA
35881   TGAACGATGC   AGAGGCAATG   CCGATGGCGA   TAGTGGGTAT   CATGTAGCCG   CTTATGCTGG
35941   AAAGAAGCAA   TAACCCGCAG   AAAAACAAAG   CTCCAAGCTC   AACAAAACTA   AGGGCATAGA
36001   CAATAACTAC   CGATGTCATA   TACCCATACT   CTCTAATCTT   GGCCAGTCGG   CGCGTTCTGC
36061   TTCCGATTAG   AAACGTCAAG   GCAGCAATCA   GGATTGCAAT   CATGGTTCCT   GCATATGATG
36121   ACAATGTCGC   CCCAAGACCA   TCTCTATGAG   CTGAAAAAGA   AACACCAGGA   ATGTAGTGGC
36181   GGAAAAGGAG   ATAGCAAATG   CTTACGATAA   CGTAAGGAAT   TATTACTATG   TAAACACCAG
36241   GCATGATTCT   GTTCCGCATA   ATTACTCCTG   ATAATTAATC   CTTAACTTTG   CCCACCTGCC
36301   TTTTAAAACA   TTCCAGTATA   TCACTTTTCA   TTCTTGCGTA   GCAATATGCC   ATCTCTTCAG
36361   CTATCTCAGC   ATTGGTGACC   TTGTTCAGAG   GCGCTGAGAG   ATGGCCTTTT   TCTGATAGAT
36421   AATGTTCTGT   TAAAATATCT   CCGGCCTCAT   CTTTTGCCCG   CAGGCTAATG   TCTGAAAATT
36481   GAGGTGACGG   GTTAAAAATA   ATATCCTTGG   CAACCTTTTT   TATATCCCTT   TTAAATTTTG
36541   GCTTAATGAC   TATATCCAAT   GAGTCAAAAA   GCTCCCCTTC   AATATCTGTT   GCCCCTAAGA
36601   CCTTTAATAT   ATCGCCAAAT   ACAGGTAGCT   TGGCTTCTAC   CTTCACCGTT   GTTCGGCCGA
36661   TGAAATGCAT   ATGCATAACA   TCGTCTTTGG   TGGTTCCCCT   CATCAGTGGC   TCTATCTGAA
36721   CGCGCTCTCC   ACTGCTTAAT   GACATTCCTT   TCCCGATTAA   AAAATCTGTC   AGATCGGATG
36781   TGGTCGGCCC   GAAAACAGTT   CTGGCAAAAC   CAATGGTGTC   GCCTTCAACA   AACAAAAAAG
36841   ATGGGAATCC   CAATGATTCG   TCATCTGCGA   GGCTGTTCTT   AATATCTTCA   ACTGAAGCTT
36901   TAGAGCGATT   TATCTTCTGA   ACCAGACTCT   TGTCATTTGT   TTTGGTAAAG   AGAAAAGTTT
36961   TTCCATCGAT   TTTATGAATA   TACAAATAAT   TGGAGCCAAC   CTGCAGGTGA   TGATTATCAG
37021   CCAGCAGAGA   ATTAAGGAAA   ACAGACAGGT   TTATTGAGCG   CTTATCTTTC   CCTTTATTTT
37081   TGCTGCGGTA   AGTCGCATAA   AAACCATTCT   TCATAATTCA   ATCCATTTAC   TATGTTATGT
37141   TCTGAGGGGA   GTGAAAATTC   CCCTAATTCG   ATGAAGATTC   TTGCTCAATT   GTTATCAGCT
37201   ATGCGCCGAC   CAGAACACCT   TGCCGATCAG   CCAAACGTCT   CTTCAGGCCA   CTGACTAGCG
37261   ATAACTTTCC   CCACAACGGA   ACAACTCTCA   TTGCATGGGA   TCATTGGGTA   CTGTGGGTTT
37321   AGTGGTTGTA   AAAACACCTG   ACCGCTATCC   CTGATCAGTT   TCTTGAAGGT   AAACTCATCA
37381   CCCCCAAGTC   TGGCTATGCA   GAAATCACCT   GGCTCAACAG   CCTGCTCAGG   GTCAACGAGA
37441   ATTAACATTC   CGTCAGGAAA   GCTTGGCTTG   GAGCCTGTTG   GTGCGGTCAT   GGAATTACCT
37501   TCAACCTCAA   GCCAGAATGC   AGAATCACTG   GCTTTTTTGG   TTGTGCTTAC   CCATCTCTCC
37561   GCATCACCTT   TGGTAAAGGT   TCTAAGCTCA   GGTGAGAACA   TCCCTGCCTG   AACATGAGAA
37621   AAAACAGGGT   ACTCATACTC   ACTTCTAAGT   GACGGCTGCA   TACTAACCGC   TTCATACATC
37681   TCGTAGATTT   CTCTGGCGAT   TGAAGGGCTA   AATTCTTCAA   CGCTAACTTT   GAGAATTTTT
37741   GCAAGCAATG   CGGCGTTATA   AGCATTTAAT   GCATTGATGC   CATTAAATAA   AGCACCAACG
37801   CCTGACTGCC   CCATCCCCAT   CTTGTCTGCG   ACAGATTCCT   GGGATAAGCC   AAGTTCATTT
37861   TTCTTTTTTT   CATAAATTGC   TTTAAGGCGA   CGTGCGTCCT   CAAGCTGCTC   TTGTGTTAAT
37921   GGTTTCTTTT   TTGTGCTCAT   ACGTTAAATC   TATCACCGCA   AGGGATAAAT   ATCTAACACC
37981   GTGCGTGTTG   ACTATTTTAC   CTCTGGCGGT   GATAATGGTT   GCATGTACTA   AGGAGGTTGT
38041   ATGGAACAAC   GCATAACCCT   GAAAGATTAT   GCAATGCGCT   TTGGGCAAAC   CAAGACAGCT
38101   AAAGATCTCG   GCGTATATCA   AAGCGCGATC   AACAAGGCCA   TTCATGCAGG   CCGAAAGATT
38161   TTTTTAACTA   TAAACGCTGA   TGGAAGCGTT   TATGCGGAAG   AGGTAAAGCC   CTTCCCGAGT
38221   AACAAAAAAA   CAACAGCATA   AATAACCCCG   CTCTTACACA   TTCCAGCCCT   GAAAAAGGGC
38281   ATCAAATTAA   ACCACACCTA   TGGTGTATGC   ATTTATTTGC   ATACATTCAA   TCAATTGTTA
38341   TCTAAGGAAA   TACTTACATA   TGGTTCGTGC   AAACAAACGC   AACGAGGCTC   TACGAATCGA
38401   GAGTGCGTTG   CTTAACAAAA   TCGCAATGCT   TGGAACTGAG   AAGACAGCGG   AAGCTGTGGG
38461   CGTTGATAAG   TCGCAGATCA   GCAGGTGGAA   GAGGGACTGG   ATTCCAAAGT   TCTCAATGCT
38521   GCTTGCTGTT   CTTGAATGGG   GGGTCGTTGA   CGACGACATG   GCTCGATTGG   CGCGACAAGT
38581   TGCTGCGATT   CTCACCAATA   AAAAACGCCC   GGCGGCAACC   GAGCGTTCTG   AACAAATCCA
```

38641 GATGGAGTTC TGAGGTCATT ACTGGATCTA TCAACAGGAG TCATTATGAC AAATACAGCA
38701 AAAATACTCA ACTTCGGCAG AGGTAACTTT GCCGGACAGG AGCGTAATGT GGCAGATCTC
38761 GATGATGGTT ACGCCAGACT ATCAAATATG CTGCTTGAGG CTTATTCGGG CGCAGATCTG
38821 ACCAAGCGAC AGTTTAAAGT GCTGCTTGCC ATTCTGCGTA AAACCTATGG GTGGAATAAA
38881 CCAATGGACA GAATCACCGA TTCTCAACTT AGCGAGATTA CAAAGTTACC TGTCAAACGG
38941 TGCAATGAAG CCAAGTTAGA ACTCGTCAGA ATGAATATTA TCAAGCAGCA AGGCGGCATG
39001 TTTGGACCAA ATAAAAACAT CTCAGAATGG TGCATCCCTC AAAACGAGGG AAAATCCCCT
39061 AAAACGAGGG ATAAAACATC CCTCAAATTG GGGGATTGCT ATCCCTCAAA ACAGGGGGAC
39121 ACAAAAGACA CTATTACAAA AGAAAAAAGA AAAGATTATT CGTCAGAGAA TTCTGGCGAA
39181 TCCTCTGACC AGCCAGAAAA CGACCTTTCT GTGGTGAAAC CGGATGCTGC AATTCAGAGC
39241 GGCAGCAAGT GGGGGACAGC AGAAGACCTG ACCGCCGCAG AGTGGATGTT TGACATGGTG
39301 AAGACTATCG CACCATCAGC CAGAAAACCG AATTTTGCTG GGTGGGCTAA CGATATCCGC
39361 CTGATGCGTG AACGTGACGG ACGTAACCAC CGCGACATGT GTGTGCTGTT CCGCTGGGCA
39421 TGCCAGGACA ACTTCTGGTC CGGTAACGTG CTGAGCCCGG CCAAACTCCG CGATAAGTGG
39481 ACCCAACTCG AAATCAACCG TAACAAGCAA CAGGCAGGCG TGACAGCCAG CAAACCAAAA
39541 CTCGACCTGA CAAACACAGA CTGGATTTAC GGGGTGGATC TATGAAAAAC ATCGCCGCAC
39601 AGATGGTTAA CTTTGACCGT GAGCAGATGC GTCGGATCGC CAACAACATG CCGGAACAGT
39661 ACGACGAAAA GCCGCAGGTA CAGCAGGTAG CGCAGATCAT CAACGGTGTG TTCAGCCAGT
39721 TACTGGCAAC TTTCCCGGCG AGCCTGGCTA ACCGTGACCA GAACGAAGTG AACGAAATCC
39781 GTCGCCAGTG GGTTCTGGCT TTTCGGGAAA ACGGGATCAC CACGATGGAA CAGGTTAACG
39841 CAGGAATGCG CGTAGCCCGT CGGCAGAATC GACCATTTCT GCCATCACCC GGGCAGTTTG
39901 TTGCATGGTG CCGGGAAGAA GCATCCGTTA CCGCCGGACT GCCAAACGTC AGCGAGCTGG
39961 TTGATATGGT TTACGAGTAT TGCCGGAAGC GAGGCCTGTA TCCGGATGCG GAGTCTTATC
40021 CGTGGAAATC AAACGCGCAC TACTGGCTGG TTACCAACCT GTATCAGAAC ATGCGGGCCA
40081 ATGCGCTTAC TGATGCGGAA TTACGCCGTA AGGCCGCAGA TGAGCTTGTC CATATGACTG
40141 CGAGAATTAA CCGTGGTGAG GCGATCCCTG AACCAGTAAA ACAACTTCCT GTCATGGGCG
40201 GTAGACCTCT AAATCGTGCA CAGGCTCTGG CGAAGATCGC AGAAATCAAA GCTAAGTTCG
40261 GACTGAAAGG AGCAAGTGTA TGACGGGCAA AGAGGCAATT ATTCATTACC TGGGGACGCA
40321 TAATAGCTTC TGTGCGCCGG ACGTTGCCGC GCTAACAGGC GCAACAGTAA CCAGCATAAA
40381 TCAGGCCGCG GCTAAAATGG CACGGGCAGG TCTTCTGGTT ATCGAAGGTA AGGTCTGGCG
40441 AACGGTGTAT TACCGGTTTG CTACCAGGGA AGAACGGGAA GGAAAGATGA GCACGAACCT
40501 GGTTTTTAAG GAGTGTCGCC AGAGTGCCGC GATGAAACGG GTATTGGCGG TATATGGAGT
40561 TAAAAGATGA CCATCTACAT TACTGAGCTA ATAACAGGCC TGCTGGTAAT CGCAGGCCTT
40621 TTTATTTGGG GGAGAGGGAA GTCATGAAAA AACTAACCTT TGAAATTCGA TCTCCAGCAC
40681 ATCAGCAAAA CGCTATTCAC GCAGTACAGC AAATCCTTCC AGACCCAACC AAACCAATCG
40741 TAGTAACCAT TCAGGAACGC AACCGCAGCT TAGACCAAAA CAGGAAGCTA TGGGCCTGCT
40801 TAGGTGACGT CTCTCGTCAG GTTGAATGGC ATGGTCGCTG GCTGGATGCA GAAAGCTGGA
40861 AGTGTGTGTT TACCGCAGCA TTAAAGCAGC AGGATGTTGT TCCTAACCTT GCCGGGAATG
40921 GCTTTGTGGT AATAGGCCAG TCAACCAGCA GGATGCGTGT AGGCGAATTT GCGGAGCTAT
40981 TAGAGCTTAT ACAGGCATTC GGTACAGAGC GTGGCGTTAA GTGGTCAGAC GAAGCGAGAC
41041 TGGCTCTGGA GTGGAAAGCG AGATGGGGAG ACAGGGCTGC ATGATAAATG TCGTTAGTTT
41101 CTCCGGTGGC AGGACGTCAG CATATTTGCT CTGGCTAATG GAGCAAAAGC GACGGGCAGG
41161 TAAAGACGTG CATTACGTTT TCATGGATAC AGGTTGTGAA CATCCAATGA CATATCGGTT
41221 TGTCAGGGAA GTTGTGAAGT TCTGGGATAT ACCGCTCACC GTATTGCAGG TTGATATCAA
41281 CCCGGAGCTT GGACAGCCAA ATGGTTATAC GGTATGGGAA CCAAAGGATA TTCAGACGCG
41341 AATGCCTGTT CTGAAGCCAT TTATCGATAT GGTAAAGAAA TATGGCACTC CATACGTCGG
41401 CGGCGCGTTC TGCACTGACA GATTAAAACT CGTTCCCTTC ACCAAATACT GTGATGACCA
41461 TTTCGGGCGA GGGAATTACA CCACGTGGAT TGGCATCAGA GCTGATGAAC CGAAGCGGCT
41521 AAAGCCAAAG CCTGGAATCA GATATCTTGC TGAACTGTCA GACTTTGAGA AGGAAGATAT
41581 CCTCGCATGG TGGAAGCAAC AACCATTCGA TTTGCAAATA CCGGAACATC TCGGTAACTG
41641 CATATTCTGC ATTAAAAAAT CAACGCAAAA AATCGGACTT GCCTGCAAAG ATGAGGAGGG
41701 ATTGCAGCGT GTTTTTAATG AGGTCATCAC GGGATCCCAT GTGCGTGACG GACATCGGGA
41761 AACGCCAAAG GAGATTATGT ACCGAGGAAG AATGTCGCTG GACGGTATCG CGAAAATGTA
41821 TTCAGAAAAT GATTATCAAG CCCTGTATCA GGACATGGTA CGAGCTAAAA GATTCGATAC

41881 CGGCTCTTGT TCTGAGTCAT GCGAAATATT TGGAGGGCAG CTTGATTTCG ACTTCGGGAG
41941 GGAAGCTGCA TGATGCGATG TTATCGGTGC GGTGAATGCA AAGAAGATAA CCGCTTCCGA
42001 CCAAATCAAC CTTACTGGAA TCGATGGTGT CTCCGGTGTG AAAGAACACC AACAGGGGTG
42061 TTACCACTAC CGCAGGAAAA GGAGGACGTG TGGCGAGACA GCGACGAAGT ATCACCGACA
42121 TAATCTGCGA AAACTGCAAA TACCTTCCAA CGAAACGCAC CAGAAATAAA CCCAAGCCAA
42181 TCCCAAAAGA ATCTGACGTA AAAACCTTCA ACTACACGGC TCACCTGTGG GATATCCGGT
42241 GGCTAAGACG TCGTGCGAGG AAAACAAGGT GATTGACCAA AATCGAAGTT ACGAACAAGA
42301 AAGCGTCGAG CGAGCTTTAA CGTGCGCTAA CTGCGGTCAG AAGCTGCATG TGCTGGAAGT
42361 TCACGTGTGT GAGCACTGCT GCGCAGAACT GATGAGCGAT CCGAATAGCT CGATGCACGA
42421 GGAAGAAGAT GATGGCTAAA CCAGCGCGAA GACGATGTAA AAACGATGAA TGCCGGGAAT
42481 GGTTTCACCC TGCATTCGCT AATCAGTGGT GGTGCTCTCC AGAGTGTGGA ACCAAGATAG
42541 CACTCGAACG ACGAAGTAAA GAACGCGAAA AAGCGGAAAA AGCAGCAGAG AAGAAACGAC
42601 GACGAGAGGA GCAGAAACAG AAAGATAAAC TTAAGATTCG AAAACTCGCC TTAAAGCCCC
42661 GCAGTTACTG GATTAAACAA GCCCAACAAG CCGTAAACGC CTTCATCAGA GAAAGAGACC
42721 GCGACTTACC ATGTATCTCG TGCGGAACGC TCACGTCTGC TCAGTGGGAT GCCGGACATT
42781 ACCGGACAAC TGCTGCGGCA CCTCAACTCC GATTTAATGA ACGCAATATT CACAAGCAAT
42841 GCGTGGTGTG CAACCAGCAC AAAAGCGGAA ATCTCGTTCC GTATCGCGTC GAACTGATTA
42901 GCCGCATCGG GCAGGAAGCA GTAGACGAAA TCGAATCAAA CCATAACCGC CATCGCTGGA
42961 CTATCGAAGA GTGCAAGGCG ATCAAGGCAG AGTACCAACA GAAACTCAAA GACCTGCGAA
43021 ATAGCAGAAG TGAGGCCGCA TGACGTTCTC AGTAAAAACC ATTCCAGACA TGCTCGTTGA
43081 AGCATACGGA AATCAGACAG AAGTAGCACG CAGACTGAAA TGTAGTCGCG GTACGGTCAG
43141 AAAATACGTT GATGATAAAG ACGGGAAAAT GCACGCCATC GTCAACGACG TTCTCATGGT
43201 TCATCGCGGA TGGAGTGAAA GAGATGCGCT ATTACGAAAA AATTGATGGC AGCAAATACC
43261 GAAATATTTG GGTAGTTGGC GATCTGCACG GATGCTACAC GAACCTGATG AACAAACTGG
43321 ATACGATTGG ATTCGACAAC AAAAAAGACC TGCTTATCTC GGTGGGCGAT TTGGTTGATC
43381 GTGGTGCAGA GAACGTTGAA TGCCTGGAAT TAATCACATT CCCCTGGTTC AGAGCTGTAC
43441 GTGGAAACCA TGAGCAAATG ATGATTGATG GCTTATCAGA GCGTGGAAAC GTTAATCACT
43501 GGCTGCTTAA TGGCGGTGGC TGGTTCTTTA ATCTCGATTA CGACAAAGAA ATTCTGGCTA
43561 AAGCTCTTGC CCATAAAGCA GATGAACTTC CGTTAATCAT CGAACTGGTG AGCAAAGATA
43621 AAAAATATGT TATCTGCCAC GCCGATTATC CCTTTGACGA ATACGAGTTT GGAAAGCCAG
43681 TTGATCATCA GCAGGTAATC TGGAACCGCG AACGAATCAG CAACTCACAA AACGGGATCG
43741 TGAAAGAAAT CAAAGGCGCG GACACGTTCA TCTTTGGTCA TACGCCAGCA GTGAAACCAC
43801 TCAAGTTTGC CAACCAAATG TATATCGATA CCGGCGCAGT GTTCTGCGGA AACCTAACAT
43861 TGATTCAGGT ACAGGGAGAA GGCGCATGAG ACTCGAAAGC GTAGCTAAAT TTCATTCGCC
43921 AAAAAGCCCG ATGATGAGCG ACTCACCACG GGCCACGGCT TCTGACTCTC TTTCCGGTAC
43981 TGATGTGATG GCTGCTATGG GGATGGCGCA ATCACAAGCC GGATTCGGTA TGGCTGCATT
44041 CTGCGGTAAG CACGAACTCA GCCAGAACGA CAAACAAAAG GCTATCAACT ATCTGATGCA
44101 ATTTGCACAC AAGGTATCGG GGAAATACCG TGGTGTGGCA AAGCTTGAAG GAAATACTAA
44161 GGCAAAGGTA CTGCAAGTGC TCGCAACATT CGCTTATGCG GATTATTGCC GTAGTGCCGC
44221 GACGCCGGGG GCAAGATGCA GAGATTGCCA TGGTACAGGC CGTGCGGTTG ATATTGCCAA
44281 AACAGAGCTG TGGGGGAGAG TTGTCGAGAA AGAGTGCGGA AGATGCAAAG GCGTCGGCTA
44341 TTCAAGGATG CCAGCAAGCG CAGCATATCG CGCTGTGACG ATGCTAATCC CAAACCTTAC
44401 CCAACCCACC TGGTCACGCA CTGTTAAGCC GCTGTATGAC GCTCTGGTGG TGCAATGCCA
44461 CAAAGAAGAG TCAATCGCAG ACAACATTTT GAATGCGGTC ACACGTTAGC AGCATGATTG
44521 CCACGGATGG CAACATATTA ACGGCATGAT ATTGACTTAT TGAATAAAAT TGGGTAAATT
44581 TGACTCAACG ATGGGTTAAT TCGCTCGTTG TGGTAGTGAG ATGAAAAGAG GCGGCGCTTA
44641 CTACCGATTC CGCCTAGTTG GTCACTTCGA CGTATCGTCT GGAACTCCAA CCATCGCAGG
44701 CAGAGAGGTC TGCAAAATGC AATCCCGAAA CAGTTCGCAG GTAATAGTTA GAGCCTGCAT
44761 AACGGTTTCG GGATTTTTTA TATCTGCACA ACAGGTAAGA GCATTGAGTC GATAATCGTG
44821 AAGAGTCGGC GAGCCTGGTT AGCCAGTGCT CTTTCCGTTG TGCTGAATTA AGCGAATACC
44881 GGAAGCAGAA CCGGATCACC AAATGCGTAC AGGCGTCATC GCCGCCCAGC AACAGCACAA
44941 CCCAAACTGA GCCGTAGCCA CTGTCTGTCC TGAATTCATT AGTAATAGTT ACGCTGCGGC
45001 CTTTTACACA TGACCTTCGT GAAAGCGGGT GGCAGGAGGT CGCGCTAACA ACCTCCTGCC
45061 GTTTTGCCCG TGCATATCGG TCACGAACAA ATCTGATTAC TAAACACAGT AGCCTGGATT

45121	TGTTCTATCA	GTAATCGACC	TTATTCCTAA	TTAAATAGAG	CAAATCCCCT	TATTGGGGGT
45181	AAGACATGAA	GATGCCAGAA	AAACATGACC	TGTTGGCCGC	CATTCTCGCG	GCAAAGGAAC
45241	AAGGCATCGG	GGCAATCCTT	GCGTTTGCAA	TGGCGTACCT	TCGCGGCAGA	TATAATGGCG
45301	GTGCGTTTAC	AAAAACAGTA	ATCGACGCAA	CGATGTGCGC	CATTATCGCC	TGGTTCATTC
45361	GTGACCTTCT	CGACTTCGCC	GGACTAAGTA	GCAATCTCGC	TTATATAACG	AGCGTGTTTA
45421	TCGGCTACAT	CGGTACTGAC	TCGATTGGTT	CGCTTATCAA	ACGCTTCGCT	GCTAAAAAAG
45481	CCGGAGTAGA	AGATGGTAGA	AATCAATAAT	CAACGTAAGG	CGTTCCTCGA	TATGCTGGCG
45541	TGGTCGGAGG	GAACTGATAA	CGGACGTCAG	AAAACCAGAA	ATCATGGTTA	TGACGTCATT
45601	GTAGGCGGAG	AGCTATTTAC	TGATTACTCC	GATCACCCTC	GCAAACTTGT	CACGCTAAAC
45661	CCAAAACTCA	AATCAACAGG	CGCCGGACGC	TACCAGCTTC	TTTCCCGTTG	GTGGGATGCC
45721	TACCGCAAGC	AGCTTGGCCT	GAAAGACTTC	TCTCCGAAAA	GTCAGGACGC	TGTGGCATTG
45781	CAGCAGATTA	AGGAGCGTGG	CGCTTTACCT	ATGATTGATC	GTGGTGATAT	CCGTCAGGCA
45841	ATCGACCGTT	GCAGCAATAT	CTGGGCTTCA	CTGCCGGGCG	CTGGTTATGG	TCAGTTCGAG
45901	CATAAGGCTG	ACAGCCTGAT	TGCAAAATTC	AAAGAAGCGG	GCGGAACGGT	CAGAGAGATT
45961	GATGTATGAG	CAGAGTCACC	GCGATTATCT	CCGCTCTGGT	TATCTGCATC	ATCGTCTGCC
46021	TGTCATGGGC	TGTTAATCAT	TACCGTGATA	ACGCCATTAC	CTACAAAGCC	CAGCGCGACA
46081	AAAATGCCAG	AGAACTGAAG	CTGGCGAACG	CGGCAATTAC	TGACATGCAG	ATGCGTCAGC
46141	GTGATGTTGC	TGCGCTCGAT	GCAAAATACA	CGAAGGAGTT	AGCTGATGCT	AAAGCTGAAA
46201	ATGATGCTCT	GCGTGATGAT	GTTGCCGCTG	GTCGTCGTCG	GTTGCACATC	AAAGCAGTCT
46261	GTCAGTCAGT	GCGTGAAGCC	ACCACCGCCT	CCGGCGTGGA	TAATGCAGCC	TCCCCCCGAC
46321	TGGCAGACAC	CGCTGAACGG	GATTATTTCA	CCCTCAGAGA	GAGGCTGATC	ACTATGCAAA
46381	AACAACTGGA	AGGAACCCAG	AAGTATATTA	ATGAGCAGTG	CAGATAGAGT	TGCCCATATC
46441	GATGGGCAAC	TCATGCAATT	ATTGTGAGCA	ATACACACGC	GCTTCCAGCG	GAGTATAAAT
46501	GCCTAAAGTA	ATAAAACCGA	GCAATCCATT	TACGAATGTT	TGCTGGGTTT	CTGTTTTAAC
46561	AACATTTTCT	GCGCCGCCAC	AAATTTTGGC	TGCATCGACA	GTTTTCTTCT	GCCCAATTCC
46621	AGAAACGAAG	AAATGATGGG	TGATGGTTTC	CTTTGGTGCT	ACTGCTGCCG	GTTTGTTTTG
46681	AACAGTAAAC	GTCTGTTGAG	CACATCCTGT	AATAAGCAGG	GCCAGCGCAG	TAGCGAGTAG
46741	CATTTTTTTC	ATGGTGTTAT	TCCCGATGCT	TTTTGAAGTT	CGCAGAATCG	TATGTGTAGA
46801	AAATTAAACA	AACCCTAAAC	AATGAGTTGA	AATTTCATAT	TGTTAATATT	TATTAATGTA
46861	TGTCAGGTGC	GATGAATCGT	CATTGTATTC	CCGGATTAAC	TATGTCCACA	GCCCTGACGG
46921	GGAACTTCTC	TGCGGGAGTG	TCCGGGAATA	ATTAAAACGA	TGCACACAGG	GTTTAGCGCG
46981	TACACGTATT	GCATTATGCC	AACGCCCCGG	TGCTGACACG	GAAGAAACCG	GACGTTATGA
47041	TTTAGCGTGG	AAAGATTTGT	GTAGTGTTCT	GAATGCTCTC	AGTAAATAGT	AATGAATTAT
47101	CAAAGGTATA	GTAATATCTT	TTATGTTCAT	GGATATTTGT	AACCCATCGG	AAAACTCCTG
47161	CTTTAGCAAG	ATTTTCCCTG	TATTGCTGAA	ATGTGATTTC	TCTTGATTTC	AACCTATCAT
47221	AGGACGTTTC	TATAAGATGC	GTGTTTCTTG	AGAATTTAAC	ATTTACAACC	TTTTTAAGTC
47281	CTTTTATTAA	CACGGTGTTA	TCGTTTTCTA	ACACGATGTG	AATATTATCT	GTGGCTAGAT
47341	AGTAAATATA	ATGTGAGACG	TTGTGACGTT	TTAGTTCAGA	ATAAAACAAT	TCACAGTCTA
47401	AATCTTTTCG	CACTTGATCG	AATATTTCTT	TAAAAATGGC	AACCTGAGCC	ATTGGTAAAA
47461	CCTTCCATGT	GATACGAGGG	CGCGTAGTTT	GCATTATCGT	TTTTATCGTT	TCAATCTGGT
47521	CTGACCTCCT	TGTGTTTTGT	TGATGATTTA	TGTCAAATAT	TAGGAATGTT	TTCACTTAAT
47581	AGTATTGGTT	GCGTAACAAA	GTGCGGTCCT	GCTGGCATTC	TGGAGGGAAA	TACAACCGAC
47641	AGATGTATGT	AAGGCCAACG	TGCTCAAATC	TTCATACAGA	AAGATTTGAA	GTAATATTTT
47701	AACCGCTAGA	TGAAGAGCAA	GCGCATGGAG	CGACAAAATG	AATAAAGAAC	AATCTGCTGA
47761	TGATCCCTCC	GTGGATCTGA	TTCGTGTAAA	AAATATGCTT	AATAGCACCA	TTTCTATGAG
47821	TTACCCTGAT	GTTGTAATTG	CATGTATAGA	ACATAAGGTG	TCTCTGGAAG	CATTCAGAGC
47881	AATTGAGGCA	GCGTTGGTGA	AGCACGATAA	TAATATGAAG	GATTATTCCC	TGGTGGTTGA
47941	CTGATCACCA	TAACTGCTAA	TCATTCAAAC	TATTTAGTCT	GTGACAGAGC	CAACACGCAG
48001	TCTGTCACTG	TCAGGAAAGT	GGTAAAACTG	CAACTCAATT	ACTGCAATGC	CCTCGTAATT
48061	AAGTGAATTT	ACAATATCGT	CCTGTTCGGA	GGGAAGAACG	CGGGATGTTC	ATTCTTCATC
48121	ACTTTTAATT	GATGTATATG	CTCTCTTTTC	TGACGTTAGT	CTCCGACGGC	AGGCTTCAAT
48181	GACCCAGGCT	GAGAAATTCC	CGGACCCTTT	TTGCTCAAGA	GCGATGTTAA	TTTGTTCAAT
48241	CATTTGGTTA	GGAAAGCGGA	TGTTGCGGGT	TGTTGTTCTG	CGGGTTCTGT	TCTTCGTTGA

```
48301   CATGAGGTTG  CCCCGTATTC  AGTGTCGCTG  ATTTGTATTG  TCTGAAGTTG  TTTTTACGTT
48361   AAGTTGATGC  AGATCAATTA  ATACGATACC  TGCGTCATAA  TTGATTATTT  GACGTGGTTT
48421   GATGGCCTCC  ACGCACGTTG  TGATATGTAG  ATGATAATCA  TTATCACTTT  ACGGGTCCTT
48481   TCCGGTGATC  CGACAGGTTA  CG
//
```